BIOLOGY

today and tomorrow

WITHOUT PHYSIOLOGY

BIOLOGY
today and tomorrow

WITHOUT PHYSIOLOGY

SECOND EDITION

CECIE STARR

CHRISTINE A. EVERS

LISA STARR

THOMSON

BROOKS/COLE

Australia • Canada • Mexico • Singapore
Spain • United Kingdom • United States

PUBLISHER Jack C. Carey

VICE-PRESIDENT, EDITOR-IN-CHIEF Michelle Julet

SENIOR DEVELOPMENT EDITOR Peggy Williams

ASSOCIATE DEVELOPMENT EDITOR Suzannah Alexander

EDITORIAL ASSISTANT Kristina Razmara

TECHNOLOGY PROJECT MANAGER Keli Amann

MARKETING MANAGER Stacy Best, Kara Kindstrom

MARKETING ASSISTANT Brian Smith

MARKETING COMMUNICATIONS MANAGER Nathaniel
  Bergson-Michelson

PROJECT MANAGER, EDITORIAL PRODUCTION Andy Marinkovich

CREATIVE DIRECTOR Rob Hugel

PRINT BUYER Karen Hunt

PERMISSIONS EDITOR Joohee Lee, Sarah Harkrader

PRODUCTION SERVICE Grace Davidson & Associates

TEXT AND COVER DESIGN Gary Head

PHOTO RESEARCHER Myrna Engler

COPY EDITOR Christy Goldfinch

ILLUSTRATORS Gary Head, ScEYEnce Studios, Lisa Starr

COMPOSITOR Lachina Publishing Services

TEXT AND COVER PRINTER QuebecorWorld—Versailles

COVER PHOTOGRAPHER Jay Barnes

A female crab flower spider (Misumena vatia) *raises its legs and shows its fangs in a defensive posture. Such spiders exploit the mutually beneficial relationship between flowers and their pollinators. A bee or fly that is lured into range by the flower's color and scent is snatched by the spider's elongated front legs and subdued by a venomous bite from the fangs.*

Printed in the United States of America
1 2 3 4 5 6 7 09 08 07 06

Library of Congress Control Number: 2005936177

ISBN 0-495-10918-5

BOOKS IN THE BROOKS/COLE BIOLOGY SERIES

Biology: The Unity and Diversity of Life, Eleventh, Starr/Taggart
Engage Online for Biology: The Unity and Diversity of Life
Biology: Concepts and Applications, Sixth, Starr
Basic Concepts in Biology, Sixth, Starr
Biology Today and Tomorrow, Second, Starr
Biology, Seventh, Solomon/Berg/Martin
Human Biology, Sixth, Starr/McMillan
Biology: A Human Emphasis, Sixth, Starr
Human Physiology, Fifth, Sherwood
Fundamentals of Physiology, Second, Sherwood
Human Physiology, Fourth, Rhoades/Pflanzer

Laboratory Manual for Biology, Fourth, Perry/Morton/Perry
Laboratory Manual for Human Biology, Morton/Perry/Perry
Photo Atlas for Biology, Perry/Morton
Photo Atlas for Anatomy and Physiology, Morton/Perry
Photo Atlas for Botany, Perry/Morton
Virtual Biology Laboratory, Beneski/Waber
Introduction to Cell and Molecular Biology, Wolfe
Molecular and Cellular Biology, Wolfe
Biotechnology: An Introduction, Second, Barnum

Introduction to Microbiology, Third, Ingraham/Ingraham
Microbiology: An Introduction, Batzing
Genetics: The Continuity of Life, Fairbanks/Anderson
Human Heredity, Seventh, Cummings
Current Perspectives in Genetics, Second, Cummings
Gene Discovery Lab, Benfey

Animal Physiology, Sherwood, Kleindorf, Yarcey
Invertebrate Zoology, Seventh, Ruppert/Fox/Barnes
Mammalogy, Fourth, Vaughan/Ryan/Czaplewski
Biology of Fishes, Second, Bond
Vertebrate Dissection, Ninth, Homberger/Walker

Plant Biology, Second, Rost/Barbour/Stocking/Murphy
Plant Physiology, Fourth, Salisbury/Ross
Introductory Botany, Berg

General Ecology, Second, Krohne
Essentials of Ecology, Third, Miller
Terrestrial Ecosystems, Second, Aber/Melillo
Living in the Environment, Fourteenth, Miller
Environmental Science, Tenth, Miller
Sustaining the Earth, Seventh, Miller
Case Studies in Environmental Science, Second, Underwood
Environmental Ethics, Third, Des Jardins
Watersheds 3—Ten Cases in Environmental Ethics, Third, Newton/Dillingham

Problem-Based Learning Activities for General Biology, Allen/Duch
The Pocket Guide to Critical Thinking, Second, Epstein

Thomson Higher Education
10 Davis Drive
Belmont, CA 94002-3098
USA

Asia (including India)
Thomson Learning
5 Shenton Way
#01-01 UIC Building
Singapore 068808

Australia/New Zealand
Thomson Learning Australia
102 Dodds Street
Southbank, Victoria 3006
Australia

Canada
Thomson Nelson
1120 Birchmount Road
Toronto, Ontario M1K 5G4

UK/Europe/Middle East/Africa
Thomson Learning
High Holborn House
50/51 Bedford Row
London WC1R 4LR
United Kingdom

CONTENTS IN BRIEF

DETAILED CONTENTS

Unit Two   Genetics

7   How Cells Reproduce

8   Observing Patterns in Inherited Traits

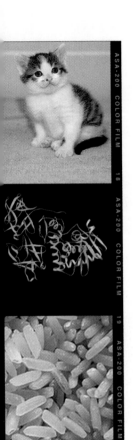

9 DNA Structure and Function

10 Gene Expression and Control

11 Studying and Manipulating Genomes

---

14 Early Life

---

15 Plant Evolution

Unit Four   Ecology

19 Community Structure and Biodiversity

20 Ecosystems

Preface

Biology opens a window on processes of nature that are so familiar that we might not even think about them. It reveals connections, sometimes subtle and sometimes not, between us and everything else on Earth. This is profound stuff that many nonmajors students are ill-prepared to grasp. Many enter college without adequate preparation in science, and often with the mistaken belief that they are not smart enough to learn biology. Should textbook writers reinforce their apprehension by squeezing all the juice out of a story, by producing a pedagogical husk? We think not. We find biology fascinating and eagerly devour news of recent discoveries. We believe that, with the proper guidance, students too can become lifelong learners who understand and are excited by the science of life.

This book is briefer than our other titles, but in it we strive to share our enthusiasm for biology with easy-to-follow writing on relevant topics. For this edition, we have shortened and streamlined many sections. We did not water down the science, but instead emphasize processes and relationships over detail.

Promote Critical Thinking Like all textbooks at this level, we walk students through examples of problem solving and experiments throughout the book. Many historical and theoretical experiments have been integrated into the text. Each chapter ends with several *Critical Thinking* questions, some illustrated. Some are more challenging than others, but all invite students to think outside the memorization box. We return again and again in chapter introductions and in the text to examples of how science is carried out. Introductory students are not scientists, however, so we do not expect them to learn the language and processes of science by intuition alone; we help them build these skills in a paced way.

Make It Brief, With Clear Explanations To keep the book length manageable, we were selective about which topics to include but were not stingy with clear explanations. If something is worth reading, why reduce it to a factoid? Factoids invite mind-numbing memorization, which does not promote critical thinking about the world and our place in it.

For instance, you can safely bet that most nonmajors simply do not want to memorize each catalytic step of crassulacean acid metabolism. They *do* want to learn about the biological basis of sex, and many female students want to know what will be going on inside them if and when they get pregnant. Good explanations can help them make their own informed decisions on many biology-related issues, including STDs, fertility drugs, prenatal diagnoses of genetic disorders, gene therapies, and abortions.

Our choices for which topics to condense, expand, or delete were not arbitrary. They reflect three decades of feedback from teachers throughout the world.

Make It Relevant Most students taking this course will not become biologists, but biological research will affect their lives in direct and often controversial ways. *What they learn today will have impact on how they make decisions tomorrow*—in the voting booth as well as in their personal lives.

Each chapter starts with an IMPACTS/ISSUES essay on a topic of current interest related to its content. For instance, the microevolution chapter opens with how the use of warfarin has favored "super rats." Essays are expanded in custom videoclips and in sidebars on relevant pages within chapters, as shown on the facing page. We return to the essay topics in exercises on the student website and in the PowerLecture, an all-in-one PowerPoint tool for instructors. We also ask the students, HOW WOULD YOU VOTE? on an application related to the essay's topic. The exercise invites them to read a selection of articles, pro and con, before voting.

---

12 PROCESSES OF EVOLUTION

IMPACTS, ISSUES

Rise of the Super Rats

Slipping in and out of the pages of human history are rats—*Rattus*—the most notorious of mammalian pests. One kind of rat or another has distributed pathogens and parasites that cause bubonic plague, typhus, and other deadly infectious diseases. The death toll from fleas that bit infected rats and then bit people has exceeded the death toll in all wars combined.

The rats themselves are far more successful. By one estimate, there is one rat for every person in urban and suburban centers of the United States. In addition to spreading diseases, rats chew their way through walls and wires of homes and cities. In any given year, they cause economic losses approaching 19 billion dollars.

For years, people have been fighting back with traps, ratproof storage facilities, and various poisons. During the 1950s, they started using baits laced with warfarin. This compound interferes with blood clotting. Rats ate the baits, then died within days after bleeding internally or losing blood through cuts or scrapes. Warfarin was extremely effective. Compared with other rat poisons, it had a lot less impact on harmless species.

In 1958, however, a Scottish researcher reported that warfarin did not work against some rats. Similar reports from other European countries followed. About twenty years later, 10 percent of the urban rats caught in the United States were warfarin resistant. *What happened?*

To find out, researchers compared warfarin-resistant rat populations with still-vulnerable rats. They traced the difference to a gene on one of the rat chromosomes. At that gene locus, a dominant allele was common in warfarin-resistant rat populations but very rare among

the vulnerable ones. "What happened" was evolution by natural selection. Warfarin was exerting selective pressure on populations of rats. The previously rare dominant allele proved to be adaptive. The lucky rats that inherited the allele survived and produced more offspring. The unlucky ones that inherited the recessive allele had no built-in defense, and died. Over time, the dominant allele's frequency increased in all rat populations exposed to the poison.

Of course, selection pressures can and often do change. When warfarin resistance increased in rat populations, people stopped using warfarin. And guess what: The dominant allele's frequency declined. Now the latest worry is the evolution of "super rats," which even more potent rodenticides can't seem to kill.

When you hear someone question whether life evolves, remember this: With respect to life, **evolution** simply means that heritable change is occurring in some line of descent. The actual mechanisms that can bring about such change are the focus of this chapter. Later chapters highlight how these mechanisms have contributed to the evolution of new species.

 How Would You Vote? Antibiotic-resistant strains of bacteria are becoming dangerously pervasive. Standard animal husbandry practice includes the repeated dosing of healthy animals with antibiotics—the same ones prescribed to people. Should this practice stop? See BiologyNow for details, then vote online.

 Key Concepts

EVOLUTIONARY VIEWS EMERGE
The world distribution of species, similarities and differences in body form, and the fossil record gave early evidence of evolution—of changes in lines of descent. Charles Darwin and Alfred Wallace had an idea of how those changes occur.

VARIATION AND ADAPTATION
An adaptation is a heritable aspect of form, function, behavior, or development that promotes survival and reproduction. It enhances the fit between the individual and prevailing conditions in its environment.

MICROEVOLUTIONARY PROCESSES
An individual does not evolve. A *population* evolves, which means its shared pool of alleles changes. Over generations, any allele may increase in its frequency among individuals, or it may become rare or lost. Mutation, genetic drift, natural selection, and gene flow change allele frequencies in a population. These processes of microevolution change the observable characteristics that define a population and, more broadly, a species.

Links to Earlier Concepts

Section 1.4 sketched out the key premises of the theory of natural selection. Here you will read about evidence that led to its formulation. You may wish to refresh your memory of protein structure (2.8), basic terms of genetics (8.1), chromosomes and crossing over (8.4, 8.5), DNA replication and repair (9.3), and gene mutation (10.5). This chapter puts the chromosomal basis of inheritance (8.4, 8.7) and continuous variation in populations (8.3) into the broader context of evolution.

Warfarin-resistant rats led to the development of anticoagulants that are more toxic and that persist longer in the environment. They are weakening and killing owls, hawks, coyotes, and other predators that have eaten poisoned rats. Between 1985 and 1999, for example, the number of barn owls with anticoagulants in their blood rose from 5 to 36 percent. And that was just one study in Great Britain.

Students throughout the country are already voting online, and they are accessing campuswide, statewide, and nationwide tallies. This interactive approach to issues reinforces the premise that an individual's actions can make a difference.

Make It Easy To Follow On each chapter's opening page is a preview of key concepts, each with a simple title. We repeat these titles at the bottom of appropriate pages as reminders of the chapter's conceptual organization.

New to this edition are LINKS TO EARLIER CONCEPTS that can help students follow the big connections within and between the chapters. The opening page lists the key sections in earlier chapters that students should be familiar with before they start the chapter. For instance, before reading about neural function, a student may wish to scan an earlier chapter's section on active transport. We repeat linking icons in the chapter's page margins as reminders of these and other relevant sections. This feature demonstrates how the concepts in the book are not separate topics, but are as closely interconnected as biology itself.

A conversational writing style eases students into the story. They can stay focused on that story without worrying about highlighting something they might be tested on. Why? We already highlight key points for them, in blue boxes at the end of each section. These HIGHLIGHTED KEY POINTS function as a *running in-text summary*. All chapters end with a *section-by-section summary* that reinforces what they have learned.

Students also can stay on track with ANIMATED DIAGRAMS. They can walk through the text's step-by-step art as a preview of major concepts, then check out the steps online. These visual learning devices are available in narrated, animated form, on *BiologyNow* and on the PowerLecture DVD. Exposing students to the material in a variety of modalities accommodates diverse learning styles and reinforces understanding.

Offer Easy-to-Use Media Tools Integrated into each end-of-chapter summary are associated media tools that will help students focus their study and understand the key concepts. With this new edition, taking advantage of these online assets is easier than ever. Students register their 1PASS ACCESS CODE at *http://1pass.thomson.com*, then log in to access all resources outlined below. An access code is packaged in each new copy of the book.

BIOLOGYNOW™ tests which topics students have not yet mastered and creates customized learning plans to focus their study and review. Responses to the diagnostic pretest questions activate a personalized learning plan. Answer incorrectly, and the plan lists the relevant text sections, figures, chapter videos, and animations that the student may review.

A *Post-Test* can be used as a self-assessment tool or submitted to an instructor. Because BiologyNow is built into iLrn, Thomson Learning's course management system, the answers and results can be fed directly into an electronic gradebook. BiologyNow can be integrated with WebCT and Blackboard, so students may log in through these systems as well.

Interactive flashcards with audio pronunciation use the definitions from this edition's highly revised glossary. All the boldface terms within the text are presented in this resource.

The *How Do I Prepare?* section of BiologyNow has tutorials in math, chemistry, and graphing, as well as a review of basic study skills.

1PASS grants access to INFOTRAC COLLEGE EDITION™, an exclusive online searchable database of more than 5,000 periodicals and close to 18 million articles. The articles are in full-text form and can be located easily and quickly with a key word search. For each chapter, the *How Would You Vote* exercise in BiologyNow references specific InfoTrac articles and websites. As students vote on each issue, the website provides a running tally by campus, state, and the country. Instructors can assign the exercises from iLrn, or students can access them through the website at *http://biology.brookscole.com/btt2*.

An online ISSUES AND RESOURCES INTEGRATOR correlates chapter sections with applications, videos, InfoTrac articles, and websites. This guide is updated each semester.

VMENTOR is an online tutorial service available from 6 AM to 12 PM Monday through Saturday. Students may interact with a live tutor in a virtual classroom by voice communication, whiteboard, and text messaging.

New college editions of the book can also include access to the AUDIOBOOK. Students can either listen to narration online or download MP3 files for use on a portable MP3 player.

1 Removal of the nucleus from a sheep egg during a cloning. At the center is an unfertilized sheep egg. A pipette has positioned the egg, which a microneedle is about to penetrate.

2 The microneedle has now emptied the sheep egg of its own nucleus, which holds the cell's DNA.

3 A nucleus from a donor cell is about to enter the enucleated egg. An electric spark will stimulate the egg to enter mitotic cell divisions. At an early stage, the ball of cells will be implanted in the womb of a surrogate mother sheep.

Figure 9.8 *Animated!* Nuclear transfer of sheep cells. In this series of photos, a microneedle replaces the nucleus of a sheep egg with a nucleus from an adult sheep cell. Newer methods involve direct transfer of nuclear DNA that has been treated with mitotic cell extracts to condense the chromosomes.

Acknowledgments

No short list can convey our thanks to the extended team that made this book not only possible, but also excellent. John Jackson and Walt Judd deserve special thanks for their detailed evaluations, perspective, and continued commitment to excellence. Dave Rintoul and his students helped us fine-tune some important last-minute details. Through focus-group participation, reviews, and class testing, he and the other instructors listed below helped us transform our first edition into the polished, cohesive textbook you see here, one that particularly addresses the needs of students who have been previously underserved.

With Jack Carey, Susan Badger, Sean Wakely, Kathie Head, Michelle Julet, and Michael Johnson, Thomson Learning proved why it is one of the world's foremost publishers; we doubt that any authors get finer support anywhere. Keli Amann managed production of all student and instructor media tools. Peggy Williams again brought her tenacity, intelligence, and humor to the developmental editing. Andy Marinkovich calmly kept us on track, and Grace Davidson made this book happen despite several hurricanes, both literal and figurative. Gary Head created great designs and graphics; Steve Bolinger did so for the media tools. Myrna Engler, Suzannah Alexander, Kristina Razmara, Karen Hunt, Stacy Best—the list goes on. Thank you to each one of you; this book would not exist without your help.

CECIE STARR, CHRIS EVERS, AND LISA STARR *November 2005*

General Advisors

DANIEL J. FAIRBANKS *Brigham Young University*
JOHN D. JACKSON *North Hennipin Community College*
WALTER JUDD *University of Florida*
E. WILLIAM WISCHUSEN *Louisiana State University*

Focus Groups, Reviewers, and Class Testing

JO DALE H. ALES *Baton Rouge Community College*
DIANNE ANDERSON *San Diego City College*
KENNETH D. ANDREWS *East Central University*
BERT ATSMA *Union County College*
ANDREW STEPHEN BALDWIN *Mesa Community College*
GREGORIO B. BEGONIA *Jackson State University*
DAVID BELT *Johnson County Community College*
PAUL A. BILLETER *College of Southern Maryland*
ANDREW BLAUSTEIN *Oregon State University*
JUDY BLUEMER *Morton College*
MARK G. BOLLONE *Tallahassee Community College*
SUSAN BOWER *Pasadena City College*
WILLIAM BROWER *Del Mar College*
STEVEN G. BRUMBAUGH *Greenriver Community College*
WILBERT BUTLER, JR. *Tallahassee Community College*
THOMAS R. CAMPBELL *Los Angeles Pierce College*
J. MICHELLE CAWTHORN *Georgia Southern University*
THOMAS T. CHEN *Santa Monica College*
THOMAS F. CHUBB *Villanova University*
JAY L. COMEAUX *Louisiana State University*
EDWARD A. DEGRAUW *Portland Community College*
JEAN DESAIX *University of North Carolina*
BRIAN DINGMANN *University of Minnesota, Crookston*

ERNEST F. DUBRUL *University of Toledo*
PARAMJIT DUGGAL *Metropolitan Community College, Maple Woods*
PAMELA K. ELF *University of Minnesota, Crookston*
ROBERT C. EVANS *Rutgers University-Camden*
W. LOGAN FINK *Pensacola Junior College*
APRIL A. FONG *Portland Community College*
EDISON R. FOWLKS *Hampton University*
TERESA LANE FULCHER *Pellissippi State Technical Community College*
RIC A. GARCIA *Clemson University*
MICHELLE GEARY *West Valley College*
MARY GOBBETT *University of Indianapolis*
MELVIN H. GREEN *University of California, San Diego*
CARLA GUTHRIDGE *Cameron University*
GALE HAIGH *McNeese State University*
JERRIE R. HANIBLE *Southeastern Louisiana University*
LINDA HARRIS-YOUNG *Motlow State Community College*
DAVID J. HENSON *Eastern Arizona College*
JULIANA HINTON *McNeese State University*
SONGQIAO (SARA) HUANG *Los Angeles Valley College*
DONNA HUFFMAN *Calhoun Community College*
KIMBERLY HURD *Bakersfield College*
DESIRÉE JACKSON *Texas Southern University*
ROSS S. JOHNSON *Virginia State University*
GLENN H. KAGEYAMA *California State Polytechnic University, Pomona*
RITA A. KARPIE *Brevard Community College*
JUDITH KELLY *Henry Ford Community College*
PAULINE LIZOTTE *Valencia Community College*
JUAN LOPEZ-BAUTISTA *University of Alabama, Tuscaloosa*
DAVID LORING *Johnson County Community College*
DANIEL MARK *Metropolitan Community College, Maple Woods*
LINDA MARTIN-MORRIS *University of Washington*
DENNIS MCDERMOT *Lake Erie College*
RON MOLLICK *Christopher Newport University*
JILL NISSEN *Montgomery College*
ALEXANDER E. OLVIDO *Virginia State University*
JOSHUA M. PARKER *Community College of Southern Nevada*
ROBERT P. PATTERSON *North Carolina State University*
JOHN S. PETERS *College of Charleston*
HAROLD PLETT *Fullerton College*
JENNIE M. PLUNKETT *San Jacinto College*
JAMES V. PRICE *Utah Valley State College*
JERRY W. PURCELL *San Antonio College*
KIRSTEN RAINES *San Jacinto College*
WENDA RIBEIRO *Thomas Nelson Community College*
PELE EVE RICH *Mt. San Jacinto College*
DAVID A. RINTOUL *Kansas State University*
FRANK A. ROMANO III *Jacksonville State University*
DAVID ROSEN *Diablo Valley College*
ETTA V. RUPPERT *Clemson University*
ERIK P. SCULLY *Towson University*
MIKE SHIFFLER *University of Alabama*
MARCIA SHOFNER *University of Maryland*
GREG A. SIEVERT *Emporia State University*
LINDA D. SMITH-STATON *Pellissippi State Technical Community College*
LARRY L. ST. CLAIR *Brigham Young University*
JACQUELINE J. STEVENS *Jackson State University*
JUDY STEWART *Community College of Southern Nevada*
ALICE B. TEMPLET *Nicholls State University*
MICHAEL T. TOLIVER *Eureka College*
JEANETTE H. TUCKER *Northern Virginia Community College*
HEATHER D. VANCE-CHALCRAFT *East Carolina University*
PATRICIA M. WALSH *University of Delaware*
JENNIFER WARNER *University of North Carolina at Charlotte*
MICHAEL WENZEL *California State University, Sacramento*
LISA A. WERNER *Pima Community College*
EMILY C. WHITELEY *Catawba Valley Community College*
MICHAEL WINDELSPECHT *Appalachian State University*
MARK WITMER *Ithaca College*
MARK WYGODA *McNeese State University*
MARTIN D. ZAHN *Thomas Nelson Community College*
ROSEMARY ZARAGOZA *University of Chicago*

Current configurations of the Earth's oceans and land masses—the geologic stage upon which life's drama continues to unfold. This composite satellite image reveals global energy use at night by the human population. Just as biological science does, it invites you to think more deeply about the world of life—and about our impact upon it.

What Am I Doing Here?

Leaf through a newspaper on any given Sunday and you may get an uneasy feeling that the world is out of control. There are stories about the Middle East, where great civilizations have come and gone. You won't find much on the amazing coral reefs in the surrounding seas, especially at the northern end of the Red Sea. Now the news is about oil and politics, bioterrorists and war.

Think back on the 1991 Persian Gulf conflict, when thick smoke from oil fires blocked out sunlight, and black rain fell. Iraqis released 460 million gallons of crude oil into the Gulf. Uncounted numbers of reef organisms and thousands of birds died.

Today Kuwaitis wonder if the oil fires have caused their increased cancer rates. They join New Yorkers who worry about developing lung problems from breathing dense, noxious dust that filled the air after the terrorist attack on the World Trade Center.

Nature, too, seems to have it in for us. The AIDS pandemic is unraveling African societies. Forests are burning fiercely. Storms, droughts, and heat waves are monstrous, polar ice caps and once-vast glaciers are melting, and the whole atmosphere is warming up.

It's enough to make you throw down the paper and long for the good old days, when things were simpler.

Of course, read up on the good old days and you'll find they weren't so good. Bioterrorists were around in 1346, when soldiers catapulted the corpses of bubonic plague victims into a walled city under siege. Infected people and rats fled the city and helped fuel the Black Death, a plague that left 25 million dead in Europe.

Nature was busy with us back then, too. In 1918, the Spanish flu raced around the world and left between 30 and 40 million people dead.

Then, as now, many felt helpless in a world that seemed out of control. Perhaps it is a human condition to think we may control something if we understand it. For millions of years, we and our ancestors have been trying to make sense of our natural world and how we fit into it. We observe it, come up with ideas, then test the ideas. This is the nature of science. But the more pieces of the puzzle we fit together, the bigger the puzzle seems to be.

You could walk away from this challenge and simply not think about it. You could let others tell you what to think. Or you could choose to develop your own understanding. Maybe you are interested in the pieces of the puzzle that affect your health, your food, or your children, should you choose to reproduce. Maybe you just find life fascinating. No matter what your focus might be, you can deepen your perspective. You can learn ways to sharpen how you interpret the natural world, including human nature. This is the gift of biology, the scientific study of life.

 How Would You Vote? The seas of the Middle East support some of the world's most spectacular coral reef ecosystems. Should the United States provide funding to help preserve the reefs? See BiologyNow for details, then vote online.

 ## Key Concepts

LIFE'S UNDERLYING UNITY
Life's organization extends from the molecular level to the biosphere. Shared features at the molecular level are the basis of life's unity.

LIFE'S DIVERSITY
Life shows spectacular diversity as well as unity. Each one of millions of kinds of organisms, past and present, has unique traits that help define it as a separate species.

EXPLAINING UNITY IN DIVERSITY
Evolutionary theories, especially the theory of natural selection, help us see a profound connection between life's underlying unity and its diversity.

HOW WE KNOW
Biologists find out about life by observing, asking questions, formulating and testing hypotheses, and reporting results in ways that others can test.

Links to Earlier Concepts

This book parallels nature's levels of organization, from atoms to the biosphere. Learning about the structure and function of atoms and molecules primes you to understand the structure of cells. Learning about protein synthesis, active transport, and other processes that keep a single cell alive helps you understand how multicelled organisms survive, because each of their cells uses the same processes. Knowing what it takes to survive helps you sense why and how organisms interact with one another and the environment. At the start of each chapter, we will be reminding you of structural and functional connections. Within chapters, you will come across keychain icons with references to relevant sections in earlier chapters.

b molecule

Two or more joined atoms of the same or different elements. The "molecules of life" are complex carbohydrates, lipids, proteins, DNA, and RNA. In today's world, only living cells make them.

c cell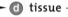

Smallest unit that can live and reproduce on its own or as part of a multicellular organism. It has an outer membrane, DNA, and other components.

d tissue

Organized cells and substances that interact in a specialized activity. Many cells (white) made this bone tissue from their own secretions.

e organ

Two or more tissues interacting in some task. A parrotfish eye, for example, is a sensory organ used in vision.

f organ system

Organs interacting physically, chemically, or both in some task. Parrotfish skin is an organ system with tissue layers, organs such as glands, and other parts.

a atom

Smallest unit of an element that still retains the element's properties. Electrons, protons, and neutrons are its building blocks. This hydrogen atom's electron zips around a proton in a spherical volume of space.

Figure 1.1 *Animated!* Increasingly complex levels of organization in nature, extending from subatomic particles to the biosphere.

1.1 Life's Levels of Organization

Biologists are interested in all forms of life, past and present. They study organisms down to their atoms and up to their interactions on a global scale. In doing so, they have discovered a great pattern of organization.

FROM SMALL TO SMALLER

Imagine yourself on the deck of a sailing ship, about to journey around the world. The distant horizon of a vast ocean beckons, and suddenly you sense that you are just one tiny part of the great scheme of things.

Now imagine one of your red blood cells can think. It realizes it is only a small part of the great scheme of your body. A string of 375 cells like itself would fit across a pin's head—and you have trillions of cells. One of the fat molecules at the red blood cell's surface is thinking about how small it is. A string of 1,200,000 molecules like itself would stretch across that pinhead.

Now a hydrogen atom in the molecule is pondering the great scheme of a fat molecule. Hydrogen atoms are so small that it would take 53,908,355 of them to reach across the pinhead! With that lone atom, you have reached the starting point of life's great pattern of organization.

FROM SMALLER TO VAST

That pattern starts at the level of atoms, which are the basic building blocks of matter. Like nonliving things, all organisms are made of atoms. Figure 1.1*a* depicts a hydrogen atom. At the next level, atoms combine into molecules (Figure 1.1*b*). Complex carbohydrates, fats and other lipids, proteins, DNA, and RNA are called the "molecules of life." In nature, only living cells can make them.

The pattern reaches the threshold of life as these molecules assemble into cells (Figure 1.1*c*). A **cell** is the smallest living unit, having the capacity to survive and reproduce on its own, given raw materials, energy

g **multicelled organism**

Individual made of different types of cells. Cells of most organisms, including this Red Sea parrotfish, are organized as tissues, organs, and organ systems.

h **population**

Group of single-celled or multicelled individuals of the same species occupying a specified area. This is a fish population in the Red Sea.

i **community**

All populations of all species occupying a specified area. This is part of a coral reef in the Gulf of Aqaba at the northern end of the Red Sea.

j **ecosystem**

A community that is interacting with its physical environment. It has inputs and outputs of energy and materials. Reef ecosystems flourish in warm, clear seawater throughout the Middle East.

k **the biosphere**

All regions of the Earth's waters, crust, and atmosphere that hold organisms. In the vast universe, Earth is a rare planet. Without its abundance of free-flowing water, there would be no life.

inputs, and suitable conditions in its environment. In multicelled species, cells are organized into tissues, organs, and organ systems. Figure 1.1*d*–g defines these levels of organization. Often there are trillions of cells, with many specialized types that interact directly and indirectly in ways that keep the whole body alive. In size, multicelled organisms range from microscopic to giant, as are redwoods and whales.

The **population** is the next level of organization. Each population is a group of the same kind of single-celled or multicelled organisms in some specified area. Fields of red poppies in a valley or schools of fish in a lake are such groups (Figure 1.1*h*).

At the next level of organization, the **community** includes all populations of all species in some specified area (Figure 1.1*i*). An example could be all of the kelps, corals, shrimps, fishes, sea anenomes, plankton, crabs, and everything else living in an underwater cave in the Red Sea. Another example would be the populations of tiny organisms that live, reproduce, and die quickly in a drop of water on the cupped petals of a flower.

The next level is the **ecosystem**, or a community interacting with its environment (Figure 1.1*j*). Finally, the **biosphere** is the highest level of life. It encompasses all regions of Earth's crust, waters, and atmosphere in which organisms live.

Bear in mind, life is more than the sum of its parts. At each successive level of organization, new properties emerge that are not inherent in any part by itself, as when living cells emerge from lifeless molecules.

This book is a journey through the globe-spanning organization of life. So take a moment to study Figure 1.1. You can use it as a road map of where each part fits into the great scheme of things.

Nature shows levels of organization, from the simple to the increasingly complex.

Life's unique characteristics originate at the molecular level. They extend through individual cells, populations, communities, ecosystems, and the biosphere.

OIL SLICK

Nearly half of the oil exports from the Persian Gulf are shipped through the Gulf of Oman, where satellite imaging revealed this vast slick formed by an oil spill. Such water pollution is damaging reefs all over the world.

1.2 Overview of Life's Unity

"Life" is not easy to define. It is too big, and it has been changing for 3.9 billion years! Even so, you can frame a definition in terms of its unity and diversity. Here is the unity part: All living things grow and reproduce with the help of DNA, energy, and raw materials. They sense and respond to their environment. But details of their traits differ among many millions of kinds of organisms. That is the diversity part—variation in traits.

DNA, THE BASIS OF INHERITANCE

You will never, ever find a rock made of nucleic acids, proteins, and complex carbohydrates and lipids. In the natural world, only living cells make these molecules. The signature molecule of life is the nucleic acid called DNA. No chunk of granite or quartz has it.

DNA holds information for building proteins from smaller molecules: the amino acids. By analogy, if you follow suitable instructions and invest enough energy in the task, you might organize a pile of a few kinds of ceramic tiles (representing amino acids) into diverse patterns (representing proteins), as in Figure 1.2.

Why are the proteins so important? Many kinds are structural materials, and others are enzymes. Enzymes are the main worker molecules in cells. They build, split, and rearrange the molecules of life in ways that keep a cell alive. Without enzymes, the information held in DNA could not be used. There would be no new organisms.

In nature, each organism inherits its DNA—and its traits—from parents. *Inheritance* means an acquisition of traits after parents transmit their DNA to offspring. Think about it. Why do baby storks look like storks and not like pelicans? Because storks inherit stork DNA, not pelican DNA.

Reproduction refers to actual mechanisms by which parents transmit DNA to offspring. For slugs, humans, trees, and other multicelled organisms, information in DNA guides *development*—the transformation of the first cell of a new individual into a multicelled adult, typically with many different tissues and organs.

ENERGY, THE BASIS OF METABOLISM

DNA is only part of the picture. Becoming alive and maintaining life requires energy—the capacity to do work. Each normal living cell has ways to obtain and convert energy from its surroundings. By the process called **metabolism**, a cell acquires and uses energy to maintain itself, grow, and make more cells.

Where does the energy come from? Nearly all of it flows from the sun into the world of life, starting with **producers**. Producers are plants and other organisms that make their own food from simple raw materials. All other organisms are **consumers**. They cannot make their own food; they must eat other organisms.

When, say, zebras browse on plants, some energy stored in plant tissues is transferred to them. Later on, energy is transferred to a lion as it devours the zebra. It gets transferred again as decomposers go to work, acquiring energy from the remains of zebras, lions, and other organisms.

Decomposers are consumers, mostly bacteria and fungi that break down sugars and other molecules to simpler materials. Some of the breakdown products are cycled back to producers as raw materials. Over time, energy that plants originally captured from the sun returns to the environment.

ENERGY AND LIFE'S ORGANIZATION

Energy happens to flow in one direction, from the environment, through producers, and then consumers. Eventually, it ends up being lost as heat (Figure 1.3). Energy transfers maintain life's organization, both in

Figure 1.2 Examples of objects built from the same materials according to different assembly instructions.

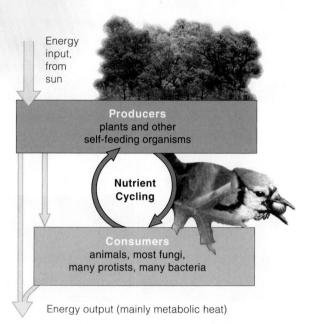

Energy input, from sun

Producers
plants and other self-feeding organisms

Nutrient Cycling

Consumers
animals, most fungi, many protists, many bacteria

Energy output (mainly metabolic heat)

Figure 1.3 *Animated!* The one-way flow of energy and the cycling of materials in the world of life.

Figure 1.4 A roaring response to signals from pain receptors, activated by a lion cub flirting with disaster.

individual organisms, and as all organisms interact throughout the biosphere. Later on, you will see how life's interconnectedness relates to problems such as food shortages, the AIDS pandemic, global warming, acid rain, cholera, and rapid losses in biodiversity.

LIFE'S RESPONSIVENESS TO CHANGE

Organisms keep their internal operating conditions within a range that their cells can tolerate. This state, called **homeostasis**, is a defining feature of life.

Only living things respond to the environment. Yet even a rock seems responsive, as when it yields to gravity and tumbles down a hill, or changes its shape slowly under the repeated batterings of wind and rain. The difference is this: Unlike rocks, living things make responses to offset changes in their environment. Such compensatory changes maintain homeostasis.

How? The answer begins with receptors. **Receptors** are structures that detect stimuli (singular, stimulus). A **stimulus** is a specific kind of energy, such as sunlight energy, chemical energy (as when a substance becomes more concentrated outside a cell than on the inside), or the mechanical energy of a bite (Figure 1.4).

Different kinds of receptors respond to different kinds of stimuli. Their responses trigger changes in cell activities. As a simple example, after you finish eating a piece of fruit, sugars leave your gut and enter your bloodstream. Think of blood and fluid around cells as an *internal* environment. The sugar from the fruit you ate changes that internal environment when it enters your bloodstream. A change in blood sugar level is a deviation from homeostasis that your body works to reverse. Cells in your pancreas (an organ) detect extra sugar in your blood and respond by secreting insulin. Many kinds of cells in your body have receptors for this hormone. When their receptors bind insulin, cells absorb sugar, removing it from the blood. After enough cells do so, the blood sugar level returns to normal. Thus, homeostasis is maintained.

Organisms reproduce based on instructions in DNA, which they inherit from their parents.

All living organisms reproduce, grow, and stay alive by way of metabolism—by ongoing energy conversions and energy transfers at the cellular level.

Organisms interact through a one-way flow of energy and a cycling of materials through the biosphere. Collectively, their interdependencies have global impact.

Organisms sense and respond to changing conditions in controlled ways. Their responses help them maintain tolerable conditions in their internal environment.

1.3 If So Much Unity, Why So Many Species?

Although unity pervades the world of life, so does diversity. Organisms differ enormously in body form, in the functions of their body parts, and in behavior.

Superimposed on life's unity—its common heritage—is tremendous diversity. Millions of different kinds of organisms, or **species**, live on Earth. Many millions more lived during the past 3.9 billion years, but their lineages disappeared; about 99.9 percent of all species have become extinct.

For centuries, scholars have tried to make sense of diversity. One, Carolus Linnaeus, devised a scheme for classifying organisms by assigning a two-part name to each newly identified species. The first part designates the genus (plural, genera). Each **genus** is one or more species grouped on the basis of a number of unique shared traits. The second part of the name refers to the particular species within the genus.

For instance, *Scarus gibbus* is a humphead parrotfish. We abbreviate a genus name once it's been spelled out in a document. Thus, we might call another species in the same genus, the midnight parrotfish, *S. coelestinus*.

Biologists are still working out how to organize the genera. Most favor a classification system that groups all organisms into three sets, which are the domains Archaea, Bacteria, and Eukarya. Eukarya encompasses protists, fungi, plants, and animals (Figures 1.5 and 1.6).

All **archaea** and **bacteria** are single cells, and all are *prokaryotic*, meaning they do not contain a nucleus (a membrane-enclosed sac that keeps DNA separated from the rest of the cell's interior.) Prokaryotes can be producers or consumers. Of all groups of organisms, theirs show the most metabolic diversity.

Archaea survive in boiling ocean water, freezing desert rocks, sulfur-rich lakes, and other habitats as harsh as those thought to have prevailed when life originated. Bacteria are sometimes called eubacteria, which means "true bacteria," to distinguish them from archaea. They are far more common than archaeans, and they live throughout the world in diverse habitats.

Bacteria   Archaea   Eukarya

Figure 1.5   Three-domain classification system.

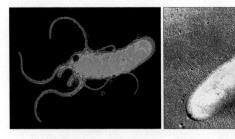

Bacteria

The most common prokaryotes; collectively, these single cells are the most metabolically diverse species on Earth.

Archaea

These prokaryotes are evolutionarily closer to eukaryotes than to bacteria. *Left*, a colony of methane-producing cells. *Right*, two species from a hydrothermal vent on the seafloor.

Figure 1.6 *Animated!*   A few representatives of life's diversity.

Plants, fungi, animals, and protists are *eukaryotic*, which means their cells start out life with a nucleus. Eukaryotes are generally larger and far more complex than prokaryotes. Differences among protist lineages are so great that this group might be subdivided into several separate groups, which would result in a major reorganization within the domain.

Plants are multicelled, photosynthetic producers. They can make their own food by using simple raw materials and sunlight as an energy source.

EXPLAINING UNITY IN DIVERSITY

Protists   Single-celled and multicelled eukaryotic species that range from the microscopic to giant seaweeds. Many biologists are now viewing the protists as many major lineages.

Plants   Generally, photosynthetic, multicelled eukaryotes, many with roots, stems, and leaves. Plants are the primary producers for ecosystems on land. Redwoods and flowering plants are examples.

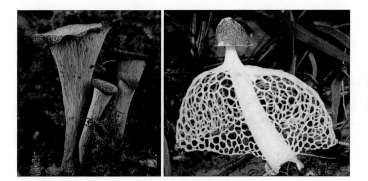

Fungi   Single-celled and multicelled eukaryotes; decomposers, parasites, and pathogens. Without decomposers, communities would become buried in their own wastes.

Animals   Multicelled eukaryotes that ingest tissues or juices of other organisms. Like this basilisk lizard, most actively move about during at least part of their life.

Most **fungi**, such as the mushrooms sold in grocery stores, are multicelled consumers with a distinctive way of feeding. They secrete enzymes that digest food outside the fungal body, and then their individual cells absorb the digested nutrients. **Animals** are multicelled consumers that ingest the tissues of other organisms. Different animals are herbivores (grazers), carnivores (meat eaters), scavengers, and parasites. All develop by a series of embryonic stages, and all actively move about during their life.

Pulling this information together, are you getting a sense of what it means when someone says that life shows unity *and* diversity?

To make the study of life's diversity more manageable, we group organisms related by descent from a shared ancestor.

We recognize three domains—Archaea, Bacteria, and Eukarya (protists, fungi, plants, and animals).

EXPLAINING UNITY IN DIVERSITY

1.4 An Evolutionary View of Diversity

How can organisms be so much alike and still show tremendous diversity? A theory of evolution by way of natural selection is one explanation.

Your traits make you and 6.4 billion other individuals members of the human population. Traits are different aspects of an organism's form, function, or behavior. For example, height and hair color are human traits. Like most traits of any natural population, there are different versions of them among individuals.

Variation in most traits arises through **mutations**, or changes in DNA. Some mutations lead to new traits that make an individual better able to find food, a mate, and so on. We call these *adaptive* traits.

An adaptive trait tends to become more common over generations, because it gives individuals a better chance to live and bear more offspring than individuals that do not have it. When different forms of a trait are becoming more or less common, evolution is under way. To biologists, **evolution** simply means heritable change in a line of descent. Mutations, the source of new traits, provide the variation that serves as the raw material for evolution.

"Diversity" refers to variations in traits that have accumulated in lines of descent. For now, think about what a great naturalist, Charles Darwin, discovered about evolution:

First, populations tend to increase in size, past the capacity of their environment to sustain them, so their members must compete for resources (food, shelter). Second, individuals of natural populations differ from one another in the details of their shared traits. Most variation has a heritable basis.

Third, when individuals differ in their ability to survive and reproduce, the traits that help them do so tend to become more common in the population over generations. This process is called **natural selection**.

Pigeon breeding is *artificial* selection. One form of a trait is favored over others in an artificial environment under contrived, manipulated conditions. Take a look at the pigeons in Figure 1.7. They differ in size, feather color, and other traits. Suppose pigeon breeders allow only the pigeons with the darkest and curliest-tipped feathers to mate. In time, pigeons with black, curly-tipped feathers make up the entire captive population.

Darwin saw that breeding practices could be an easily understood model for natural selection, a favoring of some forms of a given trait over others in nature.

Just as breeders are "selective agents" promoting reproduction of particular captive pigeons, different agents operate across the range of variation in the wild. Pigeon-eating peregrine falcons are among them (Figure 1.7). Swifter or better camouflaged pigeons are more likely to avoid peregrine falcons and live long enough to reproduce, compared with not-so-swift or more conspicuous pigeons.

> Traits are variations in form, function, or behavior that arise as a result of mutations in DNA. Some traits are more adaptive than others to prevailing conditions.
>
> Natural selection is an outcome of differences in survival and reproduction among individuals of a population that vary in one or more heritable traits. The process of evolution, or change in lines of descent, gives rise to life's diversity.

WILD ROCK DOVE

Figure 1.7 Outcome of artificial selection: just a few of the hundreds of varieties of domesticated pigeons, all descended from captive populations of wild rock doves. At right, peregrine falcons are agents of natural selection, in the wild.

1.5 The Nature of Biological Inquiry

The preceding sections introduced some big concepts. Consider approaching this or any other collection of facts with a critical attitude: "Why should I accept that they have merit?" The answer requires a look at how biologists make inferences about observations, then test their inferences against actual experience.

OBSERVATIONS, HYPOTHESES, AND TESTS

To get a sense of how science works, you might start with practices that are common in scientific research:

1. Observe some aspect of nature, carefully check what others have found out about it, then frame a question or identify a problem related to your observation.

2. Think of a **hypothesis**, a proposal that explains the cause of something that is observable, or that makes a generalization about it.

3. Using that hypothesis as a guide, make a **prediction** —a statement of what you should find in the natural world if you were to go out looking for it. This is often called the if–then process. The "if" part of a prediction is the actual hypothesis. *If* gravity does not pull objects toward the Earth, *then* it should be possible to observe apples falling up, not down, from a tree.

4. Devise one or more ways to **test** the prediction, as by making observations, conducting an experiment, or formulating a model. **Models** are theoretical, detailed descriptions of partially-understood processes.

5. If tests do not confirm your prediction, check your results. Rethink your tests, and the prediction. If the tests and prediction still stand, start again by revising your hypothesis.

6. Repeat the tests or devise new ones—the more the better, because hypotheses that withstand many tests are likely to have a higher probability of being useful.

7. Objectively analyze and report test results, and the conclusions you drew from them.

You might hear someone refer to these practices as the scientific method, as if all scientists march to the drumbeat of an absolute, fixed procedure. They do not. Many observe, describe, and report on some aspect of nature, then leave the hypothesizing to others. Some scientists stumble onto information that they are not looking for. Of course, it is not always a matter of luck. Chance seems to favor a mind that has already been prepared, by education and experience, to recognize what new information means. The scientific method is not a single way scientists do their research. It is a critical attitude about being shown rather than told, and taking a logical approach to problem solving.

ABOUT THE WORD "THEORY"

So what exactly constitutes the truth? Or proof of it? Most scientists carefully avoid the word "truth" when discussing science, preferring instead to use "accurate" in reference to data. In science, there is only evidence that supports a hypothesis.

For example, the hypothesis that the sun comes up every day has been supported by observable evidence for all of recorded history. Scientists no longer attempt to disprove that hypothesis, for the simple reason that after thousands of years of observation, no one has yet seen otherwise. The hypothesis is consistent with all of the data that has been collected, and it is now used to make different predictions. When a hypothesis meets these criteria, it has become a **scientific theory**.

You may hear someone apply the word "theory" to a speculative idea, as in the expression "only a theory." But a scientific theory differs from speculation, because it offers valid, testable predictions. Untestable ideas do not belong in science.

The predictive power of a scientific theory has been tested many times, and in many ways, and researchers have yet to find evidence that disproves it. However, a theory cannot be tested under every circumstance, so scientists do not ever assume that a theory is infallible. For example, based on the theory that the sun comes up every day, the prediction that the sun will come up again tomorrow seems like a good one. Yet, a scientist sees tomorrow as an opportunity for that prediction to fail. Like every other theory, the theory that the sun comes up every day will always be open to revision.

After more than a century of many thousands of tests, Darwin's theory of natural selection still stands, with only minor revision. It explains diverse issues, such as how river dams can alter ecosystems and why certain cancers can run in families. Such a well-tested theory is as close to the "truth" as scientists venture. As for any theory, we can say it has a high probability of not being wrong. Even so, the door is always open for scientists to look for data that might disprove it.

A scientific approach to studying nature is based on asking questions, formulating hypotheses, making predictions, testing the predictions, and objectively reporting results.

A scientific theory is a well-tested concept that can be used to explain causes of a broad range of related phenomena. Every scientific theory remains open to testing and revision.

The Persian Gulf is far saltier than most seas, and it's about to get saltier. Several new dams on the Tigris and Euphrates rivers are reducing the flow of fresh water into the Gulf to a trickle. Shown here, the Ataturk Dam on the Euphrates River, where it flows through Turkey. How will increased salinity affect life in the Gulf? It could become an agent of natural selection. It could favor individuals among the Gulf's inhabitants that are better at tolerating higher salt levels.

1.6 The Power of Experimental Tests

Natural processes are often interrelated. Researchers unravel how they work together by studying a single aspect at a time. They design experiments to observe the function, cause, or effect of that aspect in isolation.

AN ASSUMPTION OF CAUSE AND EFFECT

A scientific experiment starts with this premise: *Any aspect of nature has an underlying material cause that can be tested by controlled experiments.* This is how science is different from faith. Each has its place, but a scientific hypothesis is testable in the natural world, in stringent ways that might well disprove it.

Most aspects of nature are outcomes of interacting variables. A **variable** is a feature of an object or event that may differ over time or among the representatives of that object or event. Researchers design experiments that test one variable at a time.

They define a **control group**, which is a standard for comparison with one or more **experimental groups**. There are two kinds of control groups. One is *identical* to an experimental group, but is tested differently. For instance, consider an experiment with two identical groups of mice. The control group is given water, and the experimental group receives water with a drug in it. (In this experiment, the variable is the drug.) The other type of control group *differs* from an experimental group, but is tested the same way. For example, two different species of mice might be given the same drug. In this experiment, one of the species is the control group, and the other species is the experimental group.

EXAMPLE OF AN EXPERIMENTAL DESIGN

Here is a real-life example of an experiment. In 1996, the FDA approved Olestra®, a synthetic fat replacement made from sugar and vegetable oil, as a food additive. Potato chips were the first Olestra-laced food product on the market in the United States. Controversy raged soon after. Some people who got stomach cramps after eating Olestra chips suggested this hypothesis: Olestra causes cramps. Researchers designed an experiment to test the hypothesis. Their prediction was this: If Olestra causes cramps, then people who eat Olestra are more likely to get cramps than people who do not.

To test this prediction, they used a Chicago theater as the "laboratory." They asked more than 1,100 people between ages thirteen and thirty-eight to watch a movie and eat their fill of potato chips. Each person got an unmarked bag that held thirteen ounces of chips. The individuals who received a bag of Olestra-laced potato

Hypothesis
Olestra® causes intestinal cramps.

↓

Prediction
People who eat potato chips made with Olestra will be more likely to get intestinal cramps than those who eat potato chips made without Olestra.

Experiment	**Control Group** Eats regular potato chips	**Experimental Group** Eats Olestra potato chips
Results	93 of 529 people get cramps later (17.6%)	89 of 563 people get cramps later (15.8%)

Conclusion
Percentages are about equal. People who eat potato chips made with Olestra are just as likely to get intestinal cramps as those who eat potato chips made without Olestra. These results do not support the hypothesis.

Figure 1.8 *Animated!* Example of a typical sequence of steps taken in a scientific experiment.

chips were the experimental group. Individuals who got a bag of regular chips were the control group.

Afterward, the researchers contacted the subjects and tabulated any reports of gastrointestinal cramps. Of 563 people in the experimental group, 89 (or 15.8 percent) complained of problems. But so did 93 of 529 people (17.6 percent) in the control group, who had munched on regular chips! The experiment yielded no evidence that eating Olestra-laced potato chips, at least in one sitting, causes gastrointestinal problems. Figure 1.8 summarizes this experiment.

EXAMPLE OF A FIELD EXPERIMENT

Some biologists perform their experiments in natural settings. This one was performed by Durrell Kapan, an evolutionary biologist, in the rain forests of Ecuador.

Many toxic or bad-tasting species are vividly colored, often with distinctive patterning. After eating a few of them and suffering the consequences, predators learn to avoid individuals that display particular visual cues.

Some butterflies that are good-tasting mimic others that are bad-tasting. **Mimicry** is a case of looking like something else and confusing predators (or prey). The naturalist Fritz Müller wondered if mimicry between different species of bad-tasting butterflies would benefit both species. Young birds sample butterflies before learning to avoid eating bad-tasting ones, so if both butterfly species are sampled, then each would lose fewer individuals to predatory birds.

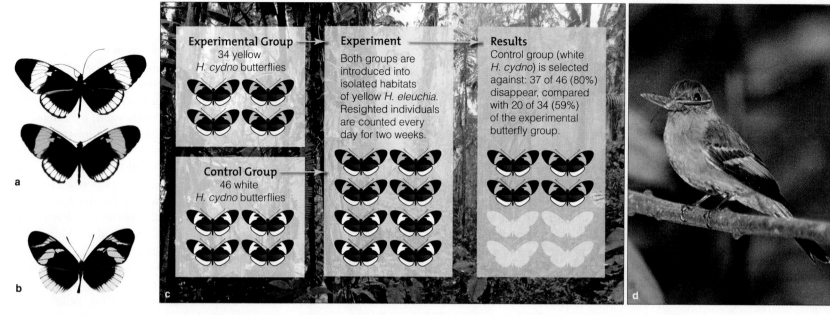

Figure 1.9 *Heliconius* butterflies. (**a**) The two forms of *H. cydno* and (**b**) yellow *H. eleuchia*. (**c**) Kapan's experiment. Both forms of *H. cydno* butterflies were captured and transferred to habitats of *H. eleuchia*. Local predatory birds (**d**), familiar with bad-tasting yellow *H. eleuchia*, avoided the yellow *H. cydno* butterflies but ate most of the white ones.

Kapan tested Müller's hypothesis by performing an experiment on some bad-tasting butterflies in Ecuador. There are two forms of *Heliconius cydno*, a bad-tasting species of butterfly. One has yellow markings on its wings; the other has white (Figure 1.9*a*).

The yellow form of *H. cydno* resembles a different unpalatable species, *H. eleuchia* (Figure 1.9*b*), that lives in another part of the forest. Kapan decided to test the hypothesis that this resemblance would benefit both species. Kapan predicted: If birds have learned to avoid *H. eleuchia* in their part of the forest, then they also will avoid imported yellow *H. cydno* butterflies.

He captured both forms of *H. cydno*. The form with white markings became the control group. The form with the yellow markings was the experimental group. He released both of the groups into isolated *H. eleuchia* habitats. For two weeks, the approximate life span of the butterflies, he counted the survivors.

Kapan found that butterflies of the control group were less likely to survive in the new forest habitat. Because they did not display the visual cue recognized by the local birds—yellow markings—they were more likely to be eaten. The experimental group did better, as the test results in Figure 1.9*c* indicate. We can expect that predatory birds recognized a familiar cue for bad taste—yellow markings—and avoided them. Besides confirming Kapan's prediction, these test results also provide evidence of natural selection in action.

In natural settings, experimenters rarely observe *all* individuals of a group. By necessity, they select subsets or samples of populations, events, and other aspects of nature. But data taken from a sample might differ from data taken from the whole. The difference is called **sampling error**. Results are less likely to be distorted (sampling error can be minimized) with large samples.

BIAS IN REPORTING RESULTS

Experimenters risk interpreting their results in terms of what they want to prove. That is why scientists design experiments to yield *quantitative* results, actual counts or some other data that can be measured and collected objectively. Data give other scientists an opportunity to confirm tests and check conclusions.

This last point gets us back to the value of thinking critically. Scientists expect one another to put aside pride or bias and test ideas in ways that may prove them wrong. If someone will not do so, others will—because science is a competitive community. It is also cooperative. Scientists share ideas, knowing it is just as useful to expose errors as to applaud insights.

Experiments simplify observations in nature by restricting a researcher's focus to one variable at a time. A variable is any feature of an object or event that may differ over time or among representatives of the object or event.

Tests are based on the premise that any aspect of nature has one or more underlying causes. Scientific hypotheses can be tested in ways that might disprove them.

The Scope and Limits of Science

Beyond the realm of scientific inquiry, some events are unexplained. Why do we exist, and for what purpose? Why do we have to die at a particular moment? Such questions lead to *subjective* answers, which come from within as an integrated outcome of all the experiences and mental connections that shape our consciousness. People differ enormously in this regard. That is why subjective answers do not readily lend themselves to scientific analysis and experiments.

This is not to say subjective answers are without value. No human society can function for long unless its individuals share a commitment to standards for making judgments, even if they are subjective. Moral, aesthetic, philosophical, and economic standards vary from one society to the next. But they all guide people in deciding what is important and good, and what is not. All attempt to give meaning to what we do.

Every so often, scientists stir up controversy when they explain something that was thought to be beyond natural explanation, or belonging to the supernatural. This is often the case when a society's moral codes are interwoven with traditional interpretations of the past. Exploring a long-standing view of the natural world from a scientific perspective might be misinterpreted as questioning morality.

As one example, centuries ago in Europe, Nikolaus Copernicus observed the motion of the planets and concluded that Earth circles the sun. Today this seems obvious. Back then, it was heresy. The prevailing belief was that the Creator made Earth—and, by extension, humans—the immovable center of the universe. One respected scholar, Galileo Galilei, studied Copernicus' model of the solar system, thought it was a good one, and said so. He was forced to retract his statement, on his knees, and put Earth back as the fixed center of things. (Word has it that he also muttered, "Even so, it *does* move.") Later, Darwin's theory of evolution also ran up against prevailing belief.

Today, as then, society has sets of standards. Those standards might be questioned when a new, logical explanation runs counter to prevailing beliefs. This does not mean scientists who raise questions are less moral, less lawful, or less spiritual than anyone else. It only means an extra standard guides their work. Their ideas about nature must be testable in the external world, in ways that can be repeated.

The external world, not internal conviction, is the testing ground for the theories generated in science.

Summary

Section 1.1 Life shows many levels of organization. All things, living and nonliving, are made of atoms. Atoms join to form molecules, and molecules of life assemble into cells. Cells are the smallest living units. An organism may be a single cell or multicellular. In most multicelled species, cells are organized as tissues and organs.

A population consists of individuals of the same species in a specified area. A community consists of all populations occupying the same area. An ecosystem is a community and its environment. The biosphere includes all regions of Earth's atmosphere, waters, and land where we find living organisms.

Biology⊘Now
Explore levels of biological organization with the interaction on BiologyNow.

Section 1.2 Life shows unity. All organisms inherit DNA instructions for building proteins and pass them on to offspring. All organisms require energy and raw

materials from the environment to grow and reproduce. All sense changes in the surroundings and respond to them in controlled ways (Table 1.1).

Biology⊘Now
Using instructions with the animation on BiologyNow, view how different objects are assembled from the same materials. Also view energy flow and materials cycling.

Section 1.3 Life shows tremendous diversity. Millions of species exist. Each is unique in some aspects of its body plan, functions, and behavior. Species that are related by descent from a common ancestor are grouped into three domains: Archaea, Bacteria, and Eukarya. Protists, plants, fungi, and animals are eukaryotes.

Biology⊘Now
Explore the characteristics of the three domains of life with the interaction on BiologyNow.

Section 1.4 Mutations change DNA and give rise to new variations of heritable traits. Natural selection occurs if a variation affects survival and reproduction. A population is evolving when adaptive forms of traits are becoming more common than less adaptive forms. Evolution means heritable change in a line of descent.

Biology⊘Now
Learn more about natural selection and evolution with Infotrac readings on BiologyNow.
Also, see the Infotrac article entitled "Will We Keep Evolving?," Ian Tattersall, Time, *April 2000.*

Section 1.5 Scientific methods are varied, but all are based on a logical approach to explaining nature. Scientists observe some aspect of nature, then develop a hypothesis about what might have caused it. They use the hypothesis to make predictions that can be tested by making more observations, building models, or doing experiments.

Scientists analyze test results, draw conclusions, and share this information with other scientists. A hypothesis that does not hold up under repeated testing is modified or discarded. A scientific theory is a long-standing hypothesis that has been supported by many tests.

Section 1.6 Most aspects of nature are complex, an outcome of many interacting variables. Researchers test the predictive power of hypotheses by performing experiments. An experiment allows a scientist to observe what happens to an experimental group when one variable is different. In every scientific experiment, an experimental group is compared with a control group. Scientists share their results so others can confirm their tests and verify their conclusions.

Section 1.7 Science cannot answer all questions. It deals only with aspects of nature that lend themselves to systematic observation, hypotheses, predictions, and experimental testing.

Table 1.1   Summary of Life's Characteristics

Shared characteristics that reflect life's unity

1. Organisms consist of one or more cells.

2. Organisms are constructed of the same kinds of atoms and molecules according to the same laws of energy.

3. Organisms engage in metabolism; they acquire and use energy and materials to survive and reproduce.

4. Organisms sense and make controlled responses to conditions in their internal and external environments.

5. Heritable information encoded in DNA gives organisms a capacity to grow and reproduce. That information also guides the development of complex multicelled organisms.

6. Characteristics that define a population can change over the generations; the population can evolve.

Foundations for life's diversity

1. Mutations (heritable changes in the structure of DNA) give rise to variation in heritable traits, including most details of body form, functioning, and behavior.

2. Diversity is the sum total of variations that have accumulated in different lines of descent over the past 3.9 billion years, by way of natural selection and other processes of evolution.

Self-Quiz

Answers in Appendix I

1. The smallest unit of life is the _____ .

2. _____ is the acquisition of traits after parents transmit their DNA to offspring.
   a. metabolism   c. homeostasis
   b. reproduction   d. inheritance

3. _____ is the capacity of cells to extract energy from sources in the environment, and use it to live, grow, and reproduce.

4. _____ is a state in which the internal environment is being maintained within a tolerable range.

5. A trait is _____ if it improves an organism's ability to survive and reproduce in a given environment.

6. Differences in heritable traits arise through _____ .

7. Researchers assign all species to one of three _____ .

8. _____ secure energy from their surroundings.
   a. producers   c. decomposers
   b. consumers   d. all of the above

9. DNA _____ .
   a. contains instructions for building proteins
   b. undergoes mutation
   c. is transmitted from parents to offspring
   d. all of the above

10. A control group is _____ .
    a. a standard against which experimental groups can be compared
    b. identical to experimental groups except for one variable only
    c. both a and b

11. Match the terms with the most suitable descriptions.
    ____ adaptive trait   a. statement of what you can expect to observe in nature
    ____ natural selection   b. proposed explanation; an educated guess
    ____ scientific theory   c. improves chances of surviving and reproducing
    ____ hypothesis   d. long-standing, much tested hypothesis used to interpret other hypotheses
    ____ prediction   e. outcome of differences in survival, reproduction among individuals of a population that differ in details of their traits

Additional questions are available on Biology ⊛Now™

Critical Thinking

1. Witnesses in a court of law are asked to "swear to tell the truth, the whole truth, and nothing but the truth." What are some of the problems inherent in this question? Can you think of a better alternative?

2. Many popular magazines publish articles on exercise, diet, and other health-related topics. Authors disagree on what is the "best" diet or the "best" supplements. What kinds of evidence do you think the articles should include so that you can decide whether to accept what they say?

3. The Olestra® potato chip experiment in Section 1.6 was a *double-blind* study: Neither the subjects of the experiment nor the researchers knew which potato chips were in which bag until after all the subjects had reported. Why do researchers perform double-blind experiments? What do think are some challenges in designing and carrying out such an experiment?

4. In 1988 Dr. Randolph Byrd and his colleagues made a study of 393 patients admitted to San Francisco General Hospital's Coronary Care Unit. In this experiment, "born-again" Christian volunteers were asked to pray daily for a patient's rapid recovery and also for the prevention of complications and death. None of the patients knew if he or she was being prayed for or not, and no volunteers or patients knew each other.

How each patient fared was classified by Byrd as "good," "intermediate," or "bad." Byrd reported that the patients who had been prayed for fared better than those who had not. His was the first experiment to document, in a scientific fashion, results in support of the prediction that prayer can have beneficial effects on the medical outcome of seriously ill patients.

Publication of these results elicited a storm of criticism from scientists who cited bias in Byrd's experimental design. For instance, Byrd himself classified the patients before observations of them were finished. Think about how bias might play a role in interpreting medical data. Why do you think this study got such a dramatic response from the scientific community?

5. A scientific theory makes a general conclusion from particular instances. The assumption is that an outcome that has been observed to happen with regularity will happen again. However, there is no way to anticipate all of the variables that may affect an outcome, as this parable illustrates.

Once there was an intelligent turkey. It lived in a pen, attended by a kind master, and had nothing to do but reflect upon the world's wonders and its regularities. The turkey observed some major regularities. Every morning began with the sky becoming light, followed by the clop, clop, clop of the master's footsteps, which was followed by the appearance of delicious food. Other things varied, but food always followed footsteps.

The sequence of events was so predictable that it became the entire basis of the turkey's theory of the goodness of the world.

One morning, after a hundred confirmations of the theory, the turkey listened for the clop, clop, clop of its master's footseps, heard them, and had its head chopped off.

The unfortunate turkey did not understand that explanations about the world only have a high or low probability of not being wrong. Some people take this uncertainty to mean that "facts are irrelevant—facts change." Should we stop doing scientific research? Why or why not?

Science or Supernatural?

About 2,000 years ago, in the mountains of Greece, the oracle of Delphi delivered prophecies from Apollo after inhaling sweet-smelling fumes that had collected in the sunken floor of her temple. Her prophecies tended to be rambling and cryptic. Why? She was babbling in a hydrocarbon-induced trance. Geologists recently found intersecting, earthquake-prone faults under the temple. Whenever the faults slipped, methane, ethane, and ethylene seeped out from the depths. All three gases are mild narcotics and hallucinogens.

methane

ethane

ethylene

Methane, ethane, and ethylene are nothing more than a few carbon and hydrogen atoms; hence the name hydrocarbons. Thanks to scientific inquiry, we now know more about them than the ancient Greeks did. For example, we know that methane was present when Earth first formed. We know that it is released into the atmosphere when volcanoes erupt, when we burn wood, peat, or fossil fuels, and when termites and cattle pass gas. We know it is one of the greenhouse gases contributing to global warming.

A legacy of the buried organic remains and wastes of countless marine organisms, methane also seeps from the ocean floor. There, the high pressure and low temperature of deep water freeze the gas as it bubbles up. A thousand billion tons of icy methane crystals lie at the bottom of the seas around the world. What will happen if the ocean warms up enough to melt those crystals is anyone's guess.

In short, knowledge about lifeless substances like methane can tell you a lot about life—past, present, and future—from Greek history, to current topics of health and disease, to physical and chemical conditions that influence the biosphere.

 How Would You Vote? *Undersea methane deposits might be developed as a vast supply of energy, but the environmental costs are unknown. Should we continue to move toward exploiting this resource? See BiologyNow for details, then vote online.*

Key Concepts

ATOMS BOND
All substances are made of one or more elements. Atoms are the smallest units of an element that still retain the element's properties. They are composed of protons, neutrons, and electrons. Bonds between atoms form molecules. Whether and how one atom will bond with another depends on the arrangements of their electrons.

WATER AND PH
Water's unique characteristics, including temperature-stabilizing effects, cohesion, and solvent properties, make life possible on Earth. Living things adapt to changing amounts of hydrogen ions and other substances dissolved in water.

BIOLOGICAL MOLECULES
Living cells combine small carbon-containing subunits into complex carbohydrates and lipids, proteins, and nucleic acids. All of these large molecules of life have a backbone of carbon atoms. Functional groups bonded to the backbone help determine the properties of a molecule.

Links to Earlier Concepts

With this chapter, we start at life's basic level of organization, so take a moment to review the simple chart in Section 1.1. You will come across an early example of how the body's built-in mechanisms return the internal environment to a homeostatic state when conditions shift beyond ranges that cells can tolerate (1.2).

Here again, you will be considering one of the consequences of mutation in DNA (1.4), this time with sickle-cell anemia as the example.

LINK TO
SECTION
1.1

2.1 Atoms and Their Interactions

A line of a million atoms would fit on the period ending this sentence. About ninety-two types of atoms make up everything in the world around you.

ATOMS AND ISOTOPES

Think about an apple. An apple, like all other matter, takes up space and responds to gravity. Now imagine using anything in your garage to demolish the apple completely: crush it, burn it, pour acid on it, plug it into an electrical socket, blow it up. When you finish, the smallest remaining bit of apple will be an atom.

Atoms are fundamental forms of matter—they can't be degraded to anything else, at least not by ordinary means. But if you use a little nuclear physics, you can separate atoms into smaller particles called protons, neutrons, and electrons. A **proton** (p^+) has a positive *charge*, or defined amount of electricity. Protons and neutrons are in the nucleus, or core region, of an atom. Zipping about that nucleus are the negatively charged **electrons** (e^-). The positive charge of a proton and the negative charge of an electron balance each other; an atom that has the same number of electrons as protons has no net electrical charge.

All matter consists of different kinds of atoms called **elements**. Figure 2.1 lists the most common elements

in humans and in Earth's crust. Each element has its own *unique* number of protons in its atoms. Hydrogen atoms have one proton, carbon atoms have six.

Atoms of the same element do not necessarily have the same number of neutrons. Atoms with the same number of protons but a different number of neutrons are **isotopes**. Carbon has three isotopes, nitrogen has two, and so on. We identify isotopes by superscript numbers on standardized symbols for elements. For example, the isotopes of carbon are ^{12}C (or carbon 12, the most common isotope of carbon, with six protons and six neutrons), ^{13}C (with six protons and seven neutrons), and ^{14}C (six protons, eight neutrons).

Many isotopes are *radio*isotopes, which means that the atoms are unstable, or radioactive. An atom that is radioactive spontaneously emits energy and particles as its nucleus disintegrates. This process, radioactive decay, transforms one element into a different, more stable element at a known rate. For instance, in about 5,700 years, about half of the atoms in a sample of ^{14}C will decay into ^{14}N (nitrogen). Measuring proportions of radioactive elements can tell us how old the Earth is (very old!). It tells us when organisms, now fossilized, originated and later vanished.

Living systems use different isotopes of an element in the same way: Carbon is carbon, regardless of how many neutrons it has. We can substitute a radioactive isotope for a stable atom in a molecule, and then feed it into a living system as a *tracer*, or probe. The energy from radioactive decay acts like a shipping label that researchers can track as it passes through the system.

For example, PET (positron-emission tomography) uses tracers to form images of body tissues. Clinicians attach a radioisotope to a molecule such as glucose and inject it into a patient. All active body cells take up glucose, radioactive or not. Some cells take up more tracer (thus becoming more radioactive) than others, indicating differences in cell activity. These differences can be detected and converted to color-coded scans, as you will see in Chapter 25.

ELECTRONS AND ENERGY LEVELS

Simple physics is great for explaining the motion of an apple falling off a tree. An electron is so tiny that it belongs to a world where simple physics just does not apply. Different forces bring about its motion. Where it can go depends on how many other electrons are buzzing around an atom's nucleus. All electrons in an atom occupy orbitals, which are volumes of space in which the electrons are likely to be at any instant.

Think of an atom as an apartment building with a nucleus in the basement, and many floors of rooms to rent to electrons. Each floor of that apartment building

Figure 2.1 You are far more than the sum of your chemical parts, which, commercially speaking, are worth just $118.63, retail price. Proportions of the most common elements making up humans and most other living things are shown here. There are a lot more *trace* elements, each less than 0.01 percent of body weight.

human body	
Oxygen	61.0%
Carbon	23.0
Hydrogen	10.0
Nitrogen	2.6
Calcium	1.4
Phosphorus	1.1
Potassium	0.2
Sulfur	0.2

Earth's crust	
Oxygen	46.0%
Silicon	27.0
Aluminum	8.2
Iron	6.3
Calcium	5.0
Magnesium	2.9
Sodium	2.3
Potassium	1.5

ATOMS BOND

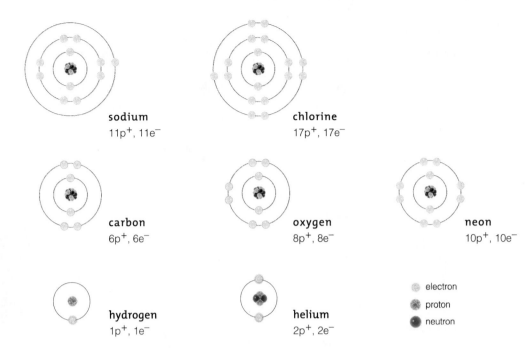

c The **third shell** shows the third energy level, a combined set of nine orbitals that have room for 18 electrons in total. Sodium has one electron in the third shell, and chlorine has seven. Both have vacancies, so both form chemical bonds.

sodium
11p⁺, 11e⁻

chlorine
17p⁺, 17e⁻

b The **second shell** shows the second energy level, representing a combined set of four orbitals with room for up to eight electrons. Carbon has six electrons, two in the first shell and four in the second shell, so it has four vacancies. Oxygen has two. Both atoms form chemical bonds. Neon, with no vacancies, does not.

carbon
6p⁺, 6e⁻

oxygen
8p⁺, 8e⁻

neon
10p⁺, 10e⁻

a The **first shell** shows the first energy level, a single orbital with room for two electrons. Hydrogen has one electron in this orbital. Hydrogen gives up its electron easily, becoming a chemically reactive free proton. A helium atom has two electrons in this orbital. Having no vacancies, helium does not usually form chemical bonds.

hydrogen
1p⁺, 1e⁻

helium
2p⁺, 2e⁻

- electron
- proton
- neutron

Figure 2.2 Shell model. Using this model, it is easy to see the vacancies in each atom's outer orbitals. Each circle represents all of the orbitals on one energy level. Larger circles correspond to higher energy levels. This model is highly simplified; a more realistic rendering would show the electrons as fuzzy clouds of probability density about ten thousand times bigger than the nucleus.

is one energy level. Each orbital is like one room that can accommodate up to two electrons. There is only one room on the first floor (one orbital at the lowest energy level), and it fills first. Hydrogen, the simplest atom, has a single electron in that room. Helium has two. Additional electrons rent second-floor rooms in larger atoms. When the second floor rooms are filled, more electrons can rent third-floor rooms, and so on. *Electrons fill orbitals at successively higher energy levels.*

The farther an electron is from the basement (the nucleus), the greater its energy. An electron in a first-floor room cannot move to the second or third floor, let alone the penthouse, unless a boost of energy gets it there. Suppose it absorbs the right amount of energy, say, from sunlight, to get excited about moving up. Move it does. If nothing fills that lower room, though, the electron will quickly go back to it, emitting extra energy as it does. Later, you will see how some cells in plants and in your eyes can capture that energy.

In the **shell model**, nested "shells" correspond to energy levels (Figure 2.2). They offer us an easy way to check for electron vacancies in various atoms. Bear in mind, atoms do not look like these flat diagrams. The shells are not three-dimensional volumes of space, and they do not show the individual electron orbitals.

Atoms that have vacancies in the outermost "shell" tend to give up, acquire, or share electrons with other atoms. This kind of electron-swapping between atoms is known as **chemical bonding** (Section 2.2). Atoms with zero vacancies rarely bond with other atoms. By

contrast, the most common atoms in organisms—such as oxygen, carbon, hydrogen, nitrogen, and calcium—have vacancies in orbitals at their outermost energy level. And they do bond with other atoms.

A union between orbitals of two or more atoms—a *chemical bond*—is the glue that makes molecules out of atoms. Some molecules consist of atoms of just one element. Molecular nitrogen (N_2), with two nitrogen atoms, is like this. So is molecular oxygen (O_2).

Molecules with unvarying proportions of two or more elements are **compounds**. A water molecule, for instance, always has one oxygen and two hydrogen atoms. The ones in a Siberian lake, a bathtub, a leaf, or anywhere else have exactly twice as many hydrogen as oxygen atoms. In a **mixture**, two or more molecules intermingle without bonding. For example, you can make a mixture by swirling water and sugar together. The proportions of elements in a mixture can vary.

vacancy

no vacancy

Electrons occupy orbitals, or defined volumes of space around an atom's nucleus. Successive orbitals correspond to levels of energy, which become higher with distance from the nucleus.

One or at most two electrons can occupy any orbital. Atoms that have vacancies in their highest-level orbitals can interact with other atoms.

A molecule is two or more atoms joined in a chemical bond. Atoms of two or more elements are bonded together in compounds. A mixture consists of intermingled molecules.

sodium atom
11 p+
11 e−
no net charge

chlorine atom
17 p+
17 e−
no net charge

electron loss

electron gain

sodium ion
11 p+
10 e−
net + charge

chloride ion
17 p+
18 e−
net − charge

an **ionic bond** forms between a positively charged sodium ion and a negatively charged chloride ion

a Ionic bonds. A sodium atom loses the lone electron in its outermost shell, becoming a positively charged sodium ion (Na+). A chlorine atom gains an electron in its outermost shell, becoming a negatively charged chloride ion (Cl−). The two ions form an ionic bond: They stay together by the mutual attraction of opposite charges. Each crystal of table salt, or NaCl, is a cubic lattice of ionically bonded sodium and chloride ions.

Figure 2.3 *Animated!* Important bonds in biological molecules.

2.2 Bonds in Biological Molecules

The distinctive properties of biological molecules start with the properties of bonds that unite their atoms.

ION FORMATION AND IONIC BONDING

An electron, recall, has a negative charge equal to a proton's positive charge. When an atom has as many electrons as protons, these charges balance each other, so the atom will have a net charge of zero.

Atoms with more electrons than protons carry a net negative charge, and atoms with more protons than electrons carry a net positive charge. An atom that has either a positive or negative charge is known as an **ion**. Ions form when atoms gain or lose electrons.

Example: An uncharged chlorine atom has seven electrons in the third shell, which is most stable when filled with eight electrons. Chlorine tends to grab an electron from other places to fill the vacancy in its third shell. That extra electron will make it a chloride ion (Cl−), with a net negative charge. A sodium atom tends to lose the lone electron in its third shell. If it

does, it becomes a sodium ion with a net positive charge (Na+), and two full shells.

Opposite charges attract. If a positively charged ion encounters a negatively charged ion, they may stay close together. A close association of ions is an **ionic bond**. Figure 2.3a shows crystals of table salt, or NaCl. In such crystals, ionic bonds constrain the ions in an orderly arrangement.

COVALENT BONDING

In an ionic bond, an atom that has gained electrons is associated with an atom that has lost electrons. What if both atoms tend to gain (or to lose) electrons? By *sharing* electrons, each atom's vacancy becomes partly filled with an electron from the other atom. Two atoms that are sharing a pair of electrons are connected by a **covalent bond**. A covalent bond is much stronger than an ionic bond (Figure 2.3b).

We depict chemical bonding patterns with structural formulas to show how atoms are physically arranged in a molecule. A single line that connects two atoms in a structural formula represents two shared electrons, or one covalent bond. Molecular hydrogen, with one covalent bond, is written H—H.

molecular hydrogen (H—H)

Two hydrogen atoms, each with one proton, share two electrons in a single covalent bond that is nonpolar.

molecular oxygen (O=O)

Two oxygen atoms, each with eight protons, share four electrons in a double covalent bond, also nonpolar.

water (H—O—H)

Two hydrogen atoms each share an electron with oxygen. The resulting two covalent bonds form a water molecule. These bonds are polar. The oxygen exerts a greater pull on the shared electrons, so it bears a slight negative charge. Each of the hydrogens bears a slight positive charge.

b Covalent bonds. Two atoms with unpaired electrons in their outermost shell become more stable by sharing electrons. Two electrons are shared in each covalent bond. When the electrons are shared equally, the covalent bond is nonpolar. If one atom exerts a slightly greater pull on the shared electrons, the covalent bond is polar.

hydrogen bond

water molecule   ammonia molecule

Two molecules interacting in one hydrogen (H) bond.

H bonds helping to hold part of two large molecules together.

Many H bonds hold DNA's two strands together. Individually, each H bond is weak, but collectively they stabilize DNA's large structure.

c Hydrogen bonds. Such bonds can form at a hydrogen atom that is already covalently bonded in a molecule. The atom's slight positive charge weakly attracts an atom with a slight negative charge that is already covalently bonded to something else. As shown above, this can happen between one of the hydrogen atoms of a water molecule and the nitrogen atom of an ammonia molecule.

Two atoms can share two electron pairs in a *double* covalent bond. Molecular oxygen (O=O) is like this. In a *triple* covalent bond, two atoms share three pairs, as they do in molecular nitrogen (N≡N). Each time you breathe in, a tremendous number of gaseous O_2 and N_2 molecules flows into your lungs.

In a *nonpolar* covalent bond, atoms share electrons equally, so there is no difference in charge between the "ends" of the bond. Molecular hydrogen (H_2) has such symmetry, as does methane (CH_4).

A covalent bond is *polar* if the atoms do not share electrons equally. One atom pulls the electrons a little more toward its end of the bond, so that part of the molecule bears a slight negative charge. This leaves a slight positive charge on the other end of the bond. A water molecule (H—O—H) has two polar covalent bonds; the oxygen carries a slight negative charge, and the hydrogens each carry a slight positive charge. We return to the polarity of water in the next section.

HYDROGEN BONDING

A hydrogen atom taking part in a polar covalent bond bears a slight positive charge, so it attracts negatively charged atoms. When the negatively charged atom is already taking part in a different covalent bond, the interaction between it and the hydrogen atom is called a **hydrogen bond** (Figure 2.3*c*). Hydrogen bonds are individually weak, so they break easily. But they also form easily, and often.

Many hydrogen bonds working together become a strong force that plays a very important role in the structure, and so the function, of biological molecules. Hydrogen bonding between different parts of a very large molecule hold it in a particular shape. Hydrogen bonds hold the two nucleotide strands of large DNA molecules together, for example.

Ions form when atoms acquire a net charge by gaining or losing electrons. Two ions of opposite charge attract each other. They can associate in an ionic bond.

In a covalent bond, atoms share electrons. When atoms share a pair of electrons equally, the bond is nonpolar. When the sharing is not equal, the bond is polar—slightly positive at one end, slightly negative at the other.

In a hydrogen bond, a covalently bound hydrogen atom bearing a slight positive charge is attracted to a covalently bound atom bearing a slight negative charge.

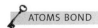
ATOMS BOND

LINK TO
SECTION
1.2

2.3 Water's Life-Giving Properties

Life originated in water. Organisms either live in it or they carry water around inside cells and tissue spaces. The properties of water that are so essential to life exist because of its unique hydrogen bonding behavior.

POLARITY OF THE WATER MOLECULE

Figure 2.4*a* shows the structure of one water molecule. In it, two hydrogen atoms have formed polar covalent bonds with an oxygen atom. The molecule has no net charge. Even so, the oxygen pulls the shared electrons more than the hydrogen atoms do. Thus, the molecule of water has an "end" with a slight negative charge, and two "ends" with slight positive charges.

The polarity of each water molecule attracts other water molecules. Hydrogen bonds form between them (Figure 2.4*b*). Water also forms hydrogen bonds with sugars and other polar molecules. That is why polar molecules are **hydrophilic** (water-loving) substances.

That same polarity repels oils and other nonpolar molecules, which are **hydrophobic** (water-dreading) substances. Shake a bottle filled with water and salad oil, then set it on a table. Soon, new hydrogen bonds replace the ones that the shaking broke. The reunited water molecules push out oil molecules, which cluster as oil droplets or as an oily film at the water's surface.

The same kinds of interactions occur at the thin, oily membrane between the water inside and outside cells. Membrane organization, and life itself, begins with hydrophilic and hydrophobic interactions. You'll be reading about membrane structure in Chapter 3.

WATER'S TEMPERATURE-STABILIZING EFFECTS

All molecules jiggle nonstop, and they move faster as they absorb heat. **Temperature** measures the energy of their motion. Extensive hydrogen bonding in water restricts the jiggles of individual water molecules. In other words, the hydrogen bonds absorb some of the energy of motion of water molecules. The result is that more heat is needed to raise the temperature of water

slight negative charge
on the oxygen atom

(−)

Overall, the molecule carries no net charge

slight positive charge
on each hydrogen atom

a

Figure 2.4 *Animated!* Water, essential for life. (**a**) Polarity of an individual water molecule.

(**b**) Hydrogen bonding pattern among water molecules in liquid water. The dashed lines signify hydrogen bonds.

(**c**) Hydrogen bonding pattern of ice. Below 0°C, every water molecule hydrogen-bonds to four others in an icy three-dimensional array. Water molecules are farther apart in ice than they would be in liquid water, because they vibrate less and so more hydrogen bonds can form. Ice floats because it has fewer molecules per unit volume (it is less dense) than liquid water.

Figure 2.5 Two spheres of hydration.

Figure 2.6 Examples of water's cohesion. (**a**) When a pebble hits liquid water and forces molecules away from the surface, the individual water molecules don't fly off every which way. They stay together in droplets. Why? Countless hydrogen bonds exert a continuous inward pull on individual molecules at the surface.

(**b**) And just how does water rise to the very top of trees? Cohesion, and evaporation from leaves, pulls it upward.

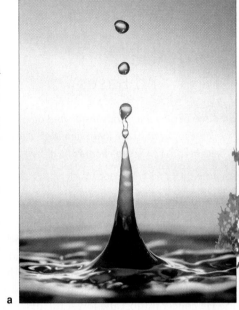

a

b

compared to other fluids. So, water is a heat reservoir, meaning that its temperature stays fairly stable even when the temperature of the environment does not.

When water temperature is below its boiling point, hydrogen bonds form as fast as they break. Energy input (heat) increases molecular motion so much that the hydrogen bonds stay broken, and molecules at the water's surface escape into air. By this process, called **evaporation**, heat energy converts liquid water to a gas. Energy input overcomes the attraction between molecules of water, which break free.

The surface temperature of water decreases during evaporation. Evaporative water loss (sweat is about 99 percent water) helps you and some other mammals cool off on hot, dry days.

Below 0°C, water molecules do not move enough to break their hydrogen bonds, and they become locked in the latticelike bonding pattern of ice (Figure 2.4c). Ice is less dense than water. During winter freezes, ice sheets may form near the surface of ponds, lakes, and streams. These "ice blankets" insulate the liquid water beneath them, helping to protect many fishes, frogs, and other aquatic organisms from freezing.

WATER'S SOLVENT PROPERTIES

Water is an excellent solvent, meaning ions and polar molecules easily dissolve in it. A dissolved substance is called a **solute**. In general, a substance is *dissolved* after water molecules cluster around ions or molecules of it and keep them dispersed in fluid.

Water molecules that cluster around a solute form a "sphere of hydration." Such spheres form around any solute in cellular fluids, a maple tree's sap, blood, the fluid in your gut, and every other fluid associated with life. Spheres of hydration keep ions dispersed in fluid. Watch it happen as you pour table salt (NaCl)

into a cup of water. After a while, the crystals dissolve as they separate into sodium ions (Na^+) and chloride ions (Cl^-). Every Na^+ attracts the oxygen ends of many water molecules, and every Cl^- attracts the hydrogen ends of many water molecules (Figure 2.5).

WATER'S COHESION

Another life-sustaining property of water is cohesion. **Cohesion** means that molecules resist separating. You see its effect as surface tension when you toss a pebble into a pond or some other body of water (Figure 2.6a). The water bounces, waves, sprays, and bubbles, but it does not fly apart into molecules. Hydrogen bonds are exerting a continuous, inward pull on the individual water molecules. This pull is so strong that they stick together in drops and pools rather than spreading out in a thin film like other liquids do.

Cohesion is in play inside organisms, too. Plants, for example, absorb nutrient-laden water while they grow. Very narrow columns of liquid water rise inside pipelines of vascular tissues, which extend from roots to leaves and other plant parts. Water evaporates from leaves as molecules break free and diffuse into the air (Figure 2.6b). The cohesive force of hydrogen bonds pulls replacements into the leaf cells, in ways you will read about in Section 18.7.

> Being polar, water molecules hydrogen-bond to one another and to other polar (hydrophilic) substances. They tend to repel nonpolar (hydrophobic) substances.
>
> Extensive hydrogen bonding between water molecules gives water unique properties that make life possible.
>
> Water has cohesion, temperature-stabilizing effects, and a capacity to dissolve many substances.

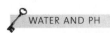

WATER AND PH

LINK TO
SECTION
1.2

2.4 Acids and Bases

Ions are dissolved in fluids inside and outside a cell. Among the most influential are hydrogen ions. They have major effects because they are chemically active.

THE PH SCALE

At any instant in liquid water, some water molecules are split into hydrogen (H^+) and hydroxide (OH^-) ions. Hydrogen ions are the basis of the **pH scale**. The scale is a way to measure the relative amount of hydrogen ions in solutions such as seawater, blood, or sap. The greater the H^+ concentration, the lower the pH. Pure water (not rainwater or tap water) always has as many H^+ as OH^- ions. This state is neutrality, or pH 7.0 (Figure 2.7).

A one unit decrease from neutrality corresponds to a tenfold increase in H^+ concentration, and an increase by one unit corresponds to a tenfold decrease in H^+ concentration. One way to get a sense of the range is to taste dissolved baking soda (pH 9), water (pH 7), and lemon juice (pH 2).

Acids lose hydrogen ions in solution. **Bases** acquire them. *Acidic* solutions, such as lemon juice, gastric fluid, and coffee, contain more H^+ than OH^-; the pH of these solutions is below 7. *Basic,* or alkaline, solutions like egg white, baking soda, and seawater contain more OH^- than H^+ and have a pH greater than 7.

Nearly all of life's chemistry occurs near pH 7. Most of your body's internal environment (tissue fluids and blood) is between pH 7.3 and 7.5. Seawater is more basic than the body fluids of the organisms living in it.

Acids and bases can be weak or strong. Weak acids, such as carbonic acid (H_2CO_3), are stingy H^+ donors. Strong acids readily give up H^+ in water. An example is the hydrochloric acid (HCl) that dissociates into H^+ and Cl^- inside your stomach. The hydrogen ions make gastric fluid very acidic, which in turn activates local protein-digesting enzymes.

HCl splashing up out of the stomach can cause *acid indigestion*. Antacids taken for this condition, including milk of magnesia, release OH^- ions that combine with H^+ to reduce the pH of stomach contents.

High concentrations of strong acids or bases can disrupt ecosystems, and make it impossible for cells to survive. Read the labels on containers of ammonia, drain cleaner, and other products commonly stored in households. Many cause severe *chemical burns*. So does sulfuric acid in car batteries. Fossil fuel burning and nitrogen fertilizers release strong acids that lower the pH of rainwater (Figure 2.8). Some regions are quite sensitive to this *acid rain*. Alterations in the pH of soil and water can harm organisms. We return to this topic in Section 30.7.

SALTS AND WATER

A **salt** is any compound that dissolves easily in water and releases ions *other than* H^+ and OH^-. It commonly forms when an acid interacts with a base. For example:

$$HCl\ (acid) + NaOH\ (base) \rightleftharpoons NaCl\ (salt) + H_2O$$

HYDROCHLORIC ACID   SODIUM HYDROXIDE   SODIUM CHLORIDE

When dissolved in water, the salt product of this reaction (NaCl) dissociates into sodium ions (Na^+) and chloride ions (Cl^-). Many salt ions are important components of cellular processes. For example, sodium, potassium, and calcium ions help nerve and muscle cells function. They also help plant cells take up water from soil.

- hydrochloric acid (HCl) — 0 — 10^0
- gastric fluid (1.0–3.0) — 1 — 10^{-1}
- lemon juice, some acid rain — 2 — 10^{-2}
- vinegar, wine, beer, oranges — 3 — 10^{-3}
- tomatoes — 4 — 10^{-4}
- bananas
- black coffee
- bread — 5 — 10^{-5}
- typical rainwater
- urine (5.0–7.0) — 6 — 10^{-6}
- milk (6.6)
- pure water ($H^+ = OH^-$) — 7 — 10^{-7}
- blood (7.3–7.5)
- egg white (8.0) — 8 — 10^{-8}
- seawater (7.8–8.3)
- baking soda
- phosphate detergents, bleach, antacids — 9 — 10^{-9}
- soapy solutions, milk of magnesia — 10 — 10^{-10}
- household ammonia (10.5–11.9) — 11 — 10^{-11}
- — 12 — 10^{-12}
- hair remover
- oven cleaner — 13 — 10^{-13}
- sodium hydroxide (NaOH) — 14 — 10^{-14}

Figure 2.7 *Animated!* The pH scale, representing concentrations of hydrogen ions in a solution. Also shown are approximate pH values of some liquids. This pH scale ranges from 0 (most acidic) to 14 (the most basic). A change of 1 on the scale means a tenfold change in H^+ concentration.

BUFFERS AGAINST SHIFTS IN PH

Cells must respond quickly to even slight shifts in pH, because enzymes and many other essential biological molecules can function properly only within a narrow range of pH. A slight deviation from that range halts cellular processes.

Most body fluids maintain a stable pH with buffers. A **buffer system** is a dynamic chemical partnership between a weak acid or base and its salt. These two related chemicals work in an equilibrium to counter slight shifts in pH. For example, if a basic substance is added to a buffered fluid, the weak acid donates some H^+ ions to the solution. Excess OH^- ions from the base combine with H^+ ions from the acid, forming water.

Carbon dioxide, a by-product of many reactions, combines with water in the blood to compose a buffer system of carbonic acid and bicarbonate ions. The acid can combine with free OH^-, forming bicarbonate (salt) ions and water:

$$OH^- + H_2CO_3 \longrightarrow HCO_3^- + H_2O$$
CARBONIC   BICARBONATE   WATER
ACID   (SALT)

The carbonate ions can combine with free H^+, forming carbonic acid:

$$HCO_3^- + H^+ \longrightarrow H_2CO_3$$
BICARBONATE   CARBONIC
ACID

Together, these reactions keep the blood pH between 7.3 and 7.5, but only up to a point. A buffer can absorb only so many excess ions. Even slightly more than that limit will cause the pH to swing widely.

The effects of buffer system failure in a biological system can be devastating. In *acute respiratory acidosis*, carbon dioxide accumulation leads to excess carbonic acid in the blood. The resulting drop in blood pH may cause a person may fall into a *coma*, a dangerous state of unconsciousness.

Coma may also occur in *alkalosis*, which is a rise in blood pH. Nausea, vomiting, and headache are among the other outcomes. If blood pH increases to just 7.8, prolonged muscle spasms called *tetany* may occur.

Ions in fluids on the inside and outside of cells have key roles in cell function. Acidic substances release hydrogen ions when dissolved, and basic substances accept them. Salts release ions other than H^+ and OH^-.

Acid–base interactions help maintain pH, which reflects the H^+ concentration in a fluid. Buffer systems keep the pH of body fluids at levels suitable for life.

Figure 2.8 Sulfur dioxide emissions from a coal-burning power plant. Some airborne pollutants, including sulfur dioxide, dissolve in water vapor to form acidic solutions that are a component of acid rain.

2.5 Molecules of Life—From Structure to Function

Under present-day conditions on Earth, only living cells can make complex carbohydrates, lipids, proteins, and nucleic acids. These are the molecules of life, and their structures hold clues to how each kind functions.

WHAT IS AN ORGANIC COMPOUND?

The molecules of life are **organic compounds**, which are compounds with covalent bonds between carbon and hydrogen atoms. The term is a holdover from a time when chemists defined "organic" substances as those made by animals and vegetables, as opposed to "inorganic" substances from other sources, like rock. Now, we can make organic compounds in laboratories. And we found out that these compounds were present on Earth even before life originated.

Vitamins, aspirin, sugars, and caffeine are organic compounds. They and most other organic compounds have at least one **functional group**: an atom (other than hydrogen) or cluster of atoms bonded to carbon. Hydrocarbons are the exception; they have carbon and hydrogen atoms only. Methane, ethane, and ethylene (shown in the chapter opener) are examples. We use a color code to depict the atoms shown at right.

 carbon (C)
 oxygen (O)
 hydrogen (H)
 nitrogen (N)
 calcium (C)
 phosphorus (P)
 potassium (K)
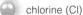 sulfur (S)
sodium (Na)
chlorine (Cl)

IT ALL STARTS WITH CARBON'S BONDING BEHAVIOR

Living organisms consist mainly of oxygen, hydrogen, and carbon. Most of the oxygen and hydrogen are in the form of water. Put water aside, and carbon makes up more than half of what is left.

Carbon's importance in life arises from its versatile bonding behavior. *Each carbon atom can covalently bond with as many as four other atoms.* Such bonds, in which two atoms share one, two, or three pairs of electrons, are relatively stable. Covalent bonds often join carbon atoms as "backbones" to which hydrogen, oxygen, and other elements are attached. The three-dimensional shapes of organic compounds start with these bonds.

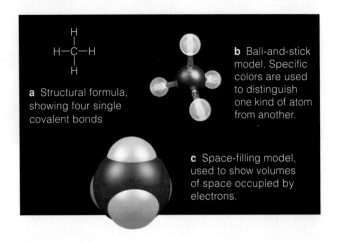

a Structural formula, showing four single covalent bonds

b Ball-and-stick model. Specific colors are used to distinguish one kind of atom from another.

c Space-filling model, used to show volumes of space occupied by electrons.

Figure 2.9 Molecular models for methane (CH_4).

Figure 2.9 shows some different ways to represent methane, the simplest organic compound. A colorless, odorless gas, methane is one carbon atom covalently bonded to four hydrogen atoms (CH_4). The ball-and-stick model conveys its bond angles and the way the atoms are distributed. The space-filling model is better to show relative sizes and surfaces.

Figure 2.10*a* is a ball-and-stick model for glucose, another organic compound. Hydrogen and oxygen are covalently bonded to a backbone of six carbon atoms. This backbone usually coils back on itself, with two carbons joined to form a ring structure (Figure 2.10*b*).

Other models look simpler. A flat structural model might show carbons but not hydrogen atoms bonded to it (Figure 2.10*c*). Some icons for a ring show only atoms that are *not* carbon or hydrogen.

We use these depictions of structure to help us visualize how molecules function. For instance, virus particles can infect a cell by attaching to its membrane proteins. Like Lego blocks, the virus surface proteins have ridges, clefts, and charged regions that match up to clefts, ridges, and charged regions of proteins at cell membrane surfaces.

FUNCTIONAL GROUPS

A functional group is an atom or cluster of atoms that imparts reactive properties to a molecule. How much can one functional group do? Consider a seemingly tiny difference—reversed functional groups—between the sex hormones shown in Figure 2.11. Early on, a vertebrate embryo appears neither male nor female; it has a set of tubes that will develop into a reproductive system. In the presence of testosterone, the tubes will become male sex organs, and male traits will develop. In the absence of testosterone, however, the tubes will become female sex organs that secrete estrogens. Those estrogens will guide the formation of female traits.

Figure 2.12 lists some major functional groups. The number, kind, and arrangement of such groups dictate features of carbohydrates, lipids, proteins, and nucleic acids. For example, hydrocarbons (organic molecules of only carbon and hydrogen atoms) are hydrophobic, or nonpolar. Fatty acids have chains of hydrocarbons, which is why lipids with fatty acid tails don't dissolve in water. Sugars, a class of compounds called alcohols, contain one or more polar *hydroxyl* (—OH) groups. Molecules of water quickly form hydrogen bonds with these groups; that is why sugars dissolve fast in water.

a Ball-and-stick model for glucose, linear structure

b Glucose ring structure

c Two kinds of simplified six-carbon rings

Figure 2.10 Some ways of representing organic compounds.

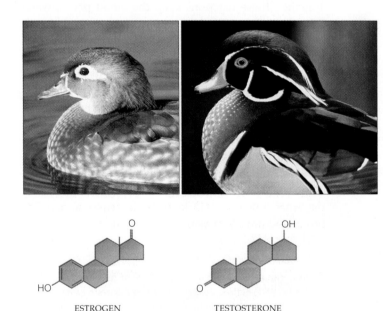

ESTROGEN

TESTOSTERONE

Figure 2.11 Observable differences in traits between female and male wood ducks (*Aix sponsa*), brought about by estrogen and testosterone. These two sex hormones have the same carbon ring structure. They differ only in the position of functional groups attached to the ring.

BIOLOGICAL MOLECULES

Hydroxyl	—OH	In alcohols (e.g., sugars, amino acids); water soluble

Methyl	H —C—H H	In fatty acid chains; insoluble in water

Carbonyl	—C—H O (aldehyde) —C— O (ketone) —CHO >CO	In sugars, amino acids, nucleotides; water soluble. An *aldehyde* if at end of a carbon backbone; a *ketone* if attached to an interior carbon of backbone

Carboxyl	—C—OH O —C—O⁻ O —COOH —COO⁻ (non-ionized) (ionized)	In amino acids, fatty acids; water soluble. Highly polar; acts as an acid (releases H⁺)

Amino	—N—H H —N—H⁺ H —NH₂ —NH₃⁺ (non-ionized) (ionized)	In amino acids and certain nucleotide bases; water soluble; acts as a weak base (accepts H⁺)

Phosphate	O⁻ —O—P—O⁻ O ⓟ icon	In nucleotides (e.g., ATP), also in DNA, RNA, many proteins, phospholipids; water soluble, acidic

Figure 2.12 *Animated!*   Common functional groups in biological molecules, with examples of their locations.

HOW DO CELLS BUILD ORGANIC COMPOUNDS?

Cells build big molecules mainly from four families of small organic compounds: simple sugars, fatty acids, amino acids, and nucleotides. These compounds are called monomers when many of the same kind bond together to form a larger molecule. Molecules with many repeating monomer units are called polymers (*mono–*, one; *poly–*, many). Starch, for example, is a polymer of glucose units. Plant cells make starch when glucose is plentiful, then break it apart when they require glucose units as a source of instant energy or to build something else.

So how do cells actually do the construction work? The **reactions** by which a cell builds, rearranges, or splits apart all organic compounds require energy and **enzymes**: proteins that make substances react much

enzyme action at functional groups

enzyme action at functional groups

a   HOH   b   OH   HO

Figure 2.13   Two metabolic reactions with roles in building organic compounds. (**a**) A condensation reaction, with two molecules being joined into a larger one. (**b**) Hydrolysis, a water-requiring cleavage reaction.

faster than they would on their own. Different kinds of enzymes mediate the following types of reactions:

1. *Functional-group transfer.* A functional group split away from one molecule is transferred to another.

2. *Electron transfer.* An electron split away from one molecule is donated to another.

3. *Rearrangement.* One type of organic compound is converted to another by a juggling of internal bonds.

4. *Condensation.* Covalent bonding between two small molecules results in a larger molecule.

5. *Cleavage.* A molecule splits into two smaller ones.

To get a sense of these cell activities, think of what happens in a **condensation reaction**. Enzymes split an —OH group from one molecule and an H atom from another. A covalent bond forms between the molecules at their exposed sites. The discarded atoms often form water, or H_2O (Figure 2.13*a*). Polymers such as starch form by way of repeated condensation reactions.

Another example: **Hydrolysis**, a type of cleavage reaction, is like condensation reversed (Figure 2.13*b*). Enzymes split molecules, then attach an —OH group and a hydrogen atom (both from a water molecule) to each of the exposed sites. With hydrolysis, cells cleave polymers into monomers when these are required for building blocks or for energy.

Complex carbohydrates, lipids, proteins, and nucleic acids are the molecules of life.

Organic compounds have diverse three-dimensional shapes and functions that start with their carbon backbone and with functional groups covalently bonded to it.

Enzyme-mediated reactions build the molecules of life mainly from smaller organic compounds—simple sugars, fatty acids, amino acids, and nucleotides.

Figure 2.14 *Animated!*
(**a**,**b**) Straight-chain and ring forms of glucose and fructose. The carbon atoms of these simple sugars are numbered in sequence, starting at the one closest to the aldehyde or ketone group. (**c**) Condensation of two monosaccharides into a disaccharide.

a Structure of glucose   **b** Structure of fructose

glucose   fructose

sucrose   + H_2O

c

a   **b**

Figure 2.15 Bonding patterns for glucose units in (**a**) starch and (**b**) cellulose. In amylose, a form of starch, a series of covalently bonded glucose units form a chain, which coils up. In cellulose, bonds form between glucose chains. The pattern stabilizes the chains, which can become tightly bundled.

2.6 The Truly Abundant Carbohydrates

Which biological molecules are most plentiful in nature?
Carbohydrates. *Most consist of carbon, hydrogen, and oxygen in a 1:2:1 ratio* $(CH_2O)_n$. *Cells use carbohydrates as structural materials and transportable or storable forms of energy. Monosaccharides, oligosaccharides, and polysaccharides are the main classes.*

THE SIMPLE SUGARS

"Saccharide" is from a Greek word that means sugar. The *mono*saccharides (one sugar unit) are the simplest carbohydrates. They all have at least two —OH groups bonded to their carbon backbone and one aldehyde or ketone group. Most dissolve easily in water. Common sugars have a backbone of five or six carbon atoms and tend to form a ring structure when dissolved.

Ribose and deoxyribose are the sugar monomers of RNA and DNA, respectively; each has five carbon atoms. Glucose has six carbons (Figure 2.14*a*). Cells use glucose as an instant energy source, as a structural unit, or as a precursor (parent molecule) to make other compounds, including vitamin C (a sugar acid) and glycerol (an alcohol with three —OH groups).

SHORT-CHAIN CARBOHYDRATES

Unlike the simple sugars, an *oligo*saccharide is a short chain of covalently bonded sugars. (*Oligo*– means a few.) The *di*saccharides consist of two sugar monomers. The lactose in milk has one glucose and one galactose unit. Sucrose, the most plentiful sugar in nature, has a glucose and a fructose unit (Figure 2.14). Table sugar is sucrose extracted from sugarcane and sugar beets. Glycolipids are lipids with oligosaccharide side chains. In later chapters, you will encounter glycolipids with essential roles in self-recognition and immunity.

COMPLEX CARBOHYDRATES

The "complex" carbohydrates, or *poly*saccharides, are straight or branched chains of many sugar monomers —often hundreds or thousands of them. Different kinds consist of one or more types of monomers. The most common polysaccharides are cellulose, glycogen, and starch. Even though all three are made of glucose, they differ a lot in their properties. Why? The answer starts with differences in bonding patterns between their glucose units, which are joined together in chains.

In starch, the covalent bonding pattern of glucose units makes the chain coil like a spiral staircase, as in Figure 2.15*a*. Many —OH groups project out from a coiled starch chain (Figure 2.16*a*). These unprotected groups are easily accessible to hydrolysis enzymes. Thus, starch can be broken apart into its glucose units quickly. Plant cells store much of their glucose in the form of starch.

a Structure of amylose, a soluble form of starch. Cells inside tree leaves briefly store excess glucose monomers as starch grains in their chloroplasts, which are tiny, membrane-bound sacs that specialize in photosynthesis.

b Structure of cellulose. In cellulose fibers, chains of glucose units stretch side by side and hydrogen-bond at –OH groups. The many hydrogen bonds stabilize the chains in tight bundles that form long fibers. Few organisms produce enzymes that can digest this insoluble material. Cellulose is a structural component of plants and plant products, such as cotton dresses and wood.

c Glycogen. Animal cells build this polysaccharide when the body has excess glucose. It is especially abundant in the liver and muscles of highly active animals, including fishes and people.

Figure 2.16 Molecular structure of starch (**a**), cellulose (**b**), and glycogen (**c**), and their typical locations in a few organisms. All three carbohydrates consist only of glucose units.

Figure 2.17 (*Top*) A tick's body covering is a protective cuticle reinforced with a polysaccharide called chitin (*right*). You may "hear" chitin when big spider legs clack across an aluminum oil pan on a garage floor.

In cellulose, many side-by-side chains of glucose monomers hydrogen-bond to each other at their —OH groups (Figure 2.15*b*). This bonding pattern stabilizes the chains into a tight bundle that resists hydrolysis. Long fibers of cellulose are a structural component of plant cell walls (Figure 2.16*b*). Like steel rods inside reinforced concrete, cellulose fibers in cell walls are tough, insoluble, and resistant to weight loads and mechanical stress, such as strong winds against stems.

In animals, glycogen is the sugar-storage equivalent of starch in plants (Figure 2.16*c*). Muscle and liver cells store a lot of it. When the sugar level in blood falls, liver cells hydrolyze glycogen into glucose monomers that enter the blood. Exercise strenuously, but briefly, and muscle cells tap glycogen for a burst of energy.

A different polysaccharide, chitin, is distinctive in that it has nitrogen-containing groups attached to its glucose units. Chitin strengthens the external skeletons and other hard body parts of a number of animals, such as crabs, earthworms, insects, spiders, and ticks of the sort shown in Figure 2.17. Chitin also stiffens the cell walls of many kinds of fungi.

The simple sugars (such as glucose), oligosaccharides, and polysaccharides (such as starch) are carbohydrates. Each cell requires carbohydrates as structural materials, and as storable or transportable packets of energy.

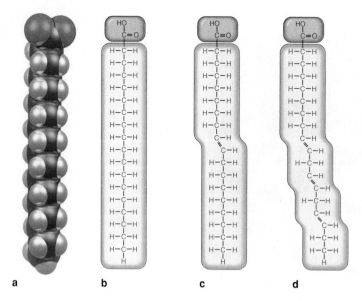

Figure 2.18 Three fatty acids. (**a,b**) Space-filling model and structural formula for stearic acid. The carbon backbone is fully saturated with hydrogen atoms. (**c**) Oleic acid, with a double bond in its backbone, is an unsaturated fatty acid. (**d**) Linoleic acid, also unsaturated, has three double bonds in its backbone.

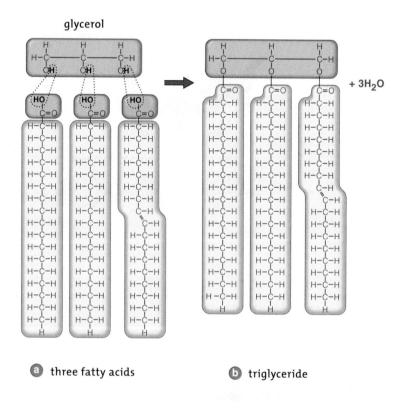

Figure 2.19 *Animated!* Condensation of (**a**) three fatty acids and one glycerol molecule into (**b**) a triglyceride.

2.7 Greasy, Fatty— Must Be Lipids

If something is greasy or oily to the touch, it may be a lipid or have lipid parts. Cells use different kinds of lipids for energy, as structural materials, and as signaling molecules. Fats, phospholipids, and waxes all have fatty acid tails. Sterols have a backbone of four carbon rings.

FATS AND FATTY ACIDS

Lipids are nonpolar hydrocarbons. They don't dissolve much in water, but they readily dissolve in nonpolar substances, as butter does in warm cream sauce.

The lipids called **fats** have one, two, or three fatty acids attached to a glycerol molecule. A **fatty acid** has a backbone of as many as thirty-six carbon atoms, a carboxyl group at one end, and hydrogen atoms at most or all of the remaining carbons (Figure 2.18).

Unsaturated fatty acids have one or more double bonds. *Saturated* fatty acids have single bonds only. Saturated fatty acids are tightly packed in animal fats, and so are solid at room temperature. Most plant fats stay liquid at room temperature, as "vegetable oils." The double bonds in unsaturated fatty acid tails form kinks that tend to prevent them from packing tightly.

"Neutral" fats, such as butter, lard, and vegetable oils, are **triglycerides**. Each triglyceride has three fatty acids stretching out like flexible tails from a glycerol unit (Figure 2.19). Triglycerides are the most abundant lipids in your body and its richest reservoir of energy. Gram for gram, the breakdown of lipids yields more than twice the energy of carbohydrate breakdown. All vertebrates store triglycerides as droplets in fat cells that make up adipose tissue.

Layers or patches of adipose tissue also insulate the body and cushion some of its parts. For instance, as in many other kinds of animals, a thick layer of triglycerides beneath the skin of Emperor penguins

a

b

hydrophilic
head

two
hydrophobic
tails

c

cell membrane section

Figure 2.20 (**a**) Space-filling model, (**b**) structural formula, and (**c**) an icon for lecithin, the most common phospholipid in animal and plant cell membranes. Phospholipids have one saturated and one unsaturated fatty acid tail.

a

b

c

Figure 2.21 (**a**) Sterol backbone. (**b**) Structural formula for cholesterol, the main sterol of animal tissues. Your liver makes enough cholesterol for your body. A fat-rich diet may lead to clogged arteries. (**c**) Honeycomb. These compartments serve as food warehouses and bee nurseries. Bees construct them from their own firm, water-repellent, waxy secretions.

insulates their bodies so they can keep warm in the extremely cold Antarctic winter months.

PHOSPHOLIPIDS

A **phospholipid** has a glycerol backbone, a hydrophilic head with a phosphate group, and two nonpolar fatty acid tails (Figure 2.20). Phospholipids are the principal component of cell membranes, which have two layers of lipids. Heads of one layer are dissolved in the cell's fluid interior; heads of the other layer are dissolved in the surroundings. Sandwiched between the two layers are all of the fatty acid tails, which are hydrophobic. Section 3.3 takes a look at membrane functions.

CHOLESTEROL AND OTHER STEROLS

Sterols are among the many lipids with no fatty acids. They differ in the number, position, and type of their functional groups, but all have a rigid backbone of four carbon rings (Figure 2.21*a*).

Sterols are key components of every eukaryotic cell membrane. The most common type in animal tissues is cholesterol (Figure 2.21*b*). Cholesterol is remodeled into compounds as diverse as steroids, bile salts, and vitamin D. Bile salts assist fat digestion in the small intestine. The sex hormones, including the two shown in Figure 2.11, are steroids. Sex hormones have roles in development and function of sexual organs, secondary sexual traits, and adult sexual behavior.

WAXES

Waxes have long-chain fatty acids tightly packed and linked to long-chain alcohols or carbon rings. All have a firm consistency; all repel water. Surfaces of plants have a cuticle that contains waxes and another lipid, cutin. A plant cuticle restricts water loss and thwarts some parasites.

Waxes also protect, lubricate, and lend pliability to skin and hair. Birds secrete waxes, fats, and fatty acids that waterproof their feathers. Bees use beeswax for honeycomb, which houses each new bee generation as well as honey (Figure 2.21*c*).

Being largely hydrocarbon, lipids can dissolve in nonpolar substances, but they resist dissolving in water.

Triglycerides, or neutral fats, have a glycerol head and three fatty acid tails. They are a major energy reservoir. Phospholipids are the main components of cell membranes.

Sterols such as cholesterol serve as membrane components and precursors of steroid hormones and other compounds. Waxes are firm yet pliable components of water-repelling and lubricating substances.

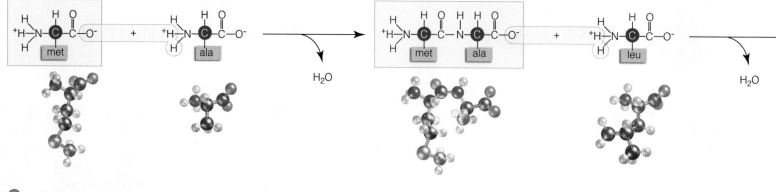

(a) Instructions encoded in DNA specify the order of amino acids to be joined in a polypeptide chain. The first amino acid is usually methionine (met). Alanine (ala) comes next in this one.

(b) In a condensation reaction, a peptide bond forms between the methionine and alanine. Leucine (leu) is next.

Figure 2.22 *Animated!* Peptide bond formation during protein synthesis. Chapter 10 offers a closer look at protein synthesis.

R group (20 kinds, each with distinct properties)

valine (val)

Figure 2.23
Generalized formula and one representative of the twenty common amino acids. The *green* boxes highlight the R groups.

2.8 Proteins—Diversity in Structure and Function

Of all large biological molecules, proteins are the most diverse. Some speed reactions; others are the stuff of spider webs, feathers, bones, hair, and other body parts. Nutritious types abound in seeds and eggs. Many move substances, help cells communicate, or defend against pathogens. Amazingly, cells build thousands of different proteins from only twenty kinds of amino acids!

WHAT IS AN AMINO ACID?

An **amino acid** is a small organic compound with an amino group ($-NH_3^+$), a carboxyl group ($-COO^-$, the acid part), a hydrogen atom, and one or more atoms called its R group. In most cases, these components are attached to the same carbon atom. The R group is a functional group that defines a particular amino acid and gives it special properties. Figures 2.22 and 2.23 show a few kinds that you will encounter later on.

LEVELS OF PROTEIN STRUCTURE

When a cell constructs a protein, it strings amino acids together, one after the other. Instructions coded in DNA specify the order in which any of the twenty kinds of amino acids will occur in a given protein. A *peptide* bond forms when a condensation reaction joins the amino group of one amino acid and the carboxyl group of the next. Each **polypeptide chain** consists of

many amino acids. Its carbon backbone (Figure 2.24*a*) contains nitrogen atoms in this regular pattern:

$$-N-C-C-N-C-C-$$

The sequence of amino acids in a polypeptide chain is its *primary* structure. Hydrogen bonding between different amino acids causes different regions of each polypeptide chain to twist, bend, loop, or fold. This is *secondary* structure. Some stretches of amino acids coil into a helical shape, a bit like a spiral staircase; other regions form sheets or loops (Figure 2.24*b*). Bear in mind, the primary structure of each type of protein is unique, but similar patterns of coils, sheets, and loops occur in most of them.

Much as an overly twisted rubber band coils back on itself, the coils, sheets, and loops of a protein fold up even more, into compact domains. A "domain" is a part of a protein that forms a functional unit. This is the third level of organization, or *tertiary* structure, of a protein. Tertiary structure is what makes a protein a working molecule (Figure 2.24*c*). For instance, barrel-shaped domains of some proteins function as tunnels through membranes.

Many proteins are two or more polypeptide chains that are bonded together or closely associated with one another. This is a protein's fourth level of organization, its *quaternary* structure. Most enzymes and many other proteins are globular, with multiple polypeptide chains folded into rounded shapes. Hemoglobin, described shortly, is an example.

Protein structure does not stop here. Enzymes often attach short, linear, or branched oligosaccharides to a

(c) A peptide bond forms between the alanine and the leucine. Tryptophan (trp) queues up.

(d) The sequence of amino acid residues in this newly forming peptide chain is now met–ala–leu–trp. The process can continue until there are hundreds or thousands of amino acids in the chain.

Figure 2.24 Protein structure. (**a**) Primary structure of a protein is its sequence of amino acids. (**b**) Hydrogen bonds (dotted lines) that connect different parts of a polypeptide chain form a protein's secondary structure, which can be coiled or sheetlike. (**c**) Coils and sheets pack together into stable, functional domains, the protein's third level of structural organization.

new polypeptide chain, making a *glyco*protein. Many glycoproteins occur at the cell surface or are secreted from cells. Lipids also get attached to many proteins. The cholesterol, triglycerides, and phospholipids that your body absorbs after a meal are transported about as components of *lipo*proteins.

Finally, some proteins array themselves by many thousands into a much larger structure. Many such conglomerate proteins are fibrous; their polypeptide chains are organized as strands or sheets. Some of them contribute to the cell's shape and organization. Others help cells and cell parts move.

A protein has primary structure, a sequence of amino acids covalently bonded as a polypeptide chain.

Local regions of a polypeptide chain become twisted and folded into helical coils, sheetlike arrays, and loops. These arrangements are the protein's secondary structure.

A polypeptide chain or parts of it become organized as structurally stable, compact, functional domains. Such domains are a protein's tertiary structure.

Many proteins show quaternary structure; they consist of two or more polypeptide chains.

(a) primary structure — one peptide group

(b) secondary structure — coil, helix — sheet

(c) tertiary structure — coiled coils — barrel

LINK TO
SECTION
1.4

heme, an iron-containing, oxygen-transporting functional group

alpha chain   alpha chain

beta chain   beta chain

Figure 2.25 (**a**) Globin. This polypeptide chain cradles heme, a functional group that contains an iron atom.

(**b**) Hemoglobin, an oxygen-transport protein in red blood cells. This is one of the proteins with quaternary structure. It consists of four globin molecules (two alphas and two betas) held together by hydrogen bonds.

2.9 Why Is Protein Structure So Important?

Cells are good at making proteins that are just what their DNA specifies. But sometimes cells make mistakes or nature sabotages the specifications, as with ultraviolet radiation attacks. The substitution of just one wrong amino acid may have far-reaching consequences.

JUST ONE WRONG AMINO ACID . . .

As blood moves through lungs, the hemoglobin in red blood cells binds oxygen gas, then gives it up in body regions where oxygen levels are low. After oxygen is released, red blood cells quickly return to the lungs and pick up more. The oxygen-binding properties of hemoglobin exist because of its structure.

Each of the four globin chains in hemoglobin folds tightly, forming a pocket that holds an iron-containing heme group (Figure 2.25*a*). During its life span, a red blood cell transports billions of oxygen molecules, all bound to 250 million or so hemoglobin molecules.

Globin comes in two slightly different forms, alpha and beta. Two of each form make up one hemoglobin molecule in adult humans (Figure 2.25*b*). Glutamate is normally the sixth amino acid in the beta globin chain, but a DNA mutation sometimes puts a different amino acid, valine, in the sixth position instead (Figure 2.26). Unlike glutamate, which carries an overall negative charge, valine has no net charge.

As a result of that one substitution, a tiny patch of the protein changes from hydrophilic to hydrophobic, which in turn causes the globin molecule's behavior to change slightly. Hemoglobin with this mutation in its beta chain is designated HbS, or sickle hemoglobin.

HbS molecules can form large, stable, rod-shaped aggregates. Red blood cells containing these aggregates become distorted into a sickle shape. Sickled cells clog tiny blood vessels, and this disrupts blood circulation.

Every human inherits two genes for beta globin, one from each of the two parents. (Genes are units of DNA that encode proteins.) Cells use both genes when they make beta globin. If one is normal and the other has the valine mutation, a person can make enough normal hemoglobin to lead a relatively normal life. But someone who inherits two mutant genes can only make the mutant HbS hemoglobin. The outcome is sickle-cell anemia, a severe genetic disorder. Figure 2.26*d* lists the far-reaching effects of sickle-cell anemia on tissues and organs.

PROTEINS UNDONE—DENATURATION

The shape of a protein defines its biological activity. A globin molecule cradles heme, an enzyme speeds some reaction, a receptor transduces an energy signal. These proteins and others cannot function unless they stay coiled, folded, and packed in a precise way. Their shape depends on many hydrogen bonds and other interactions—which heat, shifts in pH, or detergents can disrupt. At such times, polypeptide chains unwind and change shape in an event called **denaturation**.

Consider albumin, a protein in the white of an egg. When you cook eggs, the heat does not disrupt the covalent bonds of albumin's primary structure, but it does destroy the weaker hydrogen bonds that maintain the protein's shape, and so the protein unfolds. When a translucent egg white turns opaque, we know that the

BIOLOGICAL MOLECULES

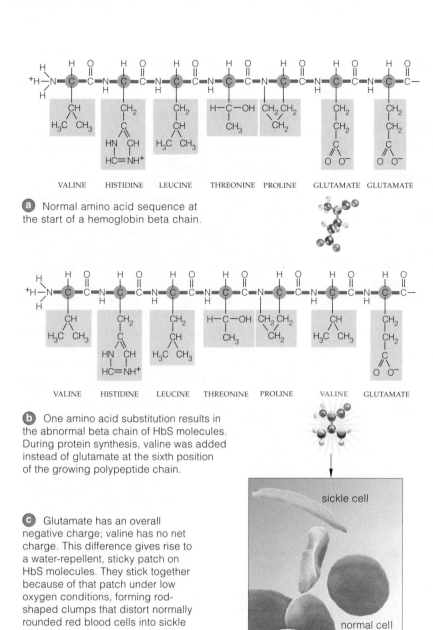

VALINE HISTIDINE LEUCINE THREONINE PROLINE GLUTAMATE GLUTAMATE

a Normal amino acid sequence at the start of a hemoglobin beta chain.

VALINE HISTIDINE LEUCINE THREONINE PROLINE VALINE GLUTAMATE

b One amino acid substitution results in the abnormal beta chain of HbS molecules. During protein synthesis, valine was added instead of glutamate at the sixth position of the growing polypeptide chain.

c Glutamate has an overall negative charge; valine has no net charge. This difference gives rise to a water-repellent, sticky patch on HbS molecules. They stick together because of that patch under low oxygen conditions, forming rod-shaped clumps that distort normally rounded red blood cells into sickle shapes. (A sickle is a farm tool that has a crescent-shaped blade.)

sickle cell

normal cell

Clumping of cells in bloodstream

↓

Circulatory problems, damage to brain, lungs, heart, skeletal muscles, gut, and kidneys

↓

Heart failure, paralysis, pneumonia, rheumatism, gut pain, kidney failure

Spleen concentrates sickle cells

↓

Spleen enlargement

↓

Immune system compromised

Rapid destruction of sickle cells

↓

Anemia, causing weakness, fatigue, impaired development, heart chamber dilation

↓

Impaired brain function, heart failure

d *Above left:* Melba Moore, celebrity spokesperson for sickle-cell anemia organizations. *Above right:* Range of symptoms for a person with two mutated genes for hemoglobin's beta chain.

Figure 2.26 *Animated!* Sickle-cell anemia's molecular basis and its main symptoms.

albumin is denatured. For a few proteins, denaturation is reversible if normal conditions return, but albumin is not one of them. There is no way to uncook an egg.

So what is the big take-home lesson? Hemoglobin, hormones, enzymes, transporters—these are the kinds of proteins that help us survive. Twists and folds in their polypeptide chains form anchors, or membrane-spanning barrels, or jaws that can grip enemy agents in the body. Mutation can alter the chains enough to block or enhance an anchoring, transport, or defensive function. Sometimes the consequences are awful. Yet changes in the sequences and functional domains also give rise to variation in traits—the raw material of evolution. *Learn about protein structure and function, and you are on your way to comprehending life in its richly normal and abnormal expressions.*

The structure of proteins dictates function. Mutations that alter a protein's structure can also alter its function, with potentially drastic consequences for an individual.

2.10 Nucleotides and the Nucleic Acids

*Certain small organic compounds called **nucleotides** are energy carriers, enzyme helpers, and messengers. Some are the building blocks for DNA and RNA. They are, in short, central to metabolism, survival, and reproduction.*

Nucleotides consist of one sugar, one nitrogen-containing base, and one or more phosphate groups. Deoxyribose or ribose is the sugar. Both sugars have a five-carbon ring structure, but ribose has an attached oxygen atom and deoxyribose does not. The bases have a single or double carbon ring structure (Figures 2.27 and 2.28).

The nucleotide **ATP** (adenosine triphosphate) has a row of three phosphate groups attached to its sugar (Figure 2.27). ATP readily transfers the outermost of these phosphate groups to other molecules inside cells. The transfer energizes acceptor molecules enough to make them reactive. ATP's phosphate-group transfers are crucial aspects of metabolism.

Other nucleotides have different metabolic roles. Some are subunits of **coenzymes**, or enzyme helpers. Coenzymes move electrons and hydrogen from one reaction site to another. NAD^+ and FAD are examples.

Still other nucleotides act as chemical messengers within and between cells. Later, you will read about one called cAMP (cyclic adenosine monophosphate).

All cells start out life and maintain themselves with instructions encoded in their **DNA** (deoxyribonucleic acid). DNA is a polymer of four kinds of nucleotide monomers: adenine, guanine, cytosine, and thymine (Figure 2.28*a*). It is one of the **nucleic acids** that store and retrieve heritable information in all cells. In all nucleic acids, a covalent bond joins the sugar of one nucleotide monomer and the phosphate group of the next (Figure 2.28*b*).

Figure 2.27 Structural formula for ATP.

Figure 2.28 (**a**) Four kinds of nucleotides that function as monomers for DNA molecules. Two of the nucleotide bases, adenine and guanine, have a double-ring structure. The two others, thymine and cytosine, have a single-ring structure. (**b**) Example of how a sequence of nucleotides is covalently bonded in a single strand of DNA or RNA.

Figure 2.29 shows how hydrogen bonds between bases join two strands of DNA along the length of a molecule. Think of every "base pairing" as one rung of a ladder, and the two sugar–phosphate backbones as the ladder's two posts. The ladder twists and turns in a regular pattern, forming a double helical coil.

The sequence of bases in DNA encodes heritable information about how to build all of the proteins that give each cell the potential to grow, maintain itself, and reproduce. Part of the sequence is unique for each species. Some parts are identical, or nearly so, among many species. We will return to DNA in Chapter 9.

Like DNA, the **RNAs** (ribonucleic acids) consist of four sugar–phosphate monomers—adenine, guanine, cytosine, and uracil (not thymine). Unlike DNA, most RNAs are single stranded. In eukaryotes, one type of RNA carries DNA's protein-building instructions from the nucleus to the cytoplasm. There, other RNAs build proteins using the instructions.

Ever since life originated, RNA has been central in all processes of retrieving and using the hereditary information stored in DNA. We return to RNA and its role in protein synthesis in Chapter 10.

> Small organic compounds called nucleotides serve as coenzymes, subunits of nucleic acids, energy carriers, and chemical messengers.
>
> The nucleic acid DNA consists of two nucleotide strands joined by hydrogen bonds and twisted as a double helix. Its nucleotide sequence contains information about how to build all of a cell's proteins.
>
> RNA is a single-stranded nucleic acid. Different RNAs have roles in the processes by which a cell retrieves and uses genetic information in DNA to build proteins.

Figure 2.29 Models for the DNA molecule, with its two strands of nucleotides hydrogen-bonded at their bases and twisted into a double helix.

covalent bonding in carbon backbone

hydrogen bonding between bases

Summary

Section 2.1 All substances consist of elements, each of which consists of one kind of atom. Atoms are fundamental forms of matter that have mass, take up space, and cannot be broken down by ordinary means like heat.

Each atom consists of positively charged protons, negatively charged electrons, and (except for hydrogen) uncharged neutrons. Protons and neutrons occupy the core region, or nucleus, of the atom.

Section 2.2 Whether a given atom will interact with others depends on the number and arrangement of its electrons. Atoms with electron vacancies in the highest energy level (outermost shell) tend to donate, accept, or share electrons with other atoms. When they do so, they become united in a chemical bond.

Atoms that lose or gain electrons are ions. In an ionic bond, ions of opposite charge attract each other and stay together. In a covalent bond, atoms share one or more pairs of electrons. Such a bond is polar when electrons are not shared equally and nonpolar when sharing is equal. Hydrogen bonds can form between a hydrogen atom and a different atom when both are already taking part in polar covalent bonds.

Biology⊜Now
Compare the types of bonds in biological molecules with the animation on Biology Now.

Section 2.3 Polar covalent bonds join two hydrogen atoms to one oxygen atom in each water molecule. The water molecule's polarity invites extensive hydrogen bonding between water molecules in liquid water and ice. Extensive hydrogen bonding is the basis of water's

Table 2.1 Summary of the Main Organic Compounds in Living Things

Category	Main Subcategories	Some Examples and Their Functions	
CARBOHYDRATES . . . contain an aldehyde or a ketone group, and one or more hydroxyl groups	**Monosaccharides** (simple sugars)	Glucose	Energy source
	Oligosaccharides (short-chain carbohydrates)	Sucrose (a disaccharide)	Most common form of sugar; the form transported through plants
	Polysaccharides (complex carbohydrates)	Starch, glycogen Cellulose	Energy storage Structural roles
LIPIDS . . . are mainly hydrocarbon; generally do not dissolve in water but do dissolve in nonpolar substances, such as other lipids	**Lipids with fatty acids** *Glycerides:* Glycerol backbone with one, two, or three fatty acid tails	Fats (e.g., butter), oils (e.g., corn oil)	Energy storage
	Phospholipids: Glycerol backbone, phosphate group, one other polar group, and (often) two fatty acids	Phosphatidylcholine	Key component of cell membranes
	Waxes: Alcohol with long-chain fatty acid tails	Waxes in cutin	Conservation of water in plants
	Lipids with no fatty acids *Sterols:* Four carbon rings; the number, position, and type of functional groups differ among sterols	Cholesterol	Component of animal cell membranes; precursor of many steroids and vitamin D
PROTEINS . . . are one or more polypeptide chains, each with as many as several thousand covalently linked amino acids	**Fibrous proteins** Long strands or sheets of polypeptide chains; often tough, water-insoluble	Keratin Collagen	Structural component of hair, nails Structural component of bone
	Globular proteins One or more polypeptide chains folded into globular shapes; many roles in cell activities	Enzymes Hemoglobin Insulin Antibodies	Great increase in rates of reactions Oxygen transport Control of glucose metabolism Tissue defense
NUCLEIC ACIDS (AND NUCLEOTIDES) . . . are chains of units (or individual units) that each consist of a five-carbon sugar, phosphate, and a nitrogen-containing base	**Adenosine phosphates**	ATP cAMP (Section 20.8)	Energy carrier Messenger in hormone regulation
	Nucleotide coenzymes	NAD^+, $NADP^+$, FAD	Transfer of electrons, protons (H^+) from one reaction site to another
	Nucleic acids Chains of thousands to millions of nucleotides	DNA, RNAs	Storage, transmission, translation of genetic information

unique properties: its ability to resist temperature changes, display internal cohesion, and easily dissolve polar or ionic substances. These properties of water make life possible.

Biology Now
Learn how water's structure gives rise to its unique properties with animation on BiologyNow.

Section 2.4 The pH of a solution indicates its H^+ (hydrogen ion) concentration. Acids have a pH that is less than 7.0; bases have a pH greater than 7.0. Buffer systems maintain pH of body fluids; most biological processes operate only within a narrow range of pH.

Biology Now
Discover the pH of common substances with the interaction on BiologyNow.

Section 2.5 Organic compounds consist primarily of carbon and hydrogen atoms. Carbon atoms bond covalently with up to four other atoms, often forming long chains or rings. Functional groups attached to the carbon backbone contribute to an organic compound's chemical properties. Enzyme-driven reactions construct carbohydrates, proteins, lipids, and nucleic acids from smaller organic subunits (Table 2.1).

Biology Now
Investigate the properties of functional groups with the interaction on BiologyNow.

Section 2.6 Carbohydrates include simple sugars, oligosaccharides, and polysaccharides. Living cells use carbohydrates as instant energy sources, transportable or storage forms of energy, and structural materials.

Biology Now
See how sucrose is formed with animation on BiologyNow.

Section 2.7 Lipids are greasy or oily compounds that tend not to dissolve in water but dissolve easily in nonpolar compounds, such as other lipids. Neutral fats (triglycerides), phospholipids, waxes, and sterols are lipids. Cells use lipids as major sources of energy and as structural materials.

Biology⑧Now
See a triglyceride form with animation on BiologyNow.

Sections 2.8, 2.9 A sequence of amino acids (a polypeptide chain) is a protein's primary structure. At higher levels of structural organization, polypeptide chains twist, coil, and bend into functional domains. Many proteins consist of two or more chains. Changes in a protein's structure may alter its function.

Biology⑧Now
Explore amino acid structure and learn about peptide bond formation with the animation on BiologyNow.

Section 2.10 ATP and other nucleotides have vital roles in metabolism. DNA and RNA are nucleic acids, each composed of four kinds of nucleotide subunits. The sequence of nucleotide bases in DNA encodes instructions for how to construct all of a cell's proteins. Different kinds of RNA molecules access and use this information to build proteins.

Figure 2.30 Lacking most basic tools and practices of modern scientists, medieval alchemists intertwined their scientific endeavors with mysticism and astrology.

12. Match the terms with their most suitable description.

_____ salt	a. long chain of amino acids
_____ protein	b. atomic nucleus component
_____ trace element	c. releases ions other than H^+ and
_____ hydrophilic	OH^- when dissolved in water
_____ covalent	d. dissolves easily in water
bond	e. double strands of nucleotides
_____ proton	f. electrons shared between atoms
_____ DNA	g. makes up less than 0.001
	percent of body weight

Additional questions are available on **Biology⑧Now™**

Self-Quiz
Answers in Appendix I

1. Is this statement true or false: Every atom consists of protons, neutrons, and electrons.

2. Atoms share electrons unequally in a(n) _____ bond.
   a. ionic
   b. nonpolar covalent
   c. polar covalent
   d. hydrogen

3. Liquid water shows _____ .
   a. polarity
   b. hydrogen-bonding capacity
   c. notable heat resistance
   d. cohesion
   e. b through d
   f. all of the above

4. Hydrogen ions (H^+) are _____ .
   a. the basis of pH values
   b. targets of certain buffers
   c. dissolved in blood
   d. all of the above

5. Name the molecules of life and the families of small organic compounds from which they are built.

6. Which of the following is a class of molecules that encompasses all of the other molecules listed?
   a. triglycerides
   b. fatty acids
   c. waxes
   d. sterols
   e. lipids
   f. phospholipids

7. Each carbon atom can share pairs of electrons with as many as _____ other atoms.
   a. one   b. two   c. three   d. four

8. _____ is a simple sugar (a monosaccharide).
   a. Glucose
   b. Sucrose
   c. Ribose
   d. Chitin
   e. both a and b
   f. both a and c

9. Unlike saturated fats, the fatty acid tails of unsaturated fats incorporate one or more _____ .
   a. single covalent bonds   b. double covalent bonds

10. _____ are to proteins as _____ are to nucleic acids.
   a. Sugars; lipids
   b. Sugars; proteins
   c. Amino acids; hydrogen bonds
   d. Amino acids; nucleotides

11. A denatured protein has lost its _____ .
   a. hydrogen bonds
   b. shape
   c. function
   d. all of the above

Critical Thinking

1. By weight, oxygen is the most abundant element in organisms, ocean water, and Earth's crust. Predict which element is the most abundant in the whole universe.

2. Medieval scientists and philosophers called alchemists (Figure 2.30) were predecessors of modern-day chemists. Many of them tried to transform lead (82 protons) into gold (79 protons). Explain why they never could have possibly succeeded.

3. David, an inquisitive three-year-old, poked his fingers into warm water in a metal pan on the stove and did not sense anything hot. Then he touched the pan itself and got a nasty burn. Explain why.

4. In the following list, identify which is the carbohydrate, the fatty acid, the amino acid, and the polypeptide:
   a. $^+NH_3$—CHR—COO^-
   b. $C_6H_{12}O_6$
   c. (glycine)$_{20}$
   d. $CH_3(CH_2)_{16}COOH$

5. A clerk in a health-food store tells you "natural" vitamin C extracts from rose hips are better than synthetic tablets of this vitamin. Given what you know about the structure of organic compounds, what would be your response? How would you design an experiment to test whether a natural and synthetic version of a vitamin differ?

6. It seems there are "good" and "bad" unsaturated fats. The double bond in *cis* fatty acid tails gives them a rigid kink. A *trans* fatty acid tail is not kinked; it can straighten.
   Some *trans* fatty acids occur naturally in beef. But most form by processes that solidify vegetable oils for margarine and shortening, synthetic substances that are widely used in prepared foods (such as cookies) and in french fries and other fast-food products. *Trans* fatty acids are linked to high levels of LDL, a "bad" form of cholesterol that can set the stage for a heart attack. Speculate on why your body might have an easier time dealing with *cis* fatty acids than *trans* fatty acids.

Animalcules and Cells Fill'd With Juices

Do you ever think of yourself as a fraction of a kilometer tall? Probably not. Yet that is how we measure cells—in micrometers, or thousandths of a millimeter, which is a thousandth of a meter, which is a thousandth of a kilometer. The bacterial cells atop the pin at right are a few micrometers "tall."

Before those cells were fixed on the head of a pin so someone could take their picture, they were living members of the most ancient lineage, the prokaryotes. Prokaryotes are the simplest of cells, structurally. Their DNA is not housed in a nucleus. Your cells, by contrast, are eukaryotic. A nucleus evolved in their single-celled ancestors, along with other internal compartments that keep metabolic activities organized.

Nearly all cells are invisible to the naked eye. No one knew about them until the seventeenth century, when the first microscopes were made. The instruments were not very complicated. Galileo Galilei, for instance, used two glass lenses inside a cylinder, but the arrangement was good enough to reveal details of an insect's eyes.

At midcentury, Robert Hooke focused a microscope on thinly sliced cork. He named the small compartments he saw *cellulae*, meaning small rooms—thus the origin of the biological term "cell." Actually, they were dead plant cell walls, but Hooke did not think of them as dead because neither he nor anyone else knew cells could be alive. He observed cells "fill'd with juices" in green plant tissues, but did not realize they were alive, either.

Given the simplicity of their instruments, it is truly amazing that the pioneers in microscopy observed as much as they did. Antoni van Leeuwenhoek, a Dutch

shopkeeper, had exceptional skill constructing lenses. By the late 1600s, he was looking at sperm, protists, some bacteria, and "many very small animalcules, the motions of which were very pleasing to behold," in scrapings of tartar from his teeth.

In the 1820s, much improved lenses brought cells into sharper focus. Robert Brown, a botanist, was the first to identify a plant cell nucleus. Later, the botanist Matthias Schleiden wondered whether each plant cell develops as an independent unit even though it is part of the plant. In 1839, the zoologist Theodor Schwann reported that cells have a life of their own even when they are part of a body. The physiologist Rudolf Virchow decided that every cell must come from one that already exists.

Microscopic analysis yielded three generalizations, which together constitute the *cell theory*. First, every organism consists of one or more cells. Second, the cell is the smallest unit of organization that still displays the properties of life. Third, the continuity of life arises directly from the growth and division of single cells.

This chapter introduces the defining features of prokaryotic and eukaryotic cells. It is not meant for memorization, but rather it is an invitation into otherwise invisible worlds.

 How Would You Vote? Researchers are in the process of modifying prokaryotes in efforts to find out just how little complexity is required for life. Should this research continue? See BiologyNow for details, then vote online.

 ## Key Concepts

BASIC CELL FEATURES
Nearly all cells are microscopic in size. They all start out life with a plasma membrane, a semifluid interior called cytoplasm, and an inner region of DNA. The plasma membrane controls the flow of specific substances into and out of cells. Two layers of phospholipid molecules are the structural basis of cell membranes. Proteins in this bilayer and those attached to its surfaces carry out many different membrane functions.

PROKARYOTIC CELLS
Compared to eukaryotic cells, prokaryotes have little internal complexity, and no nucleus. However, when taken as a group, they are the most metabolically diverse organisms. The archaea and bacteria are the only prokaryotes, but all species have prokaryotic ancestors.

EUKARYOTIC CELLS
Eukaryotic cells contain organelles, membrane-enclosed compartments that divide the cell interior into functional regions for specialized tasks. A major organelle, the nucleus, keeps the cell's DNA organized and away from cytoplasmic machinery.

Links to Earlier Concepts

Look back on your road map through the levels of organization in nature (Section 1.1). This chapter arrives at the level of living cells. You will start to see how lipids (2.7) are structurally organized into cell membranes, where DNA and RNA (2.10) reside inside cells, and where carbohydrates (2.6) are built and broken apart. You will also expand your view of how cell structure and function depend on protein structure and function (2.8–2.9). You will see how the principles of natural selection introduced in Section 1.4 are being used to make hypotheses (1.5) about how eukaryotic cells evolved.

3.1 What Is "A Cell"?

Inside your body and at its moist surfaces, trillions of cells live in interdependency. In northern forests, four-celled structures called pollen grains drift down from pine trees. In scummy pondwater, free-living single cells called bacteria and amoebas move about. How are these cells alike, and how do they differ?

Let's begin with three points of the **cell theory**. First, every organism consists of one or more cells. Second, a **cell** is the smallest unit with the properties of life; it has the capacity for metabolism, controlled responses to the environment, growth, and reproduction. Third, under conditions now prevailing in nature, only living cells give rise to new cells.

Cells differ in size, shape, and activities. Yet all are similar in three respects. They all start out life with a plasma membrane, a region of DNA, and cytoplasm (Figure 3.1).

1. **Plasma membrane**. This thin, outermost membrane separates each cell from its outside environment. The membrane lets internal metabolic reactions proceed in controllable ways, but does not isolate the cell interior. It is like a house with many windows and doors that do not open for just anyone. Water and gases cross in and out freely, but the flow of other substances across the plasma membrane is controlled.

2. **Nucleus** or **nucleoid**. In every eukaryotic cell, DNA occupies a membrane-enclosed sac called the nucleus. Inside prokaryotic cells, DNA occupies a region with no membrane. This region is called the nucleoid.

3. **Cytoplasm**. Cytoplasm is everything in between the plasma membrane and the region of DNA. It consists of a semifluid matrix and other components, including **ribosomes** (structures on which proteins are built).

Structurally, prokaryotes are the simplest of cells; nothing separates their DNA from the cytoplasm. Only the archaea and bacteria are prokaryotic cells. All other organisms—from amoebas and peach trees to puffball mushrooms and elephants—are eukaryotic. Eukaryotes have internal membranes dividing the cytoplasm into functional compartments. The compartment called the nucleus is their key defining feature.

> *All living cells have an outer plasma membrane, cytoplasm, and an internal region that contains their DNA.*
>
> *Bacteria and archaea are prokaryotic cells. Unlike eukaryotic cells, they do not have an abundance of organelles, and they do not have a nucleus.*
>
> *All other cells are eukaryotic. They have internal membranes that form organelles, particularly a nucleus.*

DNA in nucleus

cytoplasm

plasma membrane

Plant cell (eukaryotic)

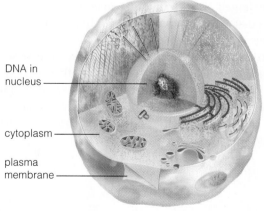

DNA in nucleus

cytoplasm

plasma membrane

Animal cell (eukaryotic)

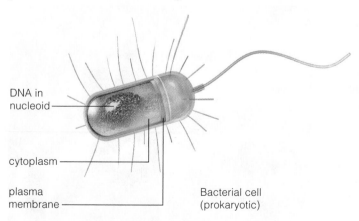

DNA in nucleoid

cytoplasm

plasma membrane

Bacterial cell (prokaryotic)

Figure 3.1 Overview of the general organization of prokaryotic cells and eukaryotic cells. The three cells are not drawn to the same scale.

Figure 3.2 *Animated!* Different types of microscopes.
(**a**) Compound light microscope. (**b**) Electron microscope.

Path of light rays (bottom to top) to eye

Ocular lens enlarges primary image formed by objective lenses.

Prism (directs rays to ocular lens)

Objective lenses (closest to specimen) form primary image.

Stage (holds microscope slide in position)

Condenser lenses focus light rays through specimen.

Illuminator

LINK TO SECTION 2.3

3.2 Most Cells Are *Really Small*

Few cells can be seen without the aid of a microscope, because physical constraints restrict their size.

TYPES OF MICROSCOPES

Like the earlier instruments, some microscopes still use visible light to illuminate objects. Light passes through a cell or other specimen, and then curved glass lenses inside the microscope bend the light and focus it into an enlarged image of the specimen (Figure 3.2). Only cells that are thin enough for light to pass through will be visible with a light microscope. Most cells will look uniformly dense unless they are *stained*, or exposed to dyes that some of their parts but not others soak up.

With the best light microscope, you can distinguish details or objects 200 billionths of a millimeter across (Figure 3.3*a,b*). Electron microscopes bring into focus objects 100,000 times smaller than that.

Electron microscopes use magnetic lenses to bend beams of electrons. A *transmission* electron microscope projects electrons through a sample and focuses them into an image of the specimen's internal components. In *scanning* electron microscopes, a beam of electrons bounces off a specimen, and is then converted into an image of the specimen's surface. Images of specimens magnified under a microscope are called micrographs. Those taken with electron microscopes have fantastic detail (Figure 3.3*c,d*).

Figure 3.3 How different microscopes reveal different aspects of the same organism, a type of green algae. All four images are at the same magnification. (**a,b**) Light micrographs. (**c**) Transmission electron micrograph. (**d**) Scanning electron micrograph. A horizontal bar on or near a micrograph provides a size reference. A micrometer (μm) is a millionth of a meter.

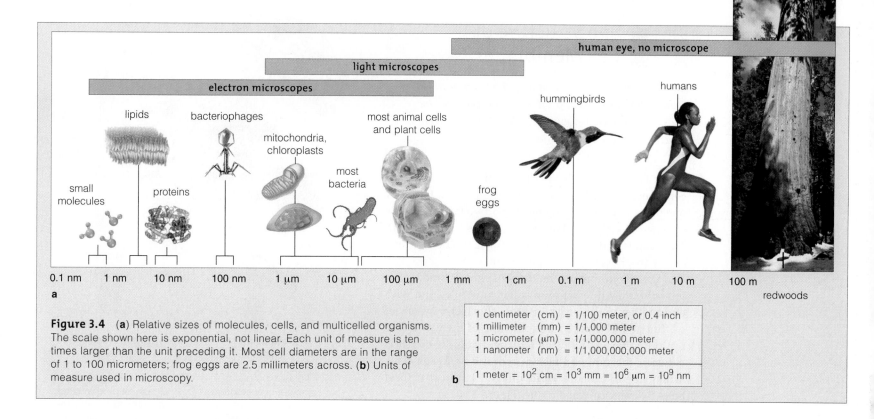

Figure 3.4 (a) Relative sizes of molecules, cells, and multicelled organisms. The scale shown here is exponential, not linear. Each unit of measure is ten times larger than the unit preceding it. Most cell diameters are in the range of 1 to 100 micrometers; frog eggs are 2.5 millimeters across. (b) Units of measure used in microscopy.

1 centimeter	(cm)	= 1/100 meter, or 0.4 inch
1 millimeter	(mm)	= 1/1,000 meter
1 micrometer	(µm)	= 1/1,000,000 meter
1 nanometer	(nm)	= 1/1,000,000,000 meter

1 meter = 10^2 cm = 10^3 mm = 10^6 µm = 10^9 nm

WHY AREN'T ALL CELLS BIG?

A few types of cells can be seen without a microscope. These cells include yolks of bird eggs, cells in the red part of watermelons, and fish eggs. These cells can get big because they are not doing much, metabolically, at maturity. Most metabolically active cells are too tiny to be seen by the unaided eye (Figure 3.4).

So why aren't all cells big? A physical relationship called the **surface-to-volume ratio** constrains increases in cell size. By this relationship, any object's volume increases with the cube of its diameter, but the surface area increases only with the square.

Apply this to a round cell. As Figure 3.5 shows, *if a cell expands in diameter during growth, then its volume will increase faster than its surface area will.* Suppose you make a round cell grow four times wider than normal. Its volume increases by 64 times, but the surface area increases by only 16 times. This means that each unit of plasma membrane must service four times as much cytoplasm as before. If a cell's diameter becomes too great, the inward flow of nutrients and the outward flow of wastes will not be able to keep up with its metabolic activity.

A large, round cell also would have trouble moving materials through its cytoplasm. Random motions of molecules can distribute materials through tiny cells. If a cell is not tiny, you can expect it to be long or thin,

diameter (cm):	0.5	1.0	1.5
surface area (cm^2):	0.79	3.14	7.07
volume (cm^3):	0.06	0.52	1.77
surface-to-volume ratio:	13.17:1	6.04:1	3.99:1

Figure 3.5 Example of the surface-to-volume ratio. This physical relationship between increases in volume and surface area restricts the size and the shape of cells.

or to have outfoldings or infoldings that increase its surface relative to its volume. *The smaller or narrower or more frilly-surfaced the cell, the more efficiently materials cross its surface and become distributed through the interior.*

Surface-to-volume constraints also shape the body plans of multicelled species. For example, small cells attach end to end in strandlike algae, so that each one interacts directly with its surroundings.

Most cells are microscopically small. Different types of microscopes reveal cells and their internal structures.

The surface-to-volume ratio constrains increases in cell size. It also influences cell shape and the body plans of multicelled organisms.

LINKS TO
SECTIONS
2.7, 2.9

3.3 | The Structure of Cell Membranes

All cell membranes consist of a lipid bilayer in which many different kinds of proteins are embedded. The membrane is a continuous boundary that selectively controls the flow of substances across it.

Think back on the phospholipids, the most abundant components of cell membranes (Section 2.7). Each has one phosphate-containing head and two fatty acid tails attached to a glycerol backbone (Figures 2.20 and 3.6*a*). The head is hydrophilic; water attracts it. The two tails are hydrophobic; water repels them.

When you immerse many phospholipid molecules in water, they interact with water molecules and with one another until they spontaneously cluster in a sheet or film at the water's surface. Some of them will line up in two layers, with the fatty acid tails sandwiched between the outward-facing hydrophilic heads. This arrangement, called a **lipid bilayer**, is the structural basis of every cell membrane (Figure 3.6*b,c*).

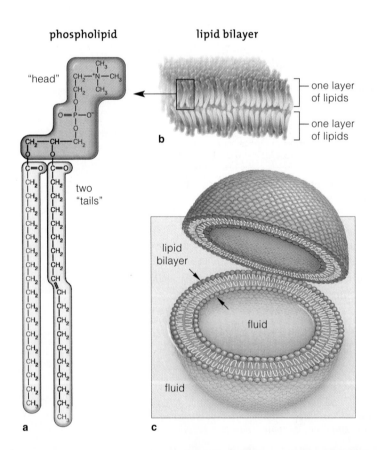

Figure 3.6 Lipid bilayer organization of cell membranes. (**a**) One of the phospholipids, the most abundant membrane components. (**b**) These lipids and others are arranged as two layers. Their hydrophobic tails are sandwiched between their hydrophilic heads in the bilayer. (**c**) Arrangement of a lipid bilayer that forms a cell.

adhesion proteins

Adhesion proteins project outward from plasma membranes of multicelled species. They help cells stick to one another, or to protein matrixes that are part of tissues.

communication proteins

Communication proteins of adjoining cells match up, forming cytoplasm-to-cytoplasm channels between cells. Signals can flow through these channels. Part of a gap junction between heart muscle cells is shown.

Figure 3.7 *Animated!* Part of a plasma membrane, which consists of a lipid bilayer and more proteins than you see here. Later chapters use the icons below these models.

The **fluid mosaic model** describes the organization of membranes (Figure 3.7). A cell membrane is a mixed composition—a *mosaic*—of phospholipids, glycolipids, sterols, and proteins. Membrane phospholipids have different kinds of heads and tails that vary in length and saturation. Unsaturated fatty acids have at least one double covalent bonds in their carbon backbone, and fully saturated fatty acids have none (Section 2.7).

Also by this model, a membrane is *fluid* because of the motions and interactions of its components. Most phospholipids drift sideways, spin on their long axes, and flex their tails, so they do not pack together into a solid layer. Most membrane phospholipids have at least one kinked (unsaturated) fatty acid tail.

Hydrogen bonds and other weak interactions help proteins associate with the phospholipids. Many of the

receptor proteins

Receptors are docks for outside substances that serve as signaling molecules. When a receptor binds one of these molecules, the cell will change its activities.

recognition proteins

A plasma membrane's recognition proteins are like molecular fingerprints. They identify each cell as *self* (belonging to one's own body or tissue) or *nonself* (foreign to the body).

passive transporters

All passive transporters are channels across a membrane. Solutes move freely through the channel to the side where they are less concentrated. Some passive transporters are open all the time, but others have molecular gates that open and close in controlled ways.

active transporters

Active transporters are channels across a membrane. They pump solutes through the channel to the side where they are more concentrated. Pumping action requires an energy input.

These two examples are a calcium pump (*blue*) and an ATP synthase (*multicolored*).

plasma membrane

membrane proteins span the bilayer, with hydrophilic parts extending beyond both of its surfaces. Others are anchored to underlying cell structures.

The lipid bilayer functions mainly as a barrier to water-soluble substances. Proteins carry out nearly all other membrane tasks. Many proteins are receptors for signals. Transport proteins move certain solutes across the bilayer. Some transport proteins require an energy input; others do not. Still other proteins are enzymes that mediate events at the membrane.

Especially among multicelled species, the plasma membrane bristles with diverse proteins. Some kinds identify the cell as belonging to the body, and others help defend it against attacks. Special proteins also help cells communicate with one another or stick together in tissues. Figure 3.7 shows some important categories of membrane proteins and gives a brief description of their functions.

The fluid mosaic model is a good starting point for thinking about cell membranes. But keep in mind that different membranes have different types of molecules. Even one bilayer's two surfaces are not exactly alike. For example, some carbohydrate side chains attached to proteins and lipids project from the cell, but not into it. Differences among plasma membranes and internal membranes reflect the roles they play in cells.

All cell membranes consist of two layers of lipids—mainly phospholipids—and diverse proteins. Hydrophobic parts of the lipids are sandwiched between hydrophilic parts, which face the cytoplasmic fluid or extracellular fluid.

All cell membranes have protein receptors, transporters, and enzymes. The plasma membrane also incorporates adhesion, communication, and recognition proteins.

LINK TO
SECTION
1.3

3.4 A Closer Look at Prokaryotic Cells

The word prokaryote means "before the nucleus." The name reminds us that bacteria and archaea originated before cells with a nucleus evolved.

Prokaryotes are the smallest known cells. As a group they are the most metabolically diverse forms of life on Earth (Figure 3.8). Different kinds exploit energy and raw materials in nearly every environment, from dry deserts to hot springs to mountain ice. There are two domains of prokaryotic cells: Bacteria and Archaea (Sections 1.3 and 14.1). Cells of both groups are alike in outward appearance, size, and where they live.

Most prokaryotes have a semirigid or rigid cell wall that surrounds the plasma membrane, supports the cell, and imparts shape to it. Prokaryotic cell walls differ from eukaryotic cell walls, but all of them are permeable to dissolved substances, which are free to move to and away from the cell's plasma membrane. Eukaryotic cell walls differ in structure, as you'll see in Section 3.8. Sticky polysaccharides often surround the prokaryotic cell wall. These help the cell attach to interesting surfaces such as river rocks, teeth, and the vagina. Many disease-causing (or pathogenic) bacteria have a jellylike protective capsule of polysaccharides around their wall.

Projecting from many prokaryotic cells are one or more **flagella** (singular, flagellum). Prokaryotic flagella are motile structures that differ from similar-looking flagella found on some kinds of eukaryotic cells; they rotate like propellers rather than waving side to side. Prokaryotic cells use the flagella to move through fluid environments (Figure 3.8*d*). Other surface projections include pili (singular, pilus). These protein filaments help the cells attach to surfaces and to one another.

Like eukaryotic cells, bacteria and archaea depend on a plasma membrane to selectively control the flow of substances into and out of the cytoplasm. Their lipid bilayer, too, bristles with diverse transporters,

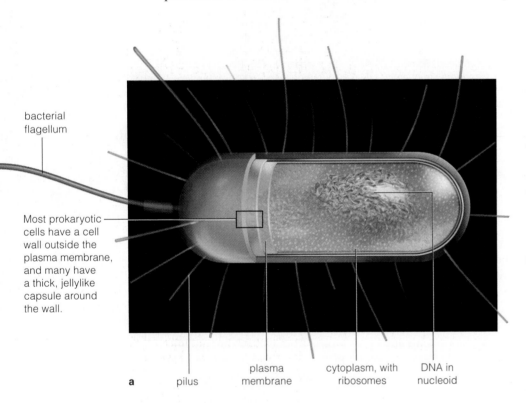

bacterial flagellum

Most prokaryotic cells have a cell wall outside the plasma membrane, and many have a thick, jellylike capsule around the wall.

a   pilus   plasma membrane   cytoplasm, with ribosomes   DNA in nucleoid

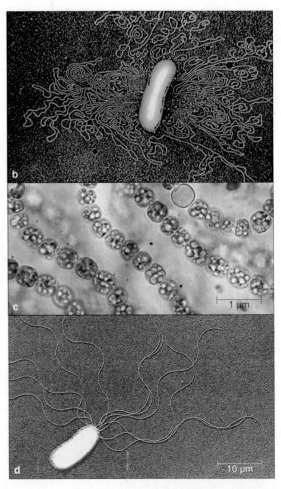

Figure 3.8 (**a**) Generalized sketch of a typical prokaryotic cell.

(**b**) Micrograph of *Escherichia coli*. Researchers manipulated this bacterial cell to release its single, circular molecule of DNA, which looks like tangled string.

(**c**) Cells of various bacterial species are shaped like balls, rods, or corkscrews. Ball-shaped cells of *Nostoc*, a photosynthetic bacterium, stick together in a thick, jellylike sheath of their own secretions.

(**d**) Like this *Pseudomonas marginalis* cell, many species have one or more bacterial flagella that propel the cell in fluid environments.

receptors, and channels. It also incorporates built-in machinery for reactions. For example, photosynthesis proceeds at the plasma membrane of many species of bacteria. Organized arrays of proteins capture light energy and convert it to chemical energy in the form of ATP, which is used to build sugars.

Given their size and simplicity, prokaryotes do not require a complex internal framework. Just under the plasma membrane, arrays of protein filaments in the cytoplasm compose a simple internal "skeleton." This protein scaffold helps shape their sturdy cell wall.

The prokaryotic cytoplasm also contains ribosomes on which polypeptide chains are built, and one DNA molecule in the form of a circle.

Prokaryotes are the simplest cells, but as a group they show the most metabolic diversity.

Bacteria and archaea are different groups of prokaryotic cells. Their DNA is not housed inside a nucleus. Most have a permeable cell wall around their plasma membrane that structurally supports and imparts shape to the cell.

Table 3.1	Common Features of Eukaryotic Cells
Organelles and Their Main Functions	
Nucleus	Localizing and protecting the cell's DNA
Endoplasmic reticulum (ER)	Routing and modifying new polypeptide chains; synthesizing lipids; other tasks
Golgi body	Modifying polypeptide chains into mature proteins; sorting and shipping proteins and lipids for secretion or for use inside cell
Vesicles	Transporting or storing a variety of substances; digesting substances and structures in the cell; other functions
Mitochondria	Making many ATP molecules with energy released from glucose breakdown
Non-Membranous Structures and Their Functions	
Ribosomes	Assembling polypeptide chains
Cytoskeleton	Imparting overall shape and internal organization to the cell; moving the cell and its internal structures

3.5 A Closer Look at Eukaryotic Cells

All cells engage in biosynthesis and energy production, but eukaryotic cells compartmentalize the tasks. Their interior is divided into organelles, each responsible for a specialized function. Ribosomes, a cytoskeleton, and other cell structures support those functions.

Like the prokaryotes, eukaryotic cells have ribosomes in the cytoplasm. Unlike them, eukaryotic cells have an intricate internal skeleton of proteins that is called a cytoskeleton. They also start out life with **organelles**: internal compartments such as the nucleus. *Eu–* means true; *karyon*, meaning kernel, refers to a nucleus. Table 3.1 lists the organelles found in most eukaryotic cells. Some specialized cells may contain additional kinds of organelles and other structures.

What advantages do compartments offer? Each has a semipermeable membrane that selectively controls the types and amounts of substances crossing it. This control maintains the microenvironment inside sacs. Compartments concentrate substances for reactions, isolate toxic or disruptive chemicals, and export others. For instance, sacs called lysosomes contain enzymes that would digest the entire cell if they escaped.

Compartmentalization is quite an organizational challenge for cells. Just as your body's organ systems

interact to keep your body running, organelles also must interact to keep a cell running. Substances move from one organelle to another, and to and from the plasma membrane, in controlled ways. As you will see later, diffusion, transport proteins, and motor proteins on tiny tracks are the movers.

Some substances travel throughout the cytoplasm along a pathway that involves a series of organelles. One series acts as a *secretory* pathway. It moves newly formed polypeptide chains from ribosomes through the organelles known as endoplasmic reticulum, then through Golgi bodies, then to the plasma membrane. Another series, the *endocytic* pathway, moves outside substances into the cell.

Substances that move along either pathway travel in tiny vesicles that bud from organelle membranes and the plasma membrane. Proteins on these vesicles help them deliver their cargo to the right destination.

All eukaryotic cells start out life with a nucleus and other organelles, as well as ribosomes and a cytoskeleton.

Specialized cells typically incorporate additional kinds of organelles and other structures.

Organelles physically separate chemical reactions. Organelles assemble, store, and move substances to and from the plasma membrane, or to specific destinations in the cytoplasm.

pore

b RNA messages are translated into polypeptide chains on ribosomes. Many chains are stockpiled in the cytoplasm or used at once. Others enter the rough ER.

chromatin   nucleolus   nuclear envelope   cytoplasm   ribosome   vesicle
(two lipid bilayers)

the cell nucleus

rough ER

a DNA instructions for making proteins are transcribed in the nucleus and moved to the cytoplasm. RNAs are the messengers and protein builders.

c Flattened sacs of rough ER form one continuous channel between the nucleus and smooth ER. Polypeptide chains that enter the channel undergo modification. They will be inserted into organelle membranes or secreted from the cell.

— DNA in nucleus
— rough ER
— smooth ER
— Golgi body

Figure 3.9 *Animated!* (**a**) The nucleus, which houses eukaryotic DNA. (**b**–**d**) Main components of the endomembrane system. Here, many proteins are processed, lipids are assembled, and both products are sorted and shipped to destinations inside the cell or to the plasma membrane for export.

THE NUCLEUS

Unlike prokaryotes, eukaryotic cells have their genetic material distributed among a number of linear DNA molecules of different lengths. The term **chromosome** refers to one double-stranded DNA molecule together with many attached protein molecules. Human body cells have forty-six chromosomes in their nucleus; frog body cells have twenty-six. **Chromatin** is the name for all of the DNA and associated proteins in a nucleus.

The nucleus has two functions. First, it isolates a cell's DNA from potentially damaging reactions in the cytoplasm. This structural separation keeps the DNA molecules organized and makes it easier for the cell to copy them before it divides. Second, it controls access to DNA through the receptors, transport proteins, and pores at its surface. This functional separation helps the cell to control the timing of protein synthesis.

Enclosing the semifluid interior of the nucleus is a **nuclear envelope**: two lipid bilayers studded with pores and transport proteins (Figure 3.9*a*). Each nucleus also

has at least one nucleolus (plural, nucleoli). This is a construction site where subunits of ribosomes are built from RNA and proteins. The subunits pass through the nuclear envelope's pores to the cytoplasm (Figure 3.9*b*). Ribosome subunits link up briefly when they assemble amino acids into polypeptide chains (Section 2.8).

THE ENDOMEMBRANE SYSTEM

Many new polypeptide chains enter the flattened sacs and tubes of the **endomembrane system**: the ER, Golgi bodies, and vesicles. Every protein destined for export or for insertion into cell membranes passes through the endomembrane system.

Endoplasmic reticulum, or **ER**, is a dynamic channel of flattened membrane sacs and tubes that starts at the nuclear envelope and extends through the cytoplasm.

Rough ER is studded with ribosomes attached to its outer surface (Figure 3.9*c*). Newly formed polypeptide chains enter it or become inserted into its membrane if (and only if) they contain a special signal sequence of amino acids. Enzymes in the rough ER channel modify polypeptide chains into final form. You can see a lot of rough ER in cells that make, store, and secrete proteins.

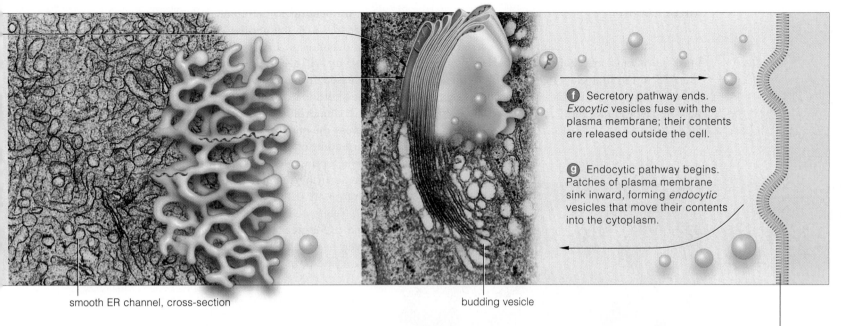

smooth ER channel, cross-section

budding vesicle

f Secretory pathway ends. *Exocytic* vesicles fuse with the plasma membrane; their contents are released outside the cell.

g Endocytic pathway begins. Patches of plasma membrane sink inward, forming *endocytic* vesicles that move their contents into the cytoplasm.

smooth ER

d Some proteins in the channel continue on, to smooth ER. Many become smooth ER enzymes or membrane proteins. The enzymes make lipids, inactivate toxins, and mediate other tasks.

Golgi body

e A Golgi body receives, processes substances that arrive in vesicles from the ER. New vesicles transport the processed substances to the plasma membrane or other parts of the cell.

plasma membrane

h Exocytic vesicles release cell products and wastes to the outside. Endocytic vesicles move nutrients, water, and other substances into the cytoplasm from outside (Section 4.7).

For example, ER-rich gland cells inside your pancreas synthesize and secrete enzymes that end up in your small intestine and help digest meals.

Smooth ER does not have ribosomes (Figure 3.9*d*). It makes lipid molecules that become part of cell membranes. It also takes part in fatty acid breakdown and degrades some kinds of toxins. In muscle cells, a type of smooth ER called sarcoplasmic reticulum has a role in contraction.

Patches of ER membrane bulge and break away as vesicles that deliver polypeptide chains and lipids to **Golgi bodies**. Golgi bodies are folded, flattened sacs that look like stacks of pancakes (Figure 3.9*e*). They attach sugar side chains to some proteins and lipids, and split others. Finished products are packaged into vesicles and then shipped to other regions of the cell, the plasma membrane, or the outside (Figure 3.9*f*).

Vesicles that bud away from the Golgi membranes include the organelles of intracellular digestion. For instance, **lysosomes** contain enzymes that break down carbohydrates, proteins, nucleic acids, and some lipids. Other types of vesicles deliver ingested bacteria, cell parts, and other materials to lysosomes for destruction. Vesicles called **peroxisomes** hold enzymes that digest

fatty acids and amino acids. An important function of peroxisomes is the breakdown of hydrogen peroxide, a toxic by-product of fatty acid metabolism. As you will see in the next chapter, enzymes convert hydrogen peroxide to water and oxygen, or use it for reactions that also break down alcohol and other toxins. Drink alcohol, and the peroxisomes of liver and kidney cells normally will degrade nearly half of it.

Some vesicles fuse to form large vacuoles. One of these is the **central vacuole** of a mature plant cell. Ions, amino acids, sugars, and toxic substances accumulate in the fluid-filled interior of a central vacuole, which expands and forces the pliable cell wall to enlarge.

The nucleus keeps eukaryotic DNA molecules organized and separated from the metabolic machinery of the cytoplasm.

In the ER and Golgi bodies of the endomembrane system, polypeptide chains are processed and lipids are assembled.

Vesicles help integrate cell activities. Different kinds store substances or transport them around the cell. Some break down nutrients and toxins.

LINK TO
SECTION
2.10

MITOCHONDRIA

ATP (Section 2.10) is the main energy carrier in cells. It jump-starts most metabolic reactions. How do cells get the ATP in the first place? One way is to break down organic compounds in the ATP-forming reactions of aerobic respiration. These reactions run to completion in **mitochondria** (singular, mitochondrion). A much-folded inner membrane divides the interior of each mitochondrion into two compartments (Figure 3.10). As you will see in Sections 5.2 and 6.3, hydrogen ions released from the breakdown of organic compounds are shuttled across the inner membrane, into the inner mitochondrial compartment. Hydrogen ion flow back across the membrane to the outer compartment drives formation of ATP. Oxygen keeps the reactions going by accepting electrons. Each time you breathe in, you are securing oxygen mainly for the mitochondria in your many trillions of cells.

Nearly all eukaryotic cells have mitochondria. Cells that use a great deal of energy, such as those of your liver, heart, and skeletal muscles, may each contain a thousand or more mitochondria.

outer membrane

outer compartment

inner compartment

inner membrane

Figure 3.10 Typical mitochondrion. This organelle specializes in producing ATP.

CHLOROPLASTS

Many plant cells have plastids, which are organelles of photosynthesis, storage, or both. Only photosynthetic eukaryotic cells contain the important plastids called **chloroplasts**. In these organelles, energy from the sun drives the formation of ATP and NADPH, which are then used to form organic compounds.

A chloroplast has two outer membranes around a semifluid interior, the stroma, which bathes an inner membrane. Often this single membrane is folded back on itself into a series of stacked, interconnected disks (Figure 3.11). Each stack is a **thylakoid**. Embedded in the thylakoid membrane are light-trapping pigments, including chlorophylls, as well as enzymes and other proteins with roles in photosynthesis. Glucose, then sucrose, starch, and other organic compounds are built from carbon dioxide and water in the stroma.

Considered individually, chloroplasts are a lot like photosynthetic bacteria. Like mitochondria, they may have evolved by way of endosymbiosis, as described in the next section.

SUMMARY OF MAJOR ORGANELLES

Before continuing with eukaryotic cell structures, take a quick look at Figures 3.12 and 3.13. They summarize the organelles we have considered so far.

two outer membranes

thylakoids (inner membrane system folded into flattened disks)

Figure 3.11 Typical chloroplast, the key defining feature of photosynthetic eukaryotic cells.

The organelles called mitochondria are the ATP-producing powerhouses of all eukaryotic cells.

Reactions that release energy from organic compounds occur in mitochondria. The reactions require oxygen.

Photosynthetic eukaryotic cells have chloroplasts, organelles that specialize in making sugars and other carbohydrates.

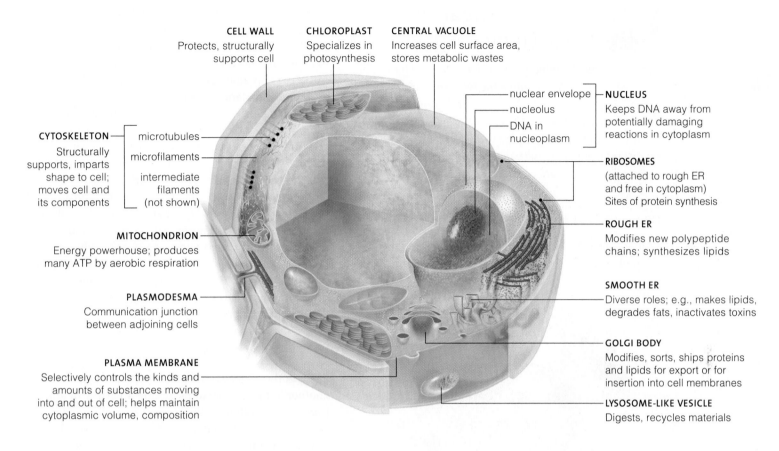

CELL WALL
Protects, structurally supports cell

CHLOROPLAST
Specializes in photosynthesis

CENTRAL VACUOLE
Increases cell surface area, stores metabolic wastes

nuclear envelope
nucleolus
DNA in nucleoplasm

NUCLEUS
Keeps DNA away from potentially damaging reactions in cytoplasm

CYTOSKELETON
Structurally supports, imparts shape to cell; moves cell and its components

microtubules
microfilaments
intermediate filaments (not shown)

RIBOSOMES
(attached to rough ER and free in cytoplasm) Sites of protein synthesis

MITOCHONDRION
Energy powerhouse; produces many ATP by aerobic respiration

ROUGH ER
Modifies new polypeptide chains; synthesizes lipids

SMOOTH ER
Diverse roles; e.g., makes lipids, degrades fats, inactivates toxins

PLASMODESMA
Communication junction between adjoining cells

GOLGI BODY
Modifies, sorts, ships proteins and lipids for export or for insertion into cell membranes

PLASMA MEMBRANE
Selectively controls the kinds and amounts of substances moving into and out of cell; helps maintain cytoplasmic volume, composition

LYSOSOME-LIKE VESICLE
Digests, recycles materials

Figure 3.12 Typical organelles and structures of plant cells.

nuclear envelope
nucleolus
DNA in nucleoplasm

NUCLEUS
Keeps DNA away from potentially damaging reactions in cytoplasm

CYTOSKELETON
Structurally supports, imparts shape to cell; moves cell and its components

microtubules
microfilaments
intermediate filaments

RIBOSOMES (attached to rough ER and free in cytoplasm) Sites of protein synthesis

ROUGH ER
Modifies new polypeptide chains; synthesizes lipids

MITOCHONDRION
Energy powerhouse; produces many ATP by aerobic respiration

SMOOTH ER
Diverse roles; e.g., makes lipids, degrades fats, inactivates toxins

CENTRIOLES
Special centers that produce and organize microtubules

GOLGI BODY
Modifies, sorts, ships proteins and lipids for export or for insertion into cell membranes

PLASMA MEMBRANE
Selectively controls the kinds and amounts of substances moving into and out of cell; helps maintain cytoplasmic volume, composition

LYSOSOME
Digests, recycles materials

Figure 3.13 Typical organelles and structures of animal cells.

LINKS TO
SECTIONS
1.4, 1.5

3.6 Where Did Organelles Come From?

Thanks to globe-hopping microfossil hunters, we have evidence of early life, including the fossil treasures shown in Figure 3.14. Today, most of the descendant species have a profusion of organelles. Where did the organelles come from? Some probably evolved by gene mutations and natural selection. For others, researchers make a case for evolution by way of endosymbiosis.

ORIGIN OF THE NUCLEUS AND ER

Prokaryotic cells, recall, do not have an abundance of organelles. Some have infoldings of plasma membrane that hold enzymes and other metabolic agents (Figure 3.15). By applying the theory of natural selection, we can hypothesize that such infoldings may also have evolved among the ancestors of eukaryotic cells. The infoldings became narrow channels in which nutrients and other substances were concentrated.

What selective advantage did the infoldings offer? More membrane meant more metabolic machinery for reactions, transport tasks, and waste disposal. Did ER channels evolve this way? Did they also protect the cell's metabolic machinery from uninvited guests?

By the latter scenario, infoldings that surrounded the DNA became a nucleus that helped protect the cell from foreign DNA (Figure 3.15). Bacteria and simple eukaryotic cells called yeasts do transfer bits of DNA between themselves. Early eukaryotic cells equipped with a nucleus could copy and use their messages of inheritance, free from competition with a potentially disruptive hodgepodge of foreign DNA.

THEORY OF ENDOSYMBIOSIS

Early in the history of life, the insertion of one cell into another led to partnerships among a variety of ancient prokaryotes. Whether they began as undigested meals or as parasites, uninvited guests evolved into what are now mitochondria, chloroplasts, and other organelles. This is one premise of a theory of **endosymbiosis**, as proposed by Lynn Margulis and others. (*Endo–* means within; *symbiosis* means living together.) One species (a guest) lives out its life inside another species (the host). The close interaction benefits one or both.

b 25 µm

a 5 µm   c 25 µm

flagellum

chloroplast

mitochondrion   nucleus   5 µm

d

Figure 3.14 A few fossilized cells: From Australia, (**a**) a strand of what may be walled prokaryotic cells 3.5 billion years old and (**b**) one of the early eukaryotes—a protist that is 1.4 billion years old. From Siberia, (**c**) aquatic eukaryotic cells about 850 million years old. (**d**) A fine example of diverse organelles—hallmarks of eukaryotic cells: *Euglena*, a single-celled protist, sliced lengthwise. It has a long flagellum, which could not completely fit in the image area at this magnification.

Figure 3.15 (**a**) Sketch of a bacterial cell (*Nitrobacter*) that lives in soil. Permanent infoldings of the plasma membrane penetrate the cytoplasm. (**b**) Model for the origin of the nuclear envelope and endoplasmic reticulum. In prokaryotic ancestors of eukaryotic cells, infoldings of the plasma membrane may have evolved into these important organelles.

DNA

a infolding of plasma membrane

b

According to this theory, eukaryotic cells evolved after the oxygen-producing pathway of photosynthesis emerged and changed the atmosphere. By 2.1 billion years ago, some populations of prokaryotic cells had adapted to the presence of oxygen and were extracting energy from glucose by way of aerobic respiration.

Some of the ancestors of eukaryotic cells certainly preyed on the aerobic bacteria. Others were perhaps hosts for parasitic infection by them. Suppose that a meal refused to be digested, or that a host persisted despite an ongoing parasitic infection. Endosymbiotic interactions would have begun at that time.

Internalized cells thrived in the sheltered, nutrient-rich environment of the host cytoplasm. The uninvited guests no longer had to spend energy on acquiring or making raw materials. The DNA regions that specified unneeded materials eventually mutated or were lost.

How did the hosts benefit? In time, they started to use some of the ATP made by their guests for growth and increased or diversified activities. Eventually they came to depend upon it. The two kinds of organisms became incapable of independent life. The guests had evolved into mitochondria.

Indeed, mitochondria of eukaryotic cells are a lot like bacteria in size and structure. Each has its own DNA and divides independently of cell division. The inner membrane of a mitochondrion also resembles the plasma membrane of a bacterial cell.

Chloroplasts, too, may have endosymbiotic origins. They would have provided a selective advantage to an aerobically respiring host by releasing needed oxygen. Chloroplasts do resemble some bacteria in the details of their metabolism and genes. They differ from one another in shape and light-absorbing pigments, like photosynthetic bacteria do. Their DNA self-replicates, and they divide independently of the cell's division.

However they arose, the hybrid cells had a nucleus, an endomembrane system, mitochondria, and in some cases, chloroplasts. They were all eukaryotic cells, the first protists. They had efficient metabolic systems, and they evolved fast. In no time at all, evolutionarily speaking, their descendants evolved into plants, fungi, and animals.

EVIDENCE OF ENDOSYMBIOSIS

Is such a theory far-fetched? A chance discovery in Jeon Kwang's laboratory suggests otherwise. In 1966, a type of rod-shaped bacteria had infected his culture of single-celled *Amoeba*. Some infected cells died right away. Others grew more slowly, and were smaller and vulnerable to starving to death. Kwang maintained the infected culture.

Five years later, amoebas stuffed with bacteria were flourishing, and actually had become symbiotic with them. Researchers discovered the symbiosis because antibiotics specific for prokaryotic cells killed infected amoebas. Also, uninfected amoebas stripped of their nucleus died after receiving a nucleus from an infected cell. Yet, more than 90 percent survived the transplant when a few bacteria tagged along with the nucleus. Later studies revealed that the infected amoebas had lost their capacity to make an essential enzyme, and were dependent on the bacteria to make it for them.

> *The nucleus and ER may have evolved through infoldings of the plasma membrane. Mitochondria, chloroplasts, and other organelles might have evolved by endosymbiosis. Such ideas may help explain how you and all other eukaryotes evolved from simple prokaryotic beginnings.*

What did the ancestors of mitochondria look like? Gene studies offer clues. *Reclinomonas americana* is the protist with the structurally simplest mitochondria. Genes of *R. americana* mitochondria resemble the genes of *Rickettsia prowazekii*, the bacteria that cause typhus. Like mitochondria, parasitic *R. prowazekii* divide only inside eukaryotic cells.

Figure 3.16 The beautiful cytoskeleton. (**a**) Microtubules (*gold*) and actin filaments (*red*) in an epithelial cell. (**b**) Actin filaments (*red*) and associated proteins (*green, blue*) help new epithelial cells such as this one migrate in tissues as human embryos are developing. (**c**) Intermediate filaments (*red*) of cultured kangaroo rat cells. The *blue* organelle inside each cell is the nucleus.

tubulin subunits assembling into a microtubule

Figure 3.17 Poisonous plants, (**a**) the autumn crocus, and (**b**) western yew. Both of these plants make toxic substances that block microtubule assembly (*above*).

3.7 The Dynamic Cytoskeleton

Like you, all eukaryotic cells have an internal structural framework—a skeleton. Unlike your skeleton, theirs has elements that are not permanently rigid; they assemble and disassemble at different times.

In between the nucleus and plasma membrane of all eukaryotic cells is a **cytoskeleton**—an interconnected system of many protein filaments. Parts of the system reinforce, organize, and move cell structures and often the whole cell. Different parts of the cytoskeleton are shown in the micrographs in Figure 3.16. Some of the cytoskeletal elements are permanent; others form at certain times in the cell's life.

Two major cytoskeletal elements, microtubules and microfilaments, occur in all eukaryotic cells. Another type, intermediate filaments, strengthens some animal cells. Prokaryotic cells do not have a cytoskeleton, but some have microfilament-like proteins that reinforce the cell body.

Microtubules are long, hollow cylinders of tubulin subunits (Figure 3.17). They organize the cell interior and form a dynamic framework that moves structures such as chromosomes to specific locations. Controls govern which microtubules grow or fall apart at any given time. Microtubules growing in one direction—say, the forward end of a prowling amoeba—might get a protein cap that keeps them intact. Those at the cell's trailing end aren't used, aren't capped, and fall apart.

Colchicine, made by the autumn crocus (*Colchicum autumnale*), blocks microtubule assembly in the cells of animals that eat the plant (Figure 3.17a). Animal cells cannot divide without microtubules. The western yew (*Taxus brevifolia*) makes taxol, a different microtubule poison (Figure 3.17b). Doctors use a synthetic version of taxol to stop the uncontrolled cell divisions of some kinds of cancers.

Microfilaments strengthen or reconfigure a cell's shape (Figure 3.16a,b). Crosslinked, bundled, and gel-like arrays of them form a **cell cortex** that underlies and reinforces the plasma membrane. Microfilaments also anchor proteins, and they help muscles contract. An animal cell divides when microfilaments around its midsection contract, pinching it in two.

Think of the bustle at a train station during a busy holiday season, and you get an idea of what goes on in cells. Microtubules and microfilaments in cells are like dynamically assembled tracks. Motor proteins are the freight engines (Figure 3.18).

Intermediate filaments are the most stable parts of some cytoskeletons (Figure 3.16c). They strengthen and help maintain structures. For example, lamins anchor the actin and myosin of contractile units inside muscle cells. Other types anchor cells in tissues.

One or at most two kinds of intermediate filaments occur in certain animal cells. Researchers use them to identify the type of cell. *Cell typing* is a useful tool in diagnosing the tissue origin of diverse cancers.

MOVING ALONG WITH MOTOR PROTEINS

All eukaryotic cells contain similar microtubules and microfilaments. In spite of the uniformity, both kinds of elements serve diverse functions. How? *Other* proteins associate with them. Among the accessory proteins are **motor proteins**. These move cell parts in a sustained, directional way when repeatedly energized by ATP.

Figure 3.19 Cilia, a flagellum, and false feet. (**a**) Cilia (*gold*) projecting from the surface of cells that line an airway to human lungs. (**b**) Human sperm, with flagellum, about to penetrate an egg. (**c**) A predatory amoeba (*Chaos carolinense*), extending two pseudopods around a single-celled green alga (*Pandorina*).

Some motor proteins move chromosomes. Others slide one microtubule over another. Some chug along tracks in nerve cells that extend from your spine to your toes. Many engines are organized in series, each moving some vesicle partway along the track before giving it up to the next in line. Motor proteins in plant cells drag the chloroplasts into new light-intercepting positions as the sun's angle changes overhead. Motor proteins also contribute to animal muscle movement, as you will see in Chapter 21.

CILIA, FLAGELLA, AND FALSE FEET

Besides moving their internal parts, many eukaryotic cells use microtubules and motor proteins to move the whole cell or to stretch out part of it. Consider flagella and **cilia** (cilium). These motile structures project from the cell surface (Figure 3.19a,b).

Compared to cilia, flagella usually are longer and less profuse. Sperm and many other cells swim with the help of their whiplike flagella. In many multicelled

organisms, ciliated cells continually stir air or fluid. Many thousands of ciliated cells line airways in your chest. Their coordinated beating propels bacteria and airborne particles away from the lungs.

Flagella and cilia move by a sliding mechanism that involves a motor protein called dynein. Paired microtubules extend the same distance into the tip of a cilium or flagellum. When ATP energizes them, short dynein arms projecting from each microtubule grab the next microtubule, tilt in a short, downward stroke, and then release. Repeated action of the dynein arms forces the microtubules to slide past each other. They don't slide far, but each *bends* a bit so that the sliding motion becomes converted to a bending motion.

Amoebas and some other types of eukaryotic cells form **pseudopods**, or "false feet" (Figure 3.19c). These temporary, irregular lobes project from the cell and are used for locomotion and prey capture. Microfilaments that elongate in the lobe also push it forward. Motor proteins on the microfilament tracks drag the plasma membrane with them.

Figure 3.18 *Animated!* A motor protein, kinesin, on a microtubule. Kinesin inches along the length of the microtubule. If the microtubule is anchored near the cell's center, the kinesin moves its freight away from the center.

A cytoskeleton of protein filaments is the basis of eukaryotic cell shape, internal structure, and movement.

Microtubules help organize the eukaryotic cell and move its parts. Microfilaments form networks that reinforce the cell surface. Intermediate filaments strengthen and maintain the shape of some animal cells.

When energized by ATP, motor proteins move along tracks of microtubules and microfilaments. They deliver specific cell components that are bound to them to new locations or into new positions.

(a) Randomly oriented cellulose strands in a growing primary wall let a cell expand in all directions. Cross-oriented strands let it lengthen only.

plasmodesma at adjoining channel across a primary cell wall connects the cytoplasm of two adjacent plant cells

(b)

plasma membrane

primary cell wall

(c)

space previously filled with cytoplasm

secondary cell wall (added in layers)

primary cell wall

(d)

(e)

Figure 3.20 *Animated!* Plant cell walls. (**a**) Microtubules orient cellulose strands, the main construction material for plant walls. Depending on the orientations, the cell will end up round or long.

(**b,c**) Cell secretions form a middle layer between the walls of adjoining cells. Channels across adjacent walls directly connect the cytoplasm of adjacent cells.

(**d**) In many plant cells, more layers are deposited on the inside of the primary wall. They strengthen the wall and maintain its shape. When the cell dies, the stiffened walls remain. (**e**) This happens in water-conducting pipelines that thread through most plant tissues. Interconnected, stiffened walls of dead cells form the tubes.

LINK TO SECTION 2.6

3.8 Cell Surface Specializations

Our survey of eukaryotic cells concludes with a look at cell walls and other specialized surface structures. Some structures are made of cell secretions. Others are clusters of proteins that connect neighboring cells.

EUKARYOTIC CELL WALLS

Single-celled eukaryotic species are directly exposed to the environment. Many have a **cell wall**, a structural component that encloses the plasma membrane. A cell wall protects and physically supports the cell. It is porous, so water and solutes easily move to and from the plasma membrane. A cell would die without these exchanges. Many single-celled protists have a cell wall around their plasma membrane. So do plant cells and many types of fungal cells.

For example, in the growing parts of multicelled plants, young cells secrete molecules of pectin and other glue-like polysaccharides, as well as cellulose. All of these materials are components of a plant cell's **primary wall** (Figure 3.20). The sticky cell wall glues abutting cells together. Thin and pliable, the wall permits the cell to enlarge.

Cells that have only a thin primary wall retain the capacity to divide or change shape as they grow and develop. Many plant cell types stop enlarging when they are mature. Mature cells secrete material on the primary wall's inner surface to form a rigid **secondary wall** that reinforces a cell's shape (Figure 3.20*d*). These secondary wall deposits are extensive, and contribute to structural support of the plant.

In woody plants, up to 25 percent of the secondary wall is made of lignin. This organic compound makes plant parts more waterproof, less susceptible to plant-attacking organisms, and stronger.

thick cuticle at
leaf surface

cell of leaf
epidermis

photosynthetic
cell inside leaf

a

b

Figure 3.21 (**a**) Section through a plant cuticle, an outer surface layer of cell secretions. (**b**) A living cell inside bone tissue, the stuff of vertebrate skeletons.

At plant surfaces exposed to outside air, waxes and other cell secretions build up and form a cuticle. This protective, semitransparent surface limits water losses from plants (Figure 3.21*a*).

MATRIXES BETWEEN ANIMAL CELLS

Animal cells have no walls, but intervening between many of them are matrixes of cell secretions and other materials absorbed from the surroundings. For instance, the cartilage at the knobby ends of leg bones consists of scattered cells and protein fibers formed from their secretions, all embedded in polysaccharides. In bone tissue, the living cells form an extensive matrix (Figure 3.21*b* and Section 21.1).

CELL JUNCTIONS

Even when a wall or some other structure imprisons a cell in its own secretions, the cell still has contact with the outside world through its plasma membrane. In multicelled species, membrane components extend into adjoining cells and the surroundings. Among the components are **cell junctions**: molecular structures through which a cell sends and receives signals and materials. Cell junctions also cement a cell to other cells and to underlying membranes.

In plants, for instance, many tiny channels cross the primary walls of adjacent living cells and interconnect their cytoplasm. Figure 3.20*b* shows one. Each channel is a plasmodesma (plural, plasmodesmata). Substances flow quickly from cell to cell across these junctions.

Three types of cell-to-cell junctions are common in most animal tissues (Figure 3.22). *Tight* junctions link cells of most tissues, including epithelia that line outer surfaces, internal cavities, and organs. Tight junctions seal abutting cells so water-soluble substances cannot pass between them. That is why gastric fluid does not

free surface of epithelial tissue (not attached to any other tissue)

examples of proteins that make up tight junctions

gap junctions

adhering junction

basement membrane

Figure 3.22 The most common types of cell junctions in animal tissues.

leak across the stomach lining and damage internal tissues. *Adhering* junctions occur in skin, the heart, and other organs that stretch. *Gap* junctions that link the cytoplasm of certain adjoining cells are open channels for a rapid flow of small molecules, most notably in heart muscle.

A variety of protist, plant, and fungal cells have a porous wall that surrounds the plasma membrane.

Young plant cells have a thin primary wall pliable enough to permit expansion. Some mature cells also deposit a lignin-reinforced secondary wall that affords structural support.

Animal cells have no walls, but they and many other cells often secrete substances that help form matrixes of tissues. Junctions form between cells of multicelled organisms.

Summary

Section 3.1 The cell is the smallest unit that still displays the properties of life. All cells start out life with an outer plasma membrane, cytoplasm, and a nucleus or nucleoid area that contains DNA.

Section 3.2 Most cells are microscopically small because a surface-to-volume ratio constrains their size. Different microscopes reveal cell shapes and structures.

Biology ⊜ Now
Learn how microscopes function with the animation on BiologyNow.

Section 3.3 Cell membranes consist mainly of lipids and proteins. A lipid bilayer gives a membrane its fluid properties and prevents water-soluble substances from freely crossing it. Proteins embedded in the bilayer or positioned at its surfaces carry out many functions.

Biology ⊜ Now
Explore the structure of cell membranes with the animation on BiologyNow.

Section 3.4 Bacteria and archaea are the prokaryotes. They have no nucleus and are the smallest, structurally simplest cells known (Table 3.2).

Biology ⊜ Now
View the structure of a typical prokaryote with the animation on BiologyNow.

Sections 3.5, 3.6 All eukaryotic cells start out life with a nucleus and other organelles, ribosomes, and a cytoskeleton. Organelles are membranous sacs that divide the cell's interior into functional compartments (Table 3.2). They physically separate chemical reactions from the rest of the cell. A nucleus isolates and protects the cell's genetic material.

In the ER and Golgi bodies of the endomembrane system, new polypeptide chains are modified and lipids are assembled. Many proteins and lipids become part of membranes or packaged inside vesicles that function in transport, storage, and other cell activities.

Mitochondria are the organelles that specialize in forming ATP by aerobic respiration. Photosynthetic eukaryotic cells have chloroplasts, which make sugars and carbohydrates. Chloroplasts and mitochondria may have evolved by endosymbiosis.

Biology ⊜ Now
See how organelles interact in a secretory pathway with the animation on BiologyNow.

Section 3.7 Eukaryotic cells have a cytoskeleton of microtubules, microfilaments, and intermediate filaments. The cytoskeleton supports and imparts shape; it also moves internal structures, motile structures, and often the whole cell.

Biology ⊜ Now
Learn about the role of motor proteins with the animation on BiologyNow.

Section 3.8 Many bacterial, protist, fungal, and plant cells have an outer, supporting wall. Animal cells do not have walls. In multicelled organisms, cell junctions form structural and functional connections between cells.

Biology ⊜ Now
Explore the structure of plant cell walls and cell junctions with the animation on BiologyNow.

Self-Quiz

Answers in Appendix I

1. Cell membranes consist mainly of a _____ .
   a. carbohydrate bilayer and proteins
   b. protein bilayer and phospholipids
   c. lipid bilayer and proteins
   d. none of the above

2. Organelles _____ .
   a. are membrane-bound compartments
   b. are typical of eukaryotic cells, not prokaryotic cells
   c. separate chemical reactions in time and space
   d. All of the above are features of organelles.

3. Cells of many protists, plants, and fungi, but not animals, commonly have _____ .
   a. mitochondria   c. ribosomes
   b. a plasma membrane   d. a cell wall

4. Is this statement true or false: The plasma membrane is the outermost component of all cells. Explain your answer.

5. Unlike eukaryotic cells, prokaryotic cells _____ .
   a. lack a plasma membrane   c. have no nucleus
   b. have RNA, not DNA   d. all of the above

6. Identify as many of the components of the cells shown at left as you can.

7. Match each cell component with its function.
   _____ mitochondrion   a. protein synthesis
   _____ chloroplast   b. initial modification of new
   _____ ribosome   polypeptide chains
   _____ rough ER   c. modification of new proteins;
   _____ Golgi body   sorting, shipping tasks
                          d. photosynthesis
                          e. formation of many molecules
                             of ATP

Additional questions are available on **Biology ⊜ Now**™

Critical Thinking

1. In a classic episode of *Star Trek*, a stupendous amoeba engulfs an entire starship. Spock blows the amoeba to bits before it reproduces and eats the universe. Think of at least one problem a biologist would identify with this scenario.

2. Your professor shows you an electron micrograph of a cell with many mitochondria and Golgi bodies. You notice the cell also contains a great deal of rough endoplasmic reticulum. What kinds of cellular activities would require such an abundance of the three kinds of organelles?

3. Peggy, a graduate student, discovers a single-celled organism swimming in a freshwater pond. Which cellular features might convince her that the cell is a bacterium? What features would suggest it is a protist?

4. *Kartagener syndrome* is a genetic disorder caused by a mutated form of the protein dynein. Affected people have chronically irritated sinuses, and mucus builds up in the airways to their lungs. Bacteria form huge populations

a **b**

Figure 3.23 Cross-section through the flagellum of a sperm cell from (**a**) a male affected by Kartagener syndrome and (**b**) an unaffected male. Notice the dynein arms that extend from the paired microtubules.

in the thick mucus. Their metabolic by-products and the inflammation they trigger combine to damage tissues.

Males affected by the syndrome can make sperm but are infertile (Figure 3.23). Some have become fathers with the help of a procedure that injects sperm cells directly into eggs. Explain how an abnormal dynein molecule could cause the observed effects.

Table 3.2 Summary of Typical Components of Prokaryotic and Eukaryotic Cells		Prokaryotic	Eukaryotic			
Cell Component	**Function**	**Bacteria, Archaea**	**Protists**	**Fungi**	**Plants**	**Animals**
Cell wall	Protection, structural support	✓*	✓*	✓	✓	None
Plasma membrane	Control of substances moving into and out of cell	✓	✓	✓	✓	✓
Nucleus	Physical separation and organization of DNA	None	✓	✓	✓	✓
DNA	Encoding of hereditary information	✓	✓	✓	✓	✓
RNA	Transcription, translation of DNA messages into polypeptide chains of specific proteins	✓	✓	✓	✓	✓
Nucleolus	Assembly of subunits of ribosomes	None	✓	✓	✓	✓
Ribosome	Protein synthesis	✓	✓	✓	✓	✓
Endoplasmic reticulum (ER)	Initial modification of many of the newly forming polypeptide chains of proteins; lipid synthesis	None	✓	✓	✓	✓
Golgi body	Final modification of proteins, lipids; sorting and packaging them for use inside cell or for export	None	✓	✓	✓	✓
Lysosome	Intracellular digestion	None	✓	✓*	✓*	✓
Mitochondrion	ATP formation	**	✓	✓	✓	✓
Photosynthetic pigments	Light–energy conversion	✓*	✓*	None	✓	None
Chloroplast	Photosynthesis; some starch storage	None	✓*	None	✓	None
Central vacuole	Increasing cell surface area; storage	None	None	✓*	✓	None
Bacterial flagellum	Locomotion through fluid surroundings	✓*	None	None	None	None
Eukaryotic flagellum or cilium	Locomotion through or motion within fluid surroundings	None	✓*	✓*	✓*	✓
Complex cytoskeleton	Cell shape; internal organization; basis of cell movement and, in many cells, locomotion	Rudimentary***	✓*	✓*	✓*	✓

*Known to be present in cells of at least some groups.
**Many groups use oxygen-requiring (aerobic) pathways of ATP formation, but mitochondria are not involved.
***Protein filaments form a simple scaffold that helps support the cell wall in at least some species.

Beer, Enzymes, and Your Liver

The next time someone asks you to have a drink, stop for a moment to think about the challenge confronting cells that keep a drinker alive—especially a heavy drinker. It makes little difference whether one swallows 12 ounces of beer, 5 ounces of wine, or 1–1/2 ounces of eighty-proof vodka. Each of these drinks has the same amount of "alcohol" or, more precisely, ethanol.

Ethanol—CH_3CH_2OH—has both water-soluble and fat-soluble components, so it is absorbed quickly by the stomach and small intestine. From there it enters the bloodstream. Most tissues in the body contain enzymes (catalase, cytochrome P-450, and alcohol dehydrogenase) that convert ethanol to acetaldehyde. Acetaldehyde can be toxic if it accumulates. In healthy people, another enzyme converts it to nontoxic acetic acid.

Almost all of the alcohol you drink ends up in the liver. But its metabolism there causes a shift in mitochondrial reactions that can be toxic to liver cells. Given the liver's central role in alcohol metabolism, heavy drinkers gamble with alcohol-induced liver diseases. Over time, there are fewer and fewer liver cells for detoxification.

What are some of the possible outcomes? Alcoholic hepatitis is a disease characterized by inflammation and destruction of liver tissue. A different disease, alcoholic cirrhosis, permanently scars the liver. In time, the liver just stops working, with devastating effects.

The liver is the largest gland in the human body, and its activity affects everything else. You would have a very hard time digesting and absorbing food without it. Your cells would have a hard time synthesizing and taking up carbohydrates, lipids, and proteins, and staying alive.

There is more to think about. The liver rids the body of toxic compounds other than acetaldehyde. It also makes plasma proteins essential for blood clotting, immunity, and maintaining fluid volume. This brings us to a self-destructive behavior called binge drinking, the idea being to consume large amounts of alcohol in a brief period.

Binge drinking is now the most serious drug problem on college campuses in the United States, though very few students recognize their own drinking problem. Consider this finding from one survey: Almost half of 17,600 students at 140 colleges and universities are binge drinkers, meaning they consumed at least four alcoholic drinks in one day, in the two weeks prior to the survey.

Binge drinking does more than damage the liver. Put aside the related 500,000 injuries from accidents, the 70,000 cases of date rape, the 400,000 cases of (whoops) unprotected sex among students in an average year. Binge drinking can kill you. Drink too much, too fast, and you can abruptly end the beating of your heart.

With this example, we turn to metabolism, the cell's capacity to acquire energy and use it to build, degrade, store, and release substances in controlled ways. At times, the activities of your cells may be the last thing you want to think about. But they help define who you are and how you live, liver and all.

 How Would You Vote? There aren't enough liver donors for all the people waiting for liver transplants. Should a person's life-style be a factor in deciding who gets one? See BiologyNow for details, then vote online.

 ## Key Concepts

HOW CELLS USE ENERGY
Energy flows into the web of life, mainly from the sun. Cells tap into this one-way flow by energy-acquiring processes, starting with photosynthesis. They convert inputs of energy to forms that can be used to drive metabolic reactions.

HOW ENZYMES WORK
Without enzymes, substances would not react fast enough to maintain life. By controlling enzyme action, cells control the inputs and outputs of substances that keep them alive.

MEMBRANES AND METABOLISM
Cells have built-in mechanisms for increasing or decreasing water and solute concentrations across the plasma membrane and internal cell membranes. These adjustments are essential for metabolic reactions and metabolic pathways.

Links to Earlier Concepts

Reflect again on the road map for life's organization (Section 1.1). Here you will gain insight into how all organisms tap into a grand, one-way flow of energy to maintain that organization (1.2). You will start thinking about how cells use electrons and protons (2.1) in ways that help make ATP (2.10). Remember what you learned about electron arrangements and hydrogen bonding (2.2)? Here you will see how the biological function of enzymes follows from their structure (2.8), and how electron arrangements contribute to cofactor activity. Remember the structure of cell membranes (3.3, 3.5) and motor proteins (3.7) as you learn more about their function.

4.1 Inputs and Outputs of Energy

We know, almost without thinking about it, that skydivers fall down instead of up, that a hot pan eventually cools off. Our sense of time moving irrevocably forward comes from these observations—which are actually observations of the tendency of energy to disperse irrevocably.

THE ONE-WAY FLOW OF ENERGY

The dictionary defines **energy** as "the capacity to do work." What exactly does that mean? Let's start with the energy part. Energy exists in many different forms. Motion, light, sound, chemical bonds, electricity, heat, and magnetism are some examples.

Energy cannot be created or destroyed. This is the **first law of thermodynamics**. Any system taken as a whole, including the universe, has a given amount of energy. That energy may be distributed in different forms.

Energy disperses spontaneously. This is the **second law of thermodynamics**. For example, a hot pan gives off heat. We will never see all of that heat flow back from the air into the pan.

Each form of energy is useful to some degree for doing work. Work may occur as energy is transferred between systems or bodies. To a biologist, this means that every organism harvests energy from the environment, then uses it for growth and maintenance. Every such event involves energy transfers, and often the conversion of one form of energy into another.

Energy transfers are never entirely efficient. This is another aspect of the second law. With every transfer, some energy dissipates, and by doing so becomes less useful for doing work. The energy of a hot pan may boil water, but only before it disperses. If you think about it, this has to be a one-way trend: Energy that is available to do work in the universe is decreasing.

Can life be one glorious pocket of resistance to this depressing flow? Energy becomes concentrated in each new organism as molecules get organized. Even so, the second law applies to life on Earth. Life's main energy source is the sun, which has been losing energy ever since it formed 5 billion years ago. Photosynthetic cells intercept light energy from the sun and convert it to chemical bond energy in sugars, starches, and other compounds. Organisms that eat plants get the stored chemical energy by breaking and rearranging chemical bonds. With each conversion, however, a bit of energy escapes into the environment, usually as heat. *Overall, energy flows in one direction.* Organisms maintain their organization only by being continually supplied with energy lost from someplace else (Figure 4.1).

ATP—THE CELL'S ENERGY CURRENCY

All cells stay alive by *coupling* energy inputs to energy outputs, mainly with adenosine triphosphate, or **ATP**. This nucleotide consists of a five-carbon sugar (ribose), a base (adenine), and three phosphate groups (Figure 4.2a). ATP readily gives up a phosphate group to other molecules and primes them to react. Such phosphate-group transfers are known as **phosphorylations**.

ATP is the currency in a cell's economy. Cells earn it by investing in energy-requiring reactions. They spend it in energy-releasing reactions that keep them alive. We use a cartoon coin to symbolize ATP.

Because ATP is the main energy carrier for so many reactions, you may infer, correctly, that cells have ways to renew it. When ATP gives up a phosphate group, ADP (adenosine diphosphate) forms. ATP forms again when ADP binds to inorganic phosphate (P_i) or to a phosphate group that was split away from a different molecule. Regenerating ATP by this **ATP/ADP cycle** helps drive most metabolic reactions (Figure 4.2b).

ENERGY LOST
Energy continuously flows from the sun.

ENERGY GAINED
Sunlight energy reaches environments on Earth. Producers secure some and convert it to stored forms of energy. They and all other organisms convert stored energy to forms that can drive cellular work.

ENERGY LOST
With each conversion, there is a one-way flow of a bit of energy back to the environment.

Figure 4.1 A one-way flow of energy into the world of life compensates for the one-way flow of energy out of it.

Figure 4.2 (a) Ball-and-stick model for the energy carrier ATP. (b) ATP couples energy-releasing reactions with energy-requiring ones. In the ATP/ADP cycle, recurring phosphate-group transfers turn ATP into ADP, and back again to ATP.

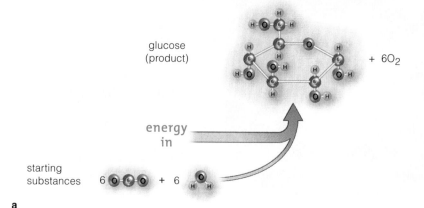

glucose
(product)

+ 6O$_2$

energy
in

starting
substances

6

+ 6

a

glucose + 6O$_2$
(starting substances)

energy
out

b

6

+ 6

products

Figure 4.3 *Animated!* Up and down the energy hills. (**a**) Endergonic reactions require a net input of energy. (**b**) Exergonic reactions end with a net release of energy.

Cells store and retrieve energy when they convert one molecule to another. Inherent in each chemical bond is a certain amount of energy. The difference between the energy of all bonds that compose a reaction's starting compounds and the energy of all bonds that compose its products is the basis of energy flow in metabolism.

By investing in a reaction in which the bonds of the products have more energy than those of the starting materials, cells store energy in the reaction's products. This kind of reaction takes a net input of energy, and is called *endergonic* (meaning energy in). Photosynthetic cells, for example, use light energy to drive formation of ATP, and then use the energy from the ATP to make glucose from carbon dioxide and water. Unlike either light or ATP, glucose is stable and storable.

We think of glucose as a high-energy compound because cells can harvest energy inherent in its bonds to build other molecules. It takes a small investment of energy to break glucose's bonds, but the formation of more stable reaction products (CO_2 and H_2O) releases a lot more energy than the amount invested. It is as if glucose is at the top of an "energy hill," and carbon dioxide and water are at the bottom (Figure 4.3). Such reactions that end in a net release of energy are called *exergonic* (meaning energy out).

Running exergonic reactions would be pointless if cells could not capture released energy. For example, a spark ignites a mixture of hydrogen and oxygen gases. Water forms with an explosive burst of energy (Figure 4.4*a*). Cells cannot capture this kind of disorganized, intense release of energy.

Cells harvest energy in manageable increments by organized, stepwise reactions. In **oxidation–reduction reactions**, one molecule gives up electrons (it becomes oxidized) to another (which becomes reduced). Redox reactions occur in **electron transfer chains**, arrays of enzymes and other molecules that pass electrons from one to the next. Think of the electrons bouncing down a staircase (Figure 4.4*b*). Energy they lose at each step is captured by cells for ATP synthesis and other tasks.

H$_2$ 1/2 O$_2$

electric
spark

Explosive
release of
energy as
heat that
cannot be
harnessed
for cellular
work

a   H$_2$O

H$_2$

① Energy input splits hydrogen into protons (H$^+$) and electrons

1/2 O$_2$

2H$^+$ **2e$^-$**

③ Some released energy is harnessed for cellular work (e.g., making ATP).

②
Electrons transferred through electron transfer chain

2e$^-$

2H$^+$ 1/2 O$_2$

④ Spent electrons and free oxygen form water.

H$_2$O

b

Figure 4.4 *Animated!* Uncontrolled versus controlled energy release. (**a**) Free hydrogen and oxygen exposed to an electric spark react and release energy all at once. (**b**) Electron transfer chains allow the same reaction to proceed in smaller, more manageable steps that can handle the released energy.

Energy is the capacity to do work. It flows in one direction, from more usable to less usable forms. Organisms maintain their organization by using energy lost from someplace else.

ATP is the main energy carrier in all living cells. It couples energy-releasing and energy-requiring reactions. ATP primes molecules to react by transferring a phosphate group to them.

The sun is the primary source of energy for the web of life. Cells store and retrieve energy by chemical reactions. They harvest energy gradually with organized, stepwise reactions.

4.2 Inputs and Outputs of Substances

How cells get energy is only one aspect of metabolism. Another is the accumulation, conversion, and disposal of materials by energy-driven reactions. Most of these reactions are part of stepwise metabolic pathways.

WHAT ARE METABOLIC PATHWAYS?

Cells use and store thousands of substances in orderly, enzyme-mediated reaction sequences called **metabolic pathways**. In *biosynthetic* pathways, small molecules become assembled into larger molecules. Biosynthetic pathways require energy inputs. Photosynthesis is the main biosynthetic pathway. *Degradative* pathways are exergonic, overall. They break down large molecules to smaller products for a net energy release. In aerobic respiration, the main degradative pathway, glucose is converted to CO_2 and H_2O.

Not all metabolic pathways are linear, a straight line from reactants to products. In cyclic pathways, the last step regenerates a reactant that is the cycle's point of entry. In branched pathways, reactants or intermediates are sent to two or more different reaction sequences.

THE NATURE OF METABOLIC REACTIONS

For any metabolic reaction, the starting substances are called *reactants*. Substances formed during a sequence of reactions are *intermediates*, and those left at the end are the *products*. ATP and other *energy carriers* activate enzymes and other molecules by making phosphate-group transfers. *Enzymes* are catalysts: They can speed specific reactions enormously. *Cofactors* are metal ions and coenzymes such as NAD^+. They help enzymes by transferring functional groups, atoms, and electrons between reaction sites. *Transport proteins* move solutes within or across membranes. Because they can change local concentrations of reactants, intermediates, and products, transport proteins and controls over them can influence the timing and direction of metabolism.

Metabolic reactions do not run only forward, from reactants to products. Most also run in reverse, with products being converted back into reactants at least to some extent. Such reversible reactions tend toward **chemical equilibrium**, when the reaction rate is about the same in both directions (Figure 4.5).

The numbers of reactant and product molecules are not usually equal at chemical equilibrium. Imagine it is like a party where people drift between two rooms. The number in each room stays the same—say, thirty in one and ten in the other—even as individuals move

back and forth. Why bother to think about this? *Each cell can bring about big changes in its activities simply by controlling a few steps of reversible metabolic pathways.*

For instance, most kinds of cells use a lot of ATP in order to function properly. They continuously convert glucose into pyruvate and ATP by a pathway of nine enzyme-mediated reactions called glycolysis (Section 6.2). When glucose supplies become too low, cells can reverse this pathway and build glucose from pyruvate and other non-carbohydrate molecules. How? Six steps of glycolysis are reversible and the other three are bypassed. Energy input from ATP drives the bypass reactions in the energetically unfavorable (or uphill) direction. If cells did not have this reverse pathway, they would not be able to compensate for episodes of starvation or intense exercise, when the blood glucose level can become dangerously low.

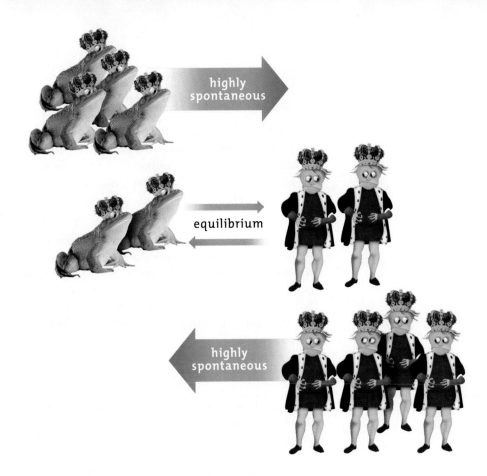

Figure 4.5 Chemical equilibrium. With a high concentration of reactant molecules (represented as wishful frogs), a reaction runs most strongly in the forward direction, to products (the princes). When the concentration of product molecules is high, it runs most strongly in reverse. At equilibrium, the rates of the forward and reverse reactions are the same.

Metabolic pathways are orderly, enzyme-mediated reaction sequences, some biosynthetic, others degradative.

All cells increase, maintain, and decrease supplies of substances by thousands of reactions. They rapidly shift activities by controlling key steps of reversible reactions.

Metabolic reactions start with reactants, typically form intermediates, and end with products. They use energy carriers (ATP), enzymes, cofactors, and transport proteins.

4.3 How Enzymes Make Substances React

What would happen if you left a cupful of glucose out in the open? Not much. Years would pass before you would see evidence of its conversion to carbon dioxide and water. Yet that same conversion takes only a few seconds in your body. Enzymes make the difference.

Enzymes are catalysts that can make reactions occur hundreds to millions of times faster than they would on their own. Enzyme molecules can work again and again; a reaction does not use them up or alter them. They recognize only certain reactants, which are called their **substrates**. Except for a few RNAs, all enzymes are proteins.

LOWERING THE ENERGY HILL

Reactions do not proceed until the reactants have a minimum amount of internal energy—the activation energy. Visualize activation energy as a barrier—a hill or a brick wall (Figure 4.6). Enzymes lower it. How? *Compared with the surrounding environment, they offer a stable microenvironment, more favorable for reaction.*

Enzyme molecules are far larger than substrates. Each has one or more **active sites**: pockets or crevices where substrates bind and where reactions occur. Part of the substrate molecule is complementary in shape, size, charge, and solubility to the active site. This fit is why each enzyme can recognize and bind its substrate among the thousands of substances inside cells.

Think back on the main types of enzyme-mediated reactions (Section 2.5). With *functional group transfers,* one molecule gives up a functional group to another. With *electron transfers*, electrons are removed from one molecule and donated to another. With *rearrangements*, a juggling of internal bonds converts one molecule to another kind. In *condensation*, two or more molecules are covalently bound into a larger molecule. Finally, with *cleavage* reactions, a larger molecule is split into smaller ones.

When we talk about activation energy, *we really are talking about the energy it takes to align reactive chemical groups, destabilize electric charges, and break bonds.* These events put a substrate into its **transition state**. In this state, substrate bonds are at the breaking point, and the reaction can run easily to product.

The binding between an enzyme and its substrate is weak and temporary (that is why the reaction does not change the enzyme). However, energy released when

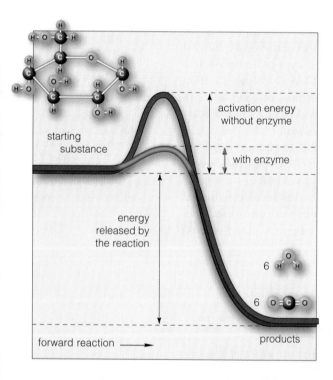

Figure 4.6 *Animated!* Activation energy: the minimum amount of internal energy that reactants must have before a reaction will run to products. An enzyme enhances the reaction rate by lowering the activation energy; by doing so it lowers the energy hill.

HOW ENZYMES WORK

one of four heme groups cradled in one of four polypeptide chains

a Hydrogen peroxide (H_2O_2) enters a cavity in catalase. It is the substrate for a reaction aided by an iron molecule in a heme group (*red*).

b A hydrogen of the peroxide is attracted to histidine, an amino acid projecting into the cavity. One oxygen binds the iron.

c This binding destabilizes the peroxide bond, which breaks. Water (H_2O) forms. In a later reaction, another H_2O_2 will pull the oxygen from iron, which will then be free to act again.

Figure 4.7 How catalase works. This enzyme has four polypeptide chains and four heme groups.

the weak bonds form stabilizes the transition state. This "binding energy" keeps enzyme and substrate together long enough for the reaction to be completed.

Four mechanisms work alone or in combination to move substrates to their transition state:

Helping substrates get together. When substrates are at low concentrations, they rarely react. Binding at an active site is an effective local boost in concentration, by as much as ten millionfold.

Orienting substrates in positions favoring reaction. On their own, substrates collide from random directions. By contrast, the weak but extensive bonds at an active site keep substrates close together.

Shutting out water. Because of its capacity to form hydrogen bonds so easily, water can interfere with the breaking and formation of chemical bonds during reactions. Some active sites are heavily populated with nonpolar amino acids. These repel water and so keep it away from the reaction.

Inducing changes in enzyme shape. Weak interactions with an enzyme may make a substrate change shape. Sometimes a substrate does not fit into an active site very well. The enzyme then stretches or squeezes the substrate into a particular shape, often next to another molecule. By optimizing its fit with the substrate, the enzyme can move the substrate to its transition state.

Substrates are not exactly complementary to active sites in an enzyme. Why? If an active site were to bind substrate too tightly, the complex would be too stable. Enzyme and substrate would stick together, and that would be a chemical, not enzymatic, reaction.

HELP FROM COFACTORS

Nearly a third of all enzymatic reactions use cofactors. **Cofactors** are non-protein compounds and metal ions that associate with certain enzymes and are required for their function. Organic cofactors such as vitamins are coenzymes, and many of them contain metal ions. Metals donate and accept electrons easily. Proximity to a metal ion can alter the arrangement of electrons in a substrate, and so bring on its transition state. That is what happens in a catalase, an enzyme with four heme cofactors. The iron atom in each heme helps catalase break down hydrogen peroxide to water and oxygen (Figure 4.7). Catalase, an **antioxidant**, also neutralizes other strong oxidizers such as *free radicals*, substances with unpaired electrons. Free radicals are dangerously reactive by-products of aerobic respiration, attacking DNA, lipids, and other biological molecules. We make less catalase as we age, so free radicals accumulate.

Activation energy is the minimum amount of energy required to convert reactants into products. Enzymes greatly increase reaction rates by lowering activation energy.

Together with cofactors, enzymes drive their substrates to a transition state, when the reaction can most easily run to completion. This happens in the enzyme's active site.

In the active site, substrates move to the transition state by mechanisms that concentrate and orient them, that exclude water, and that induce an optimal fit with the active site.

LINK TO
SECTION
2.2

IMPACTS, ISSUES

One of the mutant forms of the alcohol-degrading enzyme called aldehyde dehydrogenase is not good at breaking down acetaldehyde. It is quite common in individuals of many Asian populations, and it gets them flushed, dizzy, and nauseous soon after they drink alcohol.

NAD^+ assists this enzyme, but the body has only so many NAD^+ molecules. When this coenzyme is helping to break down alcohol, it can't enter the body's energy-producing reactions. It is diverted into fat-synthesizing reactions. That is why a fatty liver is the first stage of alcoholic liver disease.

HOW ENZYMES WORK

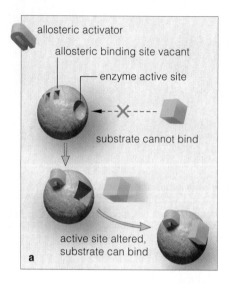

allosteric activator
allosteric binding site vacant
enzyme active site
substrate cannot bind
active site altered, substrate can bind

a

allosteric inhibitor
allosteric binding site vacant; active site can bind substrate
active site altered, can't bind substrate

b

Figure 4.8 Allosteric control over enzyme activity. (**a**) An active site is unblocked when an activator binds to a vacant allosteric site. (**b**) An active site is blocked when an inhibitor binds to a vacant allosteric site.

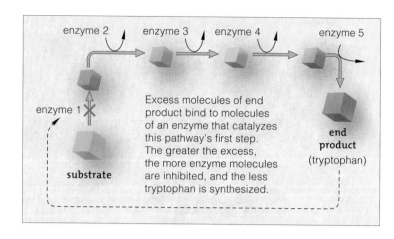

enzyme 2   enzyme 3   enzyme 4   enzyme 5

enzyme 1

substrate

Excess molecules of end product bind to molecules of an enzyme that catalyzes this pathway's first step. The greater the excess, the more enzyme molecules are inhibited, and the less tryptophan is synthesized.

end product (tryptophan)

Figure 4.9 Feedback inhibition of a metabolic pathway. Five kinds of enzymes act in sequence to convert a substrate to tryptophan.

HOW IS ENZYME ACTIVITY CONTROLLED?

LINK TO SECTION 2.4

Controls over enzymes can raise, lower, and maintain concentrations of substances. Some controls can adjust the rate of enzyme synthesis. Others activate or inhibit the enzymes that are already built.

In some cases, a molecule that acts as an activator or inhibitor reversibly binds to its own *allosteric* site, not the active site, on the enzyme (*allo–*, other; *steric*, structure). Binding alters the enzyme's shape in a way that either hides or exposes the active site (Figure 4.8). Allosteric enzymes are part of many feedback loops.

Feedback loops adjust the concentrations of many substances (Figure 4.9). Picture a bacterial cell making tryptophan and the other amino acids—the building blocks for proteins. When the cell has made enough proteins, tryptophan synthesis will continue until its accumulation causes **feedback inhibition**. This means some result of a specific activity *shuts down the activity*. Here, tryptophan stops tryptophan synthesis. Unused tryptophan binds to an allosteric site on an enzyme in the tryptophan synthesis pathway. Binding blocks the enzyme's active site, so less tryptophan is made. More of the active sites are available when few tryptophan molecules are around, and the synthesis rate increases.

EFFECTS OF TEMPERATURE, PH, AND SALINITY

Temperature is a measure of molecular motion. As it rises, it boosts reaction rates both by increasing the chance that a substrate will bump into an enzyme, and by raising a substrate's internal energy. Remember, the more energy a reactant has, the closer it is to jumping the activation energy barrier. Outside a certain range of temperature, hydrogen bonds break. Enzyme shape changes, and the active site no longer binds substrate. The reaction rate falls sharply (Figure 4.10*a,b*). Such declines typically occur in fevers above 44°C (112°F), which are usually deadly.

Enzyme function is also affected by pH and salt concentration, because they influence both hydrogen bonding and reaction chemistry. Most enzymes have a narrow range of pH and salinity under which they function properly (Figure 4.10*c*). This range is found where the enzyme occurs in nature. For instance, in the human body, most enzymes work best when the pH is between 6 and 8.

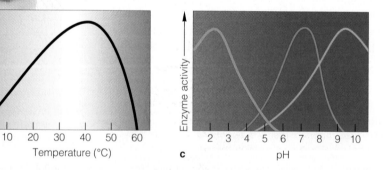

a

Enzyme activity

10 20 30 40 50 60
b   Temperature (°C)

Enzyme activity

2 3 4 5 6 7 8 9 10
c   pH

Figure 4.10 Enzymes and the environment. (**a**) Temperature outside the body affects the fur color of Siamese cats. Epidermal cells that give rise to the cat's fur produce a brownish-black pigment, melanin. Tyrosinase, an enzyme in the melanin production pathway, is heat-sensitive in the Siamese. It becomes inactive in warmer parts of the cat's body, which end up with less melanin, and lighter fur. Put this cat in booties for a few weeks and its warm feet will turn white. (**b**) How increases in temperature affect one enzyme's activity. (**c**) How rising pH values affect three kinds of enzymes.

> Enzyme action adjusts the concentrations and kinds of substances available in cells. Controls over enzymes can enhance or inhibit their activity.
>
> Enzymes work best when the cellular environment stays within limited ranges of temperature, pH, and salinity. The actual ranges differ from one type of enzyme to the next.

4.4 Diffusion and Metabolism

What determines whether a substance will move one way or another across a cell membrane? Part of the answer has to do with diffusion.

Think about the water bathing the surfaces of a cell membrane. Plenty of substances are dissolved in it, but the kinds and amounts close to the two membrane surfaces differ. The membrane set up the difference and busily maintains it. How? Its structure lets some substances but not others cross it in certain ways, at certain times. This is called **selective permeability**.

Lipids of a membrane's bilayer are mostly nonpolar, so they let small, nonpolar molecules such as O_2 and CO_2 slip across. Water molecules are polar, but some slip through gaps that open when hydrophobic tails of many lipids flex and bend. A bilayer is not permeable to ions or large, polar molecules such as glucose; these cross only with the help of proteins. Water frequently crosses with them (Figure 4.11).

LINKS TO
SECTIONS
2.3, 3.3

Membrane barriers and crossings are vital, because metabolism depends on the cell's capacity to increase, decrease, and maintain the concentrations of ions and molecules required for reactions. As well, they supply cells or organelles with raw materials, remove wastes, and collectively maintain the cell's volume and pH.

WHAT IS A CONCENTRATION GRADIENT?

A **concentration gradient** is a difference in the number per unit volume of molecules (or ions) of a substance between two regions. In the absence of other forces, molecules move into an adjoining region where they are not as concentrated. Why? They randomly collide many millions of times a second. This goes on in both regions; it just happens more often in a bigger crowd. In any given interval, more molecules are moving to the less crowded region than are moving out of it.

Diffusion is the net movement of like molecules or ions down a concentration gradient. It contributes to the movement of substances through the cytoplasm and across membranes. In multicelled species, it moves substances between body parts and between the body and its environment. For example, when oxygen builds up in leaf cells, it may diffuse into air inside the leaf, then into air outside, where its concentration is lower.

A substance tends to diffuse in a direction set by its *own* concentration gradient, not by gradients of other solutes. You can see this by dropping dye into water. Dye molecules diffuse into the region where they are less concentrated, and water molecules move into the region where *they* are less concentrated (Figure 4.12).

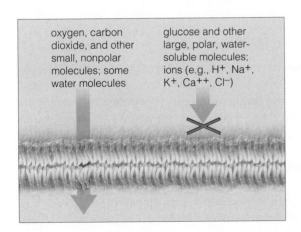

| oxygen, carbon dioxide, and other small, nonpolar molecules; some water molecules | glucose and other large, polar, water-soluble molecules; ions (e.g., H^+, Na^+, K^+, Ca^{++}, Cl^-) |

Figure 4.11 *Animated!* Selective permeability of cell membranes. Small, nonpolar molecules and some water molecules cross the lipid bilayer. Ions and large, polar, water-soluble molecules and the water dissolving them cross with the help of transport proteins.

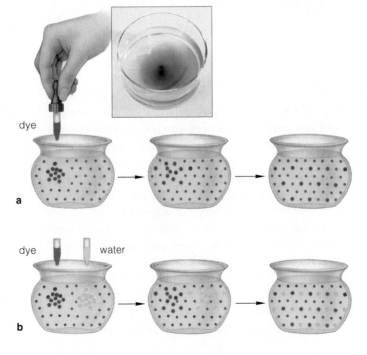

dye

a

dye   water

b

Figure 4.12 *Animated!* Two examples of diffusion. (**a**) A drop of dye enters a bowl of water. Gradually, the dye molecules become evenly dispersed through the molecules of water.

(**b**) The same thing occurs with the water molecules. Here, dye (*red*) and water (*yellow*) are added to the same bowl. Each of the substances will show a net movement down its own concentration gradient.

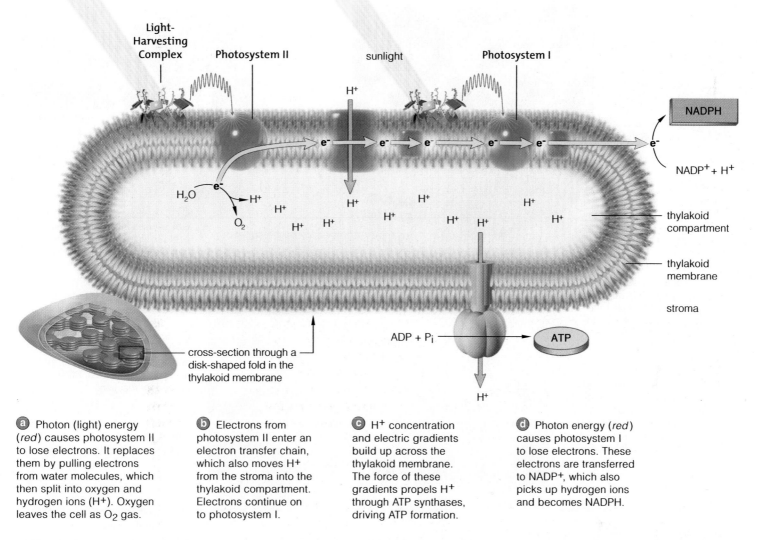

a Photon (light) energy (*red*) causes photosystem II to lose electrons. It replaces them by pulling electrons from water molecules, which then split into oxygen and hydrogen ions (H+). Oxygen leaves the cell as O₂ gas.

b Electrons from photosystem II enter an electron transfer chain, which also moves H+ from the stroma into the thylakoid compartment. Electrons continue on to photosystem I.

c H+ concentration and electric gradients build up across the thylakoid membrane. The force of these gradients propels H+ through ATP synthases, driving ATP formation.

d Photon energy (*red*) causes photosystem I to lose electrons. These electrons are transferred to NADP+, which also picks up hydrogen ions and becomes NADPH.

Figure 5.7 *Animated!* How ATP and NADPH form in the first stage of photosynthesis. The drawing represents a cross-section through one of the disk-shaped folds of the thylakoid membrane.

So where do the electrons end up? After passing through the electron transfer chain, they continue to a type I photosystem. Here, light-harvesting complexes volley energy to a special pair of chlorophylls at the photosystem's reaction center, causing them to release electrons. An intermediary molecule transfers them to NADP+—which attracts hydrogen ions at the same time. In this way, NADPH forms (Figures 5.6a and 5.7).

If too much NADPH forms, it accumulates in the stroma, and the photosystem II pathway backs up. In such a case, photosystem I may run independently of photosystem II so the cell can continue to make ATP. In a cyclic pathway, light energy frees electrons from the special chlorophyll pair in photosystem I. The released electrons enter an electron transfer chain that moves hydrogen ions into the thylakoid compartment. The resulting H+ gradient drives ATP formation. Electrons are recycled back to photosystem I at the end of this transfer chain, and so no NADPH forms. Electron flow in the cyclic pathway is shown in Figure 5.6b.

Today, both the cyclic and noncyclic pathways of electron flow operate in all photosynthetic eukaryotes. Which one dominates at a given time depends on the organism's metabolic demands for ATP and NADPH.

In the light-dependent reactions, sunlight energy drives the formation of ATP, NADPH, or both. ATP alone forms in a cyclic pathway that starts and ends with one type of photosystem.

Both ATP and NADPH form by a noncyclic pathway in which electrons flow from split water molecules, through two types of photosystems, and finally to NADP+. This is the oxygen-releasing pathway in photosynthesis.

Leaves of basswood, a typical C3 plant. Far right, basswood leaf cross section.

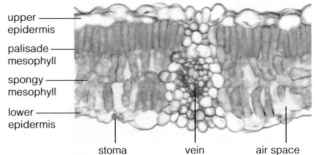

upper epidermis
palisade mesophyll
spongy mesophyll
lower epidermis

stoma   vein   air space

THESE REACTIONS PROCEED IN THE CHLOROPLAST'S STROMA

6CO₂

f It takes six turns of the Calvin–Benson cycle (six carbon atoms) to make one glucose molecule in this mesophyll cell.

e Ten of the PGAL get phosphate groups from ATP. In terms of energy, this primes them for an uphill run— for synthesis reactions that regenerate RuBP.

6 RuBP

ATP

Calvin–Benson cycle

ATP

12 PGA

ATP

NADPH

12 PGAL

1 glucose

a CO₂ in air spaces inside a leaf diffuses into a photosynthetic cell. Six times, rubisco attaches a carbon atom from CO₂ to RuBP. The resulting intermediate splits each time into two PGAs, for a total of 12 PGAs.

b Each PGA gets a phosphate group from ATP, plus hydrogen and electrons from NADPH. The resulting intermediate, PGAL, is primed for reaction.

c Two of the twelve PGAL molecules combine to form one six-carbon molecule of glucose.

d Glucose enters reactions that form carbohydrate end products of photosynthesis—mainly sucrose, starch, and cellulose.

Figure 5.8 *Animated!* Light-independent reactions of photosynthesis. The sketch is a summary of all six turns of the Calvin–Benson cycle and its product, one glucose molecule. *Brown* circles signify carbon atoms.

5.3 Light-Independent Reactions

The chloroplast is a sugar factory, and the Calvin–Benson cycle is its machinery. These cyclic, light-independent reactions are the "synthesis" part of photosynthesis.

THE SUGAR FACTORY

Sugars form in the **Calvin–Benson cycle**, a pathway that runs inside the stroma of chloroplasts (Figure 5.8). The cycle also is called light-*independent* because light is not needed to run the reaction. It uses the ATP and NADPH that formed in the light-dependent reactions. Plants get their carbon and oxygen building blocks from carbon dioxide (CO₂) in the air. Algae of aquatic habitats get them from CO₂ dissolved in water.

The enzyme **rubisco** joins a carbon from CO₂ to five-carbon ribulose bisphosphate (RuBP). The resulting unstable intermediate is the molecule that starts the Calvin–Benson cycle. It splits at once into two stable molecules of phosphoglycerate (PGA), each having a backbone of three carbons. The process of securing carbon from the environment by incorporating it into a stable organic compound is called **carbon fixation**.

Each PGA gets a phosphate group from ATP, and hydrogen and electrons from NADPH. For every six CO₂ molecules fixed, twelve phosphoglyceraldehydes (PGAL) form. Ten PGAL become rearranged in a way that regenerates RuBP. The other two combine to make a six-carbon glucose with a phosphate group attached.

Most of the glucose is converted at once to sucrose or starch by other pathways that conclude the light-independent reactions. Sucrose is a transportable form of carbohydrate in plants; starch is the main storage form. Plants use both as energy sources, and also as building blocks for other complex molecules.

DIFFERENT PLANTS, DIFFERENT PATHWAYS

If sunlight intensity, air temperature, rainfall, and soil composition never varied, photosynthesis might be the same in all plants. But environments differ, and so do details of photosynthesis. For example, you see such differences on hot days when water is scarce.

A waxy cuticle stops water loss from leaves. Water and gases move into and out of leaves only across tiny openings called **stomata** (singular, stoma), as in Figure 5.9. Stomata close on hot, dry days, so water and O_2 do not exit leaves and CO_2 does not enter them. Plants cannot make as much sugar when their photosynthetic cells are exposed to too much O_2 and too little CO_2.

That's why beans, sunflowers, and other **C3 plants** don't grow well without irrigation in hot, dry climates. Their name is a nod to the *three*-carbon PGA, the first stable intermediate to form during the Calvin–Benson cycle. When oxygen builds up in C3 leaves, rubisco binds oxygen, not CO_2. This alternate reaction yields CO_2, so the plant loses carbon instead of fixing it.

The C4 cycle evolved independently over millions of years, in many lineages. In **C4 plants**, *four*-carbon oxaloacetate forms first in reactions that run through two cell types. In mesophyll cells, the C4 cycle will fix carbon no matter how much O_2 there is. The cycle deposits CO_2 directly into bundle-sheath cells where it enters the Calvin–Benson cycle (Figure 5.9). The CO_2 level near rubisco is kept high enough to block the competing reaction. C4 plants use more ATP than C3 plants do because they have to fix carbon twice, but they can make more sugar on dry days.

CAM plants open stomata at night and fix carbon by repeated turns of a C4 cycle, and then the Calvin–Benson cycle runs on the next day (Figure 5.10). CAM stands for crassulacean acid metabolism. These plants include cacti and other succulents, which have juicy, water-storing tissues and thick surface layers. Both are adaptations to hot, dry climates. Some CAM plants survive prolonged droughts by closing their stomata even at night. They fix CO_2 released during aerobic respiration, which is enough to support slow growth.

Driven by ATP energy, the light-independent reactions make sugars with hydrogen from NADPH, and with carbon from carbon dioxide.

C3 plants, C4 plants, and CAM plants respond differently to hot, dry conditions.

Leaves of corn, a typical C4 plant

upper epidermis —
mesophyll cell —
bundle-sheath cell —
lower epidermis —
vein
stoma

a Corn leaf, cross-section

Figure 5.9 (**a**) Typical C4 leaves. (**b**) How C4 plants fix carbon in hot, dry weather, when there is too little CO_2 and too much O_2 inside leaves. A C4 cycle is common in grasses, corn, and other plants that evolved in the tropics. In their mesophyll cells, CO_2 gets fixed by an enzyme that ignores O_2. That reaction releases carbon dioxide in adjoining bundle sheath cells, where the Calvin–Benson cycle runs.

By contrast, many C3 plants evolved in moist temperate zones. The basswood tree (Figure 5.8) is one of them. On hot, dry days, it does not grow as well as a C4 plant. With stomata closed, the oxygen level rises, and rubisco uses O_2 instead of CO_2 in a different reaction. The plant loses carbon instead of fixing it.

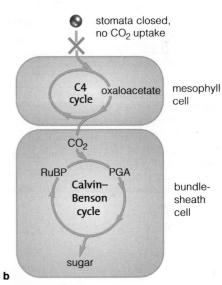

stomata closed, no CO_2 uptake
C4 cycle — oxaloacetate — mesophyll cell
CO_2
RuBP — PGA
Calvin–Benson cycle — bundle-sheath cell
sugar
b

Flat, fleshy leaves of a beavertail cactus, one of the CAM plants.

Figure 5.10 How CAM plants fix carbon in hot, dry climates. Their stomata limit water loss by opening only at night. CO_2 enters and O_2 departs. CO_2 is fixed by a C4 cycle that runs at night, then it enters the Calvin–Benson cycle during the day in the same cell.

CO_2 uptake at night only
C4 cycle — runs at night
mesophyll cell
Calvin–Benson cycle — runs during day
sugar

5.4 Pastures of The Seas

We conclude this chapter with a story to reinforce how some photosynthesizers fit into the world of living things.

Each spring, the renewed growth of photoautotrophs is evident as trees leaf out and fields turn green. At the same time, innumerable single-celled species drifting through the ocean's surface waters make a seasonal response. You cannot see them without a microscope; a row of a million cells of one species would be about one inch long. In some regions, a cup of seawater may hold 24 million cells of one species, not including any other aquatic species suspended in the cup.

Collectively, these cells form the "pastures of the seas." Most are bacteria and protists that ultimately feed nearly all other marine organisms. Their primary productivity is the start of vast aquatic food webs.

Imagine zooming in on just one small patch of "pasture" in an Antarctic sea. There, tiny shrimplike crustaceans are feeding on tinier photosynthesizers. Dense concentrations of these crustaceans—krill—are food for other animals, including fishes, penguins, seabirds, and immense blue whales. A single, mature whale is straining four tons of krill from the water today, as it has been doing for many months. And before they themselves were eaten, those four tons of krill had munched through 1,200 tons of the pasture!

Collectively, tiny marine photoautotrophs have a tremendous impact on global climate. They use about 45 billion metric tons of carbon from CO_2 each year to build organic compounds by photosynthesis.

That is a LOT of photosynthesis.

Think about all the carbon dioxide that we humans release into the atmosphere. We do this, for example, by burning fossil fuels and vast tracts of forests to clear land for farming, and even by making cement. Without help from the little marine photoautotrophs, atmospheric carbon dioxide would accumulate very quickly and accelerate global warming (Section 30.6)

Although drastic global change is a real possibility, we continue to dump many tons of industrial wastes, sewage, and fertilizers in runoff from croplands into the ocean every day. Pollutants dramatically alter the chemical composition of seawater. How will marine photoautotrophs function in this chemical brew?

The flow of energy and the cycling of carbon through the biosphere start with photoautotrophs.

Summary

Section 5.1 Photosynthesis is the main metabolic pathway by which carbon and energy enter the web of life. Photoautotrophs get carbon from carbon dioxide and energy from the sun or inorganic compounds. Animals and other heterotrophs are not self-nourishing; they get carbon and energy from organic compounds made by autotrophs or other heterotrophs.

Photosynthesis runs on energy that pigments harvest from visible light. Pigments capture light of only certain wavelengths. Other wavelengths are reflected as the characteristic color of each pigment.

Chlorophyll *a*, the main photosynthetic pigment, absorbs mainly violet and red light. Accessory pigments capture additional wavelengths of light.

Photosynthesis has two stages: light-dependent and light-independent reactions. This equation and Figure 5.11 summarize the process:

$$12H_2O + 6CO_2 \xrightarrow[\text{enzymes}]{\text{light energy}} 6O_2 + C_6H_{12}O_6 + 6H_2O$$

water   carbon   oxygen   glucose   water
         dioxide

In chloroplasts, light-dependent reactions occur at a thylakoid membrane that forms a single compartment in a semifluid interior (stroma). The light-independent reactions occur in the stroma.

Biology Now
View the sites where photosynthesis takes place with the animation on BiologyNow.

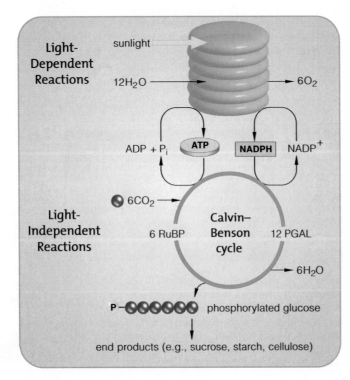

Figure 5.11 Summary of the key reactants, intermediates, and products of photosynthesis.

Section 5.2 Pigments clustered in the thylakoid membrane absorb light energy and pass it to many photosystems. Light-dependent reactions start when electrons released from photosystems enter electron transfer chains.

The released electrons can move through a noncyclic or a cyclic pathway. In the noncyclic reactions, light energy prompts photosystem II to lose electrons, which then move through an electron transfer chain and enter photosystem I. Light energy prompts photosystem I to lose electrons which, along with hydrogen ions, convert $NADP^+$ to NADPH.

In the cyclic reactions, electrons from photosystem I enter a different electron transfer chain, then are recycled back to the same photosystem. NADPH does not form.

In both pathways, electron flow through transfer chains causes H^+ to accumulate in the thylakoid compartment, forming concentration and electric gradients across the thylakoid membrane. H^+ flows in response to the gradients, through ATP synthases. The flow causes phosphate ATP to form in the stroma.

Electrons lost from photosystem I are replaced by electrons lost from photosystem II. Photosystem II replaces its lost electrons by pulling electrons from water molecules; the H^+ and O_2 also are released.

Biology⊛Now
Review the pathways by which light energy is used to form ATP with the animation on BiologyNow.

Section 5.3 The light-independent reactions are the synthesis part of photosynthesis. They occur in the stroma. The enzyme rubisco attaches carbon from CO_2 to RuBP to start the Calvin–Benson cycle. In this cyclic pathway, energy from ATP, carbon and oxygen from CO_2, and hydrogen and electrons from NADPH are used to make glucose, which quickly enters reactions that form the products of photosynthesis (mainly glucose, sucrose, and starch). It takes six turns of the Calvin–Benson cycle to fix the six CO_2 molecules required to make one six-carbon glucose molecule.

On hot dry days, plants shut stomata and oxygen can build up inside leaves. In C3 plants, this causes rubisco to attach oxygen (rather than carbon) to RuBP. The plant then loses carbon instead of fixing it. C4 plants raise the CO_2 level near rubisco by fixing carbon twice, in two cell types. CAM plants capture light energy during the day, and then fix carbon at night.

Biology⊛Now
Take a spin around the Calvin–Benson cycle with the animation on BiologyNow.
Read the InfoTrac article "Light of our Lives," Norman Miller, Geographical, January 2001.

Section 5.4 Photoautotrophs and chemoautotrophs produce the food that sustains themselves and all other organisms. Many different aerobic and anaerobic kinds influence the global cycling of oxygen, carbon, nitrogen, and other substances. Aerobic photoautotrophs live in mind-boggling numbers in the seas as well as on land.

Self-Quiz

Answers in Appendix I

1. Photosynthetic autotrophs use _____ from the air as a carbon source and _____ as their energy source.

2. Light-*dependent* reactions in plants occur at the _____ .
   a. thylakoid membrane   c. stroma
   b. plasma membrane   d. cytoplasm

3. In the light-*dependent* reactions, _____ .
   a. carbon dioxide is fixed   c. CO_2 accepts electrons
   b. ATP and NADPH form   d. sugars form

4. What accumulates inside the thylakoid compartment during the light-*dependent* reactions?
   a. glucose   c. hydrogen ions
   b. RuBP   d. carbon dioxide

5. Light-*independent* reactions proceed in the _____ .
   a. cytoplasm   b. plasma membrane   c. stroma

6. The Calvin–Benson cycle starts when _____ .
   a. light is available
   b. carbon dioxide is attached to RuBP
   c. electrons leave photosystem II

7. Match each event with its most suitable description.
   _____ NADPH formation   a. rubisco required
   _____ CO_2 fixation   b. ATP, NADPH required
   _____ PGAL formation   c. electrons cycled back
   _____ ATP formation only   to photosystem I
                             d. water molecules split

Additional questions are available on Biology⊛Now™

Figure 5.12 (**a**) A green alga (*Codium*) from shallow coastal waters. (**b**) Red alga from a tropical reef.

Critical Thinking

1. Krishna exposes plants to a radioisotope of carbon ($^{14}CO_2$), which they absorb. In which compound will he see the labeled carbon appear first if the plants are C3? C4?

2. Approximately 400 years ago, Jan Baptista van Helmont experimented on the nature of photosynthesis. He wanted to know where growing plants get the materials necessary for increases in size. He planted a tree seedling weighing 5 pounds in a barrel filled with 200 pounds of soil. He watered the tree regularly. Five years passed. Then Van Helmont weighed the tree and the soil. The tree weighed 169 pounds, 3 ounces. The soil weighed 199 pounds, 14 ounces. Because the tree gained so much weight and the soil lost so little, he concluded the tree had gained weight after absorbing the water he had added to the barrel. Given what you know about the composition of biological molecules, why was he misguided? Explain what really happened.

3. The green alga in Figure 5.12*a* lives in seawater. Its main pigments absorb red light. Its accessory pigments help harvest energy in sunlit waters, and some shield it against ultraviolet radiation. Other green algae live in ponds, in lakes, on rocks, even in snow. The red alga in Figure 5.12*b* grows on tropical reefs in clear, warm water. Its phycobilin absorbs green and blue-green wavelengths that penetrate deep water. Some of its relatives live in deep, dimly lit waters; their pigments are nearly black.

If wavelengths are such a vital source of energy, why aren't all pigments black? (*Hint*: Photoautotrophs first evolved in the seas.)

When Mitochondria Spin Their Wheels

In the early 1960s, Swedish physician Rolf Luft reflected on some odd symptoms of a young patient. The woman felt weak and too hot all the time. Even on the coldest winter days, she could not stop sweating and her skin was flushed. She was thin in spite of a huge appetite. Luft inferred that his patient's symptoms pointed to a metabolic disorder. Tests designed to detect her rate of metabolism showed that her oxygen consumption was the highest ever recorded!

Microscopic examination of a tissue sample from the patient's muscles revealed mitochondria, the cell's ATP-producing powerhouses. But there were far too many of them. The mitochondria were working hard, but producing very little ATP for their efforts.

Her disorder, now called *Luft's syndrome*, was the first to be linked directly to a defective organelle. By analogy, someone with this mitochondrial disorder functions like a city with half of its power plants shut down. Muscles, the brain, and other body parts with the highest energy demands are hurt the most.

More than a hundred other mitochondrial disorders are now known. One, a genetic disorder called *Friedreich's ataxia*, runs in families. Like siblings Leah and Joshua at *left*, affected people develop weak muscles, ataxia (loss of coordination), and visual problems. Many die young because of heart problems.

Leah and Joshua

A mutant gene and its abnormal protein product give rise to Friedreich's ataxia. The abnormal protein causes iron to accumulate inside mitochondria. Iron is required for electron transfers that drive ATP formation, but too much invites an accumulation of **free radicals**—unbound molecular fragments with an unstable number of electrons. These highly reactive fragments can attack all of the molecules of life.

Type 1 diabetes, atherosclerosis, amyotrophic lateral sclerosis (Lou Gehrig's disease), Parkinson's, Alzheimer's, and Huntington's diseases—defective mitochondria contribute to every one of these problems. So when you consider mitochondria in this chapter, do not assume they are too remote from your interests. Without them, you would not make enough ATP even to read about how they do it.

You already read about the way ATP molecules form during photosynthesis. Turn now to how all organisms make ATP by tapping into the chemical bond energy of organic compounds—especially glucose.

 How Would You Vote? *Drug development is costly. There is little incentive for private companies to target cures for ailments, such as Friedreich's ataxia, that affect relatively few individuals. Should the federal government allocate some funds to private companies that search for cures for diseases affecting a relatively small number of people? See BiologyNow for details, then vote online.*

Key Concepts

ENERGY-RELEASING PATHWAYS
All organisms release chemical bond energy of organic molecules to drive ATP formation. The main pathways start in the cytoplasm with glycolysis. Products of glycolysis can enter the aerobic respiration pathway, which uses oxygen, or anaerobic pathways such as alcoholic and lactate fermentation, which do not use oxygen.

YIELDS AND TALLIES
Only aerobic respiration ends in mitochondria. Fermentation's net energy yield per glucose molecule is always smaller than that of aerobic respiration. Molecules other than glucose can enter the aerobic pathway as alternative energy sources.

EVOLUTIONARY CONNECTIONS
Aerobic respiration and photosynthesis are connected on a global scale. Photosynthesis uses the carbon dioxide released by aerobic respiration, which uses oxygen released by photosynthesis. This connection can be traced to the evolution of novel metabolic pathways.

Links to Earlier Concepts

This chapter expands the picture of life's dependence on energy flow (Section 4.1) by showing how all organisms tap energy stored in glucose and convert it to different, transportable forms, especially ATP (4.2). Review the structure of glucose and other carbohydrates (2.6) and lipids (2.7), as well as the introduction to membrane function (3.3), mitochondria (3.5), and mechanisms of active transport (4.5). You will see another example of controlled energy release at electron transfer chains (4.1, 5.2), and you will explore global connections between photosynthesis (5.2, 5.3) and aerobic respiration.

6.1 Overview of Energy-Releasing Pathways

Organisms stay alive by taking in energy. Regardless of its source, energy must be converted to a form that can drive thousands of diverse life-sustaining reactions. The chemical bond energy of ATP serves that purpose.

The first energy-releasing metabolic pathways were operating billions of years before Earth's oxygen-rich atmosphere evolved, so we can expect that they were *anaerobic*; the reactions did not use free oxygen. Many prokaryotes still live in oxygen-free habitats, making ATP by **fermentation**, breaking down carbohydrates in pathways that do not use oxygen. Many eukaryotic cells still use fermentation.

Cells of nearly all species use **aerobic respiration**, a breakdown pathway that uses oxygen to extract energy very efficiently from glucose and other carbohydrates. Every time you inhale, you provide your aerobically-respiring cells with a fresh supply of oxygen.

Both energy-releasing pathways start with the same set of reactions in the cytoplasm. During these initial reactions, called **glycolysis**, enzymes rearrange a six-carbon glucose molecule into two pyruvate molecules. **Pyruvate** is an organic compound with a three-carbon backbone. Once glycolysis ends, the energy-releasing pathways differ. Only the aerobic pathway continues in mitochondria, where oxygen accepts electrons at the end of electron transfer chains.

By contrast, the anaerobic pathways start and end in the cytoplasm, and the final electron acceptor is not oxygen; it is some other substance.

Of energy-releasing pathways, aerobic respiration gets the most bang for the buck, typically yielding thirty-six ATP for each molecule of glucose. Compare this with two ATP per glucose yielded by anaerobic pathways (Figure 6.1). If you were a single cell, you would not require much ATP; you could depend on fermentation for all of your energy needs. Being far larger and highly active, however, you require the high energy yield of aerobic respiration. Aerobic respiration can be summarized this way:

$$C_6H_{12}O_6 + 6O_2 \longrightarrow 6CO_2 + 6H_2O + ATP$$

GLUCOSE   OXYGEN   CARBON DIOXIDE   WATER

Figure 6.1 *Animated!*
Overview of the three stages of aerobic respiration. Reactions start in the cytoplasm, and end inside mitochondria.

(a) The first stage is glycolysis. Enzymes partly break down glucose to pyruvate.

(b) In the second stage, the pyruvate is broken down to carbon dioxide.

(c) Coenzymes NAD$^+$ and FAD pick up electrons and hydrogen to become NADH and FADH$_2$.

(d) In the final stage, the NADH and FADH$_2$ give up electrons to electron transfer chains. The flow of electrons through the transfer chains results in H$^+$ gradients that drive ATP formation.

(e) At the end of the transfer chains, oxygen accepts electrons and hydrogen, forming water.

(f) From start to finish, a typical net energy yield from a single glucose molecule is thirty-six ATP.

The summary equation tells you which substances are present at the pathway's start and end. In between, however, are three reaction stages.

Again, glycolysis is the first stage. The second is a cyclic pathway called the **Krebs cycle** and a few steps preceding it. In these reactions, enzymes break down pyruvate to carbon dioxide and water. Electrons and hydrogen atoms released by the reactions are picked up by the coenzymes **NAD^+** and **FAD**, which become NADH and $FADH_2$.

Few ATP form during glycolysis or the Krebs cycle. The big energy payoff is the third stage, as NADH and $FADH_2$ give up their cargo of electrons and hydrogen to electron transfer chains—the machinery of **electron transfer phosphorylation**. Operation of these chains establishes hydrogen ion gradients, which in turn drive formation of ATP. Oxygen inside the mitochondrion accepts the electrons and hydrogen at the end of the chains, thereby forming water.

Nearly all metabolic reactions run on energy released from glucose and other organic compounds. The main energy-releasing pathways start in the cytoplasm with glycolysis, a series of reactions that break down glucose to pyruvate.

Anaerobic pathways have a small net energy yield, typically two ATP per glucose.

Aerobic respiration, an oxygen-dependent pathway, runs to completion in mitochondria. From start (glycolysis) to finish, it typically has a net energy yield of thirty-six ATP.

6.2 Glycolysis—Glucose Breakdown Starts

Let's track what happens to a glucose molecule in the first stage of aerobic respiration. Remember, the same steps happen in anaerobic energy-releasing pathways.

Several kinds of sugars can enter the enzyme-driven reactions of glycolysis. Glucose is one of them. Each molecule of glucose has six carbon atoms that form its backbone. During glycolysis, glucose is partly broken down to two molecules of pyruvate, a three-carbon compound, for a net yield of two ATP (Figure 6.2).

The first steps of glycolysis are *energy-requiring*—cells invest two ATP to begin the reactions. Let's start with a molecule of glucose that diffused into the cell through a passive transport protein. One ATP donates a phosphate group to the glucose. Phosphorylation traps the glucose inside the cell, because the resulting

glucose

Figure 6.2 *Animated!* Glycolysis occurs in the cytoplasm of all prokaryotic and eukaryotic cells. Glucose is the reactant in this example. Two pyruvate, two NADH, and four ATP form. Cells invest two ATP to start the reactions, however, so the *net* energy yield is two ATP.

Depending on the type of cell and environmental conditions, the pyruvate may enter the second set of reactions of the aerobic pathway, including the Krebs cycle. Or it may be used in other reactions, such as those of fermentation.

molecule (glucose–6–phosphate) cannot go through the glucose transporter. Glucose–6–phosphate then accepts a phosphate group from a second ATP. The resulting 6-carbon intermediate splits at once into two 3-carbon molecules of PGAL (Section 5.3).

In the first *energy-releasing* step of glycolysis, both PGAL are converted to intermediates that each give up a phosphate to ADP, which results in two ATP. Two later intermediates do the same thing. Thus, four ATP form by **substrate-level phosphorylation**. We define this metabolic event as a direct transfer of a phosphate group from a reaction substrate to another molecule—

energy-requiring steps of glycolysis

energy-releasing steps of glycolysis

a Track the six carbon atoms (*brown* circles) of glucose. Glycolysis starts with an energy investment of two ATP.

One ATP transfers a phosphate group to glucose. Then, the phosphorylated glucose accepts a phosphate group from another ATP. The intermediate splits at once into two, each three-carbon PGAL molecules.

b Two NADH form when each PGAL gives up two electrons and a hydrogen atom to NAD+.

Two intermediates each transfer a phosphate group to ADP. *Two ATP have now formed by substrate-level phosphorylation.* The original energy investment of two ATP has been recovered.

c Two intermediates form and are rearranged. Each one gives up a hydrogen atom and an —OH group. These combine as water. Two molecules called PEP form by these reactions.

d Each PEP transfers a phosphate group to ADP. *Once again, two ATP have formed by substrate-level phosphorylation.*

In sum, glycolysis has a net energy yield of two ATP for each glucose molecule. Two NADH also form during the reactions, and two molecules of pyruvate are the end products.

in this case, to ADP. Meanwhile, the two PGAL give up their electrons and hydrogens to two NAD+, which become NADH. Later these two NADH will give up their electrons and hydrogens at another reaction site, and so revert to NAD+. In other words, like all other coenzymes, NAD+ is reusable.

Four ATP have now formed. Remember two ATP were invested to start the reactions. The *net* yield of glycolysis is two ATP and two NADH.

To summarize, glycolysis converts the bond energy of glucose (a storable, stable form of energy) to the bond energy of ATP (a transportable form of energy).

The electrons and hydrogen from glucose that are picked up by NAD+ enter the next stage of reactions. And so can the end products of glycolysis—the two pyruvate molecules.

Glycolysis is a series of reactions that partially break down glucose or some other carbohydrate to two molecules of pyruvate, thereby releasing energy.

Two NADH and four ATP form. However, when we subtract the two ATP required to start the reactions, the net energy yield of glycolysis is two ATP from one glucose molecule.

a One carbon atom is stripped from pyruvate and is released, as CO_2. The fragment remaining binds with a coenzyme, forming acetyl–CoA.

Acetyl-CoA Formation

pyruvate

coenzyme A → ← NAD⁺

(CO_2) → **NADH**

—CoA

acetyl-CoA

b NADH forms when NAD⁺ picks up hydrogen and electrons. *One NADH forms during a few steps preceding the Krebs cycle.*

glucose

GLYCOLYSIS

pyruvate

KREBS CYLE

ELECTRON TRANSFER PHOSPHORYLATION

h The final steps regenerate oxaloacetate. NAD⁺ picks up hydrogen and electrons, forming NADH. *At this point in the cycle, three NADH and one $FADH_2$ have formed.*

Krebs Cycle

CoA

oxaloacetate   citrate

NADH

NAD⁺

FADH₂

FAD   NAD⁺

ATP

ADP + phosphate group

NAD⁺ → **NADH**

NAD⁺ → **NADH**

c In the first step of the Krebs cycle, acetyl–CoA transfers two carbons to oxaloacetate, forming citrate.

d In rearrangements of intermediates, another carbon atom is released as CO_2, and NADH forms as NAD⁺ picks up hydrogen and electrons.

e Another carbon atom is released as CO_2. Another NADH forms. *The three carbon atoms that entered the second-stage reactions in each pyruvate have now been released.*

Figure 6.3 *Animated!* Second stage of aerobic respiration: formation of acetyl–CoA and the Krebs cycle. These reactions occur in a mitochondrion's inner compartment. *It takes two turns of the cycle to break down two pyruvates from glycolysis.* A total of two ATP, eight NADH, two $FADH_2$, and six CO_2 molecules form.

g $FADH_2$ forms as the coenzyme FAD picks up electrons and hydrogen.

f A phosphate group is attached to ADP. At this point, one ATP has formed by substrate-level phosphorylation.

AEROBIC RESPIRATION

starts (glycolysis) in cytoplasm

finishes in mitochondrion

inner mitochondrial membrane   outer mitochondrial membrane

inner compartment   outer compartment

b

Figure 6.4 *Animated!* Functional zones inside the mitochondrion. (**a**) Micrograph of a mitochondrion, sliced lengthwise. Most eukaryotic species contain these organelles. (**b**) An inner membrane divides a mitochondrion's interior into two compartments. The second and third stages of aerobic respiration take place at this membrane.

6.3 Second and Third Stages of Aerobic Respiration

The second and third stages of the aerobic pathway unfold inside mitochondria. The two pyruvate molecules formed by glucose breakdown in glycolysis enter a series of reactions that result in the release of carbon dioxide and, later, the formation of many ATP molecules.

Suppose two pyruvate molecules formed by glycolysis leave the cytoplasm and enter a **mitochondrion** (plural, mitochondria). In this organelle, the second and third stages of the aerobic pathway are completed. Figures 6.3 and 6.4 show the reactions and where they occur.

ACETYL–CoA FORMATION AND THE KREBS CYCLE

During the second stage, what remains of the original glucose molecule is broken down completely to water and carbon dioxide. Six carbon atoms, three from each

YIELDS AND TALLIES

Figure 6.5
Animated!
Summary of electron transfers in aerobic respiration.

(a) Electrons and hydrogen from NADH and FADH$_2$ that formed during the first and second stages enter electron transfer chains.

(b) As electrons are being transferred through these chains, H$^+$ ions are moved across the inner membrane, into the outer compartment.

(c) As H$^+$ accumulates in the outer compartment, chemical and electric gradients form across the inner membrane.

(d) Hydrogen ions follow the gradients through the interior of ATP synthases, driving ATP formation from ADP and phosphate (P$_i$).

---

pyruvate, enter these reactions. Six carbon atoms depart, as six molecules of carbon dioxide that escape into the surroundings (Figure 6.3).

In the preparatory reactions, an enzyme removes one carbon atom from each pyruvate and attaches it to oxygen, forming CO$_2$. Each two-carbon fragment then combines with a coenzyme to form acetyl–CoA, a type of cofactor that is the starting material for the Krebs cycle. Two NADH also form.

Each acetyl–CoA molecule begins the Krebs cycle by transferring its two-carbon acetyl group to four-carbon oxaloacetate, which becomes citrate—a form of citric acid. (That is why the Krebs cycle is also known as the citric acid cycle.) Intermediates are rearranged to become oxaloacetate. Cells have limited amounts of oxaloacetate, and so must regenerate it to keep the Krebs cycle running. Two ATP form by substrate-level

phosphorylation. Six more NADH and two FADH$_2$ also form. With their cargo of electrons and hydrogen, these coenzymes—*eight NADH and two FADH$_2$*—are the big potential payoff for the cell.

ELECTRON TRANSFER PHOSPHORYLATION

In the third stage, ATP formation goes into high gear. It begins when coenzymes donate electrons to electron transfer chains in the inner mitochondrial membrane (Figure 6.5). The flow of electrons through the chains ultimately causes the attachment of phosphate to ADP molecules—thus the name of these reactions, *electron transfer phosphorylation.*

As electrons are passed through the transfer chains, hydrogen ions in the inner compartment are shuttled into the outer one. The accumulation of H$^+$ ions there

LINKS TO
SECTIONS 3.3, 3.5, 4.5

IMPACTS, ISSUES

The enzyme pyruvate dehydrogenase speeds conversion of pyruvate to acetyl–CoA. When mutated, this enzyme disrupts the formation of citrate in the Krebs cycle. A fetus with this severe mutation will die before birth.

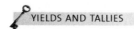

forms concentration and electric gradients across the inner membrane. These gradients motivate hydrogen ion flow back to the inner compartment, but the ions can only cross the membrane through ATP synthases (Figure 6.5). The flow of hydrogen ions through these transport proteins drives the formation of ATP.

At the end of the electron transfer chains, electrons are passed to oxygen, which then combines with H^+ to form water. *Oxygen is the final acceptor of the electrons from glucose.* In oxygen-starved cells, electrons have nowhere to go. The chain backs up with electrons all the way to NADPH, so no H^+ gradients form, and no ATP forms, either. Without oxygen, cells of complex organisms do not survive for very long, because they cannot produce enough ATP to sustain life processes.

Thirty-two ATP form in the third stage of aerobic respiration. The net harvest, including the four ATP from the preceding stages, is thirty-six ATP from one molecule of glucose.

In the second stage of aerobic respiration, two molecules of pyruvate from glycolysis are converted to acetyl–CoA, which enters the Krebs cycle. Two ATP and ten coenzymes form. All of pyruvate's carbons depart, in the form of carbon dioxide.

In aerobic respiration's third stage, coenzymes deliver their electrons and hydrogen to electron transfer chains. Electron flow through the chains sets up gradients across the inner membrane. These drive the synthesis of thirty-two ATP.

Aerobic respiration typically nets thirty-six ATP molecules from each glucose molecule metabolized.

6.4 Anaerobic Energy-Releasing Pathways

We turn now to the use of glucose as a substrate for fermentation pathways. Unlike aerobic respiration, these anaerobic pathways do not use oxygen as a final acceptor of electrons. Their final steps regenerate NAD^+.

Many fermenters are protists and bacterial species that live in marshes, ocean sediments, animal guts, canned foods, sewage treatment ponds, bogs, mud, and other oxygen-free places. Some die when they are exposed to oxygen. Bacterial species that cause botulism and many other diseases are like this. Some fermenters can tolerate the presence of oxygen. Still other kinds use oxygen, but they also can switch to fermentation when oxygen becomes scarce.

Glycolysis is the first stage of fermentation, just as it is in aerobic respiration (Figures 6.6 and 6.7). Here again, two pyruvate, two NADH, and two ATP form. But unlike aerobic respiration, fermentation does not completely break down glucose to carbon dioxide and water. It produces no additional ATP beyond the small yield from glycolysis. *The final steps in fermentation pathways regenerate the coenzyme NAD^+.*

Fermentation yields enough energy to sustain many single-celled anaerobic species. It helps some aerobic cells when the oxygen level is stressfully low. It is not enough to sustain large, multicelled organisms, this being why you will never see an anaerobic elephant.

Figure 6.6 *Animated!* Alcoholic fermentation. (**a**) Yeasts, which are single-celled fungi, use this anaerobic pathway to make ATP.

(**b**) A vintner examining a fermentation product of *Saccharomyces*. This yeast lives on sugar-rich tissues of ripe grapes.

(**c**) Carbon dioxide released from cells of *S. cerevisiae* makes bread dough rise.

(**d**) Alcoholic fermentation. An intermediate, acetaldehyde, functions as this pathway's final electron acceptor. The end product of the reactions is ethanol (ethyl alcohol).

glycolysis $C_6H_{12}O_6$

2 ATP energy input — 2 ADP — 2 NAD^+

2 NADH

4 ATP energy output — 2 pyruvate

2 ATP net

ethanol formation — 2 H_2O → 2 CO_2

2 acetaldehyde

electrons, hydrogen from NADH

2 ethanol

ALCOHOLIC FERMENTATION

In **alcoholic fermentation**, the three-carbon backbones of the two pyruvate molecules from glycolysis are split into two molecules of acetaldehyde and two of carbon dioxide. Next, the acetaldehydes accept electrons and hydrogen from NADH, thus becoming the alcohol called ethanol (Figure 6.6).

Some single-celled fungi called yeasts are famous because they use this pathway. Bakers mix one type, *Saccharomyces cerevisiae*, into dough. Fermenting yeast cells release carbon dioxide, and the dough expands (rises) as the gas forms bubbles in it. Oven heat forces the bubbles out of spaces they had occupied in the dough, and the alcohol product evaporates away.

Some wild and cultivated strains of *Saccharomyces* are used to produce wine. Crushed grapes are left in vats along with the yeast, which converts sugar in the juice to ethanol. Ethanol is toxic to microbes. When a fermenting brew's ethanol content nears 10 percent, yeast cells start dying and fermentation ends.

LACTATE FERMENTATION

In **lactate fermentation**, NADH gives up electrons and hydrogen to two pyruvate molecules from glycolysis. The transfer converts each pyruvate to lactate, a three-carbon compound (Figure 6.7).

Lactobacillus and other kinds of bacteria use lactate fermentation. Their fermenting action can spoil food, yet some species have commercial uses. For instance, bacteria that break down the glucose in milk give us dairy products like cheeses, yogurt, and buttermilk. Lactate fermentation is used to cure meats and pickle fruits and vegetables, such as sauerkraut. Lactate is an acid so it gives these foods a sour taste.

In humans, rabbits, and many other animals, some types of cells use lactate fermentation. When demands for energy are intense but brief—say, in a short race—some muscle cells use this pathway instead of aerobic respiration. Lactate fermentation works quickly but not for long. It does not produce enough ATP from glucose to sustain prolonged activity.

> *In fermentation pathways, an intermediate that forms during the reactions functions as the final acceptor of electrons originally derived from glucose.*
>
> *Alcoholic fermentation and lactate fermentation both have a net energy yield of two ATP for each glucose molecule metabolized. That ATP formed in glycolysis. The rest of the reactions regenerate NAD+, the coenzyme that keeps both of these pathways operational.*

Figure 6.7 Lactate fermentation. In this anaerobic pathway, electrons originally stripped from glucose end up in lactate, the reaction product. The reactions have a net energy yield of two ATP (from glycolysis).

6.5 Alternative Energy Sources in the Body

LINKS TO SECTIONS 2.6, 2.7

So far, you've looked at what happens after glucose molecules enter an energy-releasing pathway. Now start thinking about what cells do when they have too much or too little glucose.

THE FATE OF GLUCOSE AT MEALTIME AND IN BETWEEN MEALS

What happens to glucose at mealtime? While you and all other mammals are eating, glucose and other small organic molecules are being absorbed across the gut lining, and your blood is transporting them through the body. The rising glucose concentration in blood prompts an organ, the pancreas, to secrete insulin. This hormone stimulates cells to take up glucose faster.

Cells trap the incoming glucose by converting it to glucose–6–phosphate. This is the first intermediate of glycolysis, formed by a phosphate group transfer from ATP (Figure 6.2). When glucose is phosphorylated this way, it cannot be transported back out of the cell.

When glucose intake exceeds cellular demands for energy, ATP-producing machinery goes into high gear. Unless a cell is using ATP rapidly, the concentration of ATP in its cytoplasm rises. Then, glucose–6–phosphate from glycolysis is diverted into a different pathway,

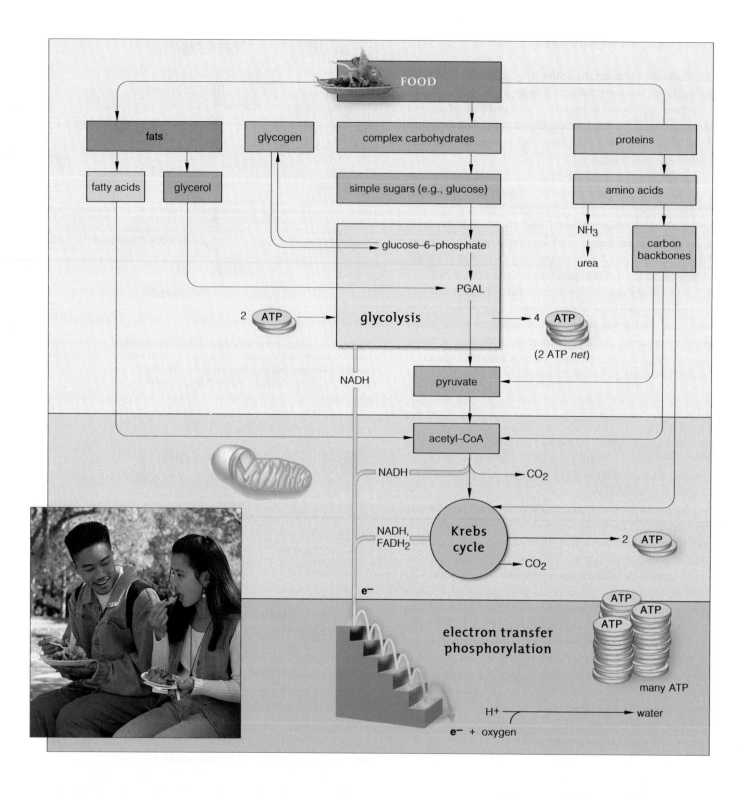

Figure 6.8 Reaction sites where a variety of organic compounds can enter the different stages of aerobic respiration. These compounds are alternative energy sources in the human body.

In humans and other mammals, complex carbohydrates, fats, and proteins cannot enter the aerobic pathway directly. The digestive system, and then individual cells, must first break apart these molecules into simpler subunits that can be degraded.

and ends up being converted to glycogen, a storage polysaccharide (Section 2.6). The glucose–to–glycogen pathway is the dominant one in muscle and liver cells, which have the largest stores of glycogen in the body.

Between meals, the blood level of glucose declines. If the decline were not countered, that would be bad news for the brain, your body's glucose hog. At any time, your brain is taking up more than two-thirds of the freely circulating glucose. Why? Its many hundreds of millions of neurons (or nerve cells) utilize this sugar as their primary energy source. Glucose-starved brain cells don't function, and so dizziness, weakness, and confusion occur.

In response to a decline in blood sugar, pancreas cells secrete glucagon, a hormone that prompts liver cells to convert stored glycogen into glucose and send it back to the blood. Only liver cells do this; muscle cells will not give it up. Blood glucose levels rise, and brain cells keep functioning. *Hormones control whether your body's cells use free glucose as an energy source or tuck it away.*

Don't let this explanation lead you to believe cells squirrel away gigantic amounts of glycogen. In adult humans, glycogen makes up only about 1 percent of the body's total energy reserves, the energy equivalent of two cups of cooked pasta. If you stop eating, you will completely deplete your liver's stash of glycogen in less than twelve hours.

Of the total energy reserves in, say, a typical adult who eats well, 78 percent (about 10,000 kilocalories) is concentrated in body fat and 21 percent in proteins.

ENERGY FROM FATS

How does a human body access its fat reservoir? A fat molecule, recall, has a glycerol head and one, two, or three fatty acid tails (Section 3.4). The body stores most fats as triglycerides, which have three tails each. Triglycerides accumulate in fat cells of adipose tissue. This tissue is strategically located beneath the skin of buttocks and other body regions.

When the blood glucose level falls, triglycerides are tapped as an energy alternative. Enzymes in fat cells cleave bonds between glycerol and fatty acids, which both enter the blood. Enzymes in the liver convert the glycerol to PGAL. And PGAL, recall, is one of the key intermediates for glycolysis (Figure 6.2). Nearly all cells of your body take up circulating fatty acids, and enzymes inside them cleave the fatty acid backbones. The fragments become converted to acetyl–CoA, which can enter the Krebs cycle.

Compared to glucose, a fatty acid tail has far more carbon-bound hydrogen atoms, so it yields more ATP. Between meals or during sustained exercise, fatty acid conversions supply about half of the ATP that muscle, liver, and kidney cells require.

What happens if you eat too many carbohydrates? Exceed the glycogen-storing capacity of your liver and muscle cells, and the excess gets converted to fats. *Too much glucose ends up as excess fat.*

ENERGY FROM PROTEINS

Some enzymes in your digestive system split dietary proteins into their amino acid subunits, which can be absorbed into the bloodstream. Cells use amino acids to make new proteins and other nitrogen-containing compounds. Even so, if you eat more protein than your body needs, amino acids will be broken down too. Their $-NH_3^+$ group is removed, and then becomes ammonia (NH_3). Depending on the type of amino acid, the carbon backbones are split to form acetyl–CoA, pyruvate, or an intermediate of the Krebs cycle. Your cells can divert any of these organic compounds into the Krebs cycle (Figure 6.8).

As you can see, maintaining and accessing energy reserves is complicated business. Controlling the use of glucose is special because it is the fuel of choice for the brain. However, providing all of your cells with energy starts with the kinds of food you eat.

> *In humans and other mammals, the entrance of glucose or other organic compounds into an energy-releasing pathway depends on the kinds and proportions of carbohydrates, fats, and proteins in the diet as well as on the type of cell.*

6.6 Connections With Photosynthesis

In this unit you read about photosynthesis and aerobic respiration—the main pathways by which cells trap, store, and release energy. Take a moment to consider how these two processes became interconnected.

All organisms require ongoing supplies of energy and organic compounds to survive, grow, and reproduce. Autotrophs get them from the physical environment. They grow, their populations grow, and they become concentrated stores of food tempting to heterotrophs.

Autotrophs are more than carbon-rich food for the biosphere. Early practitioners of the noncyclic pathway of photosynthesis filled the atmosphere with oxygen. Before then, the iron and other metals abundant on all surfaces both above and below the seas was in reduced form. All of the oxygen released by early autotrophs immediately became bound to (oxidized) these metals. Over time, all of the exposed iron rusted, as evidenced by global bands of red iron deposits below the seafloor. Eventually, there was little exposed metal that was not

CPT deficiency is a mitochondrial disorder that interferes with fatty acid metabolism. Normally, prolonged exercise, fasting, and other events that can deplete glycogen trigger fatty acid breakdown. When the body does not have enough of the enzyme known as CPT, cells break down the proteins in muscle cells instead, which causes painful muscle cramps.

LINKS TO SECTIONS 5.2, 5.3

Figure 6.9 Bubbling mud pit at Yellowstone National Park. Anaerobic archaeans live deep in the superheated water of this hot spring. They use the hydrogen sulfide gas bubbling from underlying magma as their energy source. A metabolic by-product is sulfuric acid, which dissolves adjacent rocks into mud.

oxidized. Oxygen, bubbling from the now tremendous populations of photoautotrophs, gushed unimpeded into the atmosphere.

In a wink of geologic time, maybe a few hundred thousand years, oxygen levels rose in the seas and the sky. It was too fast for most anaerobic species to adapt, and they perished in a mass extinction. Others clung to life in marginalized anaerobic habitats. We find their descendants in hydrothermal vents where molten rock and superheated water spew out from fissures in the seafloor. These archaeans get hydrogen and electrons from hydrogen sulfide gas in mineral-rich volcanic water (Figure 6.9). Other anaerobes that persist in soil get energy from nitrogen-rich wastes and the remains of other organisms.

Some ancient species survived because they had metabolic pathways to detoxify oxygen. These later became remodeled to *use* oxygen as the final electron acceptor for reactions that release energy from carbon compounds. *As a direct outcome of selection pressure by a by-product of photosynthesis, aerobic respiration evolved.*

With the evolution of aerobic respiration, the flow of carbon, and oxygen through the web of life came full circle. Why? Products of aerobic respiration—carbon dioxide and water—are precisely the raw materials that photoautotrophs use.

Such processes might seem far removed from your daily interests. But the food that nourishes you and most other heterotrophs would not be produced without them. We will return to this point in later chapters. It affords perspective on how we as a species are having a global impact on our primary sources of food and oxygen.

sunlight energy

photosynthesis

carbon dioxide, water

organic compounds, oxygen

aerobic respiration

Aerobic respiration evolved as a direct outcome of selection pressure by oxygen, a by-product of photosynthesis.

Products of aerobic respiration, carbon dioxide and water, are starting materials for photosynthesis.

Summary

Section 6.1   All organisms, including photosynthetic ones, access the chemical bond energy of glucose and other organic compounds, then use it to make ATP. They do so because ATP is a transportable form of energy that can jump-start nearly all metabolic reactions. They break apart compounds and release electrons and hydrogen, both of which have roles in ATP formation.

Glycolysis, the partial breakdown of one glucose to two pyruvate molecules, takes place in the cytoplasm of all cells. It is the first stage of all the main energy-releasing pathways, and it does not use free oxygen.

Anaerobic pathways end in the cytoplasm. The oxygen-requiring pathway called aerobic respiration ends in mitochondria. It harvests far more ATP from glucose than anaerobic pathways do.

Biology⊜Now
Get an overview of aerobic respiration with the animation on BiologyNow.

Section 6.2   It takes two ATP molecules to jump-start glycolysis. Two molecules of the coenzyme NAD^+ accept electrons and hydrogen originally from glucose, forming 2 NADH. Two pyruvate and four ATP form. The net energy yield is 2 ATP (because 2 ATP were invested up front). The end products of glycolysis are two molecules of pyruvate.

Biology⊜Now
Take a step-by-step journey through glycolysis with the animation on BiologyNow.

Section 6.3   After glycolysis, aerobic respiration goes through two more reaction stages, in mitochondria.

The second stage consists of acetyl–CoA formation and the Krebs cycle. It takes two full turns of these cyclic reactions to break down both of the pyruvates from glycolysis of one glucose molecule. In total, the second stage of aerobic respiration results in the formation of six CO_2, two ATP, eight NADH, and two $FADH_2$.

The third stage, electron transfer phosphorylation, proceeds at the inner mitochondrial membrane. There, electron transfer chains accept electrons and hydrogen from NADH and $FADH_2$. Electron flow through the transfer chains causes H^+ to accumulate in the inner compartment, so concentration and electric gradients build up across the inner membrane. H^+ follows its gradients back to the outer mitochondrial compartment, through the interior of ATP synthases. The flow drives attachment of phosphate to ADP, forming many ATP. Free oxygen picks up the electrons at the end of the transfer chains and combines with H^+, forming water.

Aerobic respiration has a typical net energy yield of 36 ATP for each glucose molecule metabolized.

Biology⊜Now
Explore a mitochondrion and observe the reactions inside. Study how each step in aerobic respiration contributes to energy harvests with animations on BiologyNow.

Section 6.4 Fermentation pathways do not require oxygen, and they take place only in the cytoplasm. They start with the formation of two ATP during glycolysis. The remaining reactions regenerate NAD^+ but produce no additional ATP. The net energy yield is only the two ATP that formed in glycolysis.

In alcoholic fermentation, pyruvate is first converted to acetaldehyde and carbon dioxide. When NADH transfers electrons and hydrogen to acetaldehyde, two ethanol molecules form and the NAD^+ is regenerated.

In lactate fermentation, NAD^+ is regenerated when electrons and hydrogen are transferred from NADH to pyruvate, which then becomes lactate.

Biology⊗Now
Compare alcoholic and lactate fermentation with the animation on BiologyNow.

Section 6.5 In the human body, simple sugars from carbohydrates, glycerol and fatty acids from fats, and carbon backbones of amino acids from proteins enter the aerobic pathway as alternative energy sources.

Biology⊗Now
Follow the breakdown of different organic molecules with the interaction on BiologyNow.

Section 6.6 Aerobic respiration arose after the noncyclic pathway of photosynthesis had enriched the atmosphere with oxygen. The two pathways are linked on a global scale. Photosynthesizers use water and carbon dioxide products of aerobic respiration when building organic compounds, and most organisms now depend on the oxygen released by photosynthesis.

Self-Quiz
Answers in Appendix I

1. Glycolysis starts and ends in the _____ .
   a. nucleus   c. plasma membrane
   b. mitochondrion   d. cytoplasm

2. Which of the following molecules does *not* form during glycolysis?
   a. NADH   c. $FADH_2$
   b. pyruvate   d. ATP

3. Aerobic respiration is completed in the _____ .
   a. nucleus   c. plasma membrane
   b. mitochondrion   d. cytoplasm

4. In the third stage of aerobic respiration, _____ is the final acceptor of electrons that originally were part of a glucose molecule.
   a. water   c. oxygen
   b. hydrogen   d. NADH

5. _____ engage in lactate fermentation.
   a. *Lactobacillus* cells   c. Sulfate-reducing bacteria
   b. Some muscle cells   d. a and b

6. In alcoholic fermentation, _____ is the final acceptor of electrons stripped from glucose.
   a. oxygen   c. acetaldehyde
   b. pyruvate   d. sulfate

7. The fermentation pathways produce no more ATP beyond the small yield from glycolysis, but the remaining reactions _____ .
   a. regenerate FAD   c. dump electrons on an
   b. regenerate NAD^+ inorganic substance
                                   (not oxygen)

8. In certain organisms and under certain conditions, _____ can be used as an energy alternative to glucose.
   a. fatty acids   c. amino acids
   b. glycerol   d. all of the above

9. Match the event with its most suitable description.
   ____ glycolysis   a. ATP, NADH, $FADH_2$,
   ____ fermentation   and CO_2 form
   ____ Krebs cycle   b. glucose to two pyruvate
   ____ electron transfer   c. NAD^+ regenerated, two
        phosphorylation   ATP net
                            d. H^+ flows through ATP
                               synthases

Additional questions are available on **Biology⊗Now™**

Critical Thinking

1. Living cells of your body never use their nucleic acids as alternative energy sources. Suggest why.

2. Suppose you start a body-building program. You are already eating plenty of carbohydrates. Now a qualified nutritionist recommends that you start a protein-rich diet that includes protein supplements. Speculate on how extra dietary proteins will be put to use, and in which tissues.

3. Each year, Canada geese lift off in precise formation from their northern breeding grounds to spend the winter months in warmer climates (Figure 6.10). Then they make the return trip in spring. As is the case for other migratory birds, their flight muscle cells are efficient at using fatty acids as an energy source. Remember, the carbon backbone of fatty acids can be cleaved into small fragments that can be converted to acetyl–CoA for entry into the Krebs cycle.

   Suppose a lesser Canada goose from Alaska's Point Barrow has been steadily flapping along for about three thousand kilometers and is approaching Klamath Falls, Oregon. It looks down and notices a rabbit sprinting like the wind from a coyote with a taste for rabbit. With a stunning burst of speed, the rabbit reaches the safety of its burrow.

   Which energy-releasing pathway predominated in muscle cells in the rabbit's legs? Why was the Canada goose relying on a different pathway for most of its journey? And why wouldn't the pathway of choice in goose flight muscle cells be much good for a rabbit making a mad dash from its enemy?

4. At very high altitudes, oxygen levels are low and climbers can suffer from altitude sickness. They are short of breath, weak, dizzy, and confused. Curiously, the early symptoms of *cyanide poisoning* are similar. This potent poison binds tightly to cytochrome, the last component in the mitochondrial electron transfer chain. When cyanide is bound to it, cytochrome can't transfer electrons to the next component of the chain. Explain why cytochrome shutdown might cause the same symptoms as altitude sickness.

Figure 6.10 Canada geese.

Henrietta's Immortal Cells

Each human starts out as a fertilized egg. By the time of birth, the body consists of about a trillion cells, each one descended from that single egg. Even in an adult, billions of cells still divide every day and replace their damaged or worn-out predecessors.

In 1951, George and Margaret Gey of Johns Hopkins University were trying to develop a way to keep human cells dividing outside the body. An "immortal" cell line could help researchers study basic life processes as well as cancer and other diseases. Using cells to study cancer would be a better alternative than experimenting with patients and risking their already vulnerable lives.

For almost thirty years, the Geys tried to grow human cells. But they could not stop the cellular descendants from dying within a few weeks. Mary Kubicek, the Geys' assistant, tried again and again to establish a lineage of cultured human cancer cells. She was about to give up, but she prepared one last sample. She named the cells

HeLa, a code consisting of the first two letters of the donor patient's first and last names. These HeLa cells started to divide, over and over, every twenty-four hours.

Sadly, cancer cells in the patient were dividing just as fast. Eight months after she had been diagnosed with cervical cancer, Henrietta Lacks, a young mother from Baltimore, was dead.

Henrietta Lacks

Although Henrietta passed away, her cells lived on in the Geys' laboratory. HeLa cells were shipped to research laboratories all over the world. The Geys used them to identify which strains of polio virus were harmful. The cells were also used to test the resulting polio vaccine. Other researchers used HeLa cells to investigate cancer, viral growth, protein synthesis, the effects of radiation, and more. Some HeLa cells even traveled into space.

Henrietta was only thirty-one when runaway cell divisions killed her. Decades later, her legacy continues to help humans through her cellular descendants that are still dividing day after day.

Understanding cell division—and, ultimately, how new individuals are put together in the image of their parents—starts with answers to three questions. *First*, what kind of information guides inheritance? *Second*, how is information copied in a parent cell before being distributed to each daughter cell? *Third*, what kinds of mechanisms actually parcel out the information to daughter cells?

It will take more than one chapter to survey the nature of cell reproduction and other mechanisms of inheritance. In this chapter, we introduce the structures and mechanisms that cells use to reproduce.

☑ *How Would You Vote?* It is illegal to sell your organs, but you can sell your cells, including eggs, sperm, and blood cells. HeLa cells are still being sold all over the world. Should the family of Henrietta Lacks share in the profits? See BiologyNow for details, then vote online.

Key Concepts

WHAT DIVIDES, AND WHEN
Cells of each species have a characteristic number of chromosomes, which differ in length and shape. Division mechanisms parcel out chromosomes and cytoplasm to daughter cells.

MITOSIS AND CYTOPLASMIC DIVISION
Mitosis, a nuclear division mechanism of eukaryotic cells, maintains the parental chromosome number. The cytoplasm divides by a separate process.

MEIOSIS AND SEXUAL REPRODUCTION
Meiosis, another nuclear division mechanism, divides the parental chromosome number by half. It occurs only in cells that give rise to gametes or spores. Later, fertilization restores the parental chromosome number.

CANCER
Built-in mechanisms control the timing and rate of cell division. On rare occasions, surveillance mechanisms fail, and cell division becomes uncontrollable. Tumor formation is the outcome.

Links to Earlier Concepts

Before reading this chapter, quickly review the introduction to the nucleus of eukaryotic cells (Section 3.5).

You will be also drawing upon your knowledge of microtubules, their assembly and disassembly, and the motor proteins associated with them (3.7).

A quick look back at plant cell walls (3.8) will help you see why plant cells do not divide by pinching their cytoplasm into two parcels, as animal cells do.

The sections on the structure of DNA (2.10), and the nature of inheritance and diversity (1.2, 1.4) gave you the framework to understand how the processes of asexual and sexual reproduction contribute to the evolution of a species.

7.1 Overview of Cell Division Mechanisms

The continuity of life depends on reproduction.
By this process, parents produce a new generation
of cells or multicelled individuals like themselves.
Cell division is the bridge between generations.

A dividing cell faces a challenge. Each of its daughters must get information encoded in the parental DNA and enough cytoplasm to start up its own operation. DNA "tells" a cell how to build proteins. Some proteins are structural materials; others are enzymes that speed up construction of organic compounds. If a cell does not inherit all of the information it needs to build proteins, it will not be able to grow or function properly.

Each daughter cell also inherits a blob of the parent cell's cytoplasm. That blob contains enough enzymes, organelles, and other metabolic machinery for the new cell to begin functioning.

MITOSIS AND MEIOSIS

Because their DNA is packaged into a single nucleus, eukaryotic cells cannot simply divide in half. They do split their cytoplasm into daughter cells, but not until *after* the DNA is copied and repackaged in two nuclei. Nuclear division occurs by either **mitosis** or **meiosis**. As you will see, although these two processes have a lot in common, their outcomes differ.

In multicelled eukaryotes, mitosis and cytoplasmic division occur only in *somatic* (body) cells. These two processes are the foundation of increases in body size, the replacement of dead cells, and tissue repair. Many organisms also reproduce by mitosis, making clones, or exact copies, of themselves.

Meiosis, a different mechanism of nuclear division, precedes the formation of gametes or spores. Meiosis is the basis of sexual reproduction. Gametes, which are mature reproductive cells (including eggs and sperm), form by the meiotic divisions of *germ* cells, which are immature reproductive cells. Spores form by meiosis in the life cycle of many protists, plants and fungi.

What about the prokaryotes—archaea and bacteria? These cells reproduce asexually by prokaryotic fission, a different mechanism. We will consider prokaryotic fission later, in Section 14.2.

KEY POINTS ABOUT CHROMOSOME STRUCTURE

Every eukaryotic cell has a characteristic number of DNA molecules, each with many attached proteins. Together, a molecule of DNA and its proteins are one

Figure 7.1 *Animated!* Scanning electron micrograph of a duplicated human chromosome. Interacting proteins coil the loops of coiled DNA. At a deeper level of organization, proteins and DNA interact in a cylindrical fiber. The smallest unit of organization of a chromosome is the nucleosome: part of a DNA molecule looped twice around a core of histones.

chromosome. A cell duplicates its chromosomes before the nucleus divides. A chromosome and its duplicate stay attached to each other as **sister chromatids** until late in the nuclear division process:

one chromatid
its sister chromatid

During the early stages of either mitosis or meiosis, each duplicated chromosome coils back on itself again and again, into a highly condensed form (Figure 7.1). This orderly coiling arises from interactions between DNA and proteins associated with it. At regular intervals, DNA "spools" twice around proteins called histones. Each spool of DNA and histones is a **nucleosome**, the chromosome's smallest unit of structural organization.

While each duplicated chromosome is condensing, it constricts where the sister chromatids are attached to each other. This constriction is called a **centromere**. Its location is different for each kind of chromosome.

So what is the point of all this organization? Tight packing helps keep chromosomes from getting tangled while they are moved and sorted into parcels *during* nuclear division. *Between* divisions, the packing keeps enzymes from accessing the cell's genetic information. Loosening specific regions allows a cell to use only the hereditary information that it requires at that time.

When a cell divides, each daughter cell gets chromosomes and some cytoplasm. Nuclear division and cytoplasmic division are two separate processes in eukaryotic cells.

One nuclear division mechanism, mitosis, is the basis of body growth, cell replacements, tissue repair, and asexual reproduction in eukaryotes.

Meiosis, another nuclear division mechanism, is the basis of gamete and spore formation in sexual reproduction.

7.2 Introducing the Cell Cycle

Start thinking about the reproduction of a cell as a cyclic sequence of events. This cell cycle is not the same as a life cycle, which is a series of stages that individuals of each species pass through during their lifetime.

A **cell cycle** is a series of events from one cell division to the next (Figure 7.2). It starts when a new daughter cell forms by way of mitosis and cytoplasmic division. It ends when that cell divides. Interphase, mitosis, and cytoplasmic division constitute one turn of the cycle.

During **interphase**, the cell increases in size, roughly doubles the number of components in its cytoplasm, and duplicates its DNA. For most cells, interphase is the longest portion of the cell cycle. Biologists divide it into three stages:

G1   Interval (*G*ap 1) of cell growth and functioning before the onset of DNA replication

S   Time of **S**ynthesis (DNA replication)

G2   Second interval (*G*ap 2), after DNA replication, when the cell prepares for division

G1, S, and G2 are code names for some events that are just amazing, considering how much DNA is stuffed into a nucleus. If you could stretch out all the DNA molecules from one of your somatic cells in a line, that line would be longer than you are tall, about 7.5 feet. DNA from one salamander cell would stretch 540 feet!

The wonder is, enzymes and other proteins in cells *selectively* access, activate, and silence information in all that DNA. They also make base-by-base copies of every DNA molecule before they divide. Most of this work is completed in interphase.

G1, S, and G2 of interphase have distinct patterns of biosynthesis. G1 is a growth phase. Cells remain in G1, sometimes for years, until they reach adequate cell size and the external environment is appropriate for replication. Cells destined to divide enter S, when they copy their DNA as well as the proteins attached to it. During G2, they make proteins that drive mitosis.

The length of the cycle is about the same for cells of the same type. However, it differs from one cell type to another. For example, every second, 2 to 3 million red blood cells form to replace worn-out ones circulating in your bloodstream. It takes 12 hours for the parent of a red blood cell to divide. Cells in the growing root tip of a bean plant divide every 19 hours, and those in the regenerating limb of an axolotl (a type of amphibian) divide every 53 hours. The cells in a sea urchin embryo can divide every 2 hours.

Once S begins, events will usually proceed at about the same rate and continue through mitosis. Because that rate is the same for all cells of a species, you might wonder if the cell cycle has built-in molecular brakes. It does. The red triangles in Figure 7.2 indicate where some of these controls exert their influence over the cell cycle. Apply the brakes that are supposed to work in G1, and the cycle stalls in G1. Lift the brakes, and the cell cycle will run to completion. Said another way, *control mechanisms govern the rate of cell division.*

Imagine a car losing its brakes just as it starts down a steep mountain road. As you will read later on in the chapter, cancer begins as controls over the cell cycle are lost. Stressful conditions often interrupt the cell cycle. For instance, when deprived of any vital nutrient, free-living cells called amoebas will remain in interphase until that nutrient is restored.

control or out of control?

INTERPHASE

G1
Interval of cell growth, before DNA replication (chromosomes unduplicated)

Each daughter cell starts interphase

S
Interval of cell growth, when DNA replication is completed (chromosomes duplicated)

Cytoplasmic division

Telophase
Anaphase
Metaphase
Prophase
MITOSIS

Interphase ends for parent cell

G2
Interval following DNA replication; cell prepares to divide

Figure 7.2 *Animated!*   Eukaryotic cell cycle, generalized. The length of each interval differs among different cell types. *Red* triangles indicate checkpoints in the cycle.

Interphase, mitosis, and cytoplasmic division constitute one turn of the cell cycle.

In interphase, a new cell increases its size, roughly doubles the number of its cytoplasmic components, and duplicates its chromosomes. The cycle ends after the cell undergoes mitosis and then divides.

7.3 Mitosis Maintains the Chromosome Number

Mitosis is a nuclear division mechanism that maintains the parental chromosome number in daughter cells.

After G2, a cell divides by mitosis to produce identical daughter cells, each with the same number and kind of chromosomes as the parent. The **chromosome number** is the sum of all chromosomes in a cell of a given type. The body cells of gorillas have 48, those of human cells have 46, and pea plant cells have 14 (Figure 7.3).

Actually, your cells have a **diploid** number ($2n$) of chromosomes; there are two of each type. Those 46 are like volumes of two sets of books numbered from 1 to 23. You have two volumes of, say, chromosome 22—*a pair of them*. Except for the different sex chromosomes (X and Y), chromosomes of a pair have the same length and shape, and carry the same hereditary instructions.

Think of them as two sets of books on how to build a house. Your father gave you one set. Your mother had her own ideas about wiring, plumbing, and so on. She gave you an alternate edition that has the same topics but says slightly different things about many of them.

With mitosis and cytoplasmic division, one diploid parent cell produces two diploid daughter cells (Figure 7.4). This does not mean that each daughter cell merely gets forty-six or forty-eight or fourteen chromosomes. If only the total mattered, then one cell might get, say, two pairs of chromosomes 22 and none of chromosome 9. But then neither cell could function like its parent *without two of each type of chromosome*.

Cells divide all of their chromosomes evenly into daughter cells by using a dynamic structure called the **mitotic spindle**. This network of microtubules grows from opposite sides of the cell during mitosis. Some of the microtubules overlap in the middle of the cell, and others attach to duplicated chromosomes.

Remember, all the chromosomes of a eukaryotic cell duplicate *before* mitosis starts. The spindle determines where each sister chromatid will end up before the cell divides. Microtubules extending from one side of the cell connect to one chromatid of every chromosome. Microtubules from the other side of the cell connect to its sister chromatid. The spindle then pulls the sister chromatids of each duplicated chromosome apart, to opposite poles of the cell.

Thus, each side of the cell gets a mass of DNA that includes one copy of each chromosome. When the cell divides, those masses of DNA will become the nuclei of the daughter cells.

Mitosis proceeds in four stages: *prophase, metaphase, anaphase,* and *telophase*. These are discussed next.

LINK TO SECTION 3.7

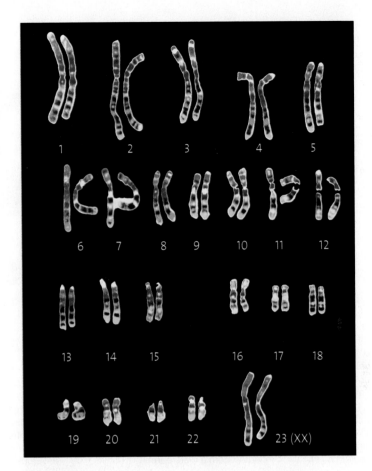

Figure 7.3 From a diploid cell of a human female, twenty-three pairs of chromosomes. The last two are a pair of sex chromosomes. Females have an XX pairing, and males have an XY pairing. The X and the Y chromosome differ in length, shape, and which genetic information they carry, but they pair up during mitosis nonetheless.

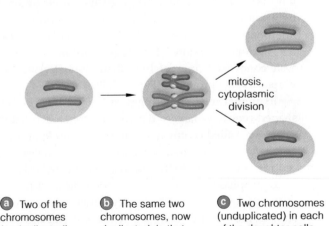

a Two of the chromosomes (unduplicated) in a parent cell at interphase

b The same two chromosomes, now duplicated, in that cell at interphase, prior to mitosis

mitosis, cytoplasmic division

c Two chromosomes (unduplicated) in each of the daughter cells, which both start out life in interphase

Figure 7.4 Preview of how mitosis maintains a parental chromosome number, one generation to the next. For clarity, the spindle is not shown here.

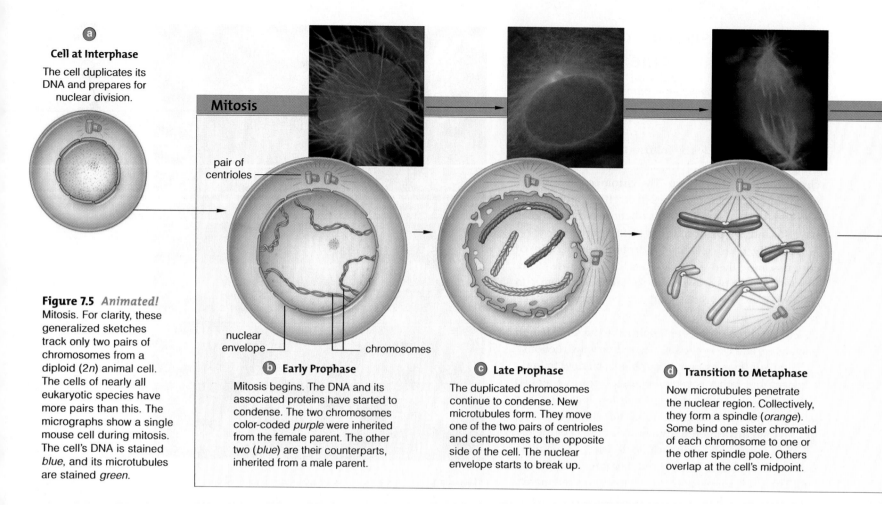

ⓐ

Cell at Interphase

The cell duplicates its DNA and prepares for nuclear division.

Mitosis

pair of centrioles

nuclear envelope

chromosomes

Figure 7.5 *Animated!* Mitosis. For clarity, these generalized sketches track only two pairs of chromosomes from a diploid (*2n*) animal cell. The cells of nearly all eukaryotic species have more pairs than this. The micrographs show a single mouse cell during mitosis. The cell's DNA is stained *blue*, and its microtubules are stained *green*.

ⓑ Early Prophase

Mitosis begins. The DNA and its associated proteins have started to condense. The two chromosomes color-coded *purple* were inherited from the female parent. The other two (*blue*) are their counterparts, inherited from a male parent.

ⓒ Late Prophase

The duplicated chromosomes continue to condense. New microtubules form. They move one of the two pairs of centrioles and centrosomes to the opposite side of the cell. The nuclear envelope starts to break up.

ⓓ Transition to Metaphase

Now microtubules penetrate the nuclear region. Collectively, they form a spindle (*orange*). Some bind one sister chromatid of each chromosome to one or the other spindle pole. Others overlap at the cell's midpoint.

PROPHASE: MITOSIS BEGINS

We can tell that a cell is in prophase, the first stage of mitosis, when its chromosomes become visible under a light microscope. Each chromosome was duplicated in interphase; each is now two sister chromatids joined at the centromere (Figure 7.5*a*). All the chromosomes will have condensed completely in their characteristic "X" shapes by late prophase (Figure 7.5*b,c*).

In animal cells, some fungi, and algae, two barrel-shaped centrioles next to the nucleus are embedded in structures called centrosomes. In certain kinds of cells, centrioles give rise to flagella or cilia. Centrioles and centrosomes duplicate themselves before prophase.

In prophase, one set of duplicated centrioles moves to the opposite side of the nucleus, then microtubules start to grow out of both sets. *These microtubules form the mitotic spindle* (Figure 7.5*d*). In cells of plants and some other species, the mitotic spindle originates from structures that are similar to centrosomes. In all cells, the areas from which spindle microtubules grow are called spindle poles.

TRANSITION TO METAPHASE

During the transition from prophase to metaphase, the nuclear envelope breaks up into many tiny, flattened vesicles. Spindle microtubules are now free to interact with the chromosomes and one another. Microtubules growing from one spindle pole bind one chromatid of each chromosome, and microtubules from the other spindle pole bind its sister.

The opposing sets of microtubules engage in a tug-of-war, growing and shrinking until they are the same length. Then, all of the duplicated chromosomes will be aligned at the spindle's equator, midway between poles. This is metaphase (Figure 7.5*e*). The alignment is crucial for the next stage.

FROM ANAPHASE THROUGH TELOPHASE

At anaphase, sister chromatids of each chromosome separate from each other and move toward opposite spindle poles. Two mechanisms bring this about. First, the microtubules attached to a chromatid shorten and

spindle microtubules

chromosomes at spindle equator, midway between spindle poles

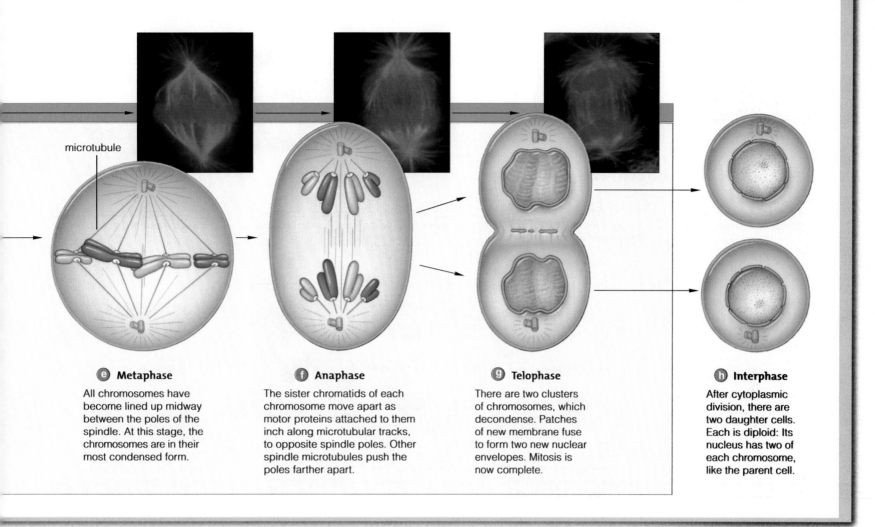

microtubule

ℯ Metaphase

All chromosomes have
become lined up midway
between the poles of the
spindle. At this stage, the
chromosomes are in their
most condensed form.

f Anaphase

The sister chromatids of each
chromosome move apart as
motor proteins attached to them
inch along microtubular tracks,
to opposite spindle poles. Other
spindle microtubules push the
poles farther apart.

g Telophase

There are two clusters
of chromosomes, which
decondense. Patches
of new membrane fuse
to form two new nuclear
envelopes. Mitosis is
now complete.

h Interphase

After cytoplasmic
division, there are
two daughter cells.
Each is diploid: Its
nucleus has two of
each chromosome,
like the parent cell.

pull it toward a spindle pole. As the spindle shortens, ATP-driven motor proteins drag the chromatids along microtubules toward the poles.

At the same time, the spindle poles themselves are being pushed farther apart. Some of the microtubules extending from spindle poles do not bind chromatids; they overlap midway between spindle poles. Motor proteins ratchet these microtubules past each other, away from the center of the cell. As they do, they push the spindle poles apart.

Sister chromatids, recall, are identical. Once they detach from each other at anaphase, each is a separate chromosome in its own right (Figure 7.5*f*). Telophase starts when the two sets of chromosomes are clustered at opposite spindle poles. Then, all the chromosomes decondense (Figure 7.5*g*).

Vesicles derived from the old nuclear envelope fuse and form patches of membrane around the clusters. Patch joins with patch until a new nuclear envelope encloses each set of chromosomes. Thus, by the end of telophase, two nuclei have formed.

If the parent cell was diploid, each cluster has two chromosomes of each type. With mitosis, remember, a new nucleus has the same chromosome number as the parent nucleus. Once the two nuclei form, telophase is over—and so is mitosis (Figure 7.5*h*).

Prior to mitosis, chromosomes are duplicated, so each chromosome consists of two sister chromatids.

In prophase, chromosomes condense. The mitotic spindle tethers sister chromatids of duplicated chromosomes to opposite spindle poles.

At metaphase, all duplicated chromosomes are aligned midway between the spindle poles.

At anaphase, microtubules move the sister chromatids of each chromosome apart, to opposite spindle poles.

At telophase, a new nuclear envelope forms around each of two clusters of decondensing chromosomes.

Thus two daughter nuclei have formed. Each has the same chromosome number as the parent cell's nucleus.

① Mitosis is over, and the spindle is disassembling.

a Animal cell division

② At the former spindle equator, a ring of microfilaments attached to the plasma membrane contracts.

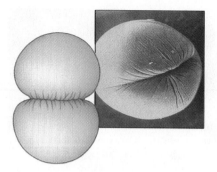

③ As the microfilament ring shrinks in diameter, it pulls the cell surface inward.

④ Contractions continue; the cell is pinched in two.

① As mitosis ends, vesicles cluster at the spindle equator. They contain materials for a new primary cell wall.

b Plant cell division

cell plate forming

② Vesicle membranes fuse. The wall material is sandwiched between two new membranes that lengthen along the plane of a newly forming cell plate.

③ Cellulose is deposited inside the sandwich. In time, these deposits will form two cell walls. Others will form the middle lamella between the walls and cement them together.

④ A cell plate grows at its margins until it fuses with the parent cell plasma membrane. The primary wall of growing plant cells is still thin. New material is deposited on it.

Figure 7.6 *Animated!* Cytoplasmic division of an animal cell (**a**) and a plant cell (**b**).

LINK TO
SECTION
3.8

animal cell pinching in two

7.4 Division of the Cytoplasm

In most cell types, the cytoplasm usually divides at some time between late anaphase and the end of telophase. The mechanism of cytoplasmic division—or, formally, cytokinesis—differs among species.

After meiosis, an animal cell's cytoplasm can pinch in two by a process called **cleavage**. Midway between the cell's two poles, a patch of plasma membrane sinks in to form a thin indentation (Figure 7.6a). This *cleavage furrow* is the first visible sign that the cytoplasm is in the process of dividing. The furrow advances until it surrounds the cell. As it does so, it deepens along the plane corresponding to the former spindle's midpoint.

What is happening? Under the plasma membrane, a thin band of actin filaments is wrapped around the cell's midsection. When energized by ATP, the actin filaments contract and slide past one another, in a way that shrinks the band's diameter. The shrinking band is attached to the plasma membrane, and thus drags the membrane inward until the cytoplasm is pinched in two. Two daughter cells form, each with a nucleus and cytoplasm enclosed in a plasma membrane.

The contractile force of actin filaments is not strong enough to pinch through the stiff cell walls of plant cells. Plant cell cytoplasm divides instead by **cell plate formation** (Figure 7.6b). Vesicles full of wall-building materials fuse to form a disk-shaped structure known as the cell plate. In time, cellulose deposits at the plate are thick enough to form a cross-wall through the cell. New plasma membrane extends across both sides of it.

The new wall grows until it bridges the cytoplasm and partitions the parent cell.

> *After mitosis, a separate mechanism partitions the cytoplasm into two daughter cells, each with a nucleus.*
>
> *A contractile ring mechanism partitions animal cells. Actin filaments form a band around the cell's midsection; as the band contracts the cytoplasm is pinched in two.*
>
> *Cell plate formation partitions plant cells. Vesicles deposit material to form a cross-wall, which connects to the parent cell wall.*

LINK TO SECTION 1.2

Figure 7.7 Examples of reproductive organs.

7.5 Meiosis and Sexual Reproduction

Meiosis is a nuclear division mechanism that is central to sexual reproduction. It reduces the parental number of chromosomes by half for forthcoming gametes.

ASEXUAL VERSUS SEXUAL REPRODUCTION

When an orchid or aphid reproduces all by itself, what sort of offspring does it get? By the process of **asexual reproduction**, all offspring inherit the same DNA from a single parent. That DNA contains all the heritable information necessary to make a new individual. Rare mutations aside, then, asexually produced individuals are *clones*, genetically identical copies of the parent.

Inheritance gets far more interesting with **sexual reproduction**, a process involving meiosis, formation of gametes, and fertilization—a union of two gametes. Unlike mitosis, meiosis sorts chromosomes into parcels *twice* prior to cytoplasmic division. Meiosis is the first step leading to the formation of gametes.

As you know, the chromosome number is the total number of chromosomes in a single cell. If that cell is diploid ($2n$), it has a *pair* of each type of chromosome. Human body cells have 46 chromosomes, or 23 pairs. Thus, we can say that human somatic cells are diploid, expressed as $2n=46$. So are the human germ cells that give rise to gametes. After meiosis, each gamete ends up with 23 chromosomes—one of each type. Meiosis has reduced the parental chromosome number by half, to a **haploid** number (n). Human gametes are haploid; they have 23 chromosomes, expressed as $n=23$.

In most multicelled eukaryotes, cells that form in specialized reproductive structures or organs give rise to gametes (Figure 7.7). Fertilization occurs as male and female gametes fuse to form a new cell that is the start of a new individual. In most sexual reproducers, this cell is diploid, with pairs of chromosomes. One of the chromosomes of each pair came from the female gamete, and one from the male gamete.

Except for a pair of nonidentical sex chromosomes, paired chromosomes have the same length, shape, and complement of genetic information. The pairs, which zipper together during meiosis, are called **homologous chromosomes** (*hom*– means alike).

A **gene** is a section of a chromosome that contains instructions, encoded in the DNA, for building some structural or functional part of an organism. Genes on homologous chromosomes match up. In other words, each pair of chromosomes has pairs of genes (Figure 7.8). If all pairs of all genes were identical, then sexual reproduction would produce only clones. Imagine if that were true—then you, everyone you know, and the entire human population would be clones.

Fortunately for us, paired genes on homologous chromosomes usually are *not* identical. Most genes are a bit different from their partners on the homologous chromosome. Why? Gene mutations make permanent changes in the molecular structure of DNA. Different mutations accumulate in the DNA of different lines of descent, so it is unlikely that two parents would have identical genes unless they were closely related.

Alleles are different molecular forms of the same gene. The differences affect thousands of traits. This is one reason why individuals of a sexually reproducing species do not all look alike. *With sexual reproduction, offspring inherit new combinations of alleles, which lead to variations in the details of their traits.*

This section introduces the cellular basis of sexual reproduction. More importantly, it starts you thinking about far-reaching effects of gene shufflings at certain stages of the process. The process introduces variations in traits among offspring that are typically acted upon by agents of natural selection. Thus, *variation in traits is a foundation for evolution.*

Figure 7.8 Homologous chromosomes. Any gene on one chromosome may be slightly different from its partner on the other chromosome, but at this magnification both of the chromosomes appear to be identical.

Meiosis I

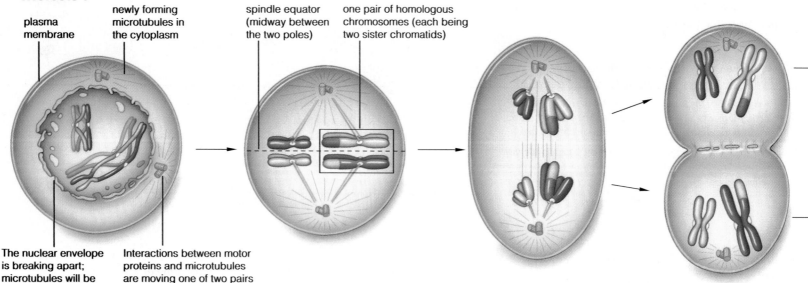

plasma membrane

newly forming microtubules in the cytoplasm

spindle equator (midway between the two poles)

one pair of homologous chromosomes (each being two sister chromatids)

The nuclear envelope is breaking apart; microtubules will be able to penetrate the nuclear region

Interactions between motor proteins and microtubules are moving one of two pairs of centrioles toward the opposite spindle pole

ⓐ Prophase I

Chromosomes have been duplicated by the end of interphase I. Now they start to condense. Each pairs with its homologue, and the two usually swap segments. This event, called *crossing over*, is indicated by the break in color on the pair of larger chromosomes. Newly forming microtubules of the spindle become attached to each sister chromatid.

ⓑ Metaphase I

Motor proteins projecting from the microtubules move the chromosomes and spindle poles apart. Chromosomes are tugged into position midway between the spindle poles. The spindle becomes fully formed by the dynamic interactions among motor proteins, microtubules, and chromosomes.

ⓒ Anaphase I

Some microtubules extend from the spindle poles and overlap at its equator. These *lengthen* and push the poles apart. Other microtubules extending from the poles *shorten* and pull each chromosome away from its homologous partner. These motions move the homologous partners to opposite poles.

ⓓ Telophase I

The cytoplasm of the cell divides at some point. There are now two haploid (*n*) cells with one of each type of chromosome that was present in the parent (2*n*) cell. *All chromosomes are still in the duplicated state.*

Figure 7.9 *Animated!* Meiosis in one type of animal cell. This nuclear division mechanism reduces the parental chromosome number in immature reproductive cells by half, to the haploid number, for forthcoming gametes. To keep things simple, we track only two pairs of homologous chromosomes. Maternal chromosomes are shaded *purple* and paternal chromosomes *blue*.

TWO DIVISIONS, NOT ONE

Meiosis is like mitosis in some ways, but the outcome is different. As in mitosis, a germ cell duplicates its DNA in interphase. The two DNA molecules and their proteins stay attached at the centromere, the notably constricted region along their length. For as long as they remain attached, they are sister chromatids:

centromere

one chromatid ⎡

its sister chromatid ⎡

one chromosome in the duplicated state

As in mitosis, the microtubules of a spindle apparatus parcel chromosomes into new nuclei by a process with several distinct stages.

With meiosis, however, *chromosomes go through two consecutive divisions that end with the formation of four haploid nuclei*. Each of these two nuclear divisions can be compared to mitosis, but interphase does not occur between divisions. In meiosis, the nucleus of a germ cell divides twice after one cycle of replication. These two divisions are called meiosis I and meiosis II:

interphase (*DNA replication before meiosis I*)	MEIOSIS I	no interphase (*no DNA replication before meiosis II*)	MEIOSIS II
	PROPHASE I		PROPHASE II
	METAPHASE I		METAPHASE II
	ANAPHASE I		ANAPHASE II
	TELOPHASE I		TELOPHASE II

In meiosis I, each duplicated chromosome aligns with its partner, *homologue to homologue*. Then, after the two

Meiosis II

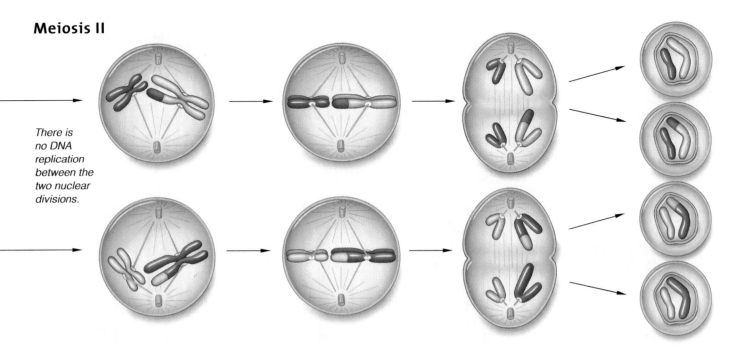

There is no DNA replication between the two nuclear divisions.

(e) Prophase II

Microtubules have moved one member of the centriole pair to the opposite spindle pole in each of two daughter cells. During prophase II, microtubules attach to the chromosomes and begin to drag them toward the middle of the cell.

(f) Metaphase II

The microtubules, motor proteins, and duplicated chromosomes interact, which positions all of the duplicated chromosomes midway between the two spindle poles.

(g) Anaphase II

The attachment between the sister chromatids of each chromosome breaks, and the two are moved to opposite spindle poles. Each former "sister" is now a chromosome on its own.

(h) Telophase II

By the time telophase II is finished, there are four daughter nuclei. When cytoplasmic division is over, each daughter cell is haploid (*n*). All chromosomes are in the unduplicated state.

---

chromosomes of every pair have lined up with each other, they are moved apart:

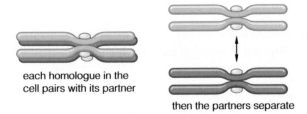

each homologue in the cell pairs with its partner

then the partners separate

The cytoplasm typically divides after each homologue has separated from its partner. Two haploid daughter cells form by this first cytoplasmic division.

Later, during meiosis II, *the two sister chromatids of each chromosome are separated from each other:*

two chromosomes (unduplicated)

one chromosome (duplicated)

Each sister chromatid has now become a chromosome in its own right. There are now four identical parcels of chromosomes, each containing one chromosome of each type. Nuclear envelopes form around the parcels, enclosing them as four nuclei. Typically the cytoplasm divides once more, so the outcome is four haploid (*n*) cells. Thus, two meiotic divisions reduced the parental chromosome number of a germ cell by half. Figure 7.9 shows the chromosomal movements in the context of the stages of meiosis.

Asexual reproduction by way of mitotic cell division results in genetically identical copies of a parent.

Sexual reproduction introduces variation in traits by bestowing novel combinations of alleles on offspring.

Meiosis is a nuclear division mechanism that reduces the chromosome number of a parental cell by half—to the haploid number (n). Cytoplasmic division typically follows.

a Both chromosomes shown here were duplicated during interphase, before meiosis. When prophase I is under way, sister chromatids of each chromosome are positioned close together.

b Each chromosome becomes zippered to its homologue, so all four chromatids are tightly aligned. The nonidentical sex chromosomes also pair up, but they zipper only in a tiny region at their ends.

c For the sake of clarity, we show the chromosomes as if they were condensed. Actually, they are in a tightly aligned form, but not condensed, during prophase I.

d The intimate contact encourages one crossover (and usually more) to happen at various intervals along the length of nonsister chromatids.

e Nonsister chromatids exchange segments at the crossover sites. They continue to condense into thicker, rodlike forms. By the start of metaphase I, they will be unzippered from each other.

f Crossing over breaks up old combinations of alleles and puts new ones together in the cell's pairs of homologous chromosomes.

Figure 7.10 *Animated!* Key events of prophase I, the first stage of meiosis. For the sake of clarity, we show only one pair of homologous chromosomes and one crossover event. Typically, more than one crossover occurs. *Blue* signifies the paternal chromosome; *purple* signifies its maternal homologue.

LINK TO SECTION 1.4

7.6 How Meiosis Puts Variation in Traits

One function of meiosis is the reduction of the parental chromosome number by half. However, two other functions are just as important. They both give sexual reproduction an evolutionary advantage over asexual reproduction.

Figure 7.9 mentioned briefly that pairs of homologous chromosomes swap pieces during prophase I. It also showed how a homologous chromosome lines up with its partner during metaphase I. Both events introduce new combinations of alleles into gametes that form at some point after meiosis. Together with chromosome shufflings that occur during fertilization, these events contribute to the variation in traits among generations of offspring in sexually reproducing species.

CROSSING OVER IN PROPHASE I

Figure 7.10*a* is a simple sketch of a pair of duplicated chromosomes, early in prophase I of meiosis. Notice how they are in long, thin, threadlike form. All of the chromosomes in a germ cell stretch out like this, and then each is drawn close to its homologue. Molecular interactions stitch homologues together point by point along their length, with little space between them. The intimate, parallel orientation favors **crossing over**, an exchange of genetic material between a chromatid of one chromosome and a chromatid of the homologous partner. During a crossing over event, two "nonsister" chromatids break at exactly the same locations along their length. The chromatids exchange corresponding segments at the breaks; they swap genes.

Gene swapping would be pointless if genes never varied. But remember, each gene can come in slightly different forms: alleles. You can predict that a number of the alleles on one chromosome will *not* be identical to their partner alleles on the homologous chromosome. Every crossover event is an opportunity to exchange different versions of heritable information.

We will look at the implications of crossing over in later chapters. For now, just remember this: *Crossing over leads to recombinations among genes of homologous chromosomes, and eventually to variation in traits among gametes, then offspring.*

METAPHASE I ALIGNMENTS

Major shufflings of intact chromosomes start during the transition from prophase I to metaphase I. Suppose this is happening right now in one of your germ cells. Crossovers have already made genetic mosaics of the chromosomes, but put this aside to simplify tracking. Just call the twenty-three chromosomes you inherited from your mother the *maternal* chromosomes, and the twenty-three homologues from your father the *paternal* chromosomes.

At metaphase I, the microtubules from both spindle poles have the duplicated chromosomes lined up at the spindle equator (Figure 7.9*b*). Have they attached all maternal chromosomes to one pole and all paternal chromosomes to the other? Maybe, but probably not. When the microtubules were growing, they connected to the first chromosome they contacted. Because the attachment was random, there is no particular pattern to the metaphase I positions of maternal and paternal chromosomes. Carry this thought one step further. During anaphase I, when a duplicated chromosome is moved away from its homologous partner, *either one* of them can end up at either spindle pole.

Think of the possibilities while tracking just three pairs of homologues. By metaphase I, these three pairs may be arranged in any one of four possible positions (Figure 7.11). This means that eight combinations (2^3) are possible for forthcoming gametes.

Cells that give rise to human gametes have twenty-three pairs of homologous chromosomes, not three. So each time a human sperm or egg forms, you can expect a total of *8,388,608* (or 2^{23}) possible combinations of maternal and paternal chromosomes!

Moreover, in each sperm or egg, many hundreds of alleles inherited from the mother might not "say" the exact same thing about hundreds of different traits as alleles inherited from the father. Are you beginning to

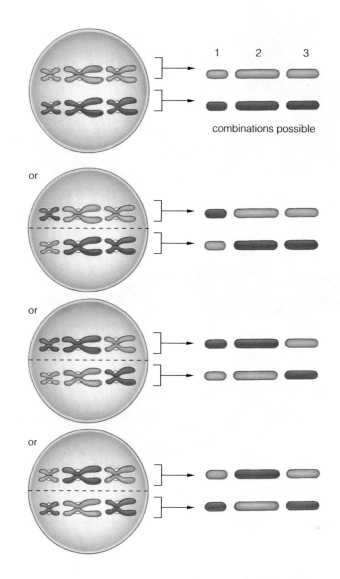

combinations possible

Figure 7.11 *Animated!* Possible outcomes for random alignment of three pairs of homologous chromosomes at metaphase I. The three types of chromosomes are labeled 1, 2, and 3. With four alignments, eight combinations of maternal (*purple*) and paternal chromosomes (*blue*) are possible in gametes.

get an idea of why such fascinating combinations of traits show up the way they do among the generations of your own family tree?

Crossing over, an interaction between a pair of homologous chromosomes, breaks up old combinations of alleles and puts new ones together during prophase I of meiosis.

The random attachment and subsequent positioning of each pair of maternal and paternal chromosomes at metaphase I lead to different combinations of maternal and paternal traits in each new generation.

Figure 7.12
(a) Generalized life cycle for most plants. (b) Generalized life cycle for animals. The zygote is the first cell to form when the nuclei of two gametes fuse at fertilization.

7.7 From Gametes to Offspring

What happens to the gametes that form after meiosis? Later chapters have specific examples. Here, just focus on where they fit into the life cycles of plants and animals.

Gametes are not all the same in their details. Human sperm have one tail, opossum sperm have two, and roundworm sperm have none. Crayfish sperm look like pinwheels. Most eggs are microscopic, and yet an ostrich egg inside its shell can weigh five pounds. A flowering plant's male gamete is just a sperm nucleus.

GAMETE FORMATION IN PLANTS

Plants on land face seasonal changes, and fertilization must coincide with spring rains and other conditions that favor growth of offspring. That is why most plant life cycles alternate between the production of spores (by sporophytes) and the production of gametes (by gametophytes). Figure 7.12a summarizes this cycle. A *sporophyte* is the plant body that bears reproductive structures. Trees and all other plants with roots, stems, and leaves are typical sporophytes. Most are diploid.

Spores form by meiosis within sporophyte tissues. A **spore** is a haploid resting structure. When spores germinate, or resume growth, they undergo mitosis to form haploid *gametophytes*, gamete-producing bodies. Gametophytes form, for example, inside stamens and carpels, the reproductive structures of flowers. During sexual reproduction, two haploid gametes will fuse to form a diploid zygote, which by mitotic cell divisions will give rise to another sporophyte. We will return to the topic of plant reproduction in later chapters.

GAMETE FORMATION IN ANIMALS

Diploid germ cells give rise to the gametes of animals (Figure 7.12b). In the male reproductive system, a germ cell develops into a primary spermatocyte. This large, immature cell enters meiosis. Four haploid cells result and develop into spermatids (Figure 7.13). These cells undergo changes, such as the formation of a tail, and become **sperm**, a type of mature male gamete.

In female animals, a germ cell becomes an **oocyte**, or immature egg. Unlike sperm, an oocyte stockpiles many cytoplasmic components, and its four daughter cells differ in size and function (Figure 7.14).

As an oocyte divides after meiosis I, one daughter cell called the secondary oocyte gets nearly all of the cytoplasm. The other cell, a first polar body, is small. Later, both of these haploid cells enter meiosis II, then cytoplasmic division. One of the secondary oocyte's daughter cells develops into a second polar body. The other gets most of the cytoplasm and develops into a gamete. The mature female gamete is called an ovum (plural, ova) or, more often, an **egg**. The three polar bodies that formed do not function as gametes; they are not rich in nutrients or plump with cytoplasm. In time they will degenerate. But their formation assures that the egg will have a haploid chromosome number. Also, by getting most of the cytoplasm, the egg holds enough metabolic machinery to support the early cell divisions of the new individual.

MORE SHUFFLINGS AT FERTILIZATION

The chromosome number characteristic of the parents is restored at **fertilization**, a time when a female and male gametes unite and their haploid nuclei fuse. If meiosis did not precede fertilization, the chromosome number would double in each generation. Doublings disrupt cell function. Why? Duplicate information on

Summary

Section 7.1 By processes of reproduction, parents produce a new generation of individuals like themselves. Cell division is the bridge between generations. When a cell divides, its daughter cells each receive a required number of DNA molecules and some cytoplasm.

A eukaryotic chromosome is a molecule of DNA together with its associated proteins. When duplicated, the chromosome consists of two sister chromatids. Until late in mitosis (or meiosis), the two sister chromatids remain attached at their centromere.

Only eukaryotic cells undergo mitosis, meiosis, or both. These nuclear division mechanisms partition the chromosomes of a parent cell into daughter nuclei. A separate mechanism divides the cytoplasm.

Mitosis is the basis of body growth of multicelled organisms, cell replacements, and tissue repair. Many eukaryotic species also reproduce asexually by mitosis. Meiosis, the basis of sexual reproduction, precedes the formation of gametes or spores.

Biology ⊜ Now
Explore the structure of a chromosome with animation on BiologyNow.

Section 7.2 The cell cycle begins when a new cell forms, and it ends when the cell reproduces by nuclear and cytoplasmic division. A cell carries out most of its functions in interphase: it increases in mass, and also duplicates each of its chromosomes.

Biology ⊜ Now
Investigate the stages of the cell cycle with the interaction on BiologyNow.

Section 7.3 The sum of all chromosomes in cells of a given type is the chromosome number. Human body cells have a diploid chromosome number of 46, or two copies of 23 types of chromosome. Mitosis maintains the chromosome number, one generation to the next.

Mitosis proceeds through prophase, metaphase, anaphase, and telophase. A mitotic spindle separates the sister chromatids of each chromosome into two daughter nuclei (Figure 7.18).

Section 7.4 Cytoplasmic division mechanisms differ. Animal cells undergo cleavage. A microfilament ring under the plasma membrane contracts, pinching the cytoplasm in two. In plant cells, a cross-wall forms in the cytoplasm and divides it.

Section 7.5 Asexual reproduction yields clones, or offspring that are genetically identical to a single parent. Sexual reproduction involves two parents that engage in meiosis, gamete formation, and fertilization. It leads to the variation in traits among offspring.

Alleles are slightly different molecular forms of the same gene. These differences affect traits. Meiosis and fertilization mix alleles in each generation of offspring. The offspring of most sexual reproducers inherit pairs of chromosomes, one from a maternal and one from a paternal parent. Except in individuals that have inherited nonidentical sex chromosomes (e.g., X with Y), the pairs are homologous (alike); each pair of chromosomes interacts during meiosis.

Mitosis

PROPHASE	METAPHASE	ANAPHASE	TELOPHASE
	Chromosomes align at spindle equator.	*Sister chromatids of chromosomes separate.*	two nuclei (2n)

Meiosis I

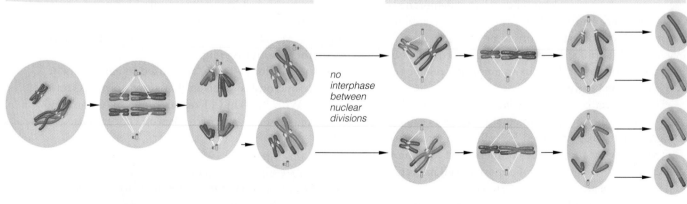

no interphase between nuclear divisions

PROPHASE I	METAPHASE I	ANAPHASE I	TELOPHASE I
Crossing over occurs between homologues.	Homologous pairs *align randomly.*	*Homologues* separate from their partner.	typically *two nuclei* (n)

Meiosis II

PROPHASE II	METAPHASE II	ANAPHASE II	TELOPHASE II
	Chromosomes align at spindle equator.	*Sister chromatids of chromosomes separate.*	four *nuclei* (n)

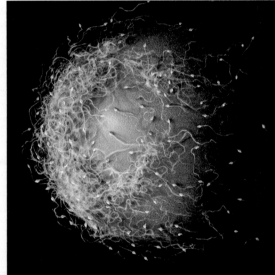

			cell differentiation, sperm formation (mature, haploid male gametes)
spermatogonium (diploid male germ cell)	primary spermatocyte (diploid)	secondary spermatocytes (haploid)	spermatids (haploid)

(a) Growth	(b) Meiosis I and cytoplasmic division	(c) Meiosis II and cytoplasmic division

Figure 7.13 *Animated!* Generalized sketch of sperm formation in animals. Figure 27.9 shows a specific example (how sperm form in human males).

		three polar bodies (haploid)
oogonium (diploid female reproductive cell)	primary oocyte (diploid)	first polar body (haploid)
		secondary oocyte (haploid)
		ovum (haploid)

(a) Growth	(b) Meiosis I and cytoplasmic division	(c) Meiosis II and cytoplasmic division

Figure 7.14 *Animated!* Animal egg formation. Eggs are far larger than sperm and larger than the three polar bodies. The painting above, based on a scanning electron micrograph, depicts human sperm surrounding an ovum.

---

extra chromosomes becomes converted into duplicate structural and functional parts of the cell, changing the finely tuned balance of its operation.

Fertilization also adds to variation among offspring. Reflect on the possibilities for humans alone. During prophase I, every human chromosome undergoes an average of two or three crossovers. In addition to the crossovers, random positioning of pairs of paternal and maternal chromosomes at metaphase I results in one of millions of possible chromosome combinations in each gamete. And of all male and female gametes that form, *which* two actually get together is a matter of chance. The sheer number of combinations that can exist at fertilization is staggering!

Distribution of random mixes of chromosomes into gametes and fertilization further contribute to the variation in traits that occur in offspring produced by sexual reproduction.

7.8 The Cell Cycle and Cancer

Growth and reproduction depend on controls over cell division and programmed cell death. On rare occasions, these controls fail in a somatic cell or reproductive cell. Cancer may be the outcome.

Take a moment to look closely at your hands. Visualize the cells making up your palms, thumbs, and fingers. Now imagine the mitotic divisions that produced the cell generations before them as you were developing, early on, inside your mother. All of those cells arose from tiny clumps of identical cells that budded from your embryonic body. Be grateful for the precision of control mechanisms that led to their formation at the right times, in the right places (Figure 7.15).

Health and survival depend on the proper timing and completion of cell cycle events. Genetic disorders arise from mistakes during duplication or distribution of even one chromosome. Uncontrolled cell divisions can destroy surrounding tissues and, ultimately, the individual. Such losses often begin in body cells that have lost control over the cell cycle. More rarely, they start in germ cells that give rise to sperm and eggs.

THE CELL CYCLE REVISITED

Millions of cells in your skin, bone marrow, gut lining, liver, and elsewhere in your body divide and replace their worn-out, dead, and dying predecessors every second of every day. The cells do not divide arbitrarily.

Figure 7.15 The paddlelike structure of a human embryo that develops into a hand by mitosis, cytoplasmic divisions, and programmed cell death. The scanning electron micrograph on the right shows individual cells.

Figure 7.16 Checkpoint proteins in action. The DNA inside this nucleus was damaged by radiation. (**a**) *Green* dots show the location of *53BP1*, and (**b**) *red* dots pinpoint the location of *BRCA1*. Both proteins have clustered around the same chromosome breaks in the same nucleus. These proteins block mitosis until the DNA breaks have been repaired.

Many mechanisms control cell growth, replication of DNA, and division. Some of them put the machinery of division to rest.

What happens when something goes wrong? For example, if sister chromatids do not separate as they should during mitosis, one daughter cell may end up with too many chromosomes, the other with too few. Chromosomal DNA can be attacked by free radicals or other chemicals, or radiation. Problems like these are frequent and inevitable.

The cell cycle has built-in checkpoints that keep the problems from getting out of control. Certain proteins monitor whether DNA gets fully replicated, whether it is damaged, and even whether nutrient concentrations are sufficient to support cell growth. This surveillance helps cells identify and correct problems.

Checkpoint proteins are the basis of mechanisms that can advance, delay, or block the cell cycle. Some of them make the cycle advance; their absence arrests it. The ones called *growth factors* invite transcription of genes that help the body grow. Other proteins inhibit cell cycle changes. When chromosomal DNA becomes damaged, several checkpoint gene products interrupt mitosis until it is repaired (Figure 7.16). If the DNA remains broken or incomplete, a cascade of signaling events induces cell death.

Sometimes a checkpoint gene mutates so that its protein product no longer functions properly. When all checkpoint mechanisms fail, the cell loses control over its replication cycle. In some cases it gets stuck in mitosis, dividing again and again with no interphase. In other cases, damaged chromosomes are replicated or cells do not die as they are supposed to, because signals calling for cell death are disabled. A growing mass of defective cellular descendants forms a tumor.

Moles and other tumors are **neoplasms**, abnormal masses of cells that lost controls over their cell cycle.

Cervical cancer cell, the kind that killed Henrietta Lacks.

Ordinary skin moles and other *benign*, noncancerous neoplasms grow very slowly, and their cells retain the surface recognition proteins that are supposed to keep cells inside a home tissue. Unless a benign neoplasm grows too large or becomes irritating, it does not pose a threat to the body.

CHARACTERISTICS OF CANCER

Cancers are abnormally growing and dividing cells of a *malignant* neoplasm. The cells disrupt surrounding tissues, both physically and metabolically. Cancer cells are grossly disfigured. They can break loose from their home tissues, slip into and out of blood vessels and lymph vessels, and invade other tissues where they do not belong (Figure 7.17).

All cancer cells display four characteristics. First, they grow and divide abnormally. Controls that limit overcrowding in tissues are lost and cell populations reach high densities. The number of tiny blood vessels that service the growing cell mass increases.

Second, the cytoskeleton and plasma membrane of cancer cells are badly altered. The plasma membrane becomes leaky and has abnormal or lost proteins. The cytoskeleton shrinks, becomes disorganized, or both. Enzyme action can shift, as in an amplified reliance on ATP formation by glycolysis.

Third, cancer cells have a weakened or nonexistent capacity for adhesion. Proteins that function in cell-cell recognition are lost or altered, so cells do not stay put. They break away and colonies of them start forming in distant tissues. This process of abnormal cell migration and tissue invasion is called *metastasis*.

Fourth, cancer cells have lethal effects. Unless they can be eradicated by surgery, chemotherapy, or other procedures, their uncontrollable divisions will put an individual on a painful road to death.

 benign tumor
 malignant tumor

ⓐ Cancer cells break away from their home tissue.

ⓑ The metastasizing cells become attached to the wall of a blood vessel or lymph vessel. They secrete digestive enzymes onto it. Then they cross the wall at the breach.

ⓒ Cancer cells creep or tumble along inside blood vessels, then leave the bloodstream the same way they got in. They start new tumors in different tissues.

Figure 7.17 *Animated!* Comparison of benign and malignant tumors. Benign tumors typically are slow-growing and stay put in their home tissue. Cells of a malignant tumor migrate abnormally through the body and establish colonies even in distant tissues.

Each year in the developed countries alone, 15 to 20 percent of all deaths result from cancer. This is not just a human problem. Cancers are known to occur in most of the animal species studied to date.

Cancer is a multistep process. But researchers have already identified many of the mutated forms of genes that contribute to it. They also are working to identify drugs that specifically target and destroy cancer cells or stop them from dividing.

HeLa cells, for instance, were used in early tests of taxol, the microtubule poison that stops spindles from forming. With this kind of research, we may one day have drugs that can put the brakes on cancer cells.

Built-in checkpoints can advance, delay, or block the cell cycle in response to environmental or internal conditions.

Loss of controls over cell division results in abnormal masses of cells called neoplasms. Malignant neoplasms are cancers; cells in them reach high densities and can metastasize.

Meiosis is central to sexual reproduction. With two nuclear divisions, it halves the chromosome number. The fusion of two gametes at fertilization restores the chromosome number (Figure 7.18).

Section 7.6 Like mitosis, meiosis uses a spindle to move and assort chromosomes. But meiosis occurs only in cells that give rise to gametes or spores. It leads to new combinations of traits in offspring.

*Non*sister chromatids of homologous chromosomes undergo crossing over during prophase I. They break at corresponding sites and exchange segments, so each gets novel combinations of alleles.

During metaphase I, and again at metaphase II, sister chromatids are randomly attached to one spindle pole or the other, so each may end up in any gamete.

Biology⊜Now
Investigate how crossing over and metaphase I alignments affect allele combinations, and learn how gametes form with the animations on BiologyNow.

Section 7.7 Life cycles vary. In animals, haploid eggs and sperm form. In plants, haploid spores, then gamete-producing bodies form. Gametes combine at fertilization, restoring the diploid chromosome number and further shuffling genetic information.

Section 7.8 Cancer is a multistep process involving altered cells that grow and divide abnormally. Such cells have lost their ability to regulate the cell cycle. Cancer cells have a weakened ability to adhere in tissues, and sometimes they migrate to new tissues.

Self-Quiz

Answers in Appendix I

1. Mitosis and cytoplasmic division function in _____ .
   a. asexual reproduction of single-celled eukaryotes
   b. growth, tissue repair, often asexual reproduction
   c. gamete formation in prokaryotes
   d. both a and b

2. A duplicated chromosome has _____ chromatid(s).
   a. one   b. two   c. three   d. four

3. A somatic cell having two of each type of chromosome has a(n) _____ chromosome number.
   a. diploid   b. haploid   c. tetraploid   d. abnormal

4. After mitosis, the chromosome number of a daughter cell is _____ the parent cell's.
   a. the same as   c. rearranged compared to
   b. one-half   d. doubled compared to

5. Only _____ is not a stage of mitosis.
   a. prophase   c. metaphase
   b. interphase   d. anaphase

6. Match each stage of mitosis with the events listed.
   ____ metaphase   a. sister chromatids move apart
   ____ prophase   b. chromosomes start to condense
   ____ telophase   c. daughter nuclei form
   ____ anaphase   d. all duplicated chromosomes are aligned at the spindle equator

7. Sexual reproduction requires _____ .
   a. meiosis   c. gamete formation
   b. fertilization   d. all of the above

8. Generally, a pair of homologous chromosomes _____ .
   a. carry the same genes   c. interact at meiosis
   b. are the same length, shape   d. all of the above

9. Meiosis _____ the parental chromosome number.
   a. doubles   c. maintains
   b. reduces   d. corrupts

10. Meiosis is a division mechanism that produces _____ .
   a. two cells   c. eight cells
   b. two nuclei   d. four nuclei

11. Pairs of duplicated, homologous chromosomes end up at opposite spindle poles during _____ .
   a. prophase I   c. anaphase I
   b. prophase II   d. anaphase II

12. Sister chromatids of each duplicated chromosome end up at opposite spindle poles during _____ .
   a. prophase I   c. anaphase I
   b. prophase II   d. anaphase II

13. Match each term with its description.
   ____ chromosome   a. different molecular forms
       number   of the same gene
   ____ alleles   b. none between meiosis I, II
   ____ metaphase   c. all chromosomes aligned
   ____ interphase   at spindle equator
                       d. all chromosomes in a cell
                          of a given type

Additional questions are available on **Biology⊜Now**™

Critical Thinking

1. Pacific yews (*Taxus brevifolius*) are among the slowest growing trees, which makes them vulnerable to extinction. People started stripping their bark and killing them when they heard that *taxol*, a chemical extracted from the bark, may work against breast and ovarian cancer. It takes bark from about six trees to treat one patient. Do some research to find out why taxol has potential as an anticancer drug and what has been done to protect the trees.

2. X-rays emitted from some radioisotopes damage DNA, especially in cells undergoing DNA replication. Humans exposed to high levels of x-rays face *radiation poisoning*. Hair loss and a damaged gut lining are early symptoms. Speculate why. Also speculate on why radiation exposure is used as a therapy to treat some cancers.

3. Why can we expect meiosis to give rise to genetic differences between parent cells and their daughter cells in fewer generations than mitosis?

4. Aphids reproduce asexually or sexually at different times of year. How might their reproductive flexibility be an adaptation that allows them to avoid predators?

5. The bdelloid rotifer lineage started at least 40 million years ago (Figure 7.19). About 360 known species of these tiny animals are found in many aquatic habitats worldwide. They show tremendous genetic diversity. Why do you think scientists were surprised to discover that all of the bdelloid rotifers are female?

Figure 7.19
Bdelloid rotifer

Menacing Mucus

Cystic fibrosis (CF) is a debilitating, frequently fatal genetic disorder. In 1989, researchers identified the mutated gene that causes it. In 2001, the American College of Obstetricians and Gynecologists suggested that all prospective parents be screened for mutated versions of the gene. The suggestion led to the first mass screening for carriers of a genetic disorder.

The gene encodes a membrane transport protein called CFTR. It helps chloride and water move into and out of cells that secrete mucus or sweat. More than 10 million people in the U.S. inherited one normal and one abnormal copy of the CFTR gene. Some of them suffer from sinus problems, but no other symptoms develop. Most do not know they carry the abnormal gene.

CF develops in anyone who inherits a mutant form of the gene from both parents. Thick, dry mucus clogs bronchial airways to their lungs (shown in the filmstrip) and makes it hard to breathe. The mucus is too thick for the ciliated cells lining the airways to sweep out, and bacteria thrive in it.

Daily routines of posture changes and thumps on the chest and back help clear the mucus, and antibiotics help control infections. Yet even with lung transplants, most patients can expect to succumb to lung failure or airway infections before their thirtieth birthday.

The severity of CF and the prevalence of carriers in the general population persuaded doctors to screen hundreds of thousands of prospective parents for the mutated gene. Many people worried that laboratory errors or confusion over the test results might prompt some parents to abort normal fetuses, but the testing continued. Since 2003, screening for CTFR mutations has been offered to all prospective parents in the U.S.

So here we are, working our way through ethical consequences of understanding human genetics. It started long before DNA, genes, or chromosomes were known, in a small garden, with a monk named Gregor Mendel. By breeding many generations of pea plants in monastery garden plots, Mendel discovered evidence of how parents bestow units of hereditary information upon their offspring.

This chapter starts out with the methods and some representative results of Mendel's experiments. His pioneering work remains a classic example of how a scientific approach can pry open important secrets about the natural world. To this day, it serves as the foundation for modern genetics.

 How Would You Vote? Advances in genetics have led to our ability to detect mutant genes that cause medical disorders in human embryos and fetuses. Should society encourage women to give birth only if their child will not develop severe medical problems? How severe? See BiologyNow for details, then vote online.

Key Concepts

MENDEL'S PATTERNS
Gregor Mendel gathered the first indirect, experimental evidence of the genetic basis of inheritance. His meticulous work tracking traits in many generations of pea plants gave him clues that heritable traits are specified in units. These units are distributed into one gamete or another according to certain patterns. The units were later identified as genes.

BEYOND MENDEL
Not all traits have clearly dominant or recessive forms. One allele of a pair may be fully or partly dominant over its partner, or codominant with it. Two or more gene pairs often influence the same trait, and some single genes influence many traits. Environmental factors can also influence the final form of a trait.

CHROMOSOMES AND INHERITANCE
Human autosomes and sex chromosomes undergo expected gene shufflings by crossing over. Many genetic disorders arise from rare changes in the chromosome number or structure.

Links to Earlier Concepts

Before starting this chapter, review the definitions for genes, alleles, chromosomes, and diploid and haploid (Sections 7.1 and 7.3).

This chapter will show you how crossing over and metaphase I alignments during meiosis (7.5, 7.6) determine patterns of inheritance.

As you read, you may wish to refer back to the introductions to experimental design (1.5, 1.6), natural selection (1.4), protein structure (2.8), and pigments (5.1).

8.1 Tracking Traits With Hybrid Crosses

We turn now to recurring inheritance patterns among humans and other sexually reproducing species.

TERMS USED IN MODERN GENETICS

In Mendel's time, no one knew about genes, meiosis, or chromosomes. As we follow his thinking, we can clarify the picture by substituting some modern terms used in inheritance studies, including those in Figure 8.1.

1. **Genes** are units of DNA information about traits, transmitted from parents to offspring. Each gene has a specific location (locus) on a chromosome.

2. Cells with a diploid chromosome number (2*n*) have pairs of genes, on pairs of homologous chromosomes.

3. Mutation alters a gene's molecular structure. It may cause a trait to change, as when a gene for flower color specifies purple and a mutated form specifies white. Different molecular forms of the same gene are **alleles**.

4. When offspring inherit a pair of *identical* alleles for a trait generation after generation, we say they belong to a true-breeding lineage. Offspring of a cross between two individuals that breed true for different forms of a trait are **hybrids**; each one has inherited *nonidentical* alleles for the trait.

5. A *homozygous* condition occurs when two alleles on homologous chromosomes are identical. A *heterozygous* condition occurs when they are not identical.

6. An allele is *dominant* when its effect on a trait masks that of any *recessive* allele paired with it. We use capital letters to signify dominant alleles and lowercase letters for recessive ones. *A* and *a* are examples.

7. Pulling this all together, a **homozygous dominant** individual has a pair of dominant alleles (*AA*) for the trait. A **homozygous recessive** individual has a pair of recessive alleles (*aa*). A **heterozygous** individual has a pair of nonidentical alleles (*Aa*).

8. *Gene expression* is the process by which information coded in a gene is converted to structural or functional parts of a cell. Expressed genes determine traits.

9. Two terms help keep the distinction clear between genes and the traits they specify. *Genotype* refers to the particular alleles that an individual carries. *Phenotype* refers to an individual's observable traits.

10. F_1 stands for first-generation offspring, and F_2 for second-generation offspring.

a A *pair of homologous chromosomes*, each in the unduplicated state (most often, one from a male parent and its partner from a female parent)

b A *gene locus* (plural, loci), the location for a specific gene on a specific type of chromosome

c A *pair of alleles* (each being a certain molecular form of a gene) at corresponding loci on a pair of homologous chromosomes

d Three *pairs of genes* (at three loci on this pair of homologous chromosomes); same thing as three pairs of alleles

LINKS TO SECTIONS 7.1, 7.3

Figure 8.1 *Animated!* A few genetic terms. Garden pea plants and other species with a diploid chromosome number have pairs of genes, on pairs of homologous chromosomes. Genes come in slightly different molecular forms called alleles. Different alleles specify different versions of the same trait. An allele at any given location on a chromosome may or may not be identical to its partner on the homologous chromosome.

MENDEL'S EXPERIMENTAL APPROACH

Reflect on cystic fibrosis, and a question comes to mind: How did we come to know such amazing things about genes? It all started with Gregor Mendel (Figure 8.2). He guessed that sperm and eggs carry distinct "units" of information about heritable traits. After tracking certain traits of pea plants generation after generation, Mendel found indirect but *observable* evidence of how parents transmit genes to offspring.

Mendel spent most of his adult life in a monastery in a city that is now part of the Czech Republic. Having grown up on a farm, he was familiar with agricultural principles and how to apply them. He kept abreast of the breeding experiments and developments described in the literature. Shortly after entering the monastery, he took courses in mathematics, physics, and botany at the University of Vienna. At the time, very few scholars showed interest in plant breeding *and* mathematics.

Shortly after his university training, Mendel began experiments with the garden pea plant, *Pisum sativum*. This plant is self-fertilizing. Sperm form in the same flower as eggs and can fertilize them. Some pea plants breed true for certain traits. This means that successive generations will be just like the parents in one or more

Figure 8.2 Gregor Mendel (1822–1884). He is considered to be the founder of modern genetics.

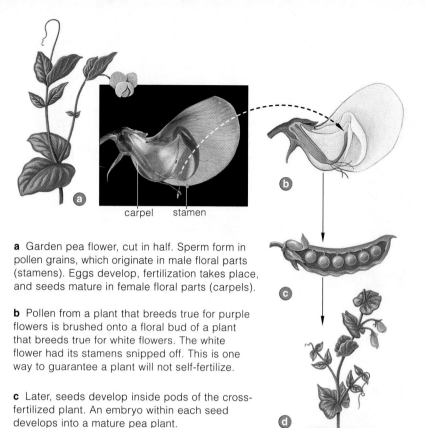

a Garden pea flower, cut in half. Sperm form in pollen grains, which originate in male floral parts (stamens). Eggs develop, fertilization takes place, and seeds mature in female floral parts (carpels).

carpel   stamen

b Pollen from a plant that breeds true for purple flowers is brushed onto a floral bud of a plant that breeds true for white flowers. The white flower had its stamens snipped off. This is one way to guarantee a plant will not self-fertilize.

c Later, seeds develop inside pods of the cross-fertilized plant. An embryo within each seed develops into a mature pea plant.

d Each new plant's flower color is indirect but observable evidence that hereditary material has been transmitted from the parent plants.

Figure 8.3 *Animated!* Garden pea plant (*Pisum sativum*), which can self-fertilize or cross-fertilize. Experimenters can control the transfer of its hereditary material from one flower to another.

homozygous dominant parent

homozygous recessive parent

(chromosomes duplicated before meiosis)

meiosis I

meiosis II

(gametes)   (gametes)

fertilization produces heterozygous offspring

Figure 8.4 One gene of a pair separates from the other gene in a monohybrid cross. Two parents that breed true for two versions of a trait produce only heterozygous offspring.

LINKS TO
SECTIONS
1.5, 1.6, 7.5

traits, as when all plants grown from the seeds of white-flowered parent plants also have white flowers.

Pea plants may be cross-fertilized by transferring pollen from one plant to another. Mendel did this with plants that bred true for *different* versions of the same trait—say, purple or white flowers, as in Figure 8.3. He thought that clearly observable differences would help him track a given trait over many generations. If there were patterns to a trait's inheritance, *then the patterns might tell him something about heredity itself.*

MENDEL'S THEORY OF SEGREGATION

Mendel hypothesized that pea plants inherit two "units" (genes) of information for a trait, one from each parent. He tested his idea with **monohybrid experiments**. In such tests, parents that breed true for two different forms of a trait are bred together, or crossed ($AA \times aa = Aa$). The experiment itself is the self-fertilization of the F_1 heterozygotes, or "monohybrids" ($Aa \times Aa$).

For one set of experiments, he crossed plants that bred true for purple flowers with plants that bred true for white flowers. All F_1 offspring had purple flowers.

Let's express this test cross in modern terms. Pea plant cells have pairs of homologous chromosomes. Assume one parent is homozygous dominant for flower color (AA), and the other parent is homozygous recessive (aa). After meiosis, all gametes of the AA parent have one A allele, and all those of the aa parent have one a allele (Figure 8.4). If any of these sperm fertilize any of these eggs, only one kind of zygote can result: Aa. All F_1 offspring will have purple flowers.

When the F_1 plants were allowed to self-fertilize, most of the F_2 offspring had purple flowers, but some of them had white flowers. After tracking offspring of his experiments with thousands of plants, Mendel was able to calculate that about three out of every four F_2 plants had purple flowers, and one out of every four had white flowers (Figure 8.5).

The ratio hinted that fertilization is a chance event having a number of possible outcomes. Mendel knew about probability, which applies to chance events *and so could help him predict the possible outcomes of genetic crosses.* **Probability** simply means this: The chance that each outcome of an event will occur is proportional to the number of ways in which that outcome can occur.

	Dominant Form	Recessive Form	
FLOWER COLOR	705 purple	224 white	3.15:1
FLOWER POSITION	651 along stem	207 at tip	3.14:1
STEM LENGTH	787 tall	227 dwarf	2.84:1

Average F$_2$ dominant-to-recessive ratio for all of the traits studied: **3:1**

Figure 8.5 Monohybrid experiments with pea plants. Mendel's counts of F$_2$ offspring having dominant or recessive hereditary "units" (alleles). The 3:1 phenotypic ratio held for other traits, including pea pod shape and color.

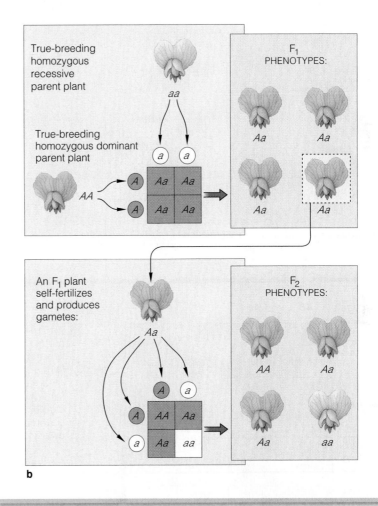

Figure 8.6 *Animated!*
(**a**) Punnett-square method of predicting the probable outcomes of genetic crosses. Circles signify gametes. *Italics* indicate dominant or recessive alleles. The possible genotypes among offspring are written inside the squares.

(**b**) Results of one of Mendel's monohybrid experiments. The ratio of dominant to recessive phenotypes in the F$_2$ (second-generation) plants was 3:1.

A **Punnett-square method**, explained and applied in Figure 8.6, shows the possibilities. If half of a plant's sperm (or eggs) are *a* and half *A*, we can expect four outcomes with each fertilization:

POSSIBLE EVENT	PROBABLE OUTCOME
sperm *A* meets egg *A*	1/4 *AA* offspring
sperm *A* meets egg *a*	1/4 *Aa*
sperm *a* meets egg *A*	1/4 *Aa*
sperm *a* meets egg *a*	1/4 *aa*

Each F$_2$ plant has 3 chances in 4 of inheriting at least one dominant allele (purple flowers). It has 1 chance in 4 of inheriting two recessive alleles (white flowers). That is a probable phenotypic ratio of 3:1.

Mendel's observed ratios were not *exactly* 3:1, but he knew about sampling error. To understand, flip a coin several times. A coin is as likely to end up heads as tails. But often it will end up heads, or tails, several times in a row. If you flip the coin only a few times, the observed ratio of heads to tails might differ a lot from the predicted ratio of 1:1. Flip it many times, and you are far more likely to approach the predicted ratio. Mendel minimized the sampling error in his results by counting many offspring.

Results from his monohybrid experiments became the basis of Mendel's theory of **segregation**, which we state here in modern terms:

MENDEL'S THEORY OF SEGREGATION *Diploid cells have pairs of genes, on pairs of homologous chromosomes. The two genes of each pair are separated from each other during meiosis, so they end up in different gametes.*

③ F₁ OUTCOME: All F₁ plants purple-flowered, tall
(**Aa**B**b** heterozygotes)

AaBb   **AaBb**

meiosis,   meiosis,
gamete formation   gamete formation

	1/4	1/4	1/4	1/4
	AB	**Ab**	**aB**	**ab**
1/4 **AB**	1/16 **AABB**	1/16 **AABb**	1/16 **AaBB**	1/16 **AaBb**
1/4 **Ab**	1/16 **AABb**	1/16 **AAbb**	1/16 **AaBb**	1/16 **Aabb**
1/4 **aB**	1/16 **AaBB**	1/16 **AaBb**	1/16 **aaBB**	1/16 **aaBb**
1/4 **ab**	1/16 **AaBb**	1/16 **Aabb**	1/16 **aaBb**	1/16 **aabb**

■ 9/16 (nine purple-flowered tall)

□ 3/16 (three purple-flowered dwarf)

□ 3/16 (three white-flowered tall)

□ 1/16 (one white-flowered dwarf)

④ Possible F₂ outcomes of cross-fertilization

Figure 8.7 Animated! A dihybrid cross. Here, two traits are tracked: flower color and plant height. A and a stand for dominant and recessive alleles for flower color. B and b stand for dominant and recessive alleles for height. The Punnett square shows all of the possible F₂ combinations.

① **AABB**
purple-flowered, tall parent (homozygous dominant)

AB × **ab**

② **aabb**
white-flowered, dwarf parent (homozygous recessive)

MENDEL'S THEORY OF INDEPENDENT ASSORTMENT

In another set of experiments, Mendel showed how *two* pairs of genes are sorted into gametes. For this he used **dihybrids**, which are heterozygous for two genes (*AaBb*). A **dihybrid experiment** is a cross between two identical dihybrid parents (*AaBb* × *AaBb*).

We can duplicate one of Mendel's dihybrid crosses. First, he bred dihybrids for flower color (alleles *A* or *a*) and for height (*B* or *b*):

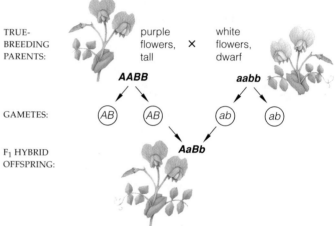

TRUE-BREEDING PARENTS: purple flowers, tall   ×   white flowers, dwarf

AABB   **aabb**

GAMETES: (AB) (AB)   (ab) (ab)

F₁ HYBRID OFFSPRING: **AaBb**

All F₁ offspring from this cross are dihybrids for flower color and height, all purple-flowered and tall (*AaBb*). Mendel then made dihybrid crosses by crossing these F₁ plants. The same ratio of phenotypes appeared in the F₂ offspring of the dihybrid crosses, time after time (Figure 8.7). To understand why Mendel's results were so predictable, think about how gene pairs are sorted into gametes during meiosis.

How two gene pairs (*Aa* and *Bb*) sort into gametes depends partly on whether the two genes are on the same chromosome. Let's assume they are not. Say one pair of homologous chromosomes carries the *A* and *a* alleles and that another pair carries the *B* and *b* alleles.

Now think about how these chromosomes become positioned at the spindle's equator during metaphase I of meiosis (Section 7.5). The chromosome with the *A* allele is positioned to move to either one of the spindle poles. Its homologue (carrying the *a* allele) will move to the opposite spindle pole. The same is true for the chromosomes that carry the *B* and *b* alleles. Following meiosis, there are four possible combinations of alleles in gametes: *AB*, *Ab*, *aB*, and *ab* (Figure 8.8).

Given the random metaphase I alignments, several combinations of alleles can occur at fertilization. Simple multiplication (four sperm types × four egg types) tells us that sixteen combinations of genotypes are possible among the F₂ offspring of a dihybrid cross.

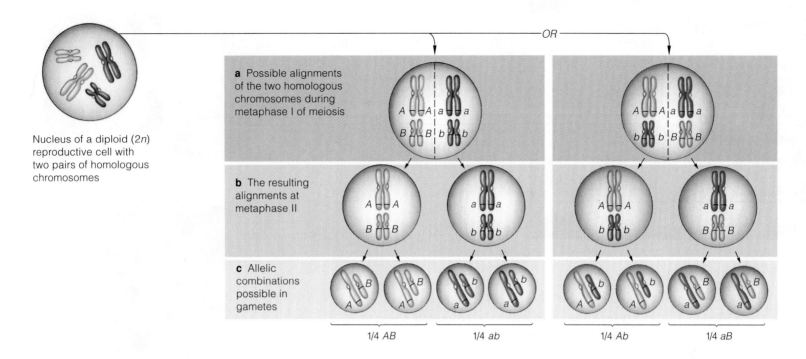

Nucleus of a diploid (2n) reproductive cell with two pairs of homologous chromosomes

OR

a Possible alignments of the two homologous chromosomes during metaphase I of meiosis

b The resulting alignments at metaphase II

c Allelic combinations possible in gametes

1/4 AB   1/4 ab   1/4 Ab   1/4 aB

Figure 8.8 An example of independent assortment during meiosis. Either chromosome of a pair may become attached to either pole of the spindle during meiosis. When just two pairs are tracked, two different metaphase I lineups are possible.

Each of these genotypes will result in a particular phenotype. Nine of the sixteen will be tall and purple-flowered, three will be dwarf purple-flowered, three will be tall white-flowered, and 1 will be dwarf white-flowered. Mendel saw this predicted phenotypic ratio, 9:3:3:1, among the F_2 offspring of his dihybrid crosses.

Mendel could predict the outcome of his dihybrid crosses, but without any knowledge of chromosomes, genes, or meiosis, he could not have understood the underlying reason for their predictability. He did not know that genes for a pea plant's traits are distributed on seven pairs of homologous chromosomes.

Mendel hypothesized, correctly, that the two "units of inheritance" which determine flower color in pea plants were sorting into offspring independently of the two "units" which determine height. Neither he, nor anyone else at the time, knew what those units were.

Mendel's hypothesis is now called the **theory of independent assortment**, stated here in modern terms: During meiosis, the genes on each pair of homologous chromosomes are assorted into gametes independently of how all other pairs are assorted.

Independent assortment and segregation give rise to genetic variation. In a monohybrid cross for one gene pair, three genotypes are possible: *AA*, *Aa*, and *aa*. With more pairs, there are more possible combinations. If the parents differ at *n* genes, each with two alleles, 3^n combinations are possible in offspring. If they differ in twenty genes, that number is nearly 3.5 billion!

The theory does not apply to *all* gene combinations, however, as you will see in the next sections.

MENDEL'S THEORY OF INDEPENDENT ASSORTMENT
In meiosis, genes on pairs on homologous chromosomes become sorted into gametes independently of how gene pairs on other chromosomes are sorted.

8.2 Not-So-Straightforward Phenotypes

Mendel focused on traits that have clearly dominant and recessive forms. However, expression of genes for most traits is not as straightforward.

ABO BLOOD TYPES—A CASE OF CODOMINANCE

In *codominance*, neither of the alleles of a gene pair in heterozygotes is dominant or recessive: both alleles are expressed fully and at the same time. The two parental forms of the trait appear together as a new phenotype.

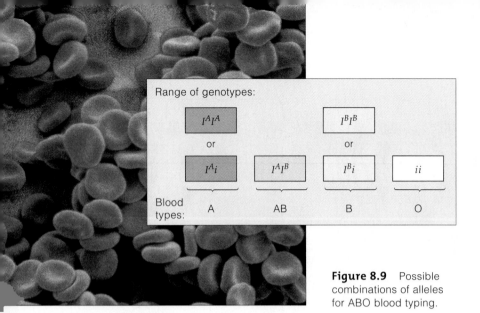

Range of genotypes:

$I^A I^A$ $I^B I^B$

or   or

$I^A i$ $I^A I^B$ $I^B i$ ii

Blood types: A   AB   B   O

Figure 8.9 Possible combinations of alleles for ABO blood typing.

homozygous parent   ×   homozygous parent

All F$_1$ offspring heterozygous for flower color:

Cross two of the F$_1$ plants, and the F$_2$ offspring will show three phenotypes in a 1:2:1 ratio:

Figure 8.10 Incomplete dominance in heterozygous (pink) snapdragons, in which an allele that affects red pigment is paired with a "white" allele.

LINKS TO
SECTIONS
2.6, 3.3, 5.1

For example, one type of glycolipid in the plasma membrane imparts an identity to your red blood cells. This glycolipid occurs in different forms. An analytical method called *ABO blood typing* reveals which of the forms a person has.

An enzyme dictates the glycolipid's final structure. Humans have three alleles that code for this enzyme. Two of them, I^A and I^B, are codominant when paired. The third, *i*, is recessive; a pairing with I^A or I^B masks its effect. Together, they are a **multiple allele system**, defined as the presence of three or more alleles of a single gene among the individuals of a population.

Which alleles do you have? If you have $I^A I^A$ or $I^A i$, your blood is type A. If you have $I^B I^B$ or $I^B i$, it is type B. If you have codominant alleles $I^A I^B$, it is AB—you have both versions of the enzyme. If you have *ii*, your blood cells do not have the glycolipid identifier, and so your blood type is O (Figure 8.9).

Receiving incompatible blood cells in a transfusion is dangerous. The immune system attacks cells with unfamiliar markers, causing them to clump or burst—with potentially lethal consequences. This transfusion reaction is the reason for blood typing. Type O blood is compatible with all other blood types, so people who have it are called universal blood donors. If you have AB blood, you can receive a transfusion of any blood type; you are called a universal recipient.

INCOMPLETE DOMINANCE

In *incomplete* dominance, one allele of a pair is partly dominant over its partner. A heterozygote displays a third phenotype, one that is a blend of the different homozygote phenotypes. For instance, a cross between true-breeding red and white snapdragons will yield *pink*-flowered F$_1$ offspring. Cross two of the F$_1$ plants

and the offspring will have red, white, or *pink* flowers (Figure 8.10). Why pink? Red snapdragon plants have two alleles for red pigment, so their flowers are red. White snapdragon plants also have two alleles for the red pigment, but neither one is functional. Flowers of these plants have no red pigment, so they are white. Heterozygote snapdragon plants have one "red" allele and one "white" allele. They make half the amount of red pigment that the red homozygotes make, so their flowers appear pink.

WHEN PRODUCTS OF TWO OR MORE GENE PAIRS INTERACT

Traits also arise from interactions among products of two or more gene pairs. One gene product may mask or alter the expression of another gene product, and some expected phenotypes may not appear at all. For example, several gene pairs govern the hair color of

	(EB)	(Eb)	(eB)	(ee)
(EB)	EEBB	EEBb	EeBB	EeBb
(Eb)	EEBb	EEBb	EeBB	EeBb
(eB)	EeBB	EeBb	eeBB	eeBb
(ee)	EeBB	EEeb	eeBB	eeBb

Figure 8.11 Coat color among Labrador retrievers, an example of differences in phenotype arising from interactions among gene pairs. Alleles *B* and *b* encode melanin. Alleles *E* and *e* govern how much of the melanin is deposited in hair.

BLACK LABRADOR YELLOW LABRADOR CHOCOLATE LABRADOR

Labrador retrievers. Their hair can be black, yellow, or brown depending on interactions between products of gene pairs. One set of alleles encodes melanin, a dark pigment. Allele *B* (black) has a stronger effect and is dominant to *b* (brown). The alleles of a different gene control how much melanin is deposited in hair. Allele *E* permits deposition. Allele *e* does not, so dogs with two recessive alleles (*ee*) have yellow hair (Figure 8.11).

Alleles at another locus (*C*) may override those two. They encode the first enzyme in a melanin-producing pathway. A *CC* or *Cc* individual makes the functional enzyme, but an individual with two recessive alleles (*cc*) does not. *Albinism*, the absence of melanin, results.

SINGLE GENES WITH A WIDE REACH

Alleles at a single locus may affect two or more traits. **Pleiotropy** occurs when the activity of one gene has multiple phenotypic effects. Cystic fibrosis (box, *right*) and sickle-cell anemia (Section 2.9) are two examples.

Figure 8.12 Flo Hyman, at left, captain of the United States volleyball team that won an Olympic silver medal in 1984. Two years later, during a game in Japan, she slid silently to the floor and died. A dime-sized weak spot in the wall of her aorta had burst. At least two college basketball stars have also died abruptly as a result of Marfan syndrome.

Marfan syndrome is another. This genetic disorder begins with certain mutations in the gene for fibrillin. Fibrillin is one of the proteins in connective tissue, the most abundant vertebrate tissue. Long, thin fibers of fibrillin, loose or cross-linked with the protein elastin, impart elasticity to the connective tissues of the heart, skin, blood vessels, skeleton, tendons, and around the eyes. Normal fibrillin fibers passively recoil after being stretched, as by the beating heart.

A mutated fibrillin gene can result in altered fibrillin. Connective tissues that lack normal fibrillin fibers form throughout the body. The outcome is Marfan syndrome, which affects 1 in 5,000 people throughout the world— men and women of all ethnicities.

Marfan mutations particularly affect the aorta, the largest blood vessel that transports blood away from the heart. Muscle cells in the aorta's thick wall do not function properly, and the aortic wall itself is not as elastic as it should be. Being under pressure, the wall eventually widens, becoming thinner and leaky. Cells infiltrate and multiply in its epithelial lining. Calcium deposits can accumulate inside. Inflamed, weakened, and thinned, the aorta may rupture suddenly during strenuous exercise.

Precisely because of its far-reaching effects, Marfan syndrome can be difficult to diagnose. Many affected people are not aware they have it. Until recent medical advances, it killed most of them before the age of fifty. Olympian Flo Hyman was one (Figure 8.12).

IMPACTS, ISSUES

The mutant gene that causes CF has pleiotropic effects. Lung infections are the greatest threat, but CF also increases the risk of gallstones, such as those shown above. Too much chloride is lost in sweat, shifting the body's salt–water balance and making the heart beat irregularly. Also, mucus thickens. This blocks secretion of digestive enzymes from the pancreas and causes nutritional problems. It also blocks ducts in the male reproductive tract. Because sperm cannot be ejaculated, sterility is the outcome.

In codominance, an allele is expressed independently of its partner on the homologous chromosome. Neither one masks the other.

In incomplete dominance, both alleles affect phenotype.

Two or more genes may affect a single phenotypic trait.

With pleiotropy, alleles at one locus affect multiple traits.

8.3 Complex Variations in Traits

For most populations or species, individuals show rich variation for many of the same traits. This variation arises from gene mutations, gene interactions, and variations in environmental conditions.

REGARDING THE UNEXPECTED PHENOTYPE

As Mendel demonstrated, phenotypic effects of one or more gene pairs often show up in predictable ratios. However, phenotypic effects of genes for some traits are quite difficult to predict.

As an example, *camptodactyly* is a rare abnormality that affects the shape and movement of fingers. Some people who carry the mutant allele for this heritable trait have immobile, bent fingers on both hands. Other people who carry the same allele have immobile, bent fingers only on one hand. Still others are not affected in any obvious way.

What causes this variation? Remember that organic compounds are synthesized in a sequence of metabolic reactions. Different enzymes mediate each of the steps. Any gene that specifies an enzyme can mutate in any number of ways. A mutant gene may result in, say, a defective version of an enzyme that fails to function, or one that does not function as effectively as the normal enzyme. Environmental conditions also can affect the performance of a crucial enzyme. Any of these factors may introduce unpredictable variation in phenotypes.

CONTINUOUS VARIATION IN POPULATIONS

Generally, individuals of each population show a range of small differences in many traits. This characteristic of populations is called **continuous variation**. It is primarily an outcome of the genes that affect a trait, and environmental factors that alter their

Figure 8.13 Part of the range of continuous variation in human eye color. Products of different gene pairs interact in producing and distributing melanin, which helps color the eye's iris. Different combinations of alleles result in small color differences. The frequency distribution for the eye-color trait is continuous over a range from black to light blue.

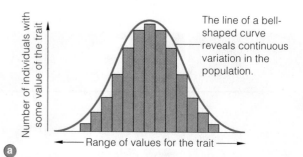

The line of a bell-shaped curve reveals continuous variation in the population.

Number of individuals with some value of the trait

Range of values for the trait

a

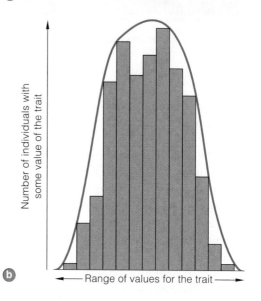

Number of individuals with some value of the trait

Range of values for the trait

b

Figure 8.14 *Animated!* Continuous variation. The number of individuals in each category is graphed against the range of measured values of a trait. (**a**) Bell-shaped curve of continuous variation in an idealized population. (**b**) A real-life example of a bell-curve for continuous variation in a trait, a graph of height distribution among the females at far right (Figure 8.15*b*).

expression. The more genes and environmental factors there are that affect a particular trait, the smaller are the differences in the range of phenotypes that result.

Eye color is one example of continuous variation. It arises from variable amounts of pigment deposited in the doughnut-shaped structure beneath the cornea, the iris. The color of the iris is the cumulative outcome of several gene products, some of which help synthesize or distribute melanin. Dark irises that are almost black have dense deposits of melanin. This pigment absorbs most wavelengths of light, so very little incoming light is reflected out of dark eyes. Melanin deposits are not as extensive in brown eyes. Light brown or hazel eyes have even less melanin (Figure 8.13).

Green, gray, or blue eyes do not have green, gray, or blue pigments. Their iris has some melanin, but not much. Thus, many or most of the wavelengths of light that do enter the eyeball are reflected out.

BEYOND MENDEL

5'3" 5'4" 5'5" 5'6" 5'7" 5'8" 5'9" 5'10" 5'11" 6'0" 6'1" 6'2" 6'3" 6'4" 6'5"

Height (feet/inches)

a

4'11" 5'0" 5'1" 5'2" 5'3" 5'4" 5'5" 5'6" 5'7" 5'8" 5'9" 5'10" 5'11"

Height (feet/inches)

b

Figure 8.15 Continuous variation in body height, one of the traits that help characterize the human population. In these two examples, professors Jon Reiskind and Greg Pryor illustrate the continuous variation in height of their biology students at the University of Florida. They divided all of the students into two groups, (**a**) males and (**b**) females. For both groups, they divided the range of possible heights incrementally, measured the students, and assigned each to the appropriate category.

How can you describe the continuous variation of a trait within a group? You might start by dividing the full range of of phenotypes for the trait—say, height—into measurable categories, such as number of inches. Then, you would count how many individuals in your group fall into each category. Finally, you would graph the data as a bar chart, such as the one in Figure 8.14*a*. This type of graph shows how different phenotypes are distributed in a group.

In Figure 8.14, the shortest bars indicate categories having the fewest individuals. The tallest bar signifies the category with the most individuals. In this case, a graph line skirting the top of all of the bars is a **bell curve**. Such bell-shaped curves are typical of any trait that shows continuous variation. Figure 8.14*b* is a bell curve that was graphed from real-life measurements of the height of many biology students at the University of Florida (Figure 8.15*b*).

ENVIRONMENTAL EFFECTS ON PHENOTYPE

LINKS TO SECTIONS 2.8, 2.9

We have mentioned, in passing, that the environment often contributes to variable gene expression among a population's individuals. Now consider a few cases.

Like a Siamese cat, the Himalayan rabbit has light hair on some of its body, and dark hair elsewhere. The Himalayan rabbit is homozygous for the c^h allele of one of the enzymes involved in melanin production. The c^h allele specifies a form of the enzyme that stops working when the temperature of the skin rises above a certain point.

Above about 33°C (91°F) the cells that give rise to a Himalayan rabbit's hairs cannot make melanin, and the resulting hairs are light. This happens in body regions that are massive enough to conserve a fair amount of metabolic heat. Ears and other slender extremities are cooler because they tend to lose metabolic heat faster.

Figure 8.16 *Animated!* Observable effect of an environmental factor that alters gene expression. A Himalayan rabbit normally has black hair only on its long ears, nose, tail, and legs—the cooler parts of its body. In one experiment, a patch of a rabbit's white fur was removed and an icepack placed over the hairless patch. Where the colder temperature had been maintained, the hairs that grew back were black.

Figure 8.16 shows one experiment demonstrating the effect of temperature on the c^h allele.

We also can identify environmental effects on the genes that govern phenotypes in plants. Consider one experiment using yarrow. This plant can grow from cuttings, so it is a useful experimental organism. Why? All cuttings from the same plant will have the same genotype, so experimenters can discount genes as the basis for the differences that appear among them.

In this classic study, cuttings (clones) from several yarrow plants were transplanted to locations at three different elevations. Researchers periodically observed the growth of the yarow plants in their new habitats. They saw that cuttings from a single plant often grew differently at the different altitudes. For example, two cuttings from one plant grew quite tall at the lowest elevation and the highest elevation, but a third cutting remained short at mid-elevation (Figure 8.17). Though the plants were genetically identical, their phenotypes differed depending on their environment.

Similarly, plant one kind of hydrangea in a garden and it may produce pink blossoms or blue blossoms. Soil acidity affects the function of gene products that dictate the color of hydrangea flowers.

What about humans? One of our genes codes for a transporter protein that moves serotonin across the plasma membrane of brain cells. Serotonin has several important effects, one of which is to counter anxiety and depression when traumatic events challenge us. For a long time, researchers have known that some people handle stress without getting too upset, while others spiral into a deep and lasting depression.

Mutation of the gene for the serotonin transporter compromises responses to stress. It is as if some of us are bicycling through life without an emotional helmet. Only when we take a fall does the phenotypic effect—depression—appear. Other genes also affect emotional states, but mutation of this one reduces our capacity to snap out of it when bad things happen.

a Mature cutting at high elevation (3,060 meters above sea level)

b Mature cutting at mid-elevation (1,400 meters above sea level)

c Mature cutting at low elevation (30 meters above sea level)

Figure 8.17 Experiment demonstrating the impact of environmental conditions on gene expression in yarrow (*Achillea millefolium*). Cuttings from the same parent plant were grown at three different elevations.

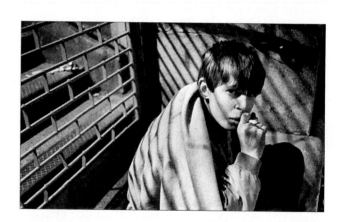

Enzymes mediate each step of most metabolic pathways. Mutations, interactions among genes, and environmental conditions may affect one or more steps. The outcome is variation in phenotypes.

For most traits, individuals of a population or species show continuous variation—a range of small differences.

Variation in traits arises not only from gene mutations and interactions, but also in response to variations in environmental conditions that each individual faces.

8.4 The Chromosomal Basis of Inheritance

You already know about the structure of chromosomes and what happens to them during meiosis. Now start correlating them with human inheritance patterns.

Figure 8.18 *Animated!* Karyotyping, a diagnostic tool that reveals an image of a single cell's diploid complement of chromosomes. This human karyotype shows 22 pairs of autosomes and both pairs of sex chromosomes—XX *or* XY.

A REST STOP ON OUR CONCEPTUAL ROAD

Before driving on into the land of human inheritance, take a few minutes to review this road map. It might give you perspective on key concepts.

A *gene*, again, is a unit of DNA information about a heritable trait. Genes are distributed among a number of chromosomes in eukaryotic cells. A gene has its own normal location, or locus, on one type of chromosome.

A cell with a diploid chromosome number (2*n*) has *pairs of homologous chromosomes*. Each chromosome of a pair is identical to its homologue in length, shape, and gene sequence. The exception is a pair of nonidentical sex chromosomes, such as X and Y.

A pair of genes on homologous chromosomes may or may not be the same, because genes mutate. For any locus, all of the slightly different molecular forms of a gene are called *alleles*.

A *wild-type* allele is the most common form of a gene in a natural population. Any less common form of the gene is a *mutant* allele.

All genes on the same chromosome are physically joined. They can become separated when *crossing over* occurs during prophase I of meiosis. By this process, the nonsister chromatids of homologous chromosomes exchange corresponding segments. Crossing over is a route to new combinations of alleles that did not exist in a parental cell.

Variable alignments of homologous chromosomes at metaphase I of meiosis are the basis of *independent assortment*, which also sorts nonparental combinations of alleles into gametes and offspring.

KARYOTYPING

Sometimes the *structure* of a chromosome can change at mitosis or meiosis. A parental chromosome *number* can change also, with variable consequences. How do we know about these changes? A diagnostic tool called *karyotyping* helps us analyze an individual's diploid complement of chromosomes (Figure 8.18).

With this procedure, a sample of cells taken from an individual is put in a growth medium that stimulates mitotic division. The medium also contains colchicine, a microtubule poison that interferes with dynamics of the mitotic spindle (Section 3.7). Thus, the cells enter

mitosis, but colchicine prevents them from dividing, so their cell cycle becomes arrested at metaphase.

The cells and the medium are transferred to a tube. Then, the cells are separated from the liquid medium with a *centrifuge*. Like a washing machine spin cycle, a centrifuge whirls tubes around a center post. The force of spinning moves the cells away from the center of rotation, and they collect at the bottom of the tube in a clump. The culture medium is removed from the tube, and then a hypotonic solution (Section 4.6) is added. The cells swell up, so the chromosomes inside of them move apart. The cells are spread on a microscope slide and stained so the chromosomes become visible under a microscope (Section 3.2).

The microscope reveals metaphase chromosomes in every cell. A micrograph of one cell is cut apart and reassembled so that the images of all the chromosomes are aligned at the centromeres and arranged according to size and shape, and length. This is the individual's **karyotype**, which is compared with a normal standard for diagnostic purposes.

AUTOSOMES AND SEX CHROMOSOMES

Like many other species, normal humans are either male or female, genetically speaking anyway. Also as in many other species, human body cells are diploid (2*n*), having pairs of homologous chromosomes. With one exception, each chromosome of a pair is identical to its homologue. One member of that nonidentical pair is a unique sex chromosome that occurs either in males or females, but not both.

LINKS TO SECTIONS 3.2, 3.7, 4.6, 7.1, 7.3, 7.5, 7.6

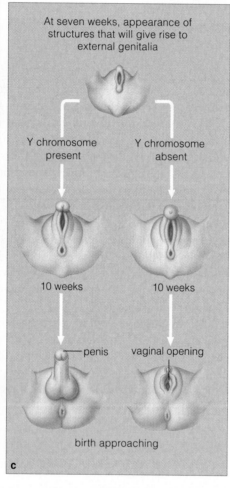

Figure 8.19 *Animated!* (**a**) Pattern of sex determination in humans. (**b**) An early human embryo appears neither male nor female. Then, ducts and other structures that develop into male *or* female reproductive organs form. In an XX embryo, ovaries form *in the absence of the SRY gene on the Y chromosome*. In an XY embryo, the gene product causes testes to form. Testes prompt the development of other male traits. (**c**) External reproductive organs in human embryos.

For instance, a diploid cell in a human female has two X chromosomes (XX). A diploid cell in a human male has one X and one Y chromosome (XY). Human X and Y chromosomes differ. In outward appearance, the considerably shorter Y chromosome appears to be a remnant of the X. The two also differ in which genes they carry. In prophase I, the X and Y chromosomes associate briefly in one small region. That small region is enough to let them interact as a homologous pair of chromosomes during meiosis.

Human X and Y chromosomes fall into the category of **sex chromosomes**, those that determine whether an individual is male or female. All other chromosomes are **autosomes**, which are the same in both sexes.

Each normal egg produced by a human female has one X chromosome. Half the sperm formed in a human male carry an X chromosome, and half carry a Y. If an X-bearing sperm fertilizes an egg, the new individual will become female. However, if the sperm carries a Y chromosome, then the new individual will become a male. The Punnett square in Figure 8.19*a* depicts sex determination in humans.

SRY, one of 330 genes on a human Y chromosome, is the master gene for sex determination in males. *SRY* expression in XY embryos triggers formation of testes, which are primary male reproductive organs. Testes make testosterone, a sex hormone that is responsible for the emergence of male sexual traits. An XX embryo has no *SRY* gene, so the primary female reproductive organs, ovaries, form instead (Figure 8.19*b,c*). Ovaries make estrogens and other sex hormones that cause the development of female sexual traits.

The human X chromosome carries 2,062 genes. Like other chromosomes, it carries some genes associated with sexual traits, such as the distribution of body fat and hair. But most of its genes deal with *nonsexual* traits, such as blood-clotting functions. Such genes can be expressed in males as well as in females. Remember, males also carry one X chromosome.

Diploid cells have pairs of genes, on pairs of homologous chromosomes. At each gene locus, the alleles (alternative forms of a gene) may be either identical or nonidentical.

Abnormal events at meiosis or mitosis can change the structure and number of chromosomes.

Autosomes are pairs of chromosomes that are the same in males and females of a species. One other pair, the sex chromosomes, differ between males and females.

The SRY gene on the human Y chromosome dictates that a new individual will develop into a male. In the absence of the Y chromosome (and the gene), a female develops.

8.5 Impact of Crossing Over on Inheritance

Crossing over between homologous chromosomes is one of the major pattern-busting events in inheritance.

We now know there are many genes on each type of autosome and sex chromosome. All the genes on one chromosome are called a linkage group. For example, the fruit fly (*Drosophila melanogaster*) has four linkage groups corresponding to its four pairs of homologous chromosomes. Indian corn (*Zea mays*) has ten linkage groups, corresponding to its ten pairs. Humans have twenty-three linkage groups, and so on.

If linked genes stayed linked all through meiosis, then there would be no surprising mixes of parental traits. You could expect parental phenotypes among, say, F₂ offspring of dihybrid crosses to show up in a predictable ratio. As early experiments with fruit flies showed, however, that ratio was predictably incorrect for many pairs of genes on the same chromosome. For example, in one experiment, 17 percent of F₂ offspring of dihybrid crosses had a combination of alleles that was found in neither parent.

Plenty of genes on the same chromosome do not stay linked during meiosis, but some do so more often than others. Why? Genes that are close together on the chromosome are separated less frequently by crossing over than genes that are far apart. Think of two genes that are on the same chromosome. *The probability that a crossover will disrupt their linkage is proportional to the distance between them.*

If genes *A* and *B* are twice as far apart as genes *C* and *D*, we would expect crossing over to disrupt the linkage between *A* and *B* much more often:

Two genes are very closely linked when the distance between them is small. Their combinations of alleles nearly always end up inside the same gamete. Linkage is more vulnerable to crossing over when the distance between them is greater (Figure 8.20). When two gene loci are far apart, crossing over is so frequent that the genes sort independently of each other into gametes.

All of the human gene linkages were identified by tracking phenotypes in families over the generations. One thing is clear from these studies: Crossovers are not rare. For most eukaryotes, meiosis cannot even be completed properly until at least one crossover occurs between each pair of homologous chromosomes.

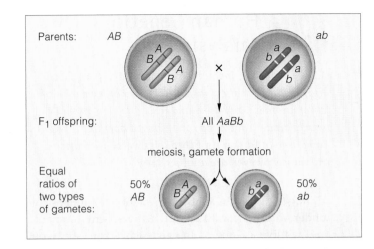

a Full linkage between two genes—no crossing over. Half the gametes have one parental genotype, and half have the other. Genes that are close together on a chromosome frequently stay together in gametes.

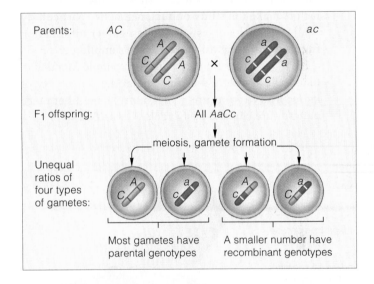

b Incomplete linkage; crossing over affected the outcome. Genes that are far apart on a chromosome are more frequently separated by crossing over than adjacent genes.

Figure 8.20 Crossover events between two gene loci influence the outcome of dihybrid crosses.

All of the genes at different locations along the length of a chromosome belong to the same linkage group. They do not all assort independently at meiosis.

Crossing over between homologous chromosomes disrupts gene linkages and results in nonparental combinations of alleles in chromosomes.

The farther apart two genes are on a chromosome, the greater will be the frequency of crossing over between them.

8.6 Human Genetic Analysis

LINK TO SECTION 1.4

Some organisms, including pea plants and fruit flies, are ideal for genetic analysis. They do not have a lot of chromosomes. They grow and reproduce fast in small spaces, under controlled conditions. It does not take a long time to track a trait through several generations. Humans, however, are another story.

Unlike fruit flies in a lab experiment, we humans live under variable conditions, in different environments, and we live as long as the geneticists who study our traits. Most of us select our own mates and reproduce if and when we want to. Most of our families are not large, which means there may not be enough offspring for geneticists to infer a pattern of inheritance. These factors make the study of human genetics particularly challenging. Geneticists will often gather information from several generations to reduce sampling error.

If a particular human trait has a simple Mendelian inheritance pattern, we can predict the probability of its recurrence in future generations. Some inheritance patterns are clues to past events (Figure 8.21).

Figure 8.21 An intriguing pattern of inheritance. Eight percent of the men in Central Asia carry nearly identical Y chromosomes, which implies descent from a shared ancestor. If so, then 16 million males living between northeastern China and Afghanistan—close to 1 of every 200 men alive today—belong to a lineage that probably started with the warrior and notorious womanizer Genghis Khan. His abundant offspring ruled an empire that stretched from China to Vienna.

Such information is often displayed in **pedigrees**, or charts of genetic connections among individuals. Standardized methods, definitions, and symbols that represent kinds of individuals are used to construct the charts (Figure 8.22*a*). Those who analyze pedigrees rely on their knowledge of probability and Mendelian inheritance patterns, which may yield clues to a trait. For example, as you will see, clues to many disorders point to a dominant or recessive allele, and even to its location on a certain autosome or sex chromosome.

Some traits are deviations from the average. Such a genetic *abnormality* is only a rare or uncommon version of a trait, as when a person is born with six fingers on each hand instead of five (Figure 8.22*b*). Whether this condition is disfiguring or interesting is a matter of opinion. On the other hand, a **genetic disorder** is an inherited condition that, sooner or later, will result in medical problems that can be quite severe. A genetic abnormality or disorder is characterized by a specific set of symptoms called a **syndrome**.

A disease also has a group of symptoms that arises from an abnormal change in how the body functions. However, a **disease** is an illness caused by infectious, dietary, or environmental factors, and does not result from a heritable mutation. It is appropriate to use the term *genetic* disease if such factors modify previously workable genes in a way that disrupts body functions.

Alleles that underlie severe genetic disorders are rare in populations because they put their bearers at

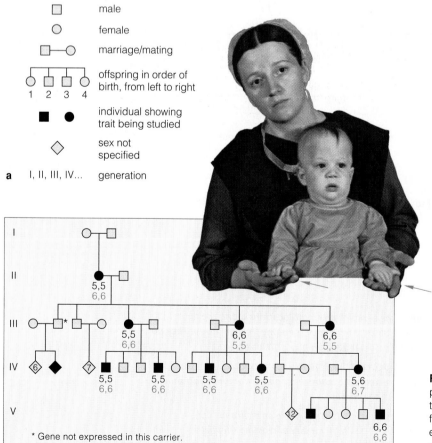

a

□ male

○ female

□─○ marriage/mating

offspring in order of birth, from left to right
1 2 3 4

■ ● individual showing trait being studied

◇ sex not specified

I, II, III, IV... generation

b * Gene not expressed in this carrier.

Figure 8.22 *Animated!* (**a**) Some standardized symbols used in pedigrees. (**b**) A pedigree for *polydactyly*, characterized by extra toes, fingers, or both. *Black* numerals signify the known number of fingers on each hand; *blue* numerals signify the number of toes on each foot. Polydactyly is one of the symptoms of Ellis–van Creveld syndrome (Section 12.7).

risk. They do not completely disappear, because rare mutations reintroduce them. Also, in heterozygotes, a normal allele that is paired with a harmful one might compensate for it. Heterozygotes can transmit mutated alleles to offspring, with effects of the sort described in the next section.

Pedigree analysis can reveal inheritance patterns. From such patterns, geneticists may determine the probability that certain genes will be passed on to children.

A genetic abnormality is a rare or less common version of an inherited trait. A genetic disorder is an inherited condition that results in mild to severe medical problems.

Figure 8.23 Pedigree for *Huntington disease*, a progressive degeneration of the nervous system. Researcher Nancy Wexler and her team constructed this extended family tree for nearly 10,000 Venezuelans. Their analysis of unaffected and affected individuals revealed that a dominant allele on human chromosome 4 is the culprit. Wexler has a special interest in the disease; it runs in her family.

8.7 Examples of Human Inheritance Patterns

Some human phenotypes arise from dominant or recessive alleles on an autosome or X chromosome, inherited in simple Mendelian patterns. Others arise from changes in chromosome structure or number.

AUTOSOMAL DOMINANT INHERITANCE

Two clues point to an autosomal dominant allele for a trait. First, the trait typically appears each generation; the allele is usually expressed even in heterozygotes. Second, if one parent is heterozygous and the other is homozygous recessive, any child of theirs will have a 50 percent chance of being heterozygous.

A few dominant alleles persist in populations even when they cause severe genetic disorders. Spontaneous mutations reintroduce some. For others, expression of a dominant allele may not interfere with reproduction, or people reproduce before symptoms are severe.

For instance, with *Huntington disease*, the nervous system deteriorates, muscle movements become more and more involuntary, and death follows (Figure 8.23). Symptoms may not start until after age thirty. Many affected individuals have already reproduced by then, which puts children at risk of inheriting the disorder. Affected people usually die in their forties or fifties.

The mutation that causes this disorder changes one of the proteins necessary for normal development of brain cells. It is one of the *expansion* mutations, which occurs as multiple repeats of the same DNA segment. The repeats disrupt gene function.

Achondroplasia is an autosomal dominant disorder that affects about 1 in 10,000 people. Commonly, the

Figure 8.24 *Animated!* Autosomal dominance. (**a**) Three males affected by achondroplasia. Verne Troyer (center) stands only two feet, eight inches tall. (**b**) One pattern for autosomal dominant inheritance. A dominant allele (*red*) is fully expressed in the carriers.

homozygous dominant condition results in stillbirth, yet heterozygotes can still reproduce. Cartilage parts of the skeleton form abnormally in achondroplasiacs. Adults have abnormally short arms and legs relative to other body parts. They are less than 4 feet, 4 inches tall. Figure 8.24 shows three such adults.

Figure 8.32 Down syndrome. (**a**) About 80 percent of trisomy 21 babies are born to mothers who are not yet 35 years old. These women are in the age categories with the highest fertility rates, and they simply have more babies than older women. The risk of having a trisomic 21 baby actually rises dramatically with the mother's age. (**b**) A boy with trisomy 21 and his brothers.

a

b

shortened body parts, loose joints, and poorly aligned bones of the hips, fingers, and toes. Skeletal muscles and reflexes are weaker than normal. Their speech and other motor skills develop slowly. With medical care, they live fifty-five years, on average.

The incidence of nondisjunction rises as mothers become older (Figure 8.32a). It may originate with the father, but less often. Trisomy 21 is one of hundreds of conditions that are detectable with prenatal diagnosis (Section 8.10). With early special training and medical intervention, affected individuals still can take part in normal activities. As a group, they tend to be cheerful and sociable people (Figure 8.32b).

CHANGES IN THE SEX CHROMOSOME NUMBER

Nondisjunction also causes alterations in the number of X and Y chromosomes, with a frequency of about 1 in 400 live births. Most often they lead to difficulties in learning and motor skills, such as a speech delay, but problems may be so subtle that the underlying cause is never diagnosed.

FEMALE SEX CHROMOSOME ABNORMALITIES *Turner syndrome* arises by inheritance of an X chromosome and no corresponding X or Y chromosome. On average, 1 in 2,500 to 10,000 newborn girls have this condition. Nondisjunction originating with the father accounts for 75 percent of Turner syndrome cases. Compared to other sex chromosome abnormalities, cases are few. Why? Probably at least 98 percent of XO zygotes abort spontaneously early in pregnancy.

Despite the near lethality, XO survivors are not as disadvantaged as the other aneuploids. They grow up well proportioned but not tall—four feet, eight inches, on average (Figure 8.33). Most do not have functional

Figure 8.33 One young girl with Turner syndrome.

ovaries and so cannot produce enough sex hormones to mature sexually. Secondary sexual traits, including breast enlargement, are affected. Any eggs that form in the ovaries disintegrate before two years of age.

Another example: A few females inherit three, four, or five X chromosomes. This *XXX condition* occurs at a frequency of about 1 in 1,000 live births. Most adults are an inch or so taller and more slender than average. They are fertile. Except for slight learning difficulties, most fall within the normal range of social behavior.

MALE SEX CHROMOSOME ABNORMALITIES One in 500 to 2,000 males has inherited one Y and two or more X chromosomes, mainly by nondisjunction. They have an XXY or, rarely, XXXY, XXXXY, or XY/XXY mosaic genotype. (Mosaic individuals have different numbers of chromosomes in different populations of their body cells.) About 67 percent of affected people inherited the extra chromosome from their mother.

The resulting *Klinefelter syndrome* develops after the onset of puberty. XXY males tend to be overweight and tall. The testes and the prostate gland usually are smaller than average. Many XXY males are within the normal range of intelligence. Some have short-term memory loss and other learning disabilities. Affected individuals make less testosterone and more estrogen than normal, with feminizing outcomes. Their sperm counts are low and hair is sparse, the voice is pitched high, and the breasts are a bit enlarged. Testosterone injections at puberty can reverse the feminized traits.

About 1 in 500 to 1,000 males has one X and two Y chromosomes. Those with this *XYY condition* tend to be taller than average, with mild mental impairment, but most are otherwise normal. At one time, XYY males were thought to be genetically predisposed to a life of crime. This misguided opinion originated with a sampling error—a study of too few cases in narrowly selected groups, including prison inmates—and flaws in experimental design (the same researchers gathered karyotypes *and* personal histories). A false report that a murderer of young nurses was XYY did not help.

In 1976 a Danish geneticist reported on a study of 4,139 tall males, twenty-six years old, who reported to their draft board. Besides giving results of physical examinations and intelligence testing, the records held clues to social and economic status, education, and any criminal convictions. Twelve males were XYY, which meant there were more than 4,000 males in the control group. The only finding was that mentally impaired, tall males who engage in criminal activity are likely to get caught—irrespective of karyotype.

The majority of XXY, XXX, and XYY children may not even be properly diagnosed. Some are dismissed unfairly as being underachievers.

> *Nondisjunction in germ cells, gametes, or early embryonic cells changes the number of autosomes or number of sex chromosomes. The change affects development and the resulting phenotypes.*
>
> *Nondisjunction at meiosis causes most sex chromosome abnormalities, which typically lead to subtle difficulties with learning, and with speech and other motor skills.*

8.10 Some Prospects in Human Genetics

With the first news of pregnancy, parents-to-be typically wonder if their baby will be normal. Quite naturally, they want their baby to be free of genetic disorders, and most babies are. What are the options when they are not?

BIOETHICAL QUESTIONS

It is a fact that we humans do not approach heritable disorders and diseases the same way. We will attack diseases with antibiotics, surgery, and other weapons. But how do we attack a heritable "enemy" that can be transmitted to offspring? Should we institute regional, national, or global programs to identify people who may carry harmful alleles? Do we tell them they are "defective" and might bestow some disorder on their children? Who decides which alleles are bad? Should society bear the cost of treating all genetic disorders before and after birth? If so, should society have a say in whether an affected embryo will be born at all, or aborted? An **abortion** is the expulsion of a pre-term embryo or fetus from the uterus. As you most likely

Figure 8.34 Amniocentesis, a prenatal diagnostic tool. A pregnant woman's doctor holds an ultrasound emitter against her abdomen while drawing a sample of amniotic fluid into a syringe. He monitors the path of the needle with an ultrasound screen, in the background. Then he directs the needle into the amniotic sac that holds the developing fetus and withdraws twenty milliliters or so of amniotic fluid. The fluid holds fetal cells and wastes that can be analyzed for genetic disorders.

have learned by now, such questions are just the tip of an ethical iceberg.

CHOICES AVAILABLE

GENETIC COUNSELING *Genetic counseling* starts with the diagnosis of parental genotypes, pedigrees, and genetic testing for known metabolic disorders. Using information gained from these tests, genetic counselors can predict the risk for disorders in future children.

If a previous child or a close relative has a genetic disorder, prospective parents may seek counseling. If one of them is a carrier for the disorder, the counselor will advise them of the chances that a child of theirs might inherit the disorder.

Even if neither prospective parent is a carrier for any genetic disorder, there is a 3 percent risk that any child will be born with a birth defect. The risk increases with the woman's age at the time of pregnancy.

PRENATAL DIAGNOSIS A genetic counselor describes available diagnostic procedures to prospective parents. Methods of *prenatal diagnosis* are used to determine whether an embryo or fetus has a genetic abnormality. (*Prenatal* means before birth. *Embryo* is the stage of prenatal human development until eight weeks after fertilization. After eight weeks, *fetus* is the correct term.)

Suppose a forty-five-year-old woman gets pregnant and worries about Down syndrome. She might choose to have *amniocentesis,* usually performed eight to twelve weeks after conception (Figure 8.34). A tiny sample of fluid inside the amnion, a membranous sac enclosing the fetus, is withdrawn. Some of the cells that the fetus has shed are suspended in the sample. These cells are analyzed for many severe genetic disorders, such as

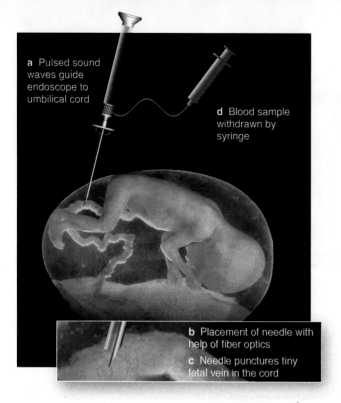

a Pulsed sound waves guide endoscope to umbilical cord

d Blood sample withdrawn by syringe

b Placement of needle with help of fiber optics

c Needle punctures tiny fetal vein in the cord

Figure 8.35
Fetoscopy for prenatal diagnosis.

Figure 8.36 Eight-cell and multicelled stages of human development.

sickle-cell anemia, Down syndrome, Turner syndrome, Kleinefelter syndrome, hemophilia, cystic fibrosis, and Duchenne muscular dystrophy.

Chorionic villi sampling (CVS) is another procedure. A clinician withdraws a few cells from the chorion, a fluid-filled, membranous sac that encloses the amnion. CVS can be performed earlier than amniocentesis can, as early as the eighth week of pregnancy.

Directly visualizing the developing fetus is possible with *fetoscopy*. An endoscope, a fiber-optic device, uses pulsed sound waves to scan the uterus and locate parts of the fetus, umbilical cord, or placenta (Figure 8.35). Fetoscopy also is used to diagnose blood cell disorders, such as sickle-cell anemia and hemophilia.

All three procedures may puncture a fetus or cause infection. If a punctured amnion does not reseal itself quickly, too much fluid leaks out, and this endangers the fetus. Amniocentesis raises the risk of miscarriage by 1 to 2 percent. After CVS, 0.3 percent of newborns have missing or deformed fingers and toes. Fetoscopy raises the risk of a miscarriage by 2 to 10 percent.

PREIMPLANTATION DIAGNOSIS This procedure relies on **in-vitro fertilization**. Sperm and eggs taken from prospective parents are mixed in a petri dish. One (or more) eggs may become fertilized. If so, mitotic cell divisions form a ball of eight cells within forty-eight hours (Figure 8.36).

All the cells in the ball have the same genes, and each is not yet committed to being specialized. Doctors carefully remove one of the cells and analyze its genes.

If the cell has no detectable genetic defects, the ball is inserted into the mother's uterus, where it may develop into an embryo. Couples who are at risk of passing on cystic fibrosis, muscular dystrophy, and other disorders often have opted for the procedure. Many *"test-tube" babies* have been born in good health.

PHENOTYPIC TREATMENTS The medical symptoms of some genetic disorders may be treated by controlling the diet, adjusting environmental factors, surgery, or hormone replacement therapy.

Dietary control works in *phenylketonuria*, or PKU. In this case, a gene specifies an enzyme that converts the amino acid phenylalanine to tyrosine. Someone who is homozygous recessive for a mutant form of the gene does not make this enzyme. Phenylalanine accumulates, and becomes diverted into other metabolic pathways. The result can be impaired brain functioning. Affected people can lead normal lives by restricting their dietary phenylalanine intake. For example, they can avoid soft drinks and other food products that contain aspartame, a sweetener that contains phenylalanine.

REGARDING ABORTION What happens when prenatal diagnosis reveals a severe problem? Do the prospective parents opt for an induced abortion? We can only say here that they must weigh awareness of the severity of the disorder against their ethical and religious beliefs. Worse, they must play out their personal tragedy on a larger stage dominated by a nationwide battle between highly vocal "pro-life" and "pro-choice" factions.

Summary

Section 8.1   Genes are heritable units of information about traits. Each gene has its own locus on one type of chromosome. Different forms of the same gene are called alleles. Diploid cells have two copies of every gene, usually one inherited from each of two parents, on homologous chromosomes.

An individual with two dominant alleles for a trait (*AA*) is homozygous dominant. A homozygous recessive has two recessive alleles (*aa*). A heterozygote has two nonidentical alleles (*Aa*) for a trait, and the dominant one may mask effects of the partner allele.

Mendel's experiments with pea plants revealed patterns of inheritance. Monohybrid experiments led to the theory of segregation, as stated here in modern terms: Pairs of genes on homologous chromosomes separate during meiosis and end up in different gametes.

Dihybrid experiments led to a theory of independent assortment, as stated here in modern terms: Pairs of genes on homologous chromosomes sort into gametes independently of those on different chromosomes. The random alignment of homologous chromosomes during metaphase I of meiosis is the basis of this outcome.

Biology⑥Now
Learn how Mendel crossed garden pea plants, and definitions of important genetic terms. Carry out monohybrid crosses and observe the results of a dihybrid cross with interactions on BiologyNow.

Section 8.2   Some alleles are not fully dominant or are codominant. Products of gene pairs often interact to influence the same trait. A single gene may have effects on two or more traits. Products of two or more pairs of genes may interact to affect a single trait.

Biology⑥Now
Explore patterns of non-Mendelian inheritance with interactions on BiologyNow.

Section 8.3   Many gene interactions give rise to continuous variation for a trait among individuals of a population. Environmental factors influence gene expression and may contribute to that variation.

Biology⑥Now
Plot the continuous distribution of height for a class with the interaction on BiologyNow.

Section 8.4   Human somatic cells are diploid (2*n*), with twenty-three pairs of homologous chromosomes. One pair are sex chromosomes—XX in females, XY in males. All other chromosomes are autosomes, which are the same in both sexes.

Biology⑥Now
See how sex is determined in humans with the interaction on BiologyNow.

Section 8.5   Genes on the same chromosome do not all assort independently during meiosis. Crossing over results in new (nonparental) combinations of alleles in

chromosomes. The farther apart two genes are on the same chromosome, the more frequently they will undergo crossing over.

Section 8.6   Geneticists often use pedigrees, or charts of genetic connections in a lineage over time, to estimate probabilities that offspring will inherit a given trait.

Section 8.7   Dominant or recessive alleles on either an autosome or X chromosome can be tracked when they are inherited in simple Mendelian patterns.

Biology⑥Now
Investigate autosomal and X-linked inheritance with the interactions on BiologyNow.

Section 8.8   On rare occasions, a chromosome's structure can change. Part of it may be duplicated, deleted, inverted, or translocated (moved to a new location on a different chromosome). Most alterations are harmful or lethal.

Section 8.9   Rarely, chromosome number can change, as by nondisjunction during meiosis. Aneuploids have one extra or one less chromosome. About half of all flowering plants, some insects, fish, and other animals are polyploid (with three or more of each chromosome).

Section 8.10   Genetic counselors review the options available for potential parents at risk of having children with a genetic disorder.

Biology⑥Now
Examine a human pedigree and also observe amniocentesis with animation on BiologyNow.

Self-Quiz   *Answers in Appendix I*

1. Alleles are _____ .
   a. different molecular forms of a gene
   b. different phenotypes
   c. self-fertilizing, true-breeding homozygotes

2. A heterozygote has a _____ for the trait being studied.
   a. pair of identical alleles
   b. pair of nonidentical alleles
   c. haploid condition, in genetic terms
   d. a and c

3. Crosses between F_1 pea plants resulting from the cross *AABB* × *aabb* lead to F_2 phenotypic ratios close to _____ .
   a. 1:2:1   b. 3:1   c. 1:1:1:1   d. 9:3:3:1

4. The probability of a crossover occurring between two genes on the same chromosome is _____ .
   a. unrelated to the distance between them
   b. increased if they are close together
   c. increased if they are far apart

5. Chromosome structure can be altered by a _____ .
   a. deletion   c. inversion   e. all of the above
   b. duplication   d. translocation

6. Genes on the same chromosome are said to be _____ .
   a. linked   c. homologous   e. all of the above
   b. alleles   d. autosomes

7. Most genes for human traits are located on _____ .
   a. the X chromosome
   b. the Y chromosome
   c. autosomes

8. Nondisjunction at meiosis can result in _____ .
   a. an inversion   c. a translocation
   b. crossing over   d. aneuploidy

9. Turner syndrome (XO) is an example of _____ .
   a. pleiotropy   c. aneuploidy
   b. polyploidy   d. gene linkage

10. Match each example with the suitable description.
   ____ dihybrid cross   a. *bb*
   ____ monohybrid cross   b. *AaBb × AaBb*
   ____ homozygous condition   c. *Aa*
   ____ heterozygous condition   d. *Aa × Aa*

11. Match the chromosome terms appropriately.
   ____ crossing over   a. number and defining
   ____ deletion   features of an individual's
   ____ nondisjunction   metaphase chromosomes
   ____ translocation   b. segment of a chromosome
   ____ karyotype   moves to a nonhomologous
   ____ linkage group   chromosome
                         c. disrupts gene linkages
                         d. can result in gametes with
                            wrong chromosome number
                         e. a chromosome segment lost
                         f. all genes on a chromosome

Additional questions are available on Biology Ⓔ Now™

Figure 8.39 Two albino organisms.

Genetics Problems   *Answers in Appendix II*

1. A certain recessive allele *c* is responsible for *albinism*, an inability to produce or deposit melanin, a brownish-black pigment, in body tissues. Humans and a number of other organisms can have this phenotype. Figure 8.39 shows two stunning examples. In each of the following cases of albinism, what are the possible genotypes of the father, the mother, and their children?
   a. Both parents have normal phenotypes; some of their children are albino and others are unaffected.
   b. Both parents are albino and have albino children.
   c. The woman is unaffected, the man is albino, and they have one albino child and three unaffected children.

2. One gene has alleles *A* and *a*. Another has alleles *B* and *b*. For each genotype, what type(s) of gametes will form? Assume that independent assortment occurs.
   a. *AABB*   c. *Aabb*
   b. *AaBB*   d. *AaBb*

3. Refer to Problem 2. What will be the genotypes of offspring from the following matings? Indicate the frequencies of each genotype among them.
   a. *AABB × aaBB*   c. *AaBb × aabb*
   b. *AaBB × AABb*   d. *AaBb × AaBb*

4. Certain dominant alleles are so essential for normal development that an individual who is homozygous recessive for a mutant recessive form cannot survive. Such recessive, *lethal alleles* can be perpetuated in the population by heterozygotes.

Figure 8.40 Manx cat, a breed that has no tail.

Consider the Manx allele (M^L) in cats. Homozygous cats (M^LM^L) die when they are still embryos inside the mother cat. In heterozygotes (M^LM), the spine develops abnormally. The cats end up with no tail (Figure 8.40).

Two M^LM cats mate. What is the probability that any one of their *surviving* kittens will be heterozygous?

5. Human females are XX and males are XY.
   a. Does a male inherit the X from his mother or father?
   b. With respect to X-linked alleles, how many different types of gametes can a male produce?
   c. If a female is homozygous for an X-linked allele, how many types of gametes can she produce with respect to that allele?
   d. If a female is heterozygous for an X-linked allele, how many types of gametes might she produce with respect to that allele?

6. A mutant allele that causes Marfan syndrome follows a pattern of autosomal dominant inheritance. What is the chance that any child will inherit it if one parent does not carry the allele and the other is heterozygous for it?

7. A dominant allele *W* confers black fur on guinea pigs. A guinea pig that is homozygous recessive (*ww*) has white fur. Fred would like to know whether his pet black-furred guinea pig is homozygous dominant (*WW*) or heterozygous (*Ww*). How might he determine his pet's genotype?

8. The somatic cells of individuals with Down syndrome usually have an extra chromosome 21; they contain forty-seven chromosomes.
   a. At which stages of meiosis I and II could a mistake alter the chromosome number?
   b. A few individuals have forty-six chromosomes, including two normal-appearing chromosomes 21 and a longer-than-normal chromosome 14. Explain how this chromosome abnormality may have arisen.

9. In the human population, mutation of two different genes on the X chromosome causes two types of X-linked hemophilia (types A and B). In a few known cases, a woman is heterozygous for both mutant alleles (one on each of her two X chromosomes). All of her sons should have either hemophilia A or B.

   However, on very rare occasions, such a woman gives birth to a son who does not have hemophilia, and his one X chromosome does not have either mutant allele. Explain how such an X chromosome could arise.

9 DNA STRUCTURE AND FUNCTION

Here, Kitty, Kitty, Kitty, Kitty, Kitty

In 1997, geneticist Ian Wilmut made headlines when he created an exact genetic copy—a **clone**—of a fully grown animal. His team used a technique called nuclear transfer. They slipped a nucleus from a sheep udder cell into an unfertilized egg that had its own nucleus gently sucked out beforehand. The hybrid egg developed into an embryo, then into a lamb, which they named Dolly.

Dolly was the first of a series of cloned animals. Researchers all over the world since have cloned cows, horses, rabbits, pigs, mules, mice, and cats, but the process remains far from routine. Not many clonings end successfully. Most nuclear transfer attempts fail to produce an embryo, and embryos that do form tend to die before birth or shortly after. Researchers made 692 nuclear transfer attempts to get one cloned guar, an endangered wild ox. The calf promptly died.

Nuclear transfer clones that survive are prone to health or development problems. Many age quickly or become unusually overweight as they age. By age five, Dolly was just like a ten-year-old sheep, fat and arthritic. Other clones are too large from birth or have enlarged organs. Cloned mice develop lung and liver problems, and almost all die prematurely. Cloned pigs have heart problems, they limp, and one never did develop a tail or, worse still, an anus.

Structural and functional problems occur in animal clones because most of the DNA in an adult cell nucleus is inactive. It has to be reprogrammed to function like the DNA of an egg, switched back on. This sounds easier than it is; most animal cloning attempts still rely heavily on trial-and-error methods.

Why do scientists keep trying to clone animals? The potential benefits are enormous. Animals with certain desirable traits could be mass-produced. Replacement cells or organs may soon cure people with degenerative diseases that are now incurable. Endangered animals may be saved from extinction; extinct animals may be brought back to life. Commercial pet cloning may be the most financially viable practical application of cloning technology yet. Several companies now are banking pet DNA for future cloning; some are making cloned cats, like the kitten in the filmstrip, for paying pet owners.

Controversies over animal cloning continue to rage. Perfecting the methods to make healthy animal clones brings us closer to cloning humans, both technically and ethically. Think about it. If cloning a cat for a grieving pet owner is acceptable, would it be also acceptable to clone a lost child for a grieving parent?

Think about these issues as you read through the rest of the chapters in this unit of the book. You will learn how cells replicate and repair their DNA, how DNA is converted to physical parts of a cell, and how researchers use these events to make clones.

 How Would You Vote? *Sickly or deformed clones are considered unfortunate but acceptable casualties of animal cloning research, but the technique also may soon provide benefits to human patients. Should animal cloning be banned? See BiologyNow for details, then vote online.*

Key Concepts

THE DNA CHASE
DNA carries instructions for making all of the structural and functional materials that cells—and bodies—use in order to function. The simplicity of its structure enables DNA to encode virtually unlimited amounts of this genetic information.

HOW DNA IS REPLICATED
Before a cell divides, enzymes and other proteins make a copy of, or replicate, the cell's DNA. During replication, double-stranded DNA molecules unwind. Each strand serves as a guide for assembling a new strand.

CLONING
What does it mean when we say that DNA holds heritable information? The answer hits home when the messages encoded in its base sequences are tapped to make an exact copy of an adult animal.

Links to Earlier Concepts

This chapter builds on your understanding of hydrogen bonding (Section 2.2), condensation reactions (2.5) and DNA structure and function (2.10). Reminding yourself about the alpha helix structure of proteins (2.8) will help you visualize the helical structure of DNA.

What you learned about chromosomes, mitosis, and meiosis (7.1, 7.3, 7.5) will let you put the molecular basis of DNA replication into a larger perspective.

Your knowledge of the cell cycle and controls over it (7.2, 7.8) will help you understand some of the technical difficulties inherent in cloning procedures.

LINK TO
SECTION
2.10

9.1 The Hunt for Fame, Fortune, and DNA

With this chapter, we turn to investigations that led to our understanding of DNA. That story is more than a march through details of its structure and function. It also reveals the scientific discovery process.

EARLY AND PUZZLING CLUES

In the 1800s, Johann Miescher, a physician, wanted to identify the composition of the nucleus. He collected white blood cells from pus-filled bandages and sperm from fish. These cells do not contain much cytoplasm, so it is easy to isolate the contents of their nuclei. He described the substance he found as large and acidic, containing nitrogen and phosphorus. This substance would later be called **deoxyribonucleic acid**, or **DNA**.

Now fast-forward to 1928. An army medical officer, Frederick Griffith, was trying to make a vaccine for pneumonia. He did not succeed, but he did isolate two forms of *Streptococcus pneumoniae*, a bacterium that can cause pneumonia. Colonies of one form had a rough appearance; colonies of the other were smooth. Griffith called the forms *R* and *S*. He inadvertently contributed to our understanding of heredity by using the different forms in four experiments (Figure 9.1).

First, he injected mice with live *R* cells. The mice did not develop pneumonia. *The R form was harmless.*

Second, he injected other mice with live *S* cells. The mice died. Blood samples from them teemed with live *S* cells. *The S form was pathogenic; it caused the disease.*

Third, he killed *S* cells by exposing them to high temperature. *Mice injected with dead S cells did not die.*

Fourth, he mixed live *R* cells with heat-killed *S* cells and injected them into mice. The mice died—and blood samples drawn from them teemed with live S cells!

What went on in the fourth experiment? Maybe heat-killed *S* cells in the mix were not really dead. But if that were so, then mice injected with heat-killed *S* cells alone (in experiment 3) would have died. Maybe the harmless *R* cells had mutated into a killer form. But if that were so, then mice injected with the *R* cells alone (experiment 1) would have died.

The simplest explanation was that heat killed the *S* cells, but it did not destroy their hereditary material—including whatever part of it that specified "infectious." Somehow, that material had been transferred from the dead *S* cells to living *R* cells, which put it to use.

The transformation was permanent and heritable. Even after a few hundred generations, the transformed *R* cell descendants were still infectious.

What *was* the hereditary material that caused the transformation? Scientists were looking in earnest, but most were thinking "proteins." Why? The molecules of inheritance had to be just as diverse as the information they encode. Virtually unlimited mixes of twenty kinds of amino acids are possible in proteins. By comparison, other molecules seemed too simple.

Griffith's results intrigued Oswald Avery, who in 1916 began transforming the *R* strain with fluid from *S* strain cultures. Something in that fluid was hereditary material. *R* cells were still transformed by *S* cell fluid after Avery added protein-digesting enzymes to it, so protein could not be the hereditary material. However, *R* cells were *not* transformed by *S* cell fluid to which Avery had added a DNA-digesting enzyme.

DNA was looking good.

CONFIRMATION OF DNA FUNCTION

By the 1950s, researchers trying to identify hereditary material were experimenting with viruses that infect bacteria, called **bacteriophages**. Like all other viruses, these infectious particles carry hereditary information about how to build new virus particles. After viruses infect a host cell, the cell starts making virus particles instead of its own components.

As researchers knew, some bacteriophages consist only of DNA and protein. Micrographs had revealed that the bacteriophage's protein coat stays on the *outer surface* of infected cells. The viruses had to be injecting genetic material into host cells, but was that material DNA, protein, or both? Figure 9.2 describes two of the landmark experiments that proved DNA is the carrier of heritable material, not protein.

Then, in 1951, a scientist named Linus Pauling and his colleagues deduced the structure of the alpha helix, a coiled motif that is part of the secondary structure of many proteins (Section 2.8). People suddenly realized that proteins had defined, organized structures.

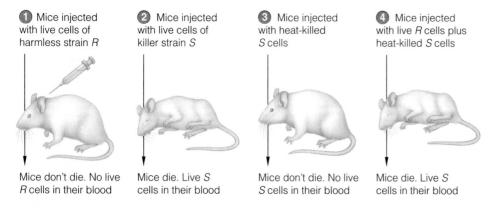

1 Mice injected with live cells of harmless strain *R*

Mice don't die. No live *R* cells in their blood

2 Mice injected with live cells of killer strain *S*

Mice die. Live *S* cells in their blood

3 Mice injected with heat-killed *S* cells

Mice don't die. No live *S* cells in their blood

4 Mice injected with live *R* cells plus heat-killed *S* cells

Mice die. Live *S* cells in their blood

Figure 9.1 *Animated!* Results of Fred Griffith's experiments with *Streptococcus pneumoniae* and laboratory mice.

THE DNA CHASE

virus particle labeled with 35S

DNA (*blue*) being injected into bacterium

a

35S remains outside cells

virus particle labeled with 32P

DNA (*blue*) being injected into bacterium

b

32P remains inside cells

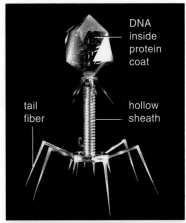

DNA inside protein coat

tail fiber

hollow sheath

micrograph of virus particles injecting DNA into an *E. coli* cell

Figure 9.2 *Animated!*   Landmark experiments that tested whether genetic material resides in bacteriophage DNA, proteins, or both. As Alfred Hershey and Martha Chase knew, sulfur (S) but not phosphorus (P) is present in proteins, and phosphorus but not sulfur is present in DNA.

(**a**) In one experiment, bacteria were infected with virus particles labeled with a radioisotope of sulfur (^{35}S). The sulfur had labeled the viral proteins. The mixture was whirled in a kitchen blender, which dislodged the viruses from the bacterial cells. Radioactive sulfur was detected mainly in the solution, not inside the cells. The viruses had not injected protein into the bacteria.

(**b**) In another experiment, bacteria were infected with virus particles labeled with a radioisotope of phosphorus (^{32}P). The phosphorus had labeled the viral DNA. When the viruses were dislodged from the bacteria, the radioactive phosphorus was detected mainly inside the bacterial cells. The viruses had injected DNA into the cells—evidence that DNA is the genetic material of this virus.

The discovery was electrifying. If someone could pry open the secrets of proteins, then why not DNA? And if DNA's structural details could be worked out, then wouldn't they be clues to its functions? *Someone would go down in history as having dissected the secret of life!*

ENTER WATSON AND CRICK

Having a shot at fame and fortune quickens the pulse of men and women in any profession, and scientists are no exception. However, science is a community effort. Individuals share not only what they find but also what they do not understand. Even if an experiment does not yield an expected result, it may turn up something that others can use or raise questions others can answer. Unexpected results, too, may be clues to secrets of the natural world.

And so scientists all over the world started sifting through results and scrambling after the prize. Among them were James Watson, a postdoctoral student from Indiana University, and Francis Crick, a researcher at Cambridge University. They spent hours arguing over everything they had read about DNA's size, shape, and bonding requirements. They fiddled with cardboard cutouts. They badgered chemists into helping them to identify bonds they might have overlooked. They built models from bits of metal connected by suitably angled "bonds" of wire.

In 1953, they built a model that fit all the pertinent biochemical rules and insights they had gleaned from other sources (Figure 9.3). They discovered the double-helix structure of the DNA molecule. As you will see in the next section, the striking simplicity of this structure is how life shows such unity at the level of molecules, and such diversity at the level of organisms.

Figure 9.3 In 1953, James Watson and Francis Crick with their newly unveiled model of the DNA molecule.

After many decades of ingenious experiments, we now know that the information required to pass parental traits to offspring is encoded in the structure of DNA.

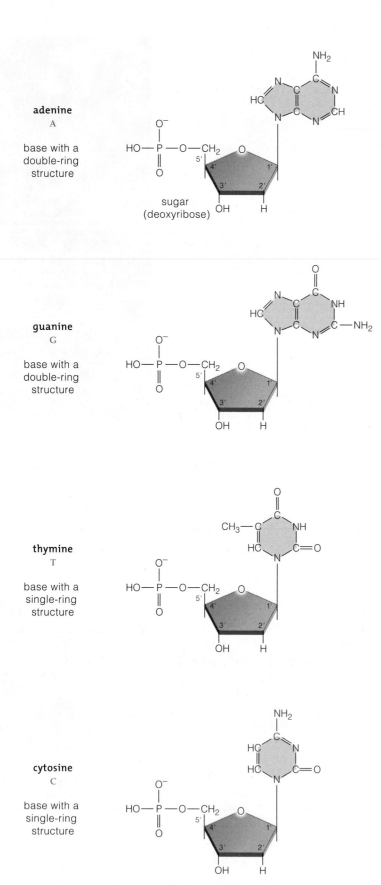

adenine
A

base with a
double-ring
structure

sugar
(deoxyribose)

guanine
G

base with a
double-ring
structure

thymine
T

base with a
single-ring
structure

cytosine
C

base with a
single-ring
structure

Figure 9.4 All chromosomes in a cell contain DNA. What does DNA contain? Four kinds of nucleotides: A, G, T, and C. Each nucleotide is named after the base structure it carries.

DNA Structure and Function

What were the critical pieces of information that led Watson and Crick to discover the structure of DNA?

DNA'S BUILDING BLOCKS

Long before the bacteriophage research, biochemists knew that DNA contains only four kinds of building blocks called nucleotides. Each **nucleotide** consists of a five-carbon sugar (deoxyribose), a phosphate group, and one of four nitrogen-containing bases:

adenine	guanine	thymine	cytosine
A	G	T	C

The structures of these four nucleotides are shown in Figure 9.4. Thymine and cytosine are the pyrimidines; their component base is a single carbon ring structure. Adenine and guanine are purines, which have larger, bulkier bases with two carbon rings.

By 1949, the biochemist Erwin Chargaff had shared with the scientific community two crucial insights into the composition of DNA. First, the amount of adenine relative to guanine differs from one species to the next. Second, the amounts of thymine and adenine in a DNA molecule are exactly the same, and so are the amounts of cytosine and guanine. We may show this as:

$$A = T \quad \text{and} \quad G = C$$

The symmetrical proportions had to mean something important. In fact, it was a tantalizing clue to how the nucleotides are arranged in a DNA molecule.

FAME AND GLORY

In 1951, Rosalind Franklin came to King's College in London to set up an x-ray crystallography laboratory. In this technique, x-rays are directed at a crystalline sample of molecules. Atoms in the molecules scatter the x-rays in a pattern that can be captured on film as an **x-ray diffraction image**. An x-ray diffraction image consists only of dots and streaks; it is not a picture of a molecule. But researchers can use the image to find clues about a molecule's structure, such as its size and shape, and the spacing between repeating patterns.

Many scientists tried to use x-ray crystallography to solve the structure of DNA, but DNA is a very large molecule, difficult to crystallize, and its shape changes depending on how it is prepared. No one yet had been able to make a clear x-ray diffraction image of DNA. Franklin was an expert at preparing samples for x-ray crystallography because of her work on the structure

THE DNA CHASE

of coal and other forms of carbon. Coal is complex and unorganized (as are also large biological molecules like DNA). Franklin had devised a new mathematical way to interpret x-ray diffraction images, and, like Pauling, had built three-dimensional molecular models.

Franklin's assignment at King's College: investigate the structure of DNA. No one told her Maurice Wilkins was already working on that same problem, just down the hall. Even the graduate student who was assigned to assist her failed to mention it. Wilkins did not know what Franklin's assignment was; he assumed she was a technician hired to do his x-ray crystallography work because he didn't know how to do it himself. And so a clash began. To Franklin, Wilkins seemed inexplicably irritable. To Wilkins, Franklin displayed an appalling lack of the deference that technicians of the era usually showed researchers.

Wilkins had a prized cache of DNA which he gave to his "technician" for x-ray crystallography studies. Franklin's meticulous preparation of that sample gave her the first clear x-ray diffraction image of DNA as it exists inside cells (Figure 9.5). Franklin gave a research presentation on what she had learned from it in 1952. DNA, she said, had two chains twisted into a double helix, with phosphate groups forming the backbone on the outside of the helix and the bases arranged in some as yet unknown way on the inside. She presented her calculations of DNA's diameter, the distance between the two chains, the pitch (angle) of the helix, and the distance between bases. She had calculated that each turn of the helix contains ten bases (Figure 9.6).

With his crystallography background, Crick would have recognized the significance of her image and her calculations—*if* he had been there. (A pair of chains in opposing directions would be the same even if flipped 180°.) Watson was in the audience but did not realize the full implications of her data.

Franklin began preparing a report on her findings. Meanwhile, Crick read a copy of her presentation, and Wilkins reviewed Franklin's best diffraction image with Watson. Watson and Crick now had all the information

Figure 9.6 *Animated!* Composite of three different models for the DNA double helix. Two sugar–phosphate backbones run in opposite directions. Think of one strand as being upside down from the other.

2-nanometer diameter overall

0.34-nanometer distance between each pair of bases

3.4-nanometer length of each full twist of the double helix

In all respects shown here, the Watson–Crick model for DNA structure is consistent with the known biochemical and x-ray diffraction data.

The pattern of base pairing (A only with T, and G only with C) is consistent with the known composition of DNA (A = T, and G = C).

Figure 9.5 Rosalind Franklin and her famous x-ray diffraction image of DNA. Franklin died at 37, of ovarian cancer that likely resulted from her intensive work with x-rays. Because the Nobel Prize is not given posthumously, she did not share in the 1962 honor that went to Watson, Crick, and Wilkins for discovery of the structure of DNA.

THE DNA CHASE

they needed to build their model of the DNA helix—one with two sugar–phosphate external chains running in opposite directions and paired bases inside.

Pauling, who had earlier discovered the alpha helix of proteins, already had been thinking "helix." So had everyone else—including Wilkins, Watson, and Crick. Later, Watson wrote, "We thought, why not try it on DNA? We were worried that *Pauling* would say, why not try it on DNA? Certainly he was a clever man. He was a hero of mine. But we beat him at his own game. I still can't figure out why."

Pauling, like most other people, had thought the nucleotide bases had to be on the outside of the helix, so they would be accessible to the enzymes that copy DNA. Franklin's data showed otherwise.

Franklin's report on her findings appeared third in a series of articles about the structure of DNA in the same magazine. Perhaps because of its placement, her report appeared only to be in support of Watson and Crick's article, which appeared first in the series.

PATTERNS OF BASE PAIRING

DNA consists of two chains, or strands, of nucleotides, coiled into a double helix. The helix is maintained by hydrogen bonding between the internally positioned bases on each strand. Only two kinds of base pairings form along the molecule's length: A—T and G—C. The simplicity of the bonding pattern belies almost infinite variation in the order of nucleotide bases. For instance, a tiny sequence of bases in DNA from a rose, a human, or any other organism might be:

The two strands of DNA match up; they are said to be complementary to each other. Although the molecular bonding pattern (A to T, and G to C) is the same in every molecule of DNA, the actual sequence of bases differs for every species. *This molecular variation is the foundation for the diversity of life.*

The pattern of base pairing between the two strands in DNA is constant for all species—A with T, and G with C. However, each species has a number of unique sequences of base pairs along the length of their DNA molecules.

9.3 DNA Replication and Repair

The discovery of DNA structure was a turning point in studies of inheritance. Suddenly it became clear how cells duplicate their DNA before they divide.

HOW IS A DNA MOLECULE DUPLICATED?

Until Watson and Crick presented their model, no one could explain **DNA replication**, or how the molecule of inheritance is duplicated before a cell divides. This is how it goes.

DNA replication uses a team of molecular workers. Replication enzymes become active along the entire length of the DNA molecule at the same time. First, enzymes pry apart the hydrogen bonds between the two nucleotide strands of a DNA molecule in several places along its length. The double helix opens up and unwinds in both directions from the unzipped sites, exposing stretches of nucleotide bases.

Then, a new strand of DNA is assembled on each parent strand according to the base-pairing rule (A to T, and G to C). **DNA polymerases** catalyze formation of two new DNA strands from free nucleotides, using the two parent strands as templates. Cells have stores of free nucleotides that can pair up with the exposed bases. The free nucleotides themselves provide energy for strand assembly. DNA polymerase splits off two of the three phosphate groups on each, and this releases energy that drives the attachments.

Last, **DNA ligases** seal any tiny gaps in the new DNA. One continuous, complementary strand of DNA winds up with the parent strand that was its template. The other new DNA strand winds up with the other parent template. The result is two DNA helixes.

The parent strands of DNA remain intact, but not together. As soon as a stretch of a new partner strand forms on the parent strand, the two twist up together into a double helix. Because the parent DNA is saved (conserved) during the replication process, half of every double-stranded DNA molecule is "old" and half is "new" (Figure 9.7). The process is called *semiconservative* replication.

MONITORING AND FIXING DNA

Errors occasionally are introduced into a new DNA molecule so that it acquires extra bases, loses bases, or no longer is complementary to the parent molecule. Some replication errors occur after the parent DNA is damaged by radiation or chemicals, but most arise simply because DNA polymerases make mistakes.

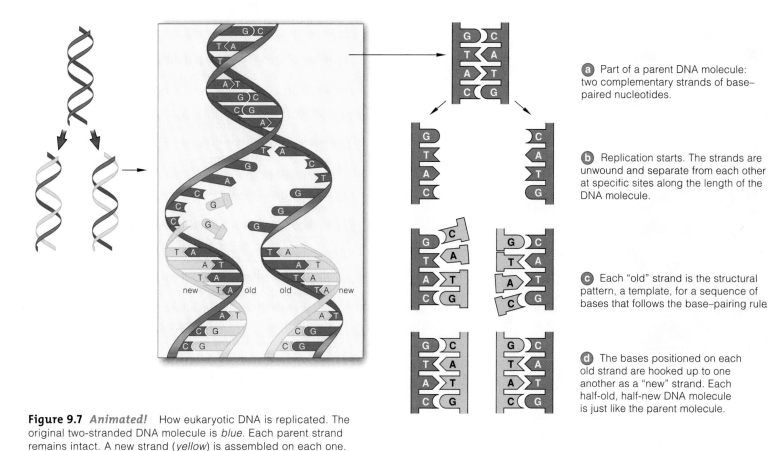

Figure 9.7 *Animated!* How eukaryotic DNA is replicated. The original two-stranded DNA molecule is *blue*. Each parent strand remains intact. A new strand (*yellow*) is assembled on each one. Hence the name, semiconservative replication.

a Part of a parent DNA molecule: two complementary strands of base–paired nucleotides.

b Replication starts. The strands are unwound and separate from each other at specific sites along the length of the DNA molecule.

c Each "old" strand is the structural pattern, a template, for a sequence of bases that follows the base–pairing rule.

d The bases positioned on each old strand are hooked up to one another as a "new" strand. Each half-old, half-new DNA molecule is just like the parent molecule.

Luckily, DNA proofreading mechanisms fix most errors as they occur. DNA polymerases themselves proofread new base pairings and correct mismatches. Other enzymes recognize and repair damaged DNA. If an error is not fixed, replication may be arrested, and the cell cycle stops. On the rare occasion that all of these processes fail, the cell acquires a mutation—a permanent alteration in its DNA sequence.

A mutation may result in the altered or diminished function of a somatic cell, and so disrupt how all of its progeny operate. This is how cancer begins. Mutations in germ cell DNA can give rise to genetic disorders in offspring. However, mutations can benefit a species: they introduce variation in traits that serve as the raw material of evolution.

DNA is replicated prior to cell division. Enzymes unwind its two strands. Each strand remains intact throughout the process—it is conserved—and enzymes assemble a new, complementary strand on each one.

DNA proofreading and repair systems fix nearly all errors and breaks, and thus help maintain the integrity of a cell's genetic information. Unrepaired errors are mutations.

LINKS TO SECTIONS 7.2, 7.8

9.4 Using DNA To Clone Mammals

Here we return to a topic introduced at the start of this chapter. Researchers can now use DNA from an adult mammal to bypass sexual reproduction, which gives rise to mixes of traits from two parents. Adult DNA cloning produces an exact genetic copy of a single adult.

"Cloning" can be a confusing word. It is a laboratory method of making multiple copies of DNA fragments. It also applies to natural and manipulated interventions in reproduction or development.

Embryo cloning occurs all the time in nature. For instance, soon after fertilized human eggs start dividing, a tiny ball of cells forms. Once in every seventy-five or so pregnancies, the ball splits, and the two parts grow and develop into identical twins. A laboratory method, "artificial twinning," simulates what happens in nature. For example, balls of cells grown from fertilized cattle eggs in petri dishes can be divided into identical-twin embryos. These are implanted into surrogate mothers, which give birth to cloned calves. Embryo cloning has been practiced for decades. However, embryo clones

① Removal of the nucleus from a sheep egg during a cloning. At the center is an unfertilized sheep egg. A pipette has positioned the egg, which a microneedle is about to penetrate.

② The microneedle has now emptied the sheep egg of its own nucleus, which holds the cell's DNA.

③ A nucleus from a donor cell is about to enter the enucleated egg. An electric spark will stimulate the egg to enter mitotic cell divisions. At an early stage, the ball of cells will be implanted in the womb of a surrogate mother sheep.

Figure 9.8 *Animated!* Nuclear transfer of sheep cells. In this series of photos, a microneedle replaces the nucleus of a sheep egg with a nucleus from an adult sheep cell. Newer methods involve direct transfer of nuclear DNA that has been treated with mitotic cell extracts to condense the chromosomes.

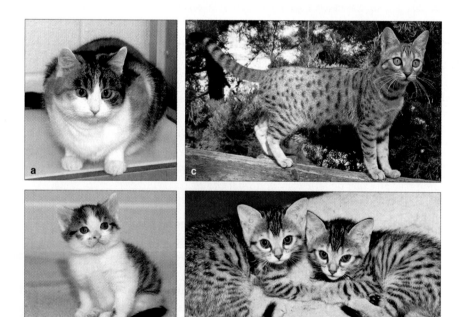

Figure 9.9 Gallery of cloned cats. (**a**) Rainbow, genetic donor for cc (**b**), the first cloned cat. Their coats differ in color because cc was cloned from a single cell from Rainbow. In that cell, the black color allele was active, but not the orange.

(**c**) DNA donor Tahini, a Bengal cat, and (**d**) Tabouli and Baba Ganoush, two of her clones. Their eye color differs; a Bengal cat's eye color changes as it matures. The three cats are household pets of the founder of Genetic Savings and Clone, a pet cloning company.

inherit DNA from two parents. If breeders are looking for a specific valued phenotype, such as plentiful milk production or mating vigor, they have to wait for the clones to grow up to see if they display the trait.

Cloning a cell from an adult animal that displays a desired phenotype is faster. There is no need to wait until clones grow up to check their phenotype—all of them will be genetically identical to the parent. But a cell from an adult will not just start dividing as if it were a newly fertilized egg starting out life. It must be tricked into rewinding its developmental clock.

All cells descended from a fertilized egg inherit the same DNA. As a new embryo develops, different cells start using different subsets of their DNA. As they do, they commit themselves and all of their descendants to be specialized in the adult. This is a one-way path. By the time a liver cell, blood cell, or other specialized cell has formed, most of its DNA is unused, turned off.

To clone an adult cell, scientists must first transform a specialized cell into an unspecialized cell by turning some of its unused DNA back on. Turning off parts of a cell's DNA is a normal process of development, but turning it back on is a different story. The DNA must be reprogrammed into directing the development of a complete individual.

Nuclear transfer is one way to do this. A researcher replaces the nucleus of an unfertilized egg with the nucleus or DNA from from an animal cell (Figure 9.8). The egg's cytoplasm provides the transplanted DNA with the appropriate environment to begin dividing. About 1 of 10 times, a cluster of embryonic cells forms that can be implanted into a surrogate mother. About 1 of every 10 implantations results in a live birth.

The technique is advancing rapidly. For instance, pet owners who have had their cats cloned report that their clones are healthy, lively, and uncannily like the DNA donor (Figure 9.9). Still, nuclear transfer is far from routine. For example, dogs are notoriously hard to clone, and for some as yet unknown reason cloned frogs remain tadpoles until they die.

The real issue, of course, is that humans—like cats, dogs, and sheep—are mammals. Not long ago, cloning mammals seemed fraught with technical problems. It still is, but as the techniques improve, making a genetic copy of a human no longer seems like science fiction.

We will return to the topic of manipulating DNA in Chapter 11.

With nuclear transfer cloning techniques, DNA of a cell taken from an adult donor can be reprogrammed to direct the development of a whole organism—one that is genetically identical to the donor.

Summary

Section 9.1 Experimental tests on bacteriophages and bacteria provided the first solid evidence that DNA is the hereditary material of all cells.

Biology ⊜ Now
Learn about experiments that revealed the function of DNA with the animation on BiologyNow.

Section 9.2 DNA consists only of nucleotides, each having a five-carbon sugar (deoxyribose), a phosphate group, and one of four kinds of nitrogen-containing bases: adenine, thymine, guanine, or cytosine.

A DNA molecule consists of two nucleotide strands coiled together into a double helix. Bases of one strand hydrogen-bond with bases of the other.

Bases of the two DNA strands pair in a constant way. Adenine pairs with thymine (A–T), and guanine with cytosine (G–C). Which base follows another along a strand varies among species. The DNA of each species incorporates some number of unique stretches of base pairs that set it apart from the DNA of all other species.

Biology ⊜ Now
Investigate the structure of DNA with the animation on BiologyNow.

Section 9.3 A cell replicates its DNA before dividing. In DNA replication, enzymes unwind a DNA molecule's double helix at several sites along its length. Then, DNA polymerases use each strand of the unwound DNA as a template to assemble new, complementary strands of DNA from free nucleotides. Two double-stranded DNA molecules result.

Proofreading mechanisms fix most base-pairing mistakes and strand breaks. These mechanisms help keep a cell's genetic information intact and accurate. Repair enzymes fix base mismatches and damaged sites in the DNA. An unrepaired error becomes a mutation.

Biology ⊜ Now
See how a DNA molecule is replicated with the animation on BiologyNow.

Section 9.4 Embryo splitting and nuclear transfers are two methods that produce clones, or individuals that are genetically identical. Cloning adult animals involves reprogramming a cell's DNA to direct the development of a new individual.

Self-Quiz
Answers in Appendix I

1. Which is *not* a nucleotide base in DNA?
   a. adenine   c. uracil   e. cytosine
   b. guanine   d. thymine   f. All are in DNA.

2. What are the base-pairing rules for DNA?
   a. A–G, T–C   c. A–U, C–G
   b. A–C, T–G   d. A–T, G–C

3. One species' DNA differs from others in its _____ .
   a. sugars   c. base sequence
   b. phosphates   d. all of the above

4. When DNA replication begins, _____ .
   a. the two DNA strands unwind from each other
   b. the two DNA strands condense for base transfers
   c. two DNA molecules bond
   d. old strands move to find new strands

5. DNA replication requires _____ .
   a. free nucleotides   c. many enzymes
   b. new hydrogen bonds   d. all of the above

6. Match the terms appropriately.
   ____ bacteriophage   a. nitrogen-containing base
   ____ clone   bonded to a sugar and one
   ____ nucleotide   or more phosphate groups
   ____ DNA ligase   b. adds nucleotides to a
   ____ DNA polymerase   growing DNA strand
                           c. only DNA and protein
                           d. seals breaks in DNA
                           e. carbon copy of donor

Additional questions are available on **Biology ⊜ Now™**

Critical Thinking

1. Matthew Meselson and Frank Stahl's experiments supported the semiconservative model of replication. These researchers obtained "heavy" DNA by growing *Escherichia coli* with ^{15}N, a radioactive isotope of nitrogen. They also prepared "light" DNA by growing *E. coli* in the presence of ^{14}N, the more common isotope. An available technique helped them identify which of the replicated molecules were heavy, light, or hybrid (one heavy strand and one light). Use pencils of two colors, one for heavy strands and one for light. Starting with a DNA molecule having two heavy strands, arrange them so that they show the formation of daughter molecules after replication in a ^{14}N-containing medium. Show the four DNA molecules that would form if the daughter molecules were replicated a second time in the ^{14}N medium. Would the resulting DNA molecules be heavy, light, or mixed?

2. During DNA replication, the template DNA forms a Y-shaped *replication fork* where it is unwinding. New DNA is assembled on both strands of the template at the same time. Look back at Figures 9.6 and 9.7. DNA polymerase only synthesizes DNA in the 5' to 3' direction (*right*). This means nucleotides can be added to each growing strand of DNA *only on its 3' end.* Which way do the polymerases move with respect to the fork as they are copying each of the parent strands near it? What has to happen as the DNA template unwinds more and more?

3. Mutations are permanent changes in base sequences of DNA and the original source of genetic variation—the raw material of evolution. Yet how can mutations accumulate, given that cells have repair systems that fix structurally altered or discontinuous DNA strands during replication?

4. In 1999, scientists discovered a woolly mammoth that had been frozen in glacial ice for the past 20,000 years. They thawed it very carefully so they could use its DNA to clone a woolly mammoth. It turns out there was not enough preserved material to work with. They plan to try again the next time a frozen woolly mammoth comes along. Speculate on the pros and cons, both technical and ethical, of cloning an extinct animal.

3' 5'

3' 5'

Ricin and Your Ribosomes

Ricin is very poisonous. The castor oil plant (*Ricinus communis*) has ricin in all of its tissues, but its oil is an ingredient in many plastics, paints, cosmetics, textiles, and adhesives. The oil, and the ricin, is concentrated in the plant's seeds (beans, *below*). The ricin is discarded when the oil is extracted.

A dose of ricin as small as a grain of salt can kill you; only plutonium and botulism toxin are more deadly. When Germany unleashed mustard gas against allied troops during World War I, both the United States and England investigated whether ricin could be used as a weapon. The research was shelved when the war ended.

Fast-forward to 1969, at the height of the Cold War between Russia and the West. A Bulgarian writer, Georgi Markov, had defected to England. As he strolled down a busy London street, an assassin jammed the tip of an umbrella into Markov's leg. The tip held a tiny ball laced with ricin. Markov died in agony three days later.

Ricin is on stage once again. During the 2001 war in Afghanistan, ricin turned up in some caves hastily

abandoned by terrorists. In 2003, police acting on an intelligence tip stormed a London apartment, where they collected laboratory glassware, bomb circuitry, castor oil beans and other raw materials, and recipes for producing ricin, cyanide, botulism toxin, and other natural poisons.

In 2004, traces of ricin were found in a United States Senate mailroom and State Department building, and in an envelope addressed to the White House. In 2005, a Florida man with castor oil beans and an AK-47 assault rifle was arrested. Jars of banana baby food laced with ground castor oil beans also made the news in 2005, as well as an Arizona man intending to settle a vendetta with his homemade ricin.

Ricin exerts its deadly effects by inactivating ribosomes, the cell's machinery for making proteins. Ricin itself is a protein. One of its two polypeptide chains, shown in the filmstrip, helps ricin insert itself into cells. The other chain wrecks part of the ribosome where amino acids are joined together. It pulls adenine subunits from the ribosome's protein-building RNA molecule. As the ribosome's three-dimensional shape unravels, protein synthesis stops. There is no antidote.

It is possible to go on with your life without knowing what a ribosome is or what it does. It is also possible to recognize that protein synthesis is not a topic invented to torture biology students. It is something that is worth appreciating for how it keeps us alive.

 How Would You Vote? *Scientists are developing a ricin vaccine even though a large-scale terrorist attack using ricin is unlikely, because it is difficult to disperse in air. Should we carry out mass immunizations with a ricin vaccine? See BiologyNow for details, then vote online.*

 ## Key Concepts

PROTEIN SYNTHESIS

DNA is a cell's master copy of genetic information, archived safely in the nucleus. RNAs are disposable copies of useful parts of the DNA archive. Three kinds of RNAs participate in building proteins. mRNA carries protein-building messages from DNA, and tRNA and rRNA translate those messages into chains of amino acids.

MUTATION

A mutation, or permanent alteration in the nucleotide sequence of DNA, may change a protein's structure and function. When the protein is an essential part of cell architecture or metabolism, the outcome may be a disruption of cellular function.

GENE CONTROLS

Controls govern when, how, and to what extent each cell uses its DNA. All cells counter changes in their external environment by adjusting transcription rates. In addition, eukaryotic cells have a long-term program of development that is based on selective gene expression.

Links to Earlier Concepts

Once again you will encounter the nucleic acids DNA and RNA (Sections 2.10, 9.2). Transcription has features in common with DNA replication, so you may wish to review Section 9.3 before you start.

In this chapter, we return again to protein structure and function (2.8), this time in the context of RNA translation. Remember the impact that mutations can have on protein structure (2.9, 8.2) as you learn about the actual mechanisms of mutation.

Your knowledge of the chromosomal basis of inheritance (Section 8.4) will help you understand gene controls.

10.1 Making and Controlling the Cell's Proteins

Two cellular processes convert DNA's protein-building information into a new protein molecule. First, part of the DNA serves as a template for making a strand of RNA. Then the RNA is used to make a protein.

If DNA is the book of protein-building information in cells, its alphabet has only four letters: the nucleotides adenine, thymine, guanine, and cytosine. How does a cell make all of its proteins from this simple alphabet? The answer starts with the order, or sequence, of the four nucleotide bases in a DNA molecule.

Genetic information is coded in the sequence itself as sets of three bases. These "genetic words" occur one after the other along the length of a chromosome. Like sentences, a linear series of genetic words can form a meaningful parcel of information. The parcel is a **gene**, a nucleotide sequence that a cell can convert to an RNA (ribonucleic acid) or protein product.

Most gene products are proteins. It takes two steps, transcription and translation, to convert a gene into a protein. In the first step, **transcription**, a gene serves as a template for making RNA. You know that the DNA double helix unwinds entirely when a cell replicates its DNA. At other times, DNA unwinds just enough to expose one gene for transcription. Then, the nucleotide base sequence of the gene is copied as a strand of RNA.

In the second step of protein synthesis, **translation**, the genetic information carried by a strand of RNA is decoded, or translated, into a sequence of amino acids that is assembled into a new polypeptide chain.

In short, DNA guides the synthesis of RNA, then RNA guides the synthesis of proteins:

$$\text{DNA} \xrightarrow{\textit{transcription}} \text{RNA} \xrightarrow{\textit{translation}} \textbf{PROTEIN}$$

A cell can regulate protein synthesis by using various control mechanisms that affect timing and duration of transcription and translation. Some of these controls are triggered if the concentrations of nutrients or other substances change. Others are activated by signaling molecules that call for changes in cell activities. These kinds of controls are the basis for homeostasis, and also for embryonic development in eukaryotic organisms.

> *The nucleotide base sequence along one of DNA's two strands guides RNA synthesis. RNA guides protein synthesis.*
>
> *Controls over transcription and translation govern which genes are expressed, at what times, and in what amounts.*

10.2 How Is RNA Transcribed From DNA?

In transcription, the first step in protein synthesis, the nucleotide sequence in a DNA region becomes a template for assembling a strand of ribonucleic acid, or RNA.

THREE CLASSES OF RNA

Transcription of most genes produces **messenger RNA**, or **mRNA**—the only class of RNA that carries *protein-building* instructions. Transcription of some other genes produces **ribosomal RNA**, or **rRNA**, a component of ribosomes. A ribosome is a large molecular structure upon which polypeptide chains are built. Transcription of other genes produces **transfer RNA**, or **tRNA**, which delivers amino acids one at a time to a ribosome. All three types of RNA participate in protein synthesis.

THE NATURE OF TRANSCRIPTION

An RNA molecule is almost, but not quite, like a short single strand of DNA. Like DNA, RNA contains four types of nucleotides. Each nucleotide has a five-carbon sugar, a ribose (not DNA's deoxyribose), a phosphate group, and a base. Three of the nucleotides—adenine, cytosine, and guanine—are the same in RNA and DNA. But in RNA, the fourth nucleotide is uracil, not thymine (Figure 10.1a,b). Like thymine, uracil can pair with adenine. This means a new RNA strand can be built according to the same base-pairing rules as DNA (Figure 10.1c).

Figure 10.1 (a) Uracil, one of the four RNA nucleotides. The other three types—adenine, guanine, and cytosine—differ only in their bases. Compare uracil to (b) thymine, a DNA nucleotide. (c) Base-pairing of DNA with RNA at transcription, compared to base-pairing during DNA replication.

gene region

newly forming
RNA transcript

DNA template
at selected
transcription
site

RNA polymerase, the enzyme
that catalyzes transcription

DNA template
winding up

DNA template
unwinding

a RNA polymerase binds the DNA at a promoter near a gene. The polymerase will begin linking RNA nucleotides (adenine, cytosine, guanine, and uracil) into a strand of RNA, in the order specified by the gene.

b All through transcription, the DNA double helix becomes unwound in front of the RNA polymerase. The newly forming RNA strand briefly winds up with its DNA template. As the RNA strand lengthens, it unwinds from the DNA (and the two DNA strands wind up again).

Figure 10.2 *Animated!* Transcription. (**a**) RNA polymerase binds to a promoter near a gene region in the DNA. (**b**) The nucleotide base sequence on one of DNA's two strands (not both) is used as a template for assembling RNA nucleotides. (**c**) The nucleotide base sequence of the transcribed RNA is complementary to that of the DNA template. (**d**) The result of transcription is a new molecule of RNA.

LINK TO
SECTION
9.3

The process of transcription *differs* from the process of DNA replication in three ways.

First, transcription produces one single-stranded molecule of RNA, not two DNA double helixes.

Second, the enzyme called **RNA polymerase**, not DNA polymerase, adds RNA nucleotides one at a time to a growing RNA molecule.

Third, a small, gene-containing part of one DNA strand is the template for transcription, not the whole DNA molecule.

A gene is at most a few thousand nucleotides long, out of millions of nucleotides on a chromosome. How does RNA polymerase know where to start? Just like a sentence, a gene contains signals that tell its readers where to begin and end. The initial capital letter of a sentence means "start," and a period at the end means "stop." The genetic signals for "start" and "stop" are special nucleotide sequences at the beginning and end of a gene. They are recognized by RNA polymerase.

Transcription is initiated when RNA polymerase binds a **promoter**, a special nucleotide sequence near a gene in the DNA. The polymerase swiftly moves along the DNA until it finds the signal for the start of the gene. Using the nucleotide sequence of the gene as a template, the polymerase joins one RNA nucleotide

after another into a strand of RNA (Figure 10.2). When it arrives at the signal for the end of the gene, the new RNA is released and transcription stops.

FINISHING UP RNA TRANSCRIPTS

In eukaryotes, most new RNA transcripts are not in a final form; they must be modified before they become functional molecules. Just as a dressmaker may snip off some loose threads or add bows on a dress before it leaves the shop, so do eukaryotic cells tailor their "pre-RNA" before it leaves the nucleus. All types of RNA undergo such modification.

For example, most eukaryotic genes contain one or more **introns**, nucleotide sequences that are removed from a pre-RNA molecule. Introns intervene between **exons**, which are sequences that remain in the RNA. Introns are transcribed into RNA along with exons, but they are snipped out before the RNA leaves the nucleus in mature form. Think of it this way: *Ex*ons are *ex*ported from the nucleus, and *in*trons stay *in* the nucleus, where they are degraded.

Exons can be snipped apart and spliced together in more than one way. Alternative splicing lets cells use the same gene to make proteins that differ slightly in

PROTEIN SYNTHESIS

direction of transcription ⟶

3' T G A G G A C T C C T C T T C 5'
A C U C C U G A G G A G A A G
5' 3'
growing RNA transcript

⟶ A C U C C U G A G G A G A A G

c RNA polymerase links RNA nucleotides, one at a time, into a strand of RNA. The order of nucleotides added to the growing strand is determined by base-pairing (U–A, G–C) with the DNA template. Many other proteins assist this process.

d At the end of the gene region, the new RNA transcript is released from its DNA template. Shown below it is a model for a transcribed strand of RNA.

form and function. Such splicing at introns is a way to maximize the information-storing capacity of a DNA molecule, hence its capacity to make diverse proteins.

Additional modification events further alter pre-mRNA, the transcripts that become mRNAs. After all the splicing events are complete, a modified guanine "cap" is attached to the start of a pre-mRNA (Figure 10.3). Later, in the cytoplasm, the cap will help bind the mRNA to a ribosome.

Then, a tail of 100 to 300 adenines is attached to the end of the pre-mRNA. Degradation enzymes begin to disassemble an mRNA immediately upon its arrival in the cytoplasm, beginning at the tail. The length of the tail paces enzyme access to the sequence part of the mRNA, and thus determines how long the mRNA will function. The tail keeps the protein-building message intact for as long as the cell needs more of that protein. Figure 10.3 summarizes mRNA processing.

In transcription, a gene sequence on one of the two strands of a DNA molecule serves as a template for RNA polymerase to assemble a single strand of RNA from free nucleotides.

In eukaryotes, each new RNA transcript, or pre-RNA, undergoes modification before leaving the nucleus.

Figure 10.3 *Animated!* How pre-mRNA transcripts are processed into final form. Inside the nucleus of eukaryotes, some or all introns are removed, and the transcript gets a modified guanine cap and a tail. The tail is a sequence of many adenines (hence the name, poly-A tail).

first base	second base				third base
	U	C	A	G	
U	phenylalanine	serine	tyrosine	cysteine	U
	phenylalanine	serine	tyrosine	cysteine	C
	leucine	serine	STOP	STOP	A
	leucine	serine	STOP	tryptophan	G
C	leucine	proline	histidine	arginine	U
	leucine	proline	histidine	arginine	C
	leucine	proline	glutamine	arginine	A
	leucine	proline	glutamine	arginine	G
A	isoleucine	threonine	asparagine	serine	U
	isoleucine	threonine	asparagine	serine	C
	isoleucine	threonine	lysine	arginine	A
	methionine (START)	threonine	lysine	arginine	G
G	valine	alanine	aspartate	glycine	U
	valine	alanine	aspartate	glycine	C
	valine	alanine	glutamate	glycine	A
	valine	alanine	glutamate	glycine	G

Figure 10.4 *Animated!* The genetic code. Codons in mRNA are nucleotide bases "read" in blocks of three. Sixty-one of the base triplets correspond to specific amino acids. Three are signals that stop translation.

The *left* vertical column lists choices for the first of three bases in an mRNA codon. The *top* horizontal row lists choices for the second codon. The *right* vertical column lists choices for the third. To give three examples, reading from left to right, the triplet U G G corresponds to tryptophan. Both U U U and U U C correspond to phenylalanine.

a

b threonine   proline   glutamate   glutamate   lysine

Figure 10.5 Example of the correspondence between genes and proteins. (**a**) An mRNA transcript of a gene region of DNA. Three nucleotide bases, equaling one codon, specify one amino acid. This series of codons (base triplets) specifies the sequence of amino acids shown in (**b**).

LINKS TO SECTIONS 2.8, 3.6

10.3 Deciphering mRNA

Like a strand of DNA, an mRNA transcript is a linear sequence of nucleotide bases. A protein-building message is encoded in consecutive groups of three bases.

Like this sentence, an mRNA is a linear sequence of words—genetic words spelled with a four nucleotide alphabet—that is readable if you can understand the language in which it is written. Researchers Marshall Nirenberg, Philip Leder, Severo Ochoa, and Gobind Korana learned the language of mRNA; they deduced the correspondence between genes and proteins.

An mRNA codes for a sequence of amino acids. To translate its nucleotide base sequence into an amino acid sequence, you have to know that mRNA bases are read *three at a time*, as triplets. Base triplets in mRNA were given the name **codons**. The order of the three bases in each codon determines which amino acid it specifies. There are sixty-four codons (Figure 10.4), many more than necessary to specify all twenty kinds of amino acids. Most amino acids are encoded by more than one codon. For instance, both GAA and GAG call for the amino acid glutamate.

The order of codons in an mRNA determines the order of amino acids in the corresponding polypeptide chain (Figure 10.5). Some codons signal the beginning and the end of the coding region. In most species, the first AUG in a transcript is the START signal for protein translation. AUG codes for methionine, so methionine is the first amino acid of most new polypeptide chains. UAA, UAG, and UGA do not call for an amino acid. They are STOP signals that block further additions of amino acids to a new chain.

The set of sixty-four different codons is the **genetic code**. For every *organism*, protein synthesis adheres to this code. One *organelle*, the mitochondrion of many species, makes its own proteins by using a few unique codons. These mitochondrial codons were a clue that led to the theory of how eukaryotic organelles arose

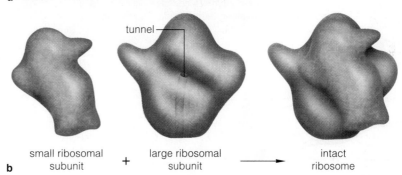

codon in mRNA transcript —

anticodon in tRNA —

amino acid

Figure 10.6   Model for a tRNA. The icon shown to the right is used in later illustrations. The "hook" at the lower end of this icon represents the binding site for a specific amino acid.

Figure 10.7   (**a**) Model of the large subunit of the bacterial ribosome. It has two rRNA molecules (*gray*) and thirty-one structural proteins (*gold*). At one end of a tunnel through the large subunit, rRNA catalyzes polypeptide chain assembly. The eukaryotic ribosome, which is larger, is similar in structure and function. (**b**) Icons for the small and large subunits of a eukaryotic ribosome.

tunnel

a

small ribosomal subunit   +   large ribosomal subunit   →   intact ribosome

b

by endosymbiosis. Mitochondria, remember, probably evolved from foreigners that stayed inside their host cells (Section 3.6).

A team of tRNAs and rRNAs translates the genetic code carried by an mRNA into a polypeptide chain. The cytoplasm of every cell contains free amino acids and tRNA molecules. A tRNA has an attachment site for one amino acid, and it also has an **anticodon**, or a base triplet that is complementary to an mRNA codon (Figure 10.6). tRNAs link the genetic code and amino acids. During translation, tRNA anticodons base-pair with codons in an mRNA, in order. The amino acid cargoes of the tRNAs thus become positioned in the order specified by the codons in the mRNA.

There are sixty-four codons but not as many kinds of tRNAs, so the tRNAs must be able to match up with more than one codon. According to base-pairing rules, adenine pairs with uracil, and cytosine with guanine. How can an anticodon pair with more than one codon? Base-pairing rules are not so strict for the third base in a codon–anticodon pair. For example, AUU, AUC, and

AUA all specify isoleucine. All three codons can base-pair with the tRNA that carries isoleucine. Freedom in codon–anticodon pairing is known as "wobble."

Interactions between tRNAs and mRNA take place at ribosomes. Ribosomes have two different subunits (Figure 10.7). Each is built from rRNA and structural proteins in the nucleus. In the cytoplasm, a large and a small ribosomal subunit join into an intact, functional ribosome when an mRNA is to be translated.

The genetic code is a set of sixty-four codons, which are nucleotide bases in mRNA that are read in sets of three. Different amino acids are specified by different codons.

mRNA carries DNA's protein-building instructions from the nucleus into the cytoplasm.

tRNAs deliver amino acids to ribosomes. Their anticodons base-pair with codons in the order specified by mRNA.

Polypeptide chains are built on ribosomes, each consisting of a large and small subunit made of rRNA and proteins.

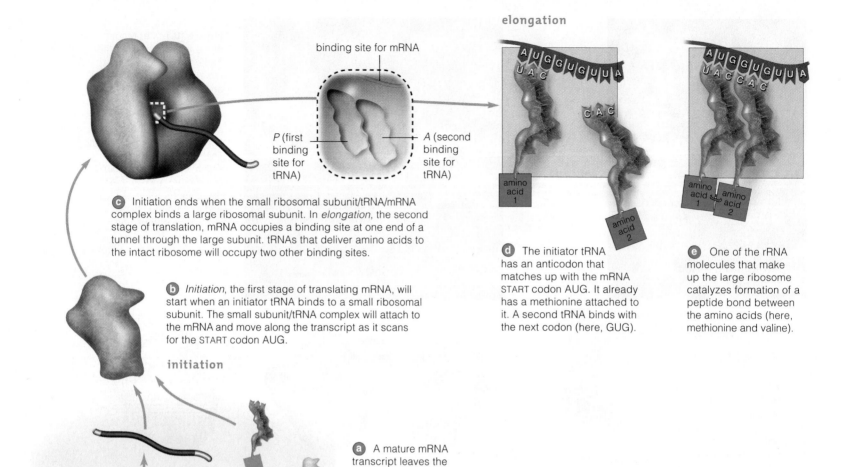

elongation

binding site for mRNA

P (first binding site for tRNA)

A (second binding site for tRNA)

c Initiation ends when the small ribosomal subunit/tRNA/mRNA complex binds a large ribosomal subunit. In *elongation*, the second stage of translation, mRNA occupies a binding site at one end of a tunnel through the large subunit. tRNAs that deliver amino acids to the intact ribosome will occupy two other binding sites.

b *Initiation*, the first stage of translating mRNA, will start when an initiator tRNA binds to a small ribosomal subunit. The small subunit/tRNA complex will attach to the mRNA and move along the transcript as it scans for the START codon AUG.

initiation

a A mature mRNA transcript leaves the nucleus through a pore in the nuclear envelope. It enters the cytoplasm, which has many free amino acids, tRNAs, and ribosomal subunits.

d The initiator tRNA has an anticodon that matches up with the mRNA START codon AUG. It already has a methionine attached to it. A second tRNA binds with the next codon (here, GUG).

e One of the rRNA molecules that make up the large ribosome catalyzes formation of a peptide bond between the amino acids (here, methionine and valine).

Figure 10.8 *Animated!* Stages of translation, the second step of protein synthesis. Here we track a mature mRNA transcript that formed inside the nucleus. It passes through pores across the nuclear envelope and enters the cytoplasm, which contains pools of many free amino acids, tRNAs, and ribosomal subunits.

LINKS TO SECTIONS 2.8, 3.5

10.4 From mRNA To Protein

DNA's hereditary information is stored intact, safely, in one place. Think of mRNAs as disposable intermediaries that deliver messages from DNA to ribosomes, which translate them into polypeptide chains.

Translation is the process of converting the genetic information carried by an mRNA into a polypeptide chain. It proceeds in the cell cytoplasm through three stages: initiation, elongation, and termination. In the first stage, *initiation*, an intact ribosome, an mRNA, and a special initiation tRNA form a cluster called an initiation complex (Figure 10.8*b,c*).

In the *elongation* stage of translation, a polypeptide chain is assembled as the mRNA passes between the two ribosomal subunits, like a thread moving through the eye of a needle. tRNAs bring amino acids to the ribosome, then bind the mRNA in the order specified by its codons. Part of an rRNA molecule in the large ribosomal subunit has enzymatic activity. It catalyzes the formation of peptide bonds between amino acids (Figure 10.8*d–f*). You already saw how peptide bonds form during protein synthesis in Figure 2.22. Peptide bonds continue to form in a series between the amino acid that was most recently attached and the next one delivered to the intact ribosome (Figure 10.8*g,h*).

During the last stage of translation, *termination*, the ribosome reaches the mRNA's STOP codon. No tRNA has an anticodon that corresponds to a STOP codon, so

f The first tRNA is released, and the ribosome moves to the next codon position.

g A third tRNA binds with the next codon (here it is UUA). The ribosome catalyzes peptide bond formation between amino acids 2 and 3.

h Steps **f** and **g** are repeated as the ribosome moves along the mRNA transcript.

termination

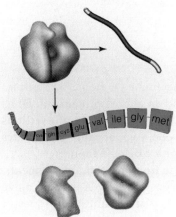

i A STOP codon moves into the area where the chain is being built. It is the signal to release the mRNA transcript from the ribosome.

j The new polypeptide chain is released from the ribosome. It joins the pool of proteins in the cytoplasm or enters the rough ER of the endomembrane system.

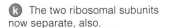

k The two ribosomal subunits now separate, also.

translation ends. Proteins called release factors bind to the ribosome, triggering enzyme activity that detaches both the mRNA and the polypeptide chain from the ribosome (Figure 10.8*i–k*).

Unfertilized eggs and other cells that will rapidly synthesize many copies of different proteins usually stockpile mRNA transcripts in their cytoplasm. In cells that are quickly using or secreting proteins, you often observe polysomes. A polysome is a cluster of many ribosomes, all translating the same mRNA transcript at the same time. Ribosomes move down the mRNA like many trains chugging along the same track. Under the electron microscope, they look like a string of pearls.

Many newly forming polypeptide chains carry out their functions in the cytoplasm. Many others have a special sequence of amino acids that allows them to

enter the ribosome-studded, flattened sacs of rough ER (Figure 3.9). In the organelles of the endomembrane system, the polypeptides will take on final form before shipment to their ultimate destinations.

Translation is initiated when a small ribosomal subunit and an initiator tRNA arrive at an mRNA's start codon. A large ribosomal subunit binds to them.

tRNAs deliver amino acids to a ribosome in the order dictated by the linear sequence of mRNA codons. A polypeptide chain lengthens as peptide bonds form between the amino acids.

Translation ends when a STOP codon triggers events that cause the polypeptide chain and the mRNA to detach from the ribosome.

a part of DNA template: T G A G G A C T C C T C T T C

mRNA transcribed from DNA: A C U C C U G A G G A G A A G

resulting amino acid sequence: THREONINE | PROLINE | GLUTAMATE | GLUTAMATE | LYSINE

b base-pair substitution in DNA: T G A G G A C A C C T C T T C

altered mRNA: A C U C C U G T G G A G A A G

altered amino acid sequence: THREONINE | PROLINE | VALINE | GLUTAMATE | LYSINE

c deletion in DNA: T G A G G A C C C T C T T C

altered mRNA: A C U C C U G G G A G A A G

altered amino acid sequence: THREONINE | PROLINE | GLYCINE | ARGININE

Figure 10.9 *Animated!*
Mutation. (**a**) Part of a gene, mRNA, and amino acid sequence of the hemoglobin beta chain.

(**b**) A base-pair substitution replaces a thymine with adenine in the DNA. When the altered mRNA transcript is translated, valine replaces glutamate as the sixth amino acid of the new polypeptide chain. The outcome is sickle hemoglobin.

(**c**) Deletion of the same thymine causes a frameshift mutation. The reading frame for the rest of the mRNA shifts, and a different protein product forms. This mutation causes thalassemia, a different type of red blood cell disorder.

LINKS TO
SECTIONS
2.8, 2.9, 8.2, 9.3

10.5 Mutated Genes and Their Protein Products

Gene sequences can change. One nucleotide might be substituted for another in the DNA sequence, an extra nucleotide may become inserted, or a nucleotide may be lost. Such changes are mutations.

COMMON MUTATIONS

In the gene mutations called **base-pair substitutions**, a base is copied incorrectly during DNA replication. A substitution causes a codon to change, and may result in a protein that incorporates a wrong amino acid or terminates when it is incomplete. There is room for error; remember, more than one codon may specify the same amino acid. For example, if one codon changes from UCU to UCC, it probably would not have dire effects because both codons specify serine. However, other mutations give rise to proteins with altered or lost functions. In the example shown in Figure 10.9b, an adenine replaced a certain thymine in the gene for beta hemoglobin. The resulting mutant gene's product has a single amino acid substitution that causes sickle-cell anemia (Section 2.9).

Figure 10.9c illustrates a different mutation, one in which a nucleotide, thymine, was *deleted*. Recall, base sequences are read in blocks of three. A deletion is one of the *frameshift* mutations, because it shifts the "three-

bases-at-a-time" reading frame. A frameshift garbles the genetic message, and an altered protein will be its outcome. Frameshift mutations result from **insertions** or **deletions**, in which one, two, or more nucleotides are inserted into DNA or are deleted from it.

Other mutations arise from transposable elements, or **transposons**, small segments of DNA that can move around in a genome. They are found in all types of organisms; about 40 percent of the human genome has been transposed. Geneticist Barbara McClintock found that some kinds of transposons move spontaneously to a new location, or even to a different chromosome. When transposons disrupt a gene, the gene's product may become altered or cease to function entirely. The unpredictability of transposons is a source of curious variations in traits. Figure 10.10 gives an example.

HOW DO MUTATIONS ARISE?

Given the very swift pace of DNA replication (about twenty nucleotide bases per second in eukaryotes, and a thousand per second in bacteria), it is not surprising that mutations happen spontaneously while DNA is being replicated. All DNA polymerases make mistakes at a predictable rate, but fix most of their own errors. Uncorrected errors may be perpetuated as mutations.

Mutations can arise if a cell is exposed to certain hazards in the environment. For example, x-rays and other forms of **ionizing radiation** have enough energy to break chromosomes into pieces (Figure 10.11). Later,

MUTATION

Figure 10.10 Barbara McClintock, who won a Nobel Prize for discovering that transposons slip into and out of different locations in DNA. The curiously nonuniform coloration of kernels in strains of Indian corn (*Zea mays*) sent her on the road to discovery. Several genes govern pigment formation and deposition in corn kernels. Mutations in these genes produce yellow, white, red, orange, blue, and purple kernels. Unstable mutations caused by transposons jumping in and out of these genes can cause streaks or spots in *individual* kernels.

Figure 10.11 Chromosomes from a human cell that was exposed to ionizing radiation. The broken pieces will be lost during DNA replication. The extent of chromosome damage in an exposed cell depends on how much radiation it absorbed.

during DNA replication, broken pieces may get lost. Ionizing radiation can also damage DNA indirectly when it penetrates living tissue, leaving a destructive trail of reactive free radicals in its wake. Free radicals also alter DNA. Doctors and dentists always use the lowest possible doses of x-rays in order to minimize damage to their patients' DNA.

Ultraviolet (UV) light is one form of **nonionizing radiation** that can boost electrons of nucleotide bases to a higher energy level. Destabilized electrons change the bonding properties of thymine and cytosine. For instance, exposure to UV light can cause two adjacent thymine bases on the same strand of DNA to bond covalently to each other. This bond is called a thymine dimer, and it wrinkles the DNA. During replication, DNA polymerase skips the wrinkled region, and thus mutations may occur. Thymine dimers are the major source of mutations that cause certain kinds of cancer, and so they are the primary reason that exposing your unprotected skin to sunlight can cause skin cancer. At least seven gene products interact as a combined DNA repair mechanism to remove them.

Natural and synthetic chemicals also accelerate the rate of mutations. Alkylating agents are one example. These chemicals transfer small hydrocarbons (such as methyl groups) to nucleotide bases in DNA. Alkylated nucleotides halt DNA replication, or mispair during it. Either event can result in a mutation. Alkylated DNA also disintegrates during enzymatic attempts at repair. Chemicals in cigarette smoke alkylate DNA.

THE PROOF IS IN THE PROTEIN

Over evolutionary time, mutations are rare. When one arises in a somatic cell of an organism that reproduces solely by sexual means, its good or bad effects will not endure beyond the life of the organism. The mutation will not be passed on to offspring.

However, if a mutation arises in the DNA of a germ cell or gamete, it may enter the evolutionary arena. The same can happen if a mutation arises in an organism or cell that reproduces asexually.

Either way, nature's test is this: An altered protein specified by a heritable mutation may have harmful, neutral, or beneficial effects on the individual's ability to function in its prevailing environment. The effects of uncountable mutations in millions of species have had spectacular evolutionary consequences in terms of diversity. That is a topic of later chapters.

A mutation is an alteration involving one or more nucleotide bases in a gene. The most common are base-pair substitutions, insertions, and deletions.

DNA polymerases make errors, but also repair most of them. Unrepaired errors are perpetuated as mutations.

Exposure to harmful radiation and certain chemicals in the environment also can cause mutations in DNA.

A protein specified by a mutated gene may have harmful, neutral, or beneficial effects.

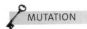

10.6 Controls Over Gene Expression

Gene expression is the conversion of the information in a DNA base sequence to functional and structural parts of a cell. When, how, and to what extent a gene is expressed depends on the type of cell and its chemical environment.

COMMON CONTROLS

Cells use some of their genes a lot, some a little, and most not at all. This regulation of gene activity allows cells to respond to changing conditions, either inside or outside the cell.

Diverse regulatory proteins adjust gene expression before, during, or after transcription and translation. They include signaling molecules such as hormones, which are small chemicals that can initiate changes in cell activities. With **negative control**, some regulatory proteins slow or stop gene expression; with **positive control**, others promote or enhance it.

Some DNA base sequences are sites of control over transcription. Gene expression is initiated, enhanced, or blocked depending on what binds to the sequence. A promoter, introduced in Section 10.2, is a sequence that marks the location of a gene for RNA polymerase to begin transcription. An *enhancer* is a binding site for proteins that increase the level of transcription.

Chemical modification is another form of control. For example, methyl groups physically block an RNA polymerase from accessing DNA. Thus, methylation prevents transcription. When the methyl groups are removed, the genes become active again. Attachment of functional groups to histones is another chemical control (Section 7.1 and Figure 10.12).

The signaling mechanisms are easier to understand in the specific context of how plants and animals work (Section 17.5, and Units Four and Five). Here, we just sample the kinds of events they set in motion.

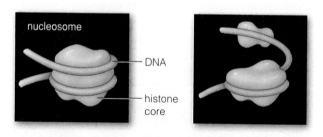

Figure 10.12 How loosening of the DNA–histone packaging in chromosomes exposes genes for transcription. Attachment of a functional group to a histone makes it loosen its grip on the DNA that is wound around it. Enzymes that are associated with transcription attach or detach the groups.

BACTERIAL CONTROL OF THE LACTOSE OPERON

Think about the dot of the letter "i." About a thousand bacterial cells would stretch side by side across that dot. Each of those microscopic cells depends just as much on gene controls as you do! When nutrients are plentiful and other conditions favor growth, bacteria divide quickly.

In bacteria, transcription occurs in the cytoplasm. Translation begins even before mRNA transcripts are fully assembled. Genes for the enzyme in a metabolic pathway often are organized along the chromosome in the same sequence as the reaction steps. All the genes in the sequence may be transcribed as one continuous strand of RNA.

With this bit of background, consider an example of how one type of bacteria adjusts transcription rates to availability of the nutrient lactose. *Escherichia coli* lives in the mammalian gut, and it dines on nutrients traveling past. Mammalian infants generally are fed milk. Milk does not contain glucose, the sugar of choice for *E. coli*. It contains lactose, a different sugar.

Once mammals are weaned, milk intake typically dwindles or stops. *E. coli* cells still take advantage of lactose if and when it becomes available by regulating the expression of three genes for lactose-metabolizing enzymes. One promoter precedes all three genes, and there are operators are on both sides of the promoter. The **operators** are binding sites for a *repressor*, a type of regulatory protein that can prevent transcription. The promoter, operators, and genes together are called the **lac operon** (Figure 10.13).

The genes for lactose-metabolizing enzymes are not used when they are not needed. In the absence of lactose, a repressor molecule binds to both operators. The binding causes the DNA region that contains the promoter to twist into a loop, as in Figure 10.13*b*. RNA polymerase cannot bind to a looped-up promoter, and so cannot transcribe the genes.

The three genes for lactose-metabolizing enzymes are used when lactose is in the gut. *E. coli* cells convert some lactose to allolactose, which binds the repressor and changes its shape. The altered repressor molecule can no longer bind to the operators. The looped DNA unwinds, and RNA polymerase transcribes the genes.

Unlike *E. coli*, humans develop *lactose intolerance*. Cells in the lining of our small intestine make and then secrete lactase, a lactose-digesting enzyme, into the gut. As many people age, however, concentrations of lactase decline. Lactose accumulates in the large intestine (colon), where it fosters population explosions of resident bacteria. As the bacteria digest the lactose, they produce carbon dioxide gas, which accumulates, distends the colon, and causes pain.

a Lactose operon: a set of operators, a promoter, three genes that specify three enzymes, and a binding site for a second messenger (*white*). A different gene upstream from the operon specifies a repressor protein that can block access to the lactose operon.

b In the absence of lactose, the repressor binds to two operators in DNA. It makes the DNA loop out in a way that blocks operon gene transcription; it stops RNA polymerase from binding to its promoter.

Figure 10.13 *Animated!* Negative control of the lactose operon. The operon's three genes code for three enzymes that allow bacteria to metabolize lactose.

c When lactose is present, some is converted to a form that binds to the repressor and alters its shape. The altered repressor can't bind to operators, so RNA polymerase is free to transcribe the operon genes.

EUKARYOTIC GENE CONTROLS

Later in the book, you will be reading about how you and other complex organisms developed from a single cell. For now, tentatively accept this premise: All cells of your body started out life with the same genes, because every one arose by mitotic cell divisions from the same fertilized egg. All of your cells transcribe many of the same genes, because most aspects of their structure and basic housekeeping activities are similar.

In other ways, however, *nearly all of your body cells became specialized in composition, structure, and function.* This process of **cell differentiation** occurs during the development of all multicelled organisms. Differences arise among cells that use different subsets of genes. Specialized tissues and organs are the result.

For example, nearly all of your cells continually transcribe the genes for glycolysis enzymes, but only immature red blood cells can transcribe the genes for hemoglobin. Only your liver cells transcribe the genes for enzymes that neutralize certain toxins. When your eyes first formed, their cells began to access the genes necessary for synthesizing crystallin. No other cells activate the genes for this protein, which helps make the transparent fibers of the lens in each eye.

By some estimates, cells rarely use more than 5 to 10 percent of their genes at a time. Gene controls keep the rest of them switched off. Which genes are active at any time depends on the type of organism, the stage of growth and development it is passing through, and the controls operating at different stages.

Controls over gene expression allow cells to respond to changing environmental conditions.

Gene expression controls also govern cell differentiation in complex eukaryotes.

Cells become specialized when they start using a unique fraction of their genes.

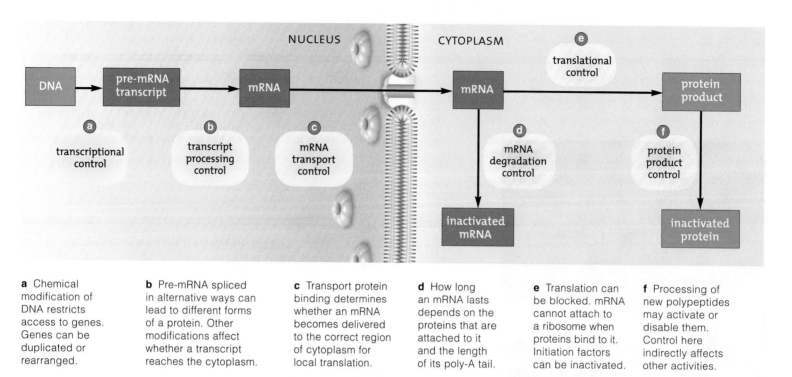

a Chemical modification of DNA restricts access to genes. Genes can be duplicated or rearranged.

b Pre-mRNA spliced in alternative ways can lead to different forms of a protein. Other modifications affect whether a transcript reaches the cytoplasm.

c Transport protein binding determines whether an mRNA becomes delivered to the correct region of cytoplasm for local translation.

d How long an mRNA lasts depends on the proteins that are attached to it and the length of its poly-A tail.

e Translation can be blocked. mRNA cannot attach to a ribosome when proteins bind to it. Initiation factors can be inactivated.

f Processing of new polypeptides may activate or disable them. Control here indirectly affects other activities.

Figure 10.14 *Animated!* Controls over eukaryotic gene expression.

We conclude this chapter with examples of two of the eukaryotic gene controls outlined in Figure 10.14.

HOMEOTIC GENES AND BODY PLANS In most species of eukaryotes, the products of **homeotic genes** guide formation of organs and limbs by turning on other genes in local areas of a developing embryo. Different homeotic genes are transcribed in specific parts of a developing embryo, so their products accumulate in local tissue regions. Body parts form as the products interact with one another and with other regulatory proteins to switch on genes in certain cells of the body.

Homeotic genes were discovered through mutations that cause cells in a *Drosophila* embryo to develop into a body part that should be somewhere else (Figure 10.15). For instance, the *antennapedia* gene is normally transcribed where a thorax and legs should form. In all other regions, cells do not transcribe *antennapedia*.

Figure 10.15*b* shows what happens when a mutation allows this gene to be wrongly transcribed in the body region that is supposed to be a head.

We know of more than a hundred homeotic genes. They control transcription by the same mechanisms in all eukaryotes, and many are interchangeable among species as evolutionarily distant as yeasts and humans. They probably evolved in the most ancient eukaryotic cells. Every homeotic gene product has a domain of about sixty amino acids that regulates transcription of other genes by binding to certain control elements in DNA. These domains differ among species only by *conservative* substitutions, in which one amino acid has replaced another with similar chemical properties. We will return to this topic in Chapter 13.

X CHROMOSOME INACTIVATION Female humans and female calico cats both have two X chromosomes in their

Figure 10.15 Gene controls that govern where body parts develop in fruit flies. (**a**) Normal fly. (**b**) A mutation that affects transcription of the homeotic gene *antennapedia* results in legs on the head. (**c**) Expression of a transcriptional control gene can induce an eye to form not only on a fruit fly's head, but also on its legs, wings, and antennae. (**d**) A mutation in one of this fly's homeotic genes caused it to develop two thoraxes.

Figure 10.16 (**a**) A condensed X chromosome, also called a Barr body (*arrow*), in the somatic cell nucleus of a human female. The X chromosome in cells of human males is not condensed this way. (**b**) Mosaic tissue effect that occurs in anhidrotic ectodermal dysplasia.

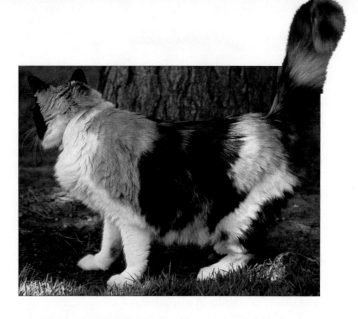

Figure 10.17 *Animated!*   Why is this female cat "calico"? In her body cells, one of the two X chromosomes has a dominant allele for the brownish-black pigment melanin. Expression of the allele on the other X chromosome results in orange fur. When this cat was an embryo, one X chromosome was inactivated at random in each cell.

Patches of different colors reflect which allele was inactivated in cells that formed a given tissue region. White patches are an outcome of an interaction that involves a different gene, the product of which blocks melanin synthesis.

diploid cells. One is stretched long and thin. The other is scrunched up, even during interphase. That scrunching is not a chromosome abnormality. It is a programmed shutdown of all but about three dozen genes on *one* of two homologous X chromosomes. This **X chromosome inactivation** happens in female embryos of all placental mammals and marsupials. Figure 10.16*a* shows a single condensed X chromosome in a nucleus at interphase. We call this condensed chromosome a **Barr body**.

One X chromosome gets inactivated when embryos are still a tiny ball of cells. In placental mammals, the shutdown is random, in that *either* chromosome could become condensed. The maternal X chromosome may be inactivated in one cell; the paternal or the maternal X chromosome may be inactivated in a cell next to it. Once that random molecular decision occurs in a cell, all of that cell's descendants make the same decision as they go on dividing to form tissues. What is the outcome? *A fully developed female has patches of tissue in which genes of the maternal X chromosome are expressed, and patches of tissue in which the genes of the paternal X chromosome are expressed.* She has become a "mosaic" for X chromosome expression (Figure 10.17).

When alleles on two homologous X chromosomes are not identical, differences occur among patches of tissues throughout the body. The mosaic effect occurs in *anhidrotic ectodermal dysplasia,* an X-linked disorder characterized by abnormalities in the skin and in the

structures derived from it, including teeth, hair, nails, and sweat glands. Human females with this disorder are heterozygous for a recessive allele that causes an absence of sweat glands. Their mosaic tissues can be observed because sweat glands are absent in patches of skin (Figure 10.16*b*). Where the glands are absent, the recessive allele is on the X chromosome that was not shut down.

In mammals, the X-chromosome shutdown has an important function. Remember, males have one X and one Y chromosome. This means females have twice as many X chromosome genes. Inactivating one of their X chromosomes balances gene expression between the sexes. The normal development of female mammalian embryos depends on this type of control mechanism, which is called **dosage compensation**.

LINK TO SECTION 8.4

Controls operate before, during, and after transcription and translation. A variety of control elements interact with one another and with DNA, RNA, and gene products.

When, how, and to what extent a gene is expressed depends on the type of cell and its functions, on the cell's chemical environment, and on signals for change.

Diverse controls over eukaryotic genes come into play at different stages of growth and development. Homeotic genes and dosage compensation are two examples.

Summary

Section 10.1 All enzymes and other proteins consist of polypeptide chains. Each chain, a linear sequence of amino acids, corresponds to a nucleotide sequence in a gene. The path from a gene to a protein has two steps: transcription and translation (Figure 10.18).

Section 10.2 Translation requires messenger RNA, transfer RNA, and ribosomal RNA. In transcription, enzymes unwind the two strands of the DNA double helix in a gene region. RNA polymerases covalently bond RNA nucleotides into a new RNA transcript, in an order complementary to the exposed nucleotides on the DNA template. Adenine, guanine, cytosine, and uracil are the RNA nucleotides.

Each new RNA molecule becomes modified before it leaves the nucleus. Its noncoding regions (exons) are removed and the remaining coding regions (introns) become spliced together as a continuous transcript.

One end of a new mRNA is capped, and the other gets a tail that helps control how long it will last in the cell.

Biology⊘Now
Learn how genes are transcribed and transcripts processed with animation on BiologyNow.

Section 10.3 Only messenger RNA (mRNA) carries DNA's protein-building information from the nucleus to the cytoplasm. Its genetic message is written in codons, sets of three nucleotides that specify amino acids. Sixty-four codons, a few of which act as START or STOP signals for translation, constitute a highly conserved genetic code.

Biology⊘Now
Explore the genetic code with the interaction on BiologyNow.

Section 10.4 Translation proceeds through three stages. In initiation, ribosomal subunits, an mRNA, and an initiator tRNA join as an initiation complex. During the elongation stage, tRNAs deliver amino acids to the intact ribosome in the order dictated by codons in the mRNA. The ribosome links the amino acids into a polypeptide chain. At termination, a STOP codon and other factors trigger release of the mRNA and the polypeptide chain, and cause the ribosome's subunits to separate from each other.

Biology⊘Now
Observe the translation of an mRNA transcript with the animation on BiologyNow.

Section 10.5 A mutation is a permanent alteration in a DNA nucleotide sequence. Mutations introduce changes in protein structure, and thus protein function. These changes may lead to small or large differences in traits among individuals of a population.

Biology⊘Now
Investigate the effects of mutation with animation on BiologyNow.

Section 10.6 Regulatory elements interact with RNA, DNA, protein products, and one another before, during, and after transcription and translation. They govern whether, when, and how a gene is expressed.

All cells respond fast to short-term shifts in nutrients and other environmental factors. Regulatory proteins adjust gene transcription rates in response. Control of bacterial operons is an example.

In eukaryotic cells, other controls also come into play during development. Selective gene expression is the basis of differentiation. All cells of multicelled organisms inherit the same genes. During development, different cell lineages begin to use different subsets of those genes. As they do, the cells diverge in structure and function, and eventually they form specialized tissues and organs.

Biology⊘Now
Review how eukaryotic gene controls influence development with animation on BiologyNow.

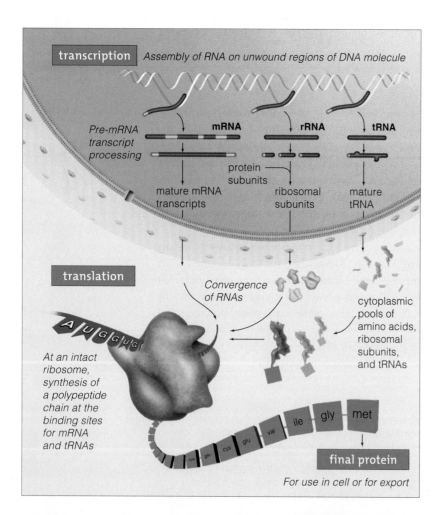

Figure 10.18 Summary of protein synthesis in eukaryotes. DNA is transcribed into RNA in the nucleus. RNA is translated into protein in the cytoplasm. Prokaryotic cells do not have a nucleus; transcription and translation occur in their cytoplasm.

Answers in Appendix I

1. DNA contains many different genes that are transcribed into different _____ .
   a. proteins
   b. mRNAs only
   c. mRNAs, tRNAs, rRNAs
   d. all of the above

2. An RNA molecule is typically _____ .
   a. a double helix
   b. single-stranded
   c. double-stranded
   d. triple-stranded

3. An mRNA molecule is produced by _____ .
   a. replication
   b. duplication
   c. transcription
   d. translation

4. Each codon calls for a specific _____ .
   a. protein
   b. polypeptide
   c. amino acid
   d. carbohydrate

5. Anticodons pair with _____ .
   a. mRNA codons
   b. DNA codons
   c. RNA anticodons
   d. amino acids

6. Cell differentiation _____ .
   a. occurs in all complex multicelled organisms
   b. requires unique genes in different cells
   c. involves selective gene expression
   d. both a and c
   e. all of the above

7. Control mechanisms adjust gene expression in response to changing _____ .
   a. nutrient availability
   b. solute concentrations
   c. signals from other cells
   d. all of the above

8. An operon typically governs _____ .
   a. prokaryotic genes
   b. eukaryotic genes
   c. genes of all types
   d. DNA replication

9. X chromosome inactivation is an example of _____ .
   a. a chromosome abnormality
   b. dosage compensation
   c. calico cats
   d. homeotic gene control

10. Homeotic genes _____ .
    a. are binding sites that flank a bacterial operon
    b. fill in body plan details of developing embryo
    c. control X chromosome inactivation
    d. both a and d

11. A cell with a Barr body is _____ .
    a. prokaryotic
    b. from a male mammal
    c. from a female mammal
    d. infected by the Barr virus

12. Match the terms with the most suitable description.
    ____ alkylating agent
    ____ chain elongation
    ____ exons
    ____ genetic code
    ____ anticodon
    ____ introns
    ____ codon

    a. coding region of mRNA transcript
    b. base triplet for amino acid
    c. second stage of translation
    d. base triplet; pairs with codon
    e. an environmental agent that induces mutation in DNA
    f. set of 64 codons for mRNA
    g. removed from a pre-mRNA transcript before translation

Additional questions are available on Biology⟨S⟩Now™

Figure 10.19 Soft skin tumors on an individual affected by the autosomal dominant disorder called neurofibromatosis.

Critical Thinking

1. Using Figure 10.4, translate this base sequence in part of an mRNA transcript into an amino acid sequence:

   5′—GGUUUCUUCAAGAGA—3′

2. *Antisense drugs* may help us fight cancer and viral diseases, including SARS. They are short mRNA strands that are complementary to mRNAs associated with these illnesses. Speculate on how these drugs work.

3. Cigarette smoke is mostly carbon dioxide, nitrogen, and oxygen. The rest contains at least fifty-five different chemicals identified as carcinogenic, or cancer-causing, by the International Agency for Research on Cancer (IARC). When these carcinogens enter the bloodstream, enzymes convert them to a series of chemical intermediates that are easier to excrete. Some of them bind irreversibly to DNA. Propose one mechanism by which smoking cigarettes causes cancer.

4. *Neurofibromatosis* is a human autosomal dominant disorder caused by mutations in the *NF1* gene. Soft, fibrous tumors form in the peripheral nervous system and skin (Figure 10.19). Because the mutant allele is dominant, an affected child usually has an affected parent. Yet in 1991, a boy developed neurofibromatosis even though his parents did not. The *NF1* gene from his father contained a transposon. Neither his father or mother had a transposon in any of the copies of their own *NF1* genes. Explain the cause of neurofibromatosis in the boy and how it arose.

5. On the right are pictured Rainbow the calico cat and cc, her clone. They were introduced in Section 9.4. One cell donated by Rainbow was cloned to form cc. Despite being an exact genetic copy of her donor, cc is not calico. Explain why cc's coat does not match Rainbow's. What other coat colors would you expect to see in different single-cell clones of Rainbow?

Golden Rice or Frankenfood?

Not long ago, the World Health Organization made an estimate that 124 million undernourished children around the world have vitamin A deficiencies. These children may become permanently blind and develop other disorders. Geneticists who wanted to help them transferred genes from daffodils into rice plants. The plants began making beta-carotene, a yellow pigment converted in the body to vitamin A. The pigment is stored in the seeds of the plants—the grains of "golden rice." Researchers predict that including these golden grains in a child's diet will prevent vitamin A deficiency

Rice is the main foodstuff for 3 billion people in impoverished countries. Getting beta-carotene into rice grains seemed the least costly way to deliver vitamin A to those who need it the most, but doing so was beyond the scope of conventional breeding practices. Rice plants just will not breed with daffodil plants.

Golden rice is a genetically modified food. Its creation was an extension of thousands of years of artificial selection and cross-breeding practices that we have been using to coax new breeds of plants and animals from wild ancestral stocks. Meatier turkeys, seedless watermelons, big juicy corn kernels from puny hard ones—the list goes on.

No one wants children to suffer or die. But many people oppose the idea of genetically modified foods, including golden rice. Maybe the unsettling part about the more recent human-directed activities is that the pace has picked up, hugely. We are getting quite good at tinkering with genetics. We do this both for pure research and also for useful, practical applications.

Some say we must never alter the DNA of anything. The concern is that we as a species simply do not have the wisdom to bring about genetic changes without causing irreparable harm. One is reminded of our very human tendency to leap before we look.

And yet, we dream of the impossible. Something about the human experience gave us a capacity to imagine wings of our own making, and that capacity carried us to the frontiers of space. Someone else dreamed of turning plain rice into more nutritious food that might keep some children from going blind.

In this unit, you began with the cell division mechanisms by which parents pass on DNA to new generations. You moved to the chromosomal and molecular basis of inheritance, then to gene controls that guide life's continuity. The sequence parallels the history of genetics. Now, you have arrived at the point in time when geneticists hold molecular keys to the kingdom of inheritance. What they are unlocking is already having an impact on life in the biosphere.

 How Would You Vote? Nutritional labeling is required on all packaged food in the United States, but genetically modified food products may be sold without labeling. Should food distributors be required to label all products made from genetically modified plants or livestock? See BiologyNow for details, then vote online.

 ## Key Concepts

DNA TECHNOLOGIES
With DNA technologies, we can identify, isolate, copy, and otherwise analyze and manipulate genes of any species. These tools have allowed researchers to decipher the linear order of all 3 billion nucleotide bases that make up the complete set of human chromosomes.

GENETIC ENGINEERING
In genetic engineering, copies of normal or modified genes are inserted into individual organisms for research and for practical applications, including medicine, agriculture, and gene therapy. Many ethical and social issues remain unresolved.

Links to Earlier Concepts

This chapter continues the story of fame and fortune that began with the search for the structure of DNA (Section 9.1). It builds on earlier discussions about the molecular structure of DNA (2.10, 9.2), DNA replication and repair (9.3), and mutations (10.5). You may wish to quickly review the nature of mRNA transcription and processing (10.2) and controls over gene transcription (10.6). You will also come across more uses for radioisotopes (2.1), and will see why inheritance patterns (7.6, 8.7) and the lactose operon (10.6) are not necessarily of obscure interest.

11.1 A Molecular Toolkit

Just as a carpenter has a toolkit for manipulating wood and metal, a molecular biologist also has a toolkit—a set of techniques for manipulating and studying DNA.

THE SCISSORS: RESTRICTION ENZYMES

In the 1950s, excitement over the discovery of DNA's structure gave way to frustration. No one could figure out how to determine the order of the nucleotides in DNA—the molecule was just too big. Identifying one particular base within the context of billions of others was an enormous technical challenge. DNA had to be divided into manageable, identifiable sections before its sequence could be determined. But how?

Then, in 1970, Hamilton Smith and his colleagues were studying *Haemophilus influenzae*, a bacteria. The bacteria were protecting themselves from infection by cutting up invading viral DNA before it could insert itself into the bacterial chromosome. This process was not random degradation; certain enzymes were making the cuts at specific nucleotide sequences.

In time, hundreds of strains of bacteria and a few eukaryotic cells yielded thousands of other enzymes like those Smith discovered. Each of these **restriction enzymes** cuts any double-stranded DNA wherever a particular nucleotide sequence occurs. For example, the enzyme EcoRI cuts between the G and the A of the sequence GAATTC. EcoRI and many other restriction enzymes make staggered cuts. These cuts leave single-stranded tails, or "sticky ends," on the cut fragments. Each tail can base-pair with the tail of any other DNA fragment that was cut by the same enzyme, because their sticky ends match up. DNA ligase appropriated from replication and repair processes is used to "glue" restriction fragments together. The result is a molecule of recombinant DNA (Figure 11.1).

CLONING VECTORS

Bacteria, recall, have one circular chromosome. Many also have **plasmids**, much smaller circles of DNA with a few extra genes (Figure 11.2*a*). Bacteria divide often, making huge populations of genetically identical cells. Before each division, replication enzymes copy both the chromosomal DNA *and* the plasmid DNA. Researchers discovered that foreign DNA inserted into a plasmid gets replicated and distributed into each daughter cell along with the plasmid. Bacteria could thus be used to copy a small fragment of DNA again and again.

A plasmid that accepts foreign DNA and slips into a host bacteria, yeast, or another kind of cell is called a **cloning vector**. Cloning vectors typically contain several unique restriction enzyme recognition sites that can be used for cloning (Figure 11.2*b*).

A cell that takes up a cloning vector with foreign DNA in it may found a huge population of descendant cells. Each will contain an identical copy of the vector and a "cloned" copy of the foreign DNA. DNA cloning

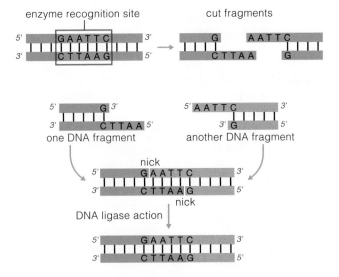

Figure 11.1 Restriction fragments are spliced together to form a recombinant DNA molecule.

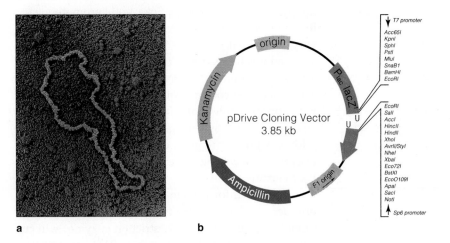

Figure 11.2 Cloning vectors. (**a**) Micrograph of a plasmid. (**b**) A commercially available cloning vector, with its useful restriction enzyme sites listed at right. This vector includes antibiotic resistance genes (*blue, purple*) and part of the bacterial lactose operon (*red*). These genes help researchers identify cells that take up recombinant DNA molecules.

a A restriction enzyme cuts a specific base sequence everywhere it occurs in DNA.

b The DNA fragments have sticky ends.

c The same enzyme cuts the same sequence in plasmid DNA.

d The plasmid DNA also has sticky ends.

e The DNA fragments and plasmid DNA are mixed with DNA ligase.

f The result? A collection of recombinant plasmids that incorporate foreign DNA fragments.

g Host cells that can divide rapidly take up the recombinant plasmids.

Figure 11.3 *Animated!* (**a–f**) Formation of recombinant DNA—in this case, a collection of chromosomal DNA fragments sealed into bacterial plasmids. (**g**) Recombinant plasmids are introduced into host cells that, by dividing, make many copies of the inserted DNA of interest.

LINK TO SECTION 10.2

is one tool in the molecular toolkit. It helps scientists to isolate and identify manageable sections of huge DNA molecules, and also to obtain enough of each bit to be able to study it (Figure 11.3).

cDNA CLONING

Figuring out where a gene is in chromosomal DNA is not by any means obvious. Eukaryotic promoters vary in sequence and in distance from a gene. Splice signals vary, and usually there is no way to tell which ones are used just by reading a nucleotide sequence. Unlike a gene in chromosomal DNA, an mRNA conveniently contains only a coding sequence, minus introns, along with appropriate signal sequences for its translation in eukaryotes. Also, when a gene is active, its transcribed mRNA is present in good quantity. For these reasons, researchers studying gene expression use mRNA as their starting material.

Genes used for protein expression in research and product development are typically cloned, but mRNA cannot be cloned directly. Restriction enzymes will cut only double-stranded DNA, so an mRNA must be first converted into double-stranded DNA before it can be spliced into a vector. A retroviral replication enzyme, **reverse transcriptase**, is used to assemble a strand of **cDNA**, or *complementary* DNA, on an mRNA template

Figure 11.4 cDNA. Reverse transcriptase assembles DNA on an mRNA template. An RNA–cDNA hybrid molecule is the result.

Then, DNA polymerase removes the mRNA as it assembles a new DNA strand on the cDNA. The result is a double-stranded DNA "copy" of an mRNA template.

(Figure 11.4). A hybrid molecule, one strand of mRNA base-paired with one strand of cDNA, is the outcome. DNA polymerase added to the mixture then strips the RNA away as it copies the first strand of cDNA into a second DNA strand. The result is a double-stranded DNA copy of the original mRNA. Because it is double-stranded, that copy may be used for cloning.

In DNA cloning, restriction enzymes cut DNA isolated from any species. DNA ligase seals the fragments into plasmids. RNA is copied into double-stranded DNA before cloning.

A plasmid cloning vector can slip into bacteria, yeast, or other host cell. Host cells divide rapidly, thereby making multiple, identical copies of the foreign DNA.

11.2 Haystacks to Needles

To study one gene, researchers must first find it among all the others in an organism's DNA. It is a bit like searching for a needle in a haystack. Once found, the gene is copied many times to make enough material for experiments.

ISOLATING GENES

A **library** is a collection of cells that host fragments of DNA from a particular organism. Collectively, all cells in a *genomic library* contain the organism's entire set of genetic material, or **genome**. A *cDNA library* is derived from mRNA, and so it represents only genes that are being expressed.

A clone that contains a particular gene of interest occurs in a library along with thousands or millions of clones that contain other genes. A probe can be used to find a single clone of interest among many others. A **probe** is a short fragment of DNA, complementary in sequence to a specific gene of interest, tagged with a detectable label such as a radioisotope or pigment.

How do researchers make a suitable probe? If they already know the desired gene sequence, they can use it to design and make a short chain of nucleotides. Or they can use some DNA from a closely related gene.

Probes distinguish one DNA sequence from all of the others in a library of clones. A probe will base-pair only with DNA that contains a complementary DNA sequence. In other words, a probe sticks to the gene of interest, but it does not stick to any other DNA. Base-pairing between any two strands of nucleic acids from different sources is called **nucleic acid hybridization**.

Researchers can pinpoint a DNA fragment that has the gene of interest, and also the clone that contains it, by detecting the label on a hybridized probe. Once the clone is found, it can be isolated from the other clones in the library. That clone will be grown in a nutrient medium to form a gigantic population of genetically identical cells, from which researchers can extract the cloned DNA in sufficient quantity for experiments. An example of how researchers isolate a bacterial clone by using a radioactive probe is shown in Figure 11.5.

PCR

Researchers can isolate and mass-produce a particular DNA fragment without cloning, by using a technique called the polymerase chain reaction, or PCR. **PCR** is a hot–and–cold cycled reaction that uses a heat-tolerant polymerase to replicate a targeted DNA fragment by the billion. It can amplify, or increase the quantity of,

a Individual cells from a library of bacterial cell clones are spread over the surface of a solid growth medium in a dish. The bacterial cells divide repeatedly to form colonies, clusters of millions of genetically identical daughter cells.

b A piece of special paper pressed onto the surface of the growth medium will bind some cells from each colony.

c The paper is soaked in a solution that ruptures the cells and releases their DNA. The DNA clings to the paper in spots mirroring the distribution of the colonies.

d A probe is added to the liquid bathing the paper. The probe hybridizes with (sticks to) only the spots of DNA that contain the targeted sequence.

e The bound probe makes a radioactive spot that will darken x-ray film. The position of the spot on the film relative to all the other spots of DNA is used to find the original colony in the dish. Cells from that colony alone are grown to isolate the cloned gene of interest.

Figure 11.5 *Animated!* Example of nucleic acid hybridization. Here, a radioactive probe helps identify a clone, in this case a bacterial colony, that contains a targeted gene.

a Primers, free nucleotides, and DNA template are mixed with heat-tolerant DNA polymerase.

b When the mixture is heated, the DNA denatures. When it is cooled, some primers hydrogen-bond to the DNA template.

c *Taq* polymerase uses the primers to initiate synthesis. It assembles a copy of the DNA template from free nucleotides. The first round of PCR is completed.

d The mixture is heated again. This denatures all the DNA into single strands. When the mixture is cooled, some of the primers hybridize with the DNA.

e *Taq* polymerase uses the primers to initiate synthesis and then copies the DNA. The second round of PCR is complete. Each successive round of synthesis can double the number of DNA molecules.

Figure 11.6 *Animated!* Two rounds of the polymerase chain reaction, or PCR. A bacterium, *Thermus aquaticus*, is the source of the *Taq* polymerase. Thirty or more cycles of PCR can yield a billionfold increase in the number of starting DNA template molecules.

an improbably tiny amount of DNA. PCR transforms a needle in a haystack, one DNA fragment mixed with millions of others, into a huge stack of needles with a little hay in it (Figure 11.6).

The starting material for PCR can be any sample that contains at least one molecule or fragment of a target DNA. It can be DNA extracted from a mixture of ten million different clones, a single sperm cell, a drop of blood at a crime scene, a mummy—essentially anything that has DNA in it.

In a PCR reaction, researchers mix the starting DNA with free nucleotides, DNA polymerase, and primers. **Primers** are synthetic single strands of DNA, usually between ten and thirty bases long. They are designed to base-pair only with nucleotide sequences on either end of the DNA sequence to be amplified.

The researchers then expose the mixture to cycles of high and low temperatures. At high temperature, the two strands of a DNA double helix separate into single strands. During a high-temperature PCR cycle, every DNA molecule unwinds and becomes single-stranded. Then, the mixture is allowed to cool. Like molecular sleuths, some of the primers find the template needle in the haystack mixture, and hybridize with it.

The high temperature needed to separate strands of DNA would destroy most DNA polymerases (Section 2.9). The special polymerase used for PCR reactions is harvested from *Thermus aquaticus*, a species of bacteria that thrives in superheated spring water (Chapter 14). The *Taq* polymerase is necessarily heat-tolerant.

Like all other DNA polymerases, *Taq* polymerase recognizes primers hybridized with template DNA as places to initiate synthesis. Synthesis proceeds along the template until the temperature cycles up and the DNA separates into single strands again. The newly synthesized DNA is a copy of the target sequence, so the number of DNA template molecules has doubled.

When the temperature cycles down again, primers rehybridize to new template DNA molecules, and the reactions start all over. With each successive round of temperature cycling, the number of copies of targeted DNA doubles. After thirty PCR cycles, the number of template molecules will have been amplified by about a billionfold.

Probes may be used to help identify a clone that hosts a DNA fragment of interest among many clones in a library. That clone is cultured, and its DNA is harvested.

The polymerase chain reaction (PCR) is a method of quickly and exponentially amplifying even a tiny amount of DNA. Compared to cloning methods, PCR more rapidly amplifies a target DNA sequence.

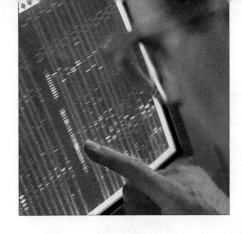

11.3 DNA Sequencing

Sequencing reveals the order of nucleotide bases in DNA. This technique uses DNA polymerase to partially replicate a DNA template. Manual sequencing methods have been replaced largely by automated techniques.

Once a targeted DNA molecule has been isolated in sufficient quantity, the order of its nucleotide bases may be determined with **DNA sequencing**—another technique in the molecular toolkit.

Researchers mix the DNA template to be sequenced with nucleotides, primer, and DNA polymerase. The reaction mixture also includes nucleotides modified to terminate DNA synthesis when added to the end of a growing strand of DNA. Also, each of the four kinds of modified nucleotide bases (G, A, T, and C) is tagged to fluoresce a different color under laser light.

As in DNA replication, the polymerase constructs a new strand of DNA by stringing nucleotides together according to the template DNA sequence. Each time, the polymerase randomly attaches either a standard *or* a modified nucleotide to the end of the new strand of DNA. If it attaches one of the modified nucleotides, synthesis of that nucleotide chain ends. After enough time, there are new strands of DNA that end at each base in the template sequence. The mixture now holds millions of DNA fragments of different lengths, each of which is labeled with one of four fluorescent pigments.

The fragments are separated by **gel electrophoresis**. With this technique, an electric field pulls the DNA fragments through a polymer matrix, a gel that is a little like Jello®. This matrix hinders the movement of larger molecules more than that of smaller ones, as a dense forest would hinder the movement of elephants more than tigers. So different sized fragments migrate at different rates through the gel according to length— larger fragments move more slowly than smaller ones. Fragments of the same length migrate through the gel at the same rate, and form distinct bands.

As the bands reach the end of the gel, they pass through a laser beam. Remember, each DNA fragment has a modified nucleotide attached to its end, one that fluoresces a certain color under laser light. Thus, each band fluoresces with the color of one kind of modified nucleotide. The order of the colored bands indicates the DNA sequence (Figure 11.7).

DNA sequencing reveals the order of nucleotides in DNA. The technique uses gel electrophoresis, which separates fragments of DNA according to their length.

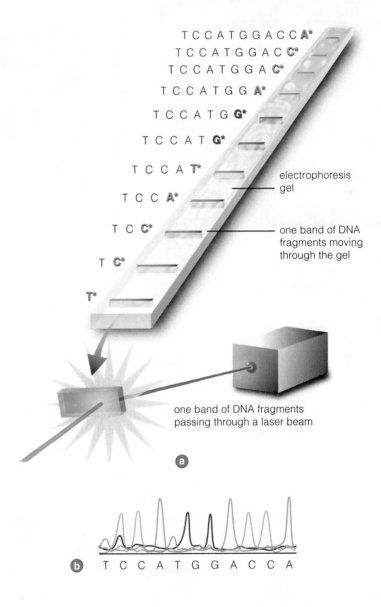

Figure 11.7 *Animated!* DNA sequencing. (**a**) DNA fragments are synthesized using a template and fluorescent nucleotides, then separated by gel electrophoresis. Fragments of the same length migrate through the gel at the same rate, and so form distinct bands. (**b**) The order of the fluorescent bands that appear in the gel indicates the template DNA sequence.

① ② ③ FROM BLOOD AT CRIME SCENE ④ ⑤ ⑥ ⑦

Figure 11.8 Comparison of the DNA fingerprint from a bloodstain left at a crime scene and from blood samples of seven suspects (the circled numbers). Can you point out which of the seven is a match?

LINKS TO SECTIONS 7.6, 8.7

11.4 First Just Fingerprints, Now DNA Fingerprints

Except for identical twins, no two people have exactly the same base sequence in their DNA. Scientists can distinguish one person from another on the basis of differences in those sequences.

Each human has a unique set of fingerprints. Like all other sexually reproducing species, each also has a **DNA fingerprint**—a unique array of DNA sequences inherited in a Mendelian pattern from parents. About 99 percent of the DNA in all humans is the same, but the other 1 percent is unique to each individual. Most unique stretches of DNA are sprinkled throughout the human genome as **tandem repeats**—many copies of the identical base sequences, positioned one after the next along the length of a chromosome.

For example, one person's DNA might contain four repeats of the nucleotides TTTTC in a certain location. Another person's DNA might have TTTTC repeated

fifteen times in the same location. One person might have five repeats of CGG, and another might have fifty. These repetitive sequences spontaneously slip into DNA during replication, and their numbers grow or shrink over generations. The mutation rate is high around tandem repeat regions.

DNA fingerprinting reveals differences in the tandem repeats among individuals. A restriction enzyme cuts genomic DNA into fragments, the sizes of which are unique to an individual, because they reflect regions of tandem repeats. Thus, the genetic differences between people can be revealed by gel electrophoresis.

As in DNA sequencing, the fragments form bands on an electrophoresis gel according to their length. The banding pattern of a person's genomic DNA fragments on an electrophoresis gel is his or her DNA fingerprint. For all practical purposes, it is identical only between identical twins, because the odds that two unrelated people would share an identical DNA fingerprint are about one in three trillion.

PCR can be used to amplify tandem-repeat regions. Differences in the sizes of amplified DNA fragments are again detected with gel electrophoresis. One drop of blood or semen, or a few cells from a hair follicle at a crime scene or on a suspect's clothing yield enough DNA to amplify with PCR.

DNA fingerprints help forensic scientists identify criminals, victims, and innocent suspects. Figure 11.8 shows some DNA fragments of genomic DNA that were separated by gel electrophoresis. Those samples of DNA were taken from seven people and from a bloodstain that was left at a crime scene. One of the DNA fingerprints matched.

Defense attorneys initially challenged the use of DNA fingerprinting as evidence in court. However, today the procedure has been firmly established as accurate and unambiguous. Now, DNA fingerprinting is routinely submitted as evidence in disputes over paternity, and it is being widely used to convict the guilty as well as exonerate the innocent. To date, such evidence has helped secure the release of more than 140 innocent people from prison.

DNA fingerprint analysis also has confirmed that human bones exhumed from a shallow pit in Siberia belonged to five individuals of the Russian imperial family, all shot to death in secrecy in 1918. It also was used to identify the remains of those who died in the World Trade Center on September 11, 2001.

Tandem repeats—many copies of short DNA sequences in a chromosome—differ from one person to the next. They are the basis of DNA fingerprint comparisons.

DNA TECHNOLOGIES

Figure 11.9
DNA sequence data. Here, a few thousand bases of the DNA sequence of the human genome.

11.5 Tinkering With the Molecules of Life

The structural and comparative analysis of genomes is yielding information about evolutionary trends as well as potential therapies for genetic diseases.

EMERGENCE OF MOLECULAR BIOLOGY

In 1953, James Watson and Francis Crick unveiled their model of the DNA double helix and ignited a global blaze of optimism about genetic research. The very book of life seemed to open up for scrutiny. In reality, it dangled just beyond reach. Major scientific breakthroughs are not very often accompanied by the simultaneous discovery of the tools to study them. New methods of doing research had to be invented before that book could become readable.

In 1972, Paul Berg and his associates were the first to make recombinant DNA by fusing fragments of DNA from one species into the genetic material from a different species. Their technique allowed them to isolate and replicate manageable subsets of DNA from any organism they wanted to study. The discipline of molecular biology was born, and suddenly everybody was worried about it.

Although scientists knew that DNA itself was not toxic, they could not predict with absolute certainty what would happen every time they fused genetic material from different organisms. Would they create new super-pathogens by accident? Could DNA from normally harmless organisms be fused to create a new form of life? What if their creation escaped into the environment and transformed other organisms?

In a remarkably immediate and responsible display of self-regulation, scientists reached a consensus on safety guidelines for doing DNA research. Adopted at once by the National Institutes of Health (NIH), the guidelines listed laboratory procedural precautions. They covered the design and use of host organisms that could survive only under the narrow range of conditions that occur in the laboratory.

A golden age of recombinant DNA research soon followed. The emphasis had shifted from the chemical and physical properties of DNA to its exact molecular structure. In 1977, Allan Maxam, Walter Gilbert, and Fred Sanger developed a method for determining the nucleotide sequence of cloned DNA fragments. The tools for reading the book of life, opened more than twenty years earlier, were now available for everyone to use.

DNA sequencing was cool, a visually rewarding, data-rich technique that soon entranced more than a few scientists (Figure 11.9). Tremendous amounts of sequence data started to amass, from diverse species. Computer technology was simultaneously advancing, but it was barely keeping pace with the demand for sequence data analysis and storage. In 1982, the NIH allocated 3 million dollars to fund the first large-scale DNA sequence database in the United States, one that was accessible to the public.

LINK TO
SECTION
9.1

THE HUMAN GENOME PROJECT

Around 1986, it seemed that everyone was arguing about sequencing the human genome. Many insisted that knowing the sequence would offer tremendous support for medicine and pure research. Others said that such a project would divert funding from other research that had greater urgency as well as a greater likelihood of success.

At the time, sequencing three billion bases seemed like an impossible task. Sequencing has a maximum resolution of about 500 bases of sequence data per run, so at least 6 million sequencing runs would be necessary. With the techniques available at the time, this would have taken at least fifty years.

But techniques were getting better every year, and more bases were being sequenced and analyzed in less time. Automated (robotic) sequencing had just been invented, as had PCR, the polymerase chain reaction. Although both techniques were still cumbersome, expensive, and far from standardized, many sensed their potential.

Figure 11.10 A few of the supercomputers used to sequence the human genome, at Celera Genomics in Maryland.

What do we do with this vast amount of data? The next step is to investigate questions about precisely what the sequence means: where all of the genes are, where they are not, what the genes encode, and what the control mechanisms are and how they operate.

GENOMICS

Research into genomes of humans and other species has converged into a new research field—**genomics**. The *structural* genomics branch deals with the actual mapping and sequencing of genomes of individuals. The *comparative* genomics branch is concerned with finding evolutionary relationships among groups of organisms. Comparative genomics researchers analyze similarities and differences among genomes.

Comparative genomics has practical applications as well as potential for research. The basic premise is that the genomes of all existing species are derived from common ancestors. For instance, pathogens share some conserved genes with their human hosts even though the lineages diverged long ago. Shared gene sequences, how they are organized, and where they differ might be clues to where our immune defenses against these pathogens are strongest or the most vulnerable.

Genomics has potential for **human gene therapy**—the transfer of one or more normal or modified genes into a person's body cells to correct a genetic defect or boost resistance to a disease. For example, some gene therapies deliver modified cells into a patient's tissue. Others use stripped-down viruses as vectors that can inject genes into a person's cells. In many cases, these therapies can alleviate the patient's symptoms even if the modified cells produce just a small amount of the gene's product.

However, a gene is not easy to manipulate within the context of a living individual, even if we know its sequence and where it lies in the genome. No one, for example, can predict where a virus-injected gene will end up in a person's chromosomes. The danger is that the insertion might disrupt other genes, particularly those which control cell division and growth (Section 11.7 gives an example). One-for-one gene swaps using genetic recombination methods are possible, but like other gene therapies they are still experimental.

LINK TO
SECTION
10.5

Waiting for faster methods seemed to be the most efficient means of sequencing the human genome, but exactly when was the technology going to be fast *enough*? Who would decide? It was during this heated debate in 1987 that several independent organizations launched their own efforts to sequence the human genome. Among them was Genome, a company started by Walter Gilbert. Gilbert announced that his company not only was going to sequence the human genome, but the company was also going to patent it.

In 1988, the NIH effectively annexed the Human Genome Project by hiring James Watson as its head and providing genome researchers with 200 million dollars per year. A consortium formed between the NIH and other institutions working on the different versions of the project.

Amid ongoing squabbles over patent issues, the bulk of the Human Genome Project continued as such in the United States until 1998, when scientist Craig Venter started Celera Genomics (Figure 11.10). Venter declared that *his* new company would be the first to complete the sequence. His challenge prompted the United States government to propel its sequencing efforts into high gear.

Sequencing of the human genome was officially completed in 2003—fifty years after the discovery of the structure of DNA. Around 99 percent of the coding regions in human DNA have been sequenced with a high degree of accuracy. To date, more than 21,000 genes have been identified. This does not mean that we know what all of those genes encode; it only means we know they are definitely genes.

In 2002, an international research team finished a draft sequence of the rice genome. The data were made available on the Internet free of charge. Rice is the staple food for more than half the world's population. It was the first crop plant to have its genome fully sequenced. This work will be used to design new rice varieties with more nutritional value, bigger crop yields, and more resistance to drought and pests.

Fifty years after the structure of DNA was discovered, the human genome sequence was completed.

We are applying knowledge about the genomes of humans and other species to reconstruct evolutionary histories and to design gene therapies.

DNA TECHNOLOGIES

11.6 Practical Genetics

Genetic engineering changes the genetic makeup of an organism, often with the intent to alter its phenotype. A gene may be transferred between species, or modified and reinserted into an organism of the same species.

DESIGNER PLANTS

As crop production expands to keep pace with human population growth, it puts unavoidable pressure on ecosystems everywhere. Irrigation leaves mineral and salt residues in soils. Tilled soil erodes, taking topsoil with it. Runoff clogs rivers, and fertilizer in it causes algae to grow so much that fish suffocate. Pesticides harm humans, other animals, and beneficial insects. Pressured to produce more food at lower cost and with less damage to the environment, some farmers are turning to genetically engineered crop plants.

Cotton plants with a built-in insecticide gene kill only the insects that eat it, so farmers that grow them do not have to use as many pesticides. Genetically modified wheat has double the yield per acre. Certain transgenic tomato plants survive in salty soils that wither other plants; they also absorb and store excess salt in their leaves, thus purifying saline soil for future crops. **Transgenic** simply refers to an organism into which DNA from another species has been inserted, as in Figure 11.11.

The cotton plants in Figure 11.11*a* were genetically engineered for resistance to a relatively short-lived herbicide. Spraying fields with this herbicide kills all of the weeds but not the engineered cotton plants. The practice means farmers can use fewer toxic chemicals

a   b

Figure 11.11 Transgenic plants. (**a**) Cotton plant (*left*), and cotton plant with a gene for herbicide resistance (*right*). Both were sprayed with weed killer.

(**b**) When biosynthesis for lignin was partially blocked in these genetically engineered aspen seedlings, cellulose production increased. Root, stem, and leaf growth were greatly enhanced. Lignin is a tough polymer that strengthens plant cell walls. Unmodified seedling is on the left.

and less of them. They also do not have to till the soil as much to control weeds, so there is less runoff.

Aspen tree seedlings in which a lignin biosynthesis pathway has been modified still make lignin, but not as much (Figure 11.11*b*). Wood from lignin-deficient trees might make it easier to manufacture paper and clean-burning fuels such as ethanol.

Engineering plants starts with cloning vectors that carry genes into plant cells. *Agrobacterium tumefaciens* is a bacterial species that infects many plants, including beans, peas, potatoes, and other vital crops. Its plasmid genes cause tumor formation on these plants; hence the name Ti plasmid (*Tumor-inducing*). The Ti plasmid is used as a vector for transferring new or modified genes into plants (Figure 11.12). Researchers excise the tumor-inducing genes, then insert a desired gene into the Ti plasmid. Plant cells cultured with the modified

a A bacterial cell contains a Ti plasmid (*purple*) that has a foreign gene (*blue*).

b The bacterium infects a plant and transfers the Ti plasmid into it. The plasmid DNA becomes integrated into one of the plant's chromosomes.

c The plant cell divides. Its descendant cells form an embryo, which may develop into a mature plant that can express the foreign gene.

A young plant expressing a fluorescent gene product

Figure 11.12 *Animated!* Gene transfer from a bacterium, *Agrobacterium tumefaciens*, to a plant cell by way of a Ti plasmid.

A new version of golden rice contains DNA from two different sources: a soil bacterium and maize. A mixture of these DNAs was introduced into rice embryos. Once inside, the foreign genes directed the production of two enzymes that catalyze the synthesis of beta-carotene in endosperm, a nutritive tissue to be used by the embryo seedling. The new golden rice has 20 times more beta-carotene than the original version.

plasmid take it up. Whole plants may be regenerated. Modified *A. tumefaciens* bacteria can deliver genes into some food crop plants, including wheat, corn, and rice. Researchers can also transfer genes into plants by way of electric shocks, chemicals, or blasts of microscopic particles coated with DNA.

To date, sixty-five genetically modified plants are approved by the USDA for general crop use in the United States, and hundreds more are in the works. Among them are berries, apples, and melons that ripen later, disease-resistant plums, papaya, grapes, carrots, and walnuts, and reduced-nicotine tobacco.

GENETICALLY ENGINEERED BACTERIA

Decoding the genetic scripts of species opens the door to great possibilities. Consider genetically engineered bacteria that produce medically valued proteins. Huge populations churn out vast quantities of desired gene products in stainless steel vats.

Diabetics, who must get daily insulin injections, are among the beneficiaries of these bacteria. At one time, insulin for their injections was extracted from pigs and cattle. In 1982, synthetic genes for human insulin were transferred into *E. coli* cells. The cells became the first large-scale, cost-effective "factory" for production of a medically important protein. Some other engineered products of bacteria are various blood-clotting factors, taxol, and human growth hormone.

Modified bacteria also hold potential for industry and for cleaning up environmental messes. Many can break down crude oil and other organic wastes into

less harmful compounds, and cycle them as nutrients through ecosystems. When sprayed on oil spills, they may avert ecological disasters of the sort that occurred in the Persian Gulf (Chapter 1). Others absorb excess phosphates and heavy metals.

Plasmid-containing species are being used in basic research, agriculture, and gene therapies. Deciphering bacterial genes is also helping researchers reconstruct the evolutionary history of life. The next two sections give a few more examples of this application.

What about the "bad" bunch, the pathogenic fungi, bacteria, and viruses? Natural selection favors mutated genes that improve a pathogen's chances of evading a host's defenses. Mutation is frequent among pathogens that reproduce fast, so designing new antibiotics and other defenses against their altered gene products is a constant challenge. If we understand their genes, we may be able to keep up with them. The story of HIV, a rapidly mutating virus that causes AIDS, hints at the magnitude of the problem (Section 23.8).

BARNYARD BIOTECH

The first mammals enlisted for experiments in genetic engineering were lab mice. Transgenic mice appeared on the research scene in 1982 when scientists built a plasmid containing a gene for rat somatotropin (growth hormone). They injected the recombinant DNA into fertilized mouse eggs, and then implanted the eggs into female mice. One-third of the resulting offspring grew much larger than their littermates. The rat gene had become integrated into their DNA and was being expressed, with impressive results.

Now, transgenic animals are commonly used for medical research. The function and regulation of many gene products have been discovered using "knockout mice," in which targeted genes are inactivated by using recombination techniques. Defects in the resulting mice give clues about the gene. Strains of mice engineered to be susceptible to human diseases give researchers the opportunity to study both the diseases and their cures without experimenting on humans.

Genetically engineered animals are also sources of pharmacological and other valuable proteins. As a few examples, different engineered goats produce proteins used to treat cystic fibrosis, heart attacks, and nerve gas exposure. Some goats produce spider silk protein for medical supplies, space equipment, and bullet-proof vests. Human antithrombin, used to treat people with blood clotting disorders, is another product of engineered goats (Figure 11.13a). Rabbits make human IL-2 (interleukin-2) a protein that stimulates immune cell divisions. Cattle produce human collagen that can be used for surgical repair of cartilage, bone, and skin.

a   b

Figure 11.13 Genetically modified animals. (**a**) Mira, a goat transgenic for human antithrombin III, an anticlotting factor. (**b**) Inquisitive transgenic pig at the Virginia Tech Swine Research facility.

11.7 Weighing the Benefits and Risks

We as a society continue to work our way through the ethical implications of DNA research even while we are applying the new techniques to medicine, industry, agriculture, and environmental remediation.

Figure 11.14 Featherless chicken developed by traditional cross-breeding methods in Israel. Such chickens survive in hot deserts where cooling systems are not an option.

Genetic engineering has also given us dairy goats with heart-healthy milk, pigs whose low phosphorus manure is easier on the environment, freeze-resistant salmon, extra-hefty sheep, low-fat pigs, and mad cow disease-resistant cows. Allergen-free cats are next.

Tinkering with the genetics of animals for the sake of human convenience does raise some serious ethical issues, particularly so because failed experiments can have gruesome results. However, is transgenic animal research not simply an extension of thousands of years of acceptable barnyard breeding practice (Figure 11.14)? The techniques have changed, but not the intent. Like our ancestors, we continue to have a vested interest in improving our livestock.

Transgenic plants help farmers grow crops more efficiently.

Transgenic animals are widely used in medical research. Some are sources of medically valued proteins. Food animals are being engineered to be more nutritious, more disease resistant, or easier to raise.

Figure 11.15 Gene therapy. Rhys Evans was born with a gene mutation that causes SCID-X1. His immune system never could fight infections. A gene transfer freed him from life in a germ-free isolation tent.

Over 15,500 serious genetic disorders affect between 3 and 5 percent of all newborns, and they cause 20 to 30 percent of all infant deaths per year. They account for about 50 percent of the mentally impaired and nearly 25 percent of all hospital admissions.

Rhys Evans (Figure 11.15) was born with a severe immune deficiency known as SCID-X1, which stems from mutations in a gene called *IL2RG*. Children who are affected by this disorder can live only in germ-free isolation tents, because they cannot fight infections.

In 1998, a viral vector was used to insert unmutated copies of *IL2RG* into cells taken from the bone marrow of eleven boys with SCID-X1. Each child's modified cells were infused back into his bone marrow. Months afterward, ten of the children left their isolation tents for good. Their immune systems had been repaired by the gene therapy. Since then, several other SCID-X1 patients, including Rhys Evans, have improved thanks to other gene therapy trials.

Three children from the initial experiment in 1998 have since developed leukemia, and one died. Their illness surprised researchers, who had predicted that any cancer related to the therapy would be extremely rare. Our understanding of how the human genome works clearly lags behind our ability to modify it.

Modifying the human genome has profound ethical implications even beyond unexpected risks. To many, correcting genetic disorders with gene therapy seems like a socially acceptable goal. Now take this idea one step further. Is it acceptable to change some genes of a normal human in order to alter or enhance traits?

The philosophy of selecting desirable human traits, called *eugenics*, is not new, but it is still controversial. Who, for example, decides which forms of a trait are the most desirable? Realistically, cures for most severe but rare genetic disorders will not be pursued because the financial payback will not be enough to cover the research. Eugenics, however, might just turn a profit. How much might potential parents pay to ensure that their children will be tall or blue-eyed or fair-skinned? Would it be okay to engineer "superhumans" with breathtaking strength or intelligence? How about an injection that would help you lose that extra weight, and keep it off permanently? The borderline between interesting and abhorrent is not the same for everyone.

transgenic mouse
expressing green
fluorescent protein

In a survey conducted not long ago in the United States, more than 40 percent of those interviewed said it would be fine to use gene therapy to make smarter or better looking babies. In one poll of British parents, 18 percent would use genetic enhancement to prevent their children from being aggressive, and 10 percent would use it to keep them from being homosexual.

KNOCKOUT CELLS AND ORGAN FACTORIES

Each year, about 75,000 people are on waiting lists for an organ transplant, but human donors are in short supply. There is talk of harvesting organs from pigs, because pig organs function very much like ours do. Transferring an organ from one species into another is called **xenotransplantation**.

The human immune system battles anything that it recognizes as "nonself." It rejects a pig organ at once. Researchers have successfully engineered pigs lacking immunity-inducing molecules on the surface of their cells. The human immune system might not recognize and attack organs from these pigs. Tissues and organs from such animals could benefit millions of people, including those who suffer from diabetes, Parkinson's disease, kidney failure, AIDS, and hepatitis.

Critics of xenotransplantation are concerned that, among other things, pig to human transplants would put selective pressure on pig viruses to cross species and infect humans. These concerns are not unfounded. In 1918, an influenza pandemic killed twenty million people worldwide. It originated with a strain of avian flu that mutated to become infectious in humans.

REGARDING "FRANKENFOOD"

Genetically engineered foods are widespread in the United States. In 2003, 81 percent of soybean crops and 40 percent of corn crops were modified to withstand weedkillers. Such crops are no longer fed only to farm animals. At least 70 percent of all processed foods in grocery stores—like tofu, cereal, soy sauce, vegetable oils, breads, frozen pizzas, hot dogs, soft drinks, and beer—may contain some ingredients from genetically engineered plants.

Read the research and form your own opinions. The alternative is to be swayed by media hype (the term "Frankenfood," for instance), or by potentially biased reports from other groups that might have a different agenda (such as herbicide manufacturers).

Our ability to manipulate genomes is outpacing attempts to work through some of its bioethical implications.

Summary

Section 11.1 Recombinant DNA technology uses restriction enzymes that cut DNA into fragments. The fragments may be spliced into cloning vectors by using DNA ligase. Recombinant plasmids are taken up by host cells, such as bacteria, to make multiple, identical copies of the foreign DNA. mRNA can be cloned after it is copied into double-stranded DNA.

Section 11.2 A gene library is a mixed collection of host cells that have taken up cloned DNA. A particular gene can be isolated from a library by using a probe, a short stretch of DNA that can base-pair with the gene and that is traceable with a radioactive or pigment label. Probes help researchers identify one particular clone among millions of others. Base-pairing between strands of DNA or RNA from different sources is called nucleic acid hybridization.

The polymerase chain reaction (PCR) is a way to rapidly copy particular pieces of DNA. A DNA template is mixed with nucleotides, primers, and a heat-resistant DNA polymerase. Each cycle of PCR proceeds through a series of temperature changes that doubles the number of DNA molecules. After thirty cycles, PCR amplification can result in a billionfold increase in the number of DNA molecules.

Biology⊗Now
Learn how researchers isolate and copy genes with the interaction on BiologyNow.

Section 11.3 DNA sequencing reveals the order of nucleotides in DNA fragments. As DNA polymerase is copying a template DNA, progressively longer fragments stop growing when one of four kinds of modified, fluorescent nucleotides becomes attached.

Electrophoresis separates the resulting labeled DNA fragments into bands according to length. The order of the colored bands as they migrate through the gel reflects which fluorescent base was added to the end of each fragment, and so indicates the template DNA nucleotide sequence.

Biology⊗Now
Investigate DNA sequencing with the animation on BiologyNow.

Section 11.4 Tandem repeats are multiple copies of a short DNA sequence that follow one another along a chromosome. The number and distribution of tandem repeats, unique in each person, can be revealed by gel electrophoresis; they form a DNA fingerprint.

Biology⊗Now
Observe the process of DNA fingerprinting with the animation on BiologyNow.

Section 11.5 Discovery of DNA's double helical structure sparked interest in deciphering its genetic messages. A global race to complete the sequence of

GENETIC ENGINEERING

the human genome spurred rapid development of new techniques to study and manipulate DNA. The entire human genome has been sequenced and is now being analyzed in detail.

Genomics, the study of human and other genomes, is shedding light on evolutionary relationships and has practical uses. Human gene therapy transfers normal or modified genes into body cells with the intent to correct genetic defects.

Section 11.6 Genetic engineering is the directed modification of the genetic makeup of an organism, often with intent to modify its phenotype. Researchers insert normal or modified genes from one organism into another of the same or different species. Gene therapies also reinsert altered genes into individuals.

Genetically engineered bacteria produce medically valued proteins. Transgenic crop plants help farmers produce food more efficiently. Genetic engineering of animals allows commercial production of human proteins, as well as research into genetic disorders.

Biology ❂ Now
See how the Ti plasmid is used to genetically engineer plants with the animation on BiologyNow.

Section 11.7 Gene therapy and the modification of animals for xenotransplantation are examples of developing technologies. As with any new technology, potential benefits must be weighed against potential risks, including ecological and social repercussions.

Self-Quiz   *Answers in Appendix I*

1. _____ is the transfer of normal genes into body cells to correct a genetic defect.
   a. Reverse transcription   c. PCR
   b. Nucleic acid hybridization   d. Gene therapy

2. DNA is cut at specific sites by _____ .
   a. DNA polymerase   c. restriction enzymes
   b. DNA probes   d. reverse transcriptase

3. Fill in the blank: A _____ is a small circle of bacterial DNA that is not part of the bacterial chromosome.

4. By reverse transcription, a molecule of _____ is assembled on a(n) _____ template.
   a. mRNA; DNA   c. DNA; ribosome
   b. cDNA; mRNA   d. protein; mRNA

5. By gel electrophoresis, fragments of DNA can be separated according to _____ .
   a. sequence   b. length   c. species

6. DNA sequencing uses _____ .
   a. standard and modified nucleotides
   b. primers and DNA polymerase
   c. gel electrophoresis and a laser beam
   d. all of the above

7. _____ can be used to insert genes into human cells.
   a. PCR   c. Xenotransplantation
   b. Modified viruses   d. cDNA

8. Match the terms with the most suitable description.
   ____ DNA fingerprint   a. selecting "desirable" traits
   ____ Ti plasmid   b. uses heat stable polymerase
   ____ PCR   c. used in some gene transfers
   ____ nucleic acid   d. a person's unique collection
       hybridization   of tandem repeats
   ____ eugenics   e. base pairing of nucleotide
                         sequences from different
                         DNA or RNA sources

Additional questions are available on **Biology ❂ Now™**

Critical Thinking

1. Lunardi's Market put out a bin of nice tomatoes having vine-ripened redness, flavor, and texture. A sign identified them as genetically engineered produce. Most shoppers selected unmodified tomatoes in the adjacent bin even though those tomatoes were pale pink, mealy-textured, and tasteless. Which tomatoes would you pick? Why?

2. Scientists at Oregon Health Sciences University produced ANDi, the first transgenic primate, by inserting a jellyfish gene into the fertilized egg of a rhesus monkey. (The gene encodes a protein that fluoresces green.) The long-term goal of this project is not to make fluorescent green monkeys, but the transfer of human genes into organisms with genomes most like ours. Transgenic primates could be studied to gain insight into human genetic disorders, which might lead to cures for those who are affected and vaccines for those at risk. Would it be ethical to transfer a monkey gene into a human embryo to cure a genetic defect? Or to bestow immunity against a potentially fatal disease such as AIDS?

3. Animal viruses can mutate so that they infect humans, occasionally with disastrous results. In 1918, an influenza pandemic that apparently originated with a strain of avian flu left in its wake 50 million people dead worldwide. Researchers recently isolated samples of that virus, the influenza A-H1N1 strain, from bodies of infected people that were fortuitously preserved in Alaskan permafrost since 1918. From the samples, the scientists reconstructed the DNA sequence of the entire viral genome, and from that sequence they reconstructed the actual virus. 39,000 times more infectious than recent influenza strains, the reconstructed A-H1N1 virus is 100 percent lethal in mice.

Understanding how the A-H1N1 strain works in a living context can help us defend ourselves against others that may be like it. For example, researchers are using the reconstructed virus to discover which of its mutations made it so infectious and deadly in humans. This research is of immediate importance. A deadly new strain of avian influenza in Asia has some mutations in common with the A-H1N1 strain. Perhaps most pertinent, researchers can test the effectiveness of antiviral drugs and vaccines on the reconstructed virus, and develop new ones.

Critics of the A-H1N1 reconstruction are concerned that if this virus escapes the containment facilities (this is not without precedent), it might cause another pandemic. Worse, terrorists could use the published DNA sequence and methods to make the virus for nefarious purposes. Do you think this research makes us more or less safe?

Rise of the Super Rats

Slipping in and out of the pages of human history are rats—*Rattus*—the most notorious of mammalian pests. One kind of rat or another has distributed pathogens and parasites that cause bubonic plague, typhus, and other deadly infectious diseases. The death toll from fleas that bit infected rats and then bit people has exceeded the death toll in all wars combined.

The rats themselves are far more successful. By one estimate, there is one rat for every person in urban and suburban centers of the United States. In addition to spreading diseases, rats chew their way through walls and wires of homes and cities. In any given year, they cause economic losses approaching 19 billion dollars.

For years, people have been fighting back with traps, ratproof storage facilities, and various poisons. During the 1950s, they started using baits laced with warfarin. This compound interferes with blood clotting. Rats ate the baits, then died within days after bleeding internally or losing blood through cuts or scrapes. Warfarin was extremely effective. Compared with other rat poisons, it had a lot less impact on harmless species.

In 1958, however, a Scottish researcher reported that warfarin did not work against some rats. Similar reports from other European countries followed. About twenty years later, 10 percent of the urban rats caught in the United States were warfarin resistant. *What happened?*

To find out, researchers compared warfarin-resistant rat populations with still-vulnerable rats. They traced the difference to a gene on one of the rat chromosomes. At that gene locus, a dominant allele was common in warfarin-resistant rat populations but very rare among the vulnerable ones. "What happened" was evolution by natural selection. Warfarin was exerting selective pressure on populations of rats. The previously rare dominant allele proved to be adaptive. The lucky rats that inherited the allele survived and produced more offspring. The unlucky ones that inherited the recessive allele had no built-in defense, and died. Over time, the dominant allele's frequency increased in all rat populations exposed to the poison.

Of course, selection pressures can and often do change. When warfarin resistance increased in rat populations, people stopped using warfarin. And guess what: The dominant allele's frequency declined. Now the latest worry is the evolution of "super rats," which even more potent rodenticides can't seem to kill.

When you hear someone question whether life evolves, remember this: With respect to life, **evolution** simply means that heritable change is occurring in some line of descent. The actual mechanisms that can bring about such change are the focus of this chapter. Later chapters highlight how these mechanisms have contributed to the evolution of new species.

 How Would You Vote? Antibiotic-resistant strains of bacteria are becoming dangerously pervasive. Standard animal husbandry practice includes the repeated dosing of healthy animals with antibiotics—the same ones prescribed to people. Should this practice stop? See BiologyNow for details, then vote online.

 Key Concepts

EVOLUTIONARY VIEWS EMERGE
The world distribution of species, similarities and differences in body form, and the fossil record gave early evidence of evolution—of changes in lines of descent. Charles Darwin and Alfred Wallace had an idea of how those changes occur.

VARIATION AND ADAPTATION
An adaptation is a heritable aspect of form, function, behavior, or development that promotes survival and reproduction. It enhances the fit between the individual and prevailing conditions in its environment.

MICROEVOLUTIONARY PROCESSES
An individual does not evolve. A *population* evolves, which means its shared pool of alleles changes. Over generations, any allele may increase in its frequency among individuals, or it may become rare or lost. Mutation, genetic drift, natural selection, and gene flow change allele frequencies in a population. These processes of microevolution change the observable characteristics that define a population and, more broadly, a species.

Links to Earlier Concepts

Section 1.4 sketched out the key premises of the theory of natural selection. Here you will read about evidence that led to its formulation. You may wish to refresh your memory of protein structure (2.8), basic terms of genetics (8.1), chromosomes and crossing over (8.4, 8.5), DNA replication and repair (9.3), and gene mutation (10.5). This chapter puts the chromosomal basis of inheritance (8.4, 8.7) and continuous variation in populations (8.3) into the broader context of evolution.

12.1 Early Beliefs, Confounding Discoveries

Prevailing beliefs can influence how we interpret clues to natural processes and their observable outcomes.

LINK TO SECTION 1.7

QUESTIONS FROM BIOGEOGRAPHY

At one time Europeans viewed nature as a great Chain of Being extending from the "lowest" forms of life to humans, and on to spiritual beings. Each kind of being, or **species** as it was called, was one separate link in the chain. All the links had been designed and forged at the same time at one center of creation. They had not changed since. Once all the links were discovered and described, the meaning of life would be revealed.

Then Europeans embarked on their globe-spanning explorations and discovered the world is a lot bigger than Europe. Tens of thousands of unique plants and animals were brought back home to be catalogued as more links in the chain. But a few naturalists moved beyond cataloguing the species and started identifying *patterns* in where species live, and how species might or might not be related. These people were pioneers in **biogeography**, the study of patterns in the geographic distribution of species and communities. They were the first to ponder the relationship between ecology and evolution.

Some patterns were intriguing. For example, many plants and animals are found only on islands in the middle of the ocean and in other remote places. Many species that are strikingly similar live far apart, and vast expanses of open ocean or impassable mountain ranges keep them separated, in isolation.

Consider the three long-necked, long-legged, and flightless birds in Figure 12.1a–c. They live on different continents. Why are they alike? The two plants shown in Figure 12.1d,e have such differences in reproductive parts that they cannot be close relatives. They live on different continents. Yet both have spines, tiny leaves, and short fleshy stems. Why are *they* so much alike?

Curiously, the flightless birds all live in the same kind of environment about the same distance from the equator. They sprint about in flat, open grasslands of dry climates, and they raise their long necks to keep an eye on predators in the distance. Both plant species also live about the same distance from the equator in the same kind of environment—a desert—where water is seasonally scarce. Both store water in their fleshy stems. Their stems have a very thick, water-conserving cuticle. Both species have rows of sharp spines that deter thirsty, hungry animals. If all birds and plants arose in one place, then how did such similar kinds end up in such distant, remote places?

QUESTIONS FROM COMPARATIVE MORPHOLOGY

Questions about similarities and differences in body plans among groups of organisms led to another field of inquiry, now called **comparative morphology**. For example, the bones of a human arm, a whale flipper, and a bat wing differ in size, shape, and function. As the next chapter explains, they have similar positions in the body and the same tissues arranged in much the same patterns. These parts develop in much the same way in embryos. The scholars who deduced all of this wondered: Why are some animals that are so different in some features so much alike in others?

Figure 12.1 Species that resemble one another, strikingly so, even though they are native to distant geographic realms. (**a**) South American rhea, (**b**) Australian emu, and (**c**) African ostrich. All three types of birds live in similar habitats. They are unlike most birds in several traits, most notably in their long, muscular legs and their inability to fly. (**d**) A spiny cactus native to the hot deserts of the American Southwest. (**e**) A spiny spurge native to southwestern Africa.

By one hypothesis, basic body plans are so perfect there was no need to come up with a new one for each organism at the time of its creation. Yet if that were so, then how could there be parts having no function? For example, some ancient whales had ankle bones, but whales do not walk (Figure 12.2). Why the bones? Our coccyx is like some tailbones in many other mammals. We do not have a tail. Why do we have parts of one?

QUESTIONS ABOUT FOSSILS

About the same time, geologists were mapping layers of rock exposed by erosion or quarrying. As you will read later on, they added to the confusion when they found fossils in the same kinds of layers in different parts of the world. **Fossils** came to be recognized as stone-hard evidence of earlier forms of life. Figure 12.2 shows an example.

A puzzle: Many deep layers held fossils of simple marine life. Some layers above them contained fossils that were similar, but more intricate. In higher layers, the fossils looked like modern species. What did these sequences in complexity among fossils of a given type mean? Were they evidence of lines of descent?

Taken as a whole, the findings from biogeography, comparative morphology, and geology did not fit with prevailing beliefs. Scholars floated novel hypotheses. If a simultaneous dispersal of all species from a center of creation was unlikely, *then perhaps species originated in more than one place.* If species had not been created in a perfect state—and fossil sequences and "useless" body parts among species implied that they had not—*then perhaps species had become modified over time.*

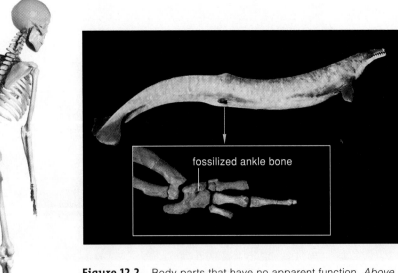

Figure 12.2 Body parts that have no apparent function. *Above*, reconstruction of an ancient whale (*Basilosaurus*), with a head as long as a sofa. This marine predator was fully aquatic, so it did not use its hindlimbs to support body weight as you do—yet it had ankle bones. We use our ankles, but not our coccyx bones.

coccyx

ankle bone

fossilized ankle bone

SQUEEZING NEW EVIDENCE INTO OLD BELIEFS

Georges Cuvier was among those trying to make sense of the growing evidence for change. For years he had compared fossils with living organisms. He was aware of the abrupt changes in the fossil record and was the first to recognize that they indicated times of mass extinctions. By Cuvier's hypothesis, a single time of creation had populated the entire world, which was an unchanging stage for the human drama. Monstrous earthquakes, floods, and other major catastrophes did happen, and many people died. Each time, survivors repopulated the world. By Cuvier's reckoning, there were no new species. Naturalists simply had not yet found all of the fossils. His hypothesis was favored for a long time. It even became elevated to the rank of theory, one that later became known as **catastrophism**.

Jean Lamarck had a different idea: The environment modifies traits of individuals in ways that are passed on to offspring. Outside pressures and internal needs cause permanent changes in body form and function, and offspring inherit the acquired changes. Life, created long ago in a simple state, gradually improved—up the Chain of Being. Apply his idea to modern giraffes. Say they had a short-necked ancestor. Pressed by the need to find food, it kept stretching its neck to reach leaves beyond the reach of other animals. Stretching made the neck lengthen permanently. The longer neck was inherited by offspring that stretched their necks, too. So generations of animals stretching to reach ever loftier leaves led to the modern giraffe.

The environment *is* a factor in evolution. However, Lamarck's idea, like others proposed at the time, has not been supported by rigorous tests.

VOYAGE OF THE *BEAGLE*

In 1831, in the midst of the confusion, Charles Darwin was twenty-two years old and wondering what to do with his life. Ever since he was eight, he had wanted to hunt, fish, collect shells, or simply watch insects and birds—anything but sit in school. Later, at his father's insistence, he tried to study medicine in college. The crude, painful procedures used on patients at that time sickened him, so he packed for Cambridge. He earned a degree in theology, but nature still beckoned him. A botanist, John Henslow, sensed Darwin's real interests. Henslow arranged for him to be appointed as ship's naturalist aboard a ship, the *Beagle*.

The *Beagle* (Figure 12.3) was about to embark on a five-year voyage that would take Darwin around the world. During the Atlantic crossing, he collected and studied marine life. At stops along coasts and islands, he observed diverse species in environments ranging

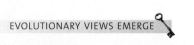

from shores to mountains. He read Henslow's parting gift, a book called *Principles of Geology* by Charles Lyell. He started circling the question of evolving life, now on the minds of many respected scholars.

Lyell and other geologists were proposing a radical hypothesis. Catastrophes, they said, had no more effect on Earth history than did subtle processes of change. They studied how long it takes for rain, the pounding surf, and other forces of nature to sculpt the surface. They considered how long it takes layers of sediments to be deposited in beds of rivers and seas. If deposition occurred just as slowly in the past, then it must have taken millions of years for thick stacks to form. And what about catastrophes? Massive floods, more than a hundred huge earthquakes, and twenty or so volcanic eruptions strike in a typical year. Catastrophes, they said, are just not that unusual.

Their view of gradual, uniformly repetitive change became a theory of **uniformity**. It directly challenged a prevailing view that the Earth was only about 6,000 years old. Traditionalists thought people had recorded everything that happened in those thousands of years. In all that time, no one had ever mentioned seeing a species evolve. By Lyell's calculations, geologic forces must have taken millions of years to sculpt the present landscape. *Was that not enough time for species to evolve in diverse ways?*

OLD BONES AND ARMADILLOS

When Darwin returned to England, he talked to other naturalists about possible evidence that life evolves. By carefully studying detailed notes from his voyage, he came up with possibilities.

In Argentina, for example, he had observed fossils of glyptodonts, now extinct. Of all animals on Earth, only living armadillos are like them (Figure 12.4). Of all places on Earth, armadillos live only in the same places where glyptodonts had lived. If the two kinds of animals had been created at the same time, lived in the same place, and were so much alike, why is only one still alive? Would it not be reasonable to assume that glyptodonts were early ancestors of armadillos? Many of their shared traits might have been retained over many thousands of generations. Other traits may have been modified in the armadillo branch of a family tree. *Descent with modification* seemed possible. What, then, could be the driving force for evolution?

A KEY INSIGHT — VARIATION IN TRAITS

While Darwin assessed his notes, an influential essay by Thomas Malthus, a clergyman and economist, made him consider a topic of social interest. Malthus had

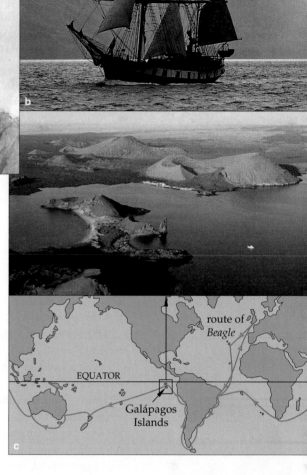

Figure 12.3 (a) Charles Darwin. (b) Replica of the *Beagle* sailing off a rugged coastline of South America. On one of his trips, Darwin ventured into the Andes. He found fossils of marine organisms in rock layers 3.6 kilometers above sea level.

(c) The Galápagos Islands are isolated in the ocean far to the west of Ecuador. They arose by volcanic action on the seafloor about 5 million years ago. Winds or ocean currents probably dispersed species to the new oceanic islands. All of the species native to the islands are their descendants.

Figure 12.4 (a) From Texas, a modern armadillo, about a foot long excluding the tail. (b) A Pleistocene glyptodont, which was about as big as a Volkswagen Beetle, and now extinct. Glyptodonts shared unusual traits and a restricted distribution with the existing armadillos. Yet the two kinds of animals are widely separated in time. Their similarities were a clue that helped Darwin develop a theory of evolution by natural selection.

LINK TO
SECTION
1.4

correlated population size with famine, disease, and war. Humans, Malthus claimed, run out of food, living space, and other resources because they reproduce too much. The larger a population gets, the more people reproduce. Resources dwindle, and struggles intensify. Many people starve, get sick, or engage in war and other forms of competition for remaining resources.

Darwin deduced that *any* population has a capacity to produce more individuals than the environment can support. Even one sea star can release 2,500,000 eggs per year, but the seas do not fill up with sea stars.

Darwin also reflected on species he had observed during his voyage. Individuals of those species were not alike in their details. They varied in size, color, and other traits. *It dawned on Darwin that variations in traits affect an individual's ability to secure resources in the environment—and thus to survive and reproduce.*

Did the Galápagos Islands reveal evidence of this? Between the islands and the South American coast are 900 kilometers of open ocean. The islands offer diverse habitats on their rocky shores, deserts, and mountains. Nearly all of their species live nowhere else, although they do resemble species on the mainland.

Did wind and water deliver colonizing species to the Galápagos? If so, then different species living on different islands might be island-hopping descendants of the first colonizers.

Darwin thought about the Galápagos finch species (Figure 12.5). Could he correlate their variations with different environmental challenges? Imagine yourself in his place, watching a population of large-billed, seed-cracking birds. A few birds with a stronger bill crack seeds too tough for the others. If most of the seeds that form during a given season have hard coats, then the strong-billed birds will have a competitive edge and a better chance of surviving and producing offspring. If the trait is heritable, then the advantage will help their descendants, also. If environmental factors continue to "select" the most adaptive version of this trait, then the population may end up being mostly strong-billed birds. *And a population is evolving when forms of heritable traits change over the generations.*

Figure 12.5 Four of thirteen finch species on the Galápagos Islands. (**a**) *Geospiza conirostris* and (**b**) *G. scandens*. Both eat cactus flowers and fruits. Other finches eat cactus seeds. (**c**) *Certhidea olivacea*, a tree-dwelling finch, uses its slender bill to probe for food. (**d**) *Camarhynchus pallidus* eats wood-boring insects such as termites. It learns to break cactus spines and twigs to suitable lengths, then hold the "tools" and use them to probe bark for insects. Compare Figure 13.17, which shows bills of the spectacularly diverse Hawaiian honeycreepers.

Let's trace Darwin's observations and conclusions about a key evolutionary process, using modern terms:

1. *Observation:* Natural populations have an inherent reproductive capacity to increase in numbers through successive generations.

2. *Observation:* No population can indefinitely grow in size, because its individuals run out of food, living space, and other resources.

3. *Inference:* Sooner or later, individuals will wind up competing for dwindling resources.

4. *Observation:* Individuals share a pool of heritable information about traits, encoded in genes.

5. *Observation:* Mutations have given rise to alleles, or slightly different molecular forms of genes, which are a source of differences in phenotypic details.

6. *Inferences:* Some phenotypes are better than others at helping an individual survive and reproduce. They increase an individual's **fitness**—its success in a given environment, as measured by its relative contribution to future generations. Over successive generations, the alleles for such phenotypes increase in frequency.

7. *Conclusions:* **Natural selection** is an outcome of the differences in reproduction among individuals of a population that vary in details of their shared traits. Environmental agents of selection act on the range of variation, and the population may evolve as a result.

Anticipating that his view would be controversial, Darwin searched for flaws in his reasoning. He took too long. More than ten years after he wrote up but did not publish his theory, another respected naturalist—Alfred Wallace—came up with the same theory. At a scientific meeting in 1858, the theory was presented and attributed to both. Wallace was still in the field, so he knew nothing about the meeting, and Darwin did not attend it. The next year, Darwin finally published *On the Origin of Species*, which has detailed evidence in support of the theory.

You may have heard that Darwin's book fanned an intellectual firestorm, but most scholars were quick to accept evolution. It was, however, fiercely debated that evolution occurs by natural selection. Decades passed before evidence from genetics led to the widespread acceptance of the theory of natural selection.

As Darwin and Wallace perceived, natural selection is the outcome of differences in survival and reproduction among individuals of a population that show variation in their shared traits. It can lead to increased fitness—that is, to increased reproductive success in a given environment.

12.2 The Nature of Adaptation

Observable traits are not always easy to correlate with specific conditions in an organism's environment.

"Adaptation" is one of those words that have different meanings in different contexts. An individual plant or animal often can quickly adjust its form, function, and behavior. Junipers in inhospitably windy places grow less tall than junipers of the same species that grow in more sheltered places. A sudden clap of thunder may make you lurch the first time you hear it, but then you may get used to the sound over time and ignore it. These are examples of *short-term* adaptations, because they last only as long as the individual does.

Over the long term, an **adaptation** is some heritable aspect of form, function, behavior, or development that improves the odds for surviving and reproducing in a given environment. It is an *outcome* of microevolution —natural selection especially—an enhancement of the fit between the individual and prevailing conditions.

SALT-TOLERANT TOMATOES

As an example of long-term adaptation, compare how tomato species handle salty water. Tomatoes evolved in Ecuador, Peru, and the Galápagos Islands. The type found most often in markets, *Lycopersicon esculentum*, has eight close relatives in the wild. If you mix ten grams of table salt with sixty milliliters of water, then pour it into the soil around *L. esculentum* roots, the plant will wilt severely in less than thirty minutes (Figure 12.6*a*). Even when the soil has only 2,500 parts per million of salt, this species grows poorly. Yet the Galápagos tomato (*L. cheesmanii*) survives in seawater-washed soils (Figure 12.6*b*). We know its salt tolerance is a heritable adaptation. How? Crosses between the wild and commercial species produce a small, edible hybrid. This hybrid tolerates irrigation water that is two parts fresh and one part salty. It is gaining interest in places where fresh water is scarce and where salts have built up in croplands.

It may take modification of only a few traits to get new salt-tolerant plants. Revving up a single gene for a sodium/hydrogen ion transporter lets tomato plants use salty water and still bear edible fruits.

ADAPTATION TO WHAT?

You can safely assume a polar bear (*Ursus maritimis*) is adapted to the Arctic, and that its form and function would be a flop in a desert. But it is not always easy to ascertain a direct relationship between an adaptation and the environment. A prevailing environment might be very different from the one in which a trait evolved.

Consider the llama. It is native to the cloud-piercing peaks of the Andes range in western South America (Figure 12.7*a*). It lives 4,800 meters (16,000 feet) above sea level. Compared to humans at lower elevations, its lungs have more air sacs and blood vessels. The llama heart has larger chambers, so it pumps larger volumes of blood. Llamas do not have to make extra blood cells, as people do when they move permanently from the lowlands to high elevations. (Extra cells make blood "stickier," so the heart has to pump harder.) But the most publicized adaptation is this: Llama hemoglobin is better than ours at latching on to oxygen. It picks up oxygen in the lungs far more efficiently.

Superficially, at least, the oxygen-binding affinity of llama hemoglobin appears to be an adaptation to thin air at high altitudes. But is it? Apparently not.

Llamas belong to the same family as the dromedary camels (Figure 12.7*b*). Their shared camelid ancestors evolved in the Eocene grasslands and deserts of North America. Later, the ancestors of the camels and llamas went their separate ways. Camel forerunners moved into Asia's low-elevation grasslands and deserts by a land bridge, which later submerged when the sea level rose. Llama forerunners moved a different way—down the Isthmus of Panama, and on into South America.

Intriguingly, a dromedary camel's hemoglobin also shows a high oxygen-binding capacity. So if the trait arose in a shared ancestor, then in what respect was it adaptive at *low* elevations if it is also adaptive at *high* elevations? We know that camels and llamas did not just happen to evolve to be similar. They are very close kin,

Figure 12.6 (**a**) Severe, rapid wilting of one commercial tomato plant (*Lycopersicon esculentum*) that absorbed salty water. (**b**) Galápagos tomato plant, *L. cheesmanii*, which stores most absorbed salts in its leaves, not in its fruits.

Figure 12.7 The Llama (**a**) and the camel (**b**) share a common ancestor.

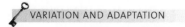

LINKS TO
SECTIONS
8.3, 8.4, 8.5, 10.5

and their most recent ancestors lived in very different environments with different oxygen concentrations.

Who knows why the trait was originally favored? Eocene climates were alternately warm and cool, and hemoglobin's oxygen-binding capacity does go down as temperatures go up. Did it prove adaptive during a long-term shift in climate? Or were its effects neutral at first? What if a small ancestral population became homozygous for the mutant gene simply by chance?

Use all of these "what-ifs" as a reminder to think critically about the connections between an organism's form and function. Identifying connections takes more than intuition.

> *A long-term adaptation is any heritable aspect of form, function, behavior, or development that contributes to the fit between the individual and its environment.*
>
> *An adaptive trait improves the odds of surviving and reproducing, or at least it did so under conditions that prevailed when genes coding for the trait first evolved.*

12.3 Individuals Don't Evolve, Populations Do

Evolution starts with changes in the gene pool of a population. Mutation is the source of new alleles, and sexual reproduction spreads them around.

VARIATION IN POPULATIONS

As Darwin and Wallace both perceived, *individuals do not evolve; populations do.* By definition, a **population** is a group of individuals of the same species occupying a given area. To understand how a population evolves, start with variation in the features that characterize it.

Individuals of a population share the same body plan. Some birds have two eyes, two wings, feathers, three toes forward and one toe back, and so on. These are *morphological* traits (*morpho–*, meaning form). Cells

and body parts of all individuals work much the same way in metabolism, growth, and reproduction. These are *physiological* traits, relating to how the body works in its environment. Individuals respond the same way to basic stimuli, as when babies instinctively imitate adult facial expressions. These are *behavioral* traits.

However, the individuals of a population also show variation in the details of their shared traits. You know this just by thinking about the variations in the color and patterning of pigeon feathers or butterfly wings or snail shells (Figure 12.8). Almost every trait of any species may vary, sometimes dramatically.

Many traits, such as those of Gregor Mendel's pea plants, show *qualitative* differences. They have two or more distinct forms (morphs), such as purple or white flowers. Having just two is called dimorphism; having three or more is called polymorphism. Do not forget that for many traits, individuals of a population show *quantitative* differences, a range of incrementally small variations in a specified trait (Section 8.3).

THE "GENE POOL"

Genes encode information about heritable traits. All of a population's individuals inherit the same number and kinds of genes (except for a pair of nonidentical sex chromosomes). Together, they and their offspring represent a **gene pool**—a pool of genetic resources.

For sexual reproducers, nearly all of the genes in the shared pool have two or more slightly different molecular forms, or **alleles**. An individual might or might not inherit identical alleles for any trait. This is the source of variations in *phenotype*—differences in details of shared traits. Whether you have red, black, brown, or blond hair depends upon which alleles you inherited from your two parents.

You read about the inheritance of alleles in earlier chapters. Here we summarize the key events involved:

1. Gene mutation (produces new alleles)

2. Crossing over events in meiosis I (introduce novel combinations of alleles in chromosomes)

3. Independent assortment at meiosis I (puts mixes of maternal and paternal chromosomes in gametes)

4. Fertilization (combines alleles from two parents)

5. Change in chromosome number or structure (loss, duplication, or repositioning of genes)

Only mutation *creates* new alleles. Other events shuffle *existing* alleles into different combinations—but what a shuffle! There are $10^{116,446,000}$ possible combinations of human alleles. Not even 10^{10} people are alive today. Unless you have an identical twin, it is very unlikely that any other person with your exact genetic makeup has ever lived, or ever will.

STABILITY AND CHANGE IN ALLELE FREQUENCIES

Researchers typically track **allele frequencies**, or the abundance of certain alleles in a population. They start from a theoretical reference point known as **genetic equilibrium**. A population at that point is *not* evolving with respect to the allele frequencies being studied.

You will read more about genetic equilibrium in Section 12.4. For now, know that five conditions are necessary for genetic equilibrium to occur, and those conditions will almost never prevail at the same time. *Mutations* are rare but inevitable, and *natural selection*, *gene flow*, and *genetic drift* can drive a population out of genetic equilibrium. **Microevolution** refers to the small-scale changes in allele frequencies that originate from those four processes, as discussed next.

Figure 12.8 Fabulous polymorphism evident among individuals of the snail populations on Caribbean islands.

MUTATIONS REVISITED

Mutations, again, are heritable changes in DNA that usually give rise to altered gene products. They are the original source of all alleles. We cannot predict when or in which individual they will occur. Yet within each species, mutations accumulate at a predictable rate. In humans, that rate is around one mutation for every replication cycle—or about one in three billion bases. That means one out of every four gametes, on average, will carry a mutation.

Mutations may give rise to structural, functional, or behavioral alterations that diminish an individual's chances of surviving and reproducing. Even one small biochemical change can be devastating. For instance, skin, bones, tendons, lungs, blood vessels, and many of the other vertebrate organs incorporate collagen. When the collagen gene mutates in a way that affects collagen function, drastic changes in the body occur. A mutation that severely changes a phenotype usually causes death; it is called a **lethal mutation**.

With respect to natural selection, a mutation can be neutral if it has no effect on survival or reproduction. It neither helps nor hurts the individual. If you carry a mutant gene that keeps your earlobes attached to your head instead of swinging freely, this in itself should not stop you from surviving and reproducing as well as anybody else. Because a neutral mutation has no effect on survival and reproduction, natural selection does not alter its frequency in the population.

Every so often, a mutation proves useful. A mutant gene product that affects growth may make a corn plant grow larger or faster and so give it the best access to sunlight and nutrients. A neutral mutation may prove

helpful after a condition in the environment changes. Even when it bestows only a small advantage, chance events or natural selection might preserve the mutant gene in the DNA and favor its representation in the next generation.

Mutations are so rare, they usually have little or no immediate effect on the population's allele frequencies. But both beneficial and neutral mutations have been accumulating in different lineages for billions of years. Through all that time, they have functioned as the raw material for evolutionary change, for the staggering range of biodiversity, past and present. Actually, the reason you do not look like a bacterium or an avocado or an earthworm—or even your neighbors down the street—started with mutations that arose at different times, in different lines of descent.

Each population or species is characterized by certain morphological, physiological, and behavioral traits. The details of these traits vary among its individuals, and most of the variation has a heritable basis.

Different combinations of alleles among individuals of a population give rise to variations in phenotype—that is, to differences in the details of their shared structural, functional, and behavioral traits.

In sexually reproducing species, a population's individuals share a pool of genetic resources—a gene pool.

Mutation alone creates new alleles. Whether an allele proves neutral, beneficial, or harmful depends on the mutation and on prevailing environmental conditions.

490 *AA* butterflies
dark-blue wings

490 *AA* butterflies
dark-blue wings

490 *AA* butterflies
dark-blue wings

420 *Aa* butterflies
medium-blue wings

420 *Aa* butterflies
medium-blue wings

420 *Aa* butterflies
medium-blue wings

90 *aa* butterflies
white wings

90 *aa* butterflies
white wings

90 *aa* butterflies
white wings

Starting Population   **Next Generation**   **Next Generation**

Figure 12.9 *Animated!* Proportions of butterflies with dark-blue, medium-blue, and white wings in a hypothetical population that remains at genetic equilibrium.

The frequencies of *A* and *a* must add up to 1.0. For example, if *A* occupies half of all the loci for this gene in the population, then *a* must occupy the other half (0.5 + 0.5 = 1.0). If *A* occupies 90 percent of all the loci, then *a* must occupy 10 percent (0.9 + 0.1 = 1.0).

No matter what the proportions,

$$p + q = 1.0$$

At meiosis, alleles of each gene pair segregate and end up in separate gametes. So the proportion of gametes having the *A* allele is *p*, and the proportion having the *a* allele is *q*. Check this Punnett square for genotypes that are possible in the next generation (*AA*, *Aa*, and *aa*):

	p ⒶA	q ⓐa
p ⒶA	AA (p^2)	Aa (pq)
q ⓐa	Aa (pq)	aa (q^2)

The frequencies add up to 1: $p^2 + 2pq + q^2 = 1.0$.

Let's say the population contains 1,000 butterflies, and each one produces two gametes:

490 *AA* individuals produce 980 *A* gametes
420 *Aa* individuals produce 420 *A* and 420 *a* gametes
 90 *aa* individuals produce 180 *a* gametes

The frequency of *A* and *a* among the 2,000 gametes is

$$A = \frac{980 + 420}{2,000 \text{ alleles}} = \frac{1,400}{2,000} = 0.7 = p$$

$$a = \frac{180 + 420}{2,000 \text{ alleles}} = \frac{600}{2,000} = 0.3 = q$$

At fertilization, the gametes combine at random and give rise to the next generation. If the population stays constant at 1,000 individuals, you now have 490 *AA*, 420 *Aa*, and 90 *aa* individuals. Because the frequencies of alleles for dark-blue, medium-blue, and white wings are the same as they were in the original gametes, they will give rise to the same phenotypic frequencies you observed in the second generation (Figure 12.9).

As long as the five conditions hold, the pattern will also hold. When they do not, one or more evolutionary forces are at work, and allele frequencies will change.

> *When traits do not appear in proportions predicted from the Hardy–Weinberg rule, one or more conditions stated in the rule are not holding. The hunt can begin for one or more specific evolutionary forces driving the change.*

LINK TO SECTION 8.1

12.4 When Is A Population *Not* Evolving?

How do researchers know whether or not a population is evolving? They can start by tracking deviations from the baseline of genetic equilibrium.

Gene pools can be stable only if five conditions are met. There is no mutation, the population is infinitely large, the population is reproductively isolated from all other populations of the species, mating is random, and all individuals survive and produce the same number of offspring. As you can imagine, these conditions rarely, if ever, occur at once in nature. Nonetheless, they serve as a hypothetical starting point.

Allele frequencies for any gene in a shared pool will stay stable unless the population is evolving. Godfrey Hardy and Wilhelm Weinberg developed a formula that tracks allele frequencies in a population of any sexually reproducing species. Their formula can be used to check whether the population is in genetic equilibrium.

Butterflies are sexually reproducing organisms with pairs of genes on homologous chromosomes. Start by assuming the population is at genetic equilibrium for a single pair of alleles that influence wing color. Allele *A* is associated with dark-blue wings, *a* with white, and *Aa* with medium-blue. The proportions of genotypes for wing color at genetic equilibrium are:

$$p^2 (AA) + 2pq(Aa) + q^2 (aa) = 1.0$$

where *p* and *q* are frequencies of alleles *A* and *a*. This formula is what is known as the *Hardy–Weinberg rule*.

Natural Selection Revisited

Natural selection operates on the variation in traits that characterize a population. It affects which individuals survive and reproduce under existing environmental conditions, one generation to the next.

DIRECTIONAL SELECTION

In cases of **directional selection**, allele frequencies that give rise to a range of variation in phenotype tend to shift in a consistent direction. The shift is a response to directional change in the environment or to one or more new conditions in it. A new mutation that proves beneficial also may cause the directional shift. Either way, forms at one end of the phenotypic range become more common than midrange forms (Figure 12.10).

ROCK POCKET MICE Directional selection is at work among rock pocket mice (*Chaetodipus intermedius*) of the Sonoran Desert in Arizona. Of more than eighty genes known to affect coat color in mice, researchers found one gene that governs a difference between two populations of this mouse species.

Rock pocket mice are small mammals that spend the day in underground burrows. At night they forage for seeds, scampering over rock outcroppings in the desert. Individuals of the largest population of mice live among tawny-colored granite rocks. Their tawny fur camouflages them from predators (Figure 12.11*a*).

A smaller population of pocket mice lives in the same region, but these mice scamper over dark basalt rocks of ancient lava flows. The individuals of this population have dark fur, and they are camouflaged from predators on the dark rocks (Figure 12.11*b*). We can expect that local night-flying predatory birds are selective agents that affect fur color. For instance, owls have an easier time seeing mice with fur that does not match the rock color.

Researchers used genetic data on laboratory mice to formulate a hypothesis about coat color differences in the two populations of wild mice. They predicted that a gene mutation causes the difference.

DNA was collected from dark-colored pocket mice at one lava flow, and from light-colored mice living on granite outcroppings. Analysis of both sets of DNA showed that the DNA sequence of the coat color gene in all dark-fur mice differed by four nucleotides from that of their light-furred neighbors. In the population of dark mice, the allele frequencies had evolved in a consistent direction as a result of selection pressure, so dark fur became more common.

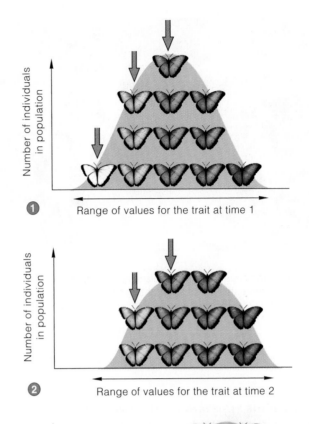

① Range of values for the trait at time 1

② Range of values for the trait at time 2

③ Range of values for the trait at time 3

(y-axis: Number of individuals in population)

Figure 12.10
Animated!
Example of directional selection, using the phenotypic variation within a population of butterflies. The bell-shaped curve spans a range of continuous variation in wing color. *Medium-blue*, the most common form, is between extreme forms of the trait—*white* and *dark purple*. *Orange* arrows signify which forms are being selected against over time.

Figure 12.11 (**a,b**) The two color morphs of rock pocket mice. (**c**) A lava basalt area in the Sonoran Desert of Arizona.

Figure 12.12 The super-rat threat. Rats infesting the Philippines have destroyed more than 20 percent of that country's 80,000 hectares of rice fields. Rice is the main food source for most people of Southeast Asia.

Antibiotics also have been overprescribed, often for simple infections that most people can fight on their own. Standard farming practices have traditionally included preventative doses of antibiotics for all farm animals, whether or not they are sick.

The overabundance of antibiotics in the human arena and the agricultural environment has put great selection pressure on bacterial populations. Remember how quickly bacteria grow and divide; their evolution proceeds at a much accelerated rate compared with ours. The intestinal bacteria *Escherichia coli* can divide once every 17 minutes, so an average two-week course of antibiotics potentially can exert selective pressure on over a thousand generations.

The genetic variation in bacterial gene pools allows some cells to survive this antibacterial onslaught even as others do not. Resistant strains are becoming the norm, particularly in hospitals, and because of this are becoming a greater threat to human health each year. Even as doctors and researchers scramble to find new antibiotics, this trend does not bode well for the many millions of people who contract cholera, tuberculosis, and other bacterial diseases each year.

PESTICIDE RESISTANCE The overuse of pesticides can cause directional selection, as it did for the super rats (Figure 12.12). Typically, a heritable aspect of body form, physiology, or behavior helps a few individuals survive the first pesticide doses. The most resistant individuals are favored. If resistance has a heritable basis, it becomes more common in each generation. The chemicals act as agents of selection, favoring the most resistant forms! Today, 450 species of pests resist one or more pesticides. Worse, pesticides also kill natural predators of the pests. When freed from their natural constraints, populations of resistant individuals overgrow, and crop damage is greater than before. This outcome of directional selection is called *pest resurgence*.

Pesticide use is decreasing where crops have been genetically engineered to resist pests. However, these plants are exerting selection pressure on populations of pests. Traits that help any pests survive the engineered defenses tend to be favored.

ANTIBIOTIC RESISTANCE When your grandparents were young, tuberculosis, pneumonia, and scarlet fever caused one-fourth of the annual deaths in the United States. Since the 1940s, we have been relying on natural and synthetic antibiotics to fight such bacterial diseases. Natural **antibiotics** are toxins that some microorganisms in soil release to kill bacterial competitors for nutrients (Section 1.6). Streptomycins, for instance, block protein synthesis in target cells. Penicillins disrupt formation of covalent bonds that hold a bacterial cell wall together, and by weakening the wall they cause the cell to rupture.

SELECTION AGAINST OR IN FAVOR OF EXTREME PHENOTYPES

Figure 12.13 shows outcomes of two more modes of natural selection. One works against phenotypes at the fringes of a range of variation. The other favors them.

STABILIZING SELECTION With **stabilizing selection**, intermediate forms of a trait in a population are favored and alleles for the extreme forms are not. This mode of selection tends to preserve the intermediate phenotypes in the population (Figure 12.13).

As an example, prospects are not good for human babies who weigh far more or far less than average at birth (Figure 12.14). Newborns weighing less than 5.51 pounds, if they survive, tend to suffer from high blood pressure, diabetes, and heart disease as adults.

A different example is stabilizing selection on the body mass of sociable weavers of the African savanna. These birds cooperate in constructing and using large communal nests in regions where trees and other good nesting sites are scarce. For seven years, researchers captured, measured, tagged, released, and recaptured birds during the breeding seasons. Their studies show that body mass in these birds is a trade-off between the risks of starvation and predation. Birds that have intermediate mass have the selective advantage. Lean birds do not store enough fat to avoid starvation, and we can expect that the fat ones are more attractive to predators and not as good at escaping.

MICROEVOLUTIONARY PROCESSES

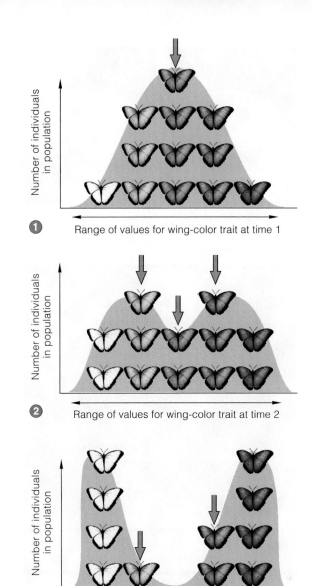

Figure 12.13
Animated!
Selection against or in favor of extreme phenotypes, with butterfly populations as examples. *Left*, stabilizing selection and *right*, disruptive selection. *Orange* arrows show forms of the trait being selected against.

DISRUPTIVE SELECTION With **disruptive selection**, forms at both ends of the range of variation are favored and intermediate forms are selected against.

Consider the black-bellied seedcracker (*Pyrenestes ostrinus*) of Cameroon, in West Africa. Both females and males of these African finches have large or small bills—but none of the birds have medium-sized bills. The pattern holds throughout the geographic range of this species. It is like everyone in Texas being four feet *or* six feet tall, with no one in between.

If the pattern is unrelated to geography or gender, what causes it? Seed-cracking ability must directly influence survival. If only two bill sizes persist, then disruptive selection might be eliminating birds with intermediate-size bills. Which factors affect feeding

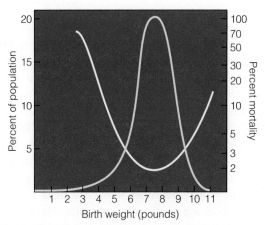

Figure 12.14 Weight distribution for 13,730 human newborns (*yellow* curve) correlated with death rate (*white* curve).

lower bill 12 mm wide — lower bill 15 mm wide

Figure 12.15 Disruptive selection in African finch populations. Selection pressures favor birds with bills that are either 12 *or* 15 millimeters wide. The size difference is correlated with competition for scarce food resources during the local dry season.

Figure 12.16 Outcome of sexual selection. This male bird of paradise (*Paradisaea raggiana*) is engaged in a flashy courtship display. He caught the eye (and, perhaps, sexual interest) of the smaller, less colorful female. The males compete fiercely for females, which are the selective agents. (Why do you suppose drab-colored females have been favored?)

performance? Cameroon's swamp forests flood during the wet season, and fires sparked by lightning burn throughout the dry season. Two kinds of fire-resistant, grasslike plants called sedges dominate these forests. One sedge has hard seeds and the other, soft. When finches reproduce, hard *and* soft seeds are abundant.

All birds prefer soft seeds for as long as they can get them. Birds with small bills are better at picking up soft, tiny seeds; and birds with large bills are better at cracking larger, hard ones. When the dry season peaks, soft seeds and other foods are hard to find, and the birds compete fiercely for seeds. The scarcity of both types of seeds during recurring periods of drought has had a disruptive effect on bill size in the seedcracker

population. Birds with intermediate sizes are selected against, and now all bills are either 12 *or* 15 millimeters wide (Figure 12.15).

In these seedcrackers, bills of a particular size have a genetic basis. In experimental crosses between two birds with the two optimal bill sizes, all offspring had a bill of one size or the other.

With directional selection, allele frequencies underlying a range of variation tend to shift in a consistent direction in response to directional change in the environment.

With stabilizing selection, intermediate phenotypes are favored and extreme phenotypes at both ends of the range of variation are eliminated.

With disruptive selection, intermediate forms of traits are selected against; extreme forms in the range of variation are favored.

12.6 Maintaining Variation in a Population

Natural selection theory helps explain diverse aspects of nature, including male–female differences as well as a balance between sickle-cell anemia and malaria.

SEXUAL SELECTION

The individuals of many sexually reproducing species show a distinct male or female phenotype, or **sexual dimorphism** (*dimorphos*, having two forms). Often the males are larger and flashier than females. Courtship rituals and male aggression are common.

These adaptations take energy and time away from survival activities. Why do they persist? The answer is **sexual selection**: a form of natural selection in which the genetic winners are the ones that can outreproduce others of the population. The most adaptive traits help individuals defeat same-sex rivals for mates or are the ones most attractive to the opposite sex.

By choosing mates, one gender acts as an agent of selection on its own species. For example, the females of some species shop among a clustering of males, which differ in appearance and courtship behavior. The selected males, as well as the females making the selection, pass on their alleles to the next generation.

Flashy structures and behaviors are correlated with species in which males have little or nothing to do with raising offspring. The female chooses a male by observing signs of his health and vigor, which might improve her odds of producing healthy and vigorous offspring (Figure 12.16).

Physician searching for *Plasmodium*
larvae in Southeast Asia

a

b

☐ less than 1 in 1,600
☐ 1 in 400–1,600
☐ 1 in 180–400
☐ 1 in 100–180
☐ 1 in 64–100
■ more than 1 in 64

Figure 12.17 (**a**) Distribution of malaria cases in Africa, Asia, and the Middle East in the 1920s, before the start of programs to control mosquitoes, the vector for *Plasmodium*. (**b**) Distribution and frequency of the allele that causes sickle-cell anemia. Notice the close correlation between the maps.

SICKLE-CELL ANEMIA—LESSER OF TWO EVILS?

With *balancing* selection, two or more alleles of a gene are being maintained at relatively high frequencies—at least one percent—in the population. Their persistence is called **balanced polymorphism** (*polymorphos*, having many forms). Allele frequencies might shift, but often they return to the same values over the long term. We see this balance when conditions favor heterozygotes. In some way, their nonidentical alleles for a trait give them higher fitness compared to homozygotes, which, recall, have identical alleles for the trait.

Consider the human heterozygotes that carry an allele for Hb^S. The Hb^S allele, remember, codes for a mutant version of hemoglobin, an oxygen-transporting protein in the blood. Homozygotes (Hb^S/Hb^S) develop *sickle-cell anemia*, a severe disorder (Section 2.9).

The frequency of Hb^S is highest in the tropical and subtropical parts of Asia and Africa. Often, Hb^S/Hb^S homozygotes will die in their teens or early twenties. Yet, in these same regions, heterozygotes (Hb^A/Hb^S) make up nearly a third of the human population! Why is this combination of alleles being maintained at such high frequency?

The balancing act is most pronounced in areas that have the highest incidence of *malaria* (Figure 12.17 and Section 14.3). A mosquito transmits the parasitic agent of malaria, *Plasmodium*, to human hosts. The protozoan multiplies in the liver and, later, in red blood cells. The infected cells rupture and release new parasites during severe, recurring bouts of infection.

It turns out that Hb^A/Hb^S heterozygotes are more likely to survive an infection of malaria than people who produce only the normal hemoglobin (Hb^A/Hb^A homozygotes). There are several potential reasons for this. For one example, in heterozygotes, infected cells sickle under normal environmental conditions. Their abnormal shape brings them to the attention of the host's immune system, which will destroy them along with their parasitic cargo. In Hb^A/Hb^A homozygotes, infected cells do not sickle, and so the parasite remains hidden from the immune system.

Also, heterozygotes have one normal hemoglobin allele. Although they are not completely healthy, they produce enough normal hemoglobin to support body functions. That is why heterozygotes are more likely to survive long enough to reproduce, compared to very sick Hb^S/Hb^S homozygotes.

So the persistence of the "harmful" Hb^S allele is a matter of relative evils. Natural selection favors the Hb^A/Hb^S combination in malaria-ridden areas because heterozygotes show more resistance to the disease. In such environments, the combination has more survival value than either homozygotic condition (Hb^S/Hb^S or Hb^A/Hb^A). Malaria has been a selective force for many thousands of years in tropical and subtropical areas of Asia, the Middle East, and Africa.

With sexual selection, some version of a trait simply gives the individual an advantage in reproductive success. Sexual dimorphism is one outcome of sexual selection.

Balanced polymorphism is a state in which natural selection is maintaining two or more alleles over the generations at frequencies greater than one percent.

LINKS TO
SECTIONS
2.9, 8.1, 8.7

IMPACTS, ISSUES

Rat populations show balanced polymorphism as a result of selection for a dominant allele that confers resistance to warfarin. This allele also promotes increases in the amount of vitamin K that all mammals require. In the absence of warfarin, rats with the dominant allele are less fit than those without it. In the presence of warfarin, the relative fitnesses are reversed.

MICROEVOLUTIONARY PROCESSES

a The size of twelve populations of beetles was maintained at 10 breeding individuals per generation for twenty generations. Allele b^+ was lost and b became fixed in one population. Notice that alleles can be fixed or lost even in the absence of selection.

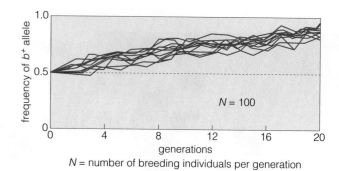

b The size of twelve populations was maintained at 100 individuals per generation for twenty generations. Allele b did not become fixed. Drift was far less in each generation than it was in the small populations tracked in (**a**).

Figure 12.18 *Animated!* The effect of genetic drift on allele frequencies in small and large populations. The starting frequency of the mutant allele (b^+) was 0.5.

LINK TO
SECTION
8.1

12.7 Genetic Drift—The Chance Changes

Random changes in allele frequencies drift toward a loss of genetic diversity. The drifting has greatest impact on small populations.

CHANCE EVENTS AND POPULATION SIZE

Genetic drift is a random change in allele frequencies over time, brought about by chance alone. Researchers measure it in terms of probability rules. *Probability* is the chance that something will happen relative to the number of times it could happen (Section 8.1). We can measure relative frequency as a fraction on a scale of 0 to 1—or 0 to 100 percent of the time. For instance, if ten million people enter a drawing for a month-long vacation in Paris, all expenses paid, each has an equal chance of winning: 1/10,000,000, or a very improbable 0.00001 percent.

By one probability rule, you are less likely to see an event occurring with expected probability if that event happens only rarely. For example, each time you flip a coin, there is a 50 percent chance it will land heads up. With 10 flips, the proportion of times that heads land up will probably deviate greatly from 50 percent. With 1,000 flips, large deviations are less likely.

We can apply the same rule to populations. Because population sizes are not infinite, there will be random

changes in allele frequencies. These random changes tend to have minor impact on large populations, but greatly increase the odds that an allele will become dramatically more or less prevalent if a population is very small.

Steven Rich and his coworkers used small and large populations of the flour beetle (*Tribolium castaneum*) to study genetic drift. They started with beetles that bred true for the allele b^+ (the superscript plus signifies the wild-type allele) and other beetles that bred true for the mutant allele b. They hybridized individuals from both groups to get a population of F_1 heterozygotes (b^+b), which they divided into sets of twelve. Different sets consisted of 10, 20, 50, and 100 randomly selected male and female beetles, and the subpopulation sizes were maintained for twenty generations.

Figure 12.18 shows two of the test results. Drift was greatest in the sets of 10 beetles and least in the sets of 100 beetles. Notice the loss of b^+ from one of the small populations (one graph line ends at 0 in Figure 12.18*a*). Only allele b remained. When all of the individuals of a population have become homozygous for one allele only at a locus, we say that **fixation** has occurred.

Thus, *random change in allele frequencies leads to the homozygous condition and a loss of genetic diversity over time.* This is genetic drift's outcome in all populations; it simply happens faster in small populations. Once alleles from the parent population have become fixed, their frequencies will not change again unless new alleles are introduced.

BOTTLENECKS AND THE FOUNDER EFFECT

Genetic drift is pronounced when a few individuals rebuild a population or start a new one. This happens after a **bottleneck**, a drastic reduction in population size brought about by selection pressure or a calamity. Suppose contagious disease, habitat loss, or hunting nearly wipes out a population. Even when a moderate number of individuals survive the bottleneck, allele frequencies will have been altered at random.

By the 1890s, northern elephant seals were on the edge of extinction, with only twenty known survivors. Hunting restrictions implemented since have allowed the population to recover to about 170,000 individuals. Each is homozygous for all of the genes analyzed yet.

Genetic outcomes can be similarly unpredictable after a few individuals leave a population and establish a new colony elsewhere. This form of bottlenecking is called the **founder effect**. The allele frequencies in the founders may differ quite dramatically from those in the original population. Genetic diversity can be much reduced relative to the original gene pool, as when a lone seed founds a population on a remote island in the middle of the ocean (Figure 12.19).

GENETIC DRIFT AND INBRED POPULATIONS

Genetic drift is pronounced in an inbred population. **Inbreeding** is mating between close relatives, which share many identical alleles. Inbreeding can lower the overall genetic diversity of a population by increasing the frequency of the homozygous state.

Most human societies forbid or discourage incest (inbreeding between parents and children or between siblings). But inbreeding among other close relatives is common in small communities that are geographically or culturally isolated from a larger population. The Old Order Amish of Pennsylvania, for instance, are a moderately inbred group having distinct genotypes. One outcome of Amish inbreeding is a relatively high frequency of the recessive allele that causes *Ellis–van Creveld syndrome*. Affected individuals have extra toes, fingers, or both, and short limbs. Section 8.6 shows one individual. The allele might have been rare when the few founders immigrated to Pennsylvania. Now, about 1 in 8 of the population are heterozygous for the allele, and 1 in 200 are homozygous for it.

> *Genetic drift is the random change in allele frequencies over the generations, brought about by chance alone. The magnitude of its effect is greatest in small populations, such as a population that makes it through a bottleneck.*

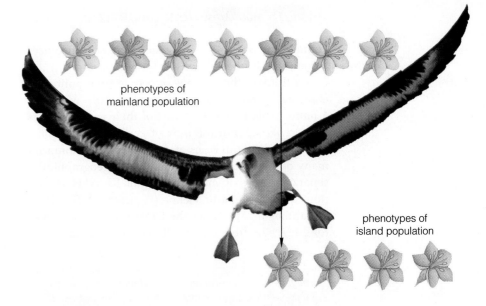

phenotypes of mainland population

phenotypes of island population

Figure 12.19 Founder effect. This wandering albatross carries seeds, stuck to its feathers, from the mainland to a remote island. By chance, most of the seeds carry an allele for orange flowers that are rare in the original population. Without further gene flow or selection for color, genetic drift will fix the allele on the island.

Figure 12.20 Blue jay, a mover of acorns that helps keep genes flowing between separate oak populations.

12.8 Gene Flow—Keeping Populations Alike

Individuals, and their alleles, move into and out of populations, and this physical flow counters changes introduced by other microevolutionary processes.

Individuals of the same species do not always stay put. A population will lose alleles whenever an individual permanently leaves it, an act called *emigration*. It gains alleles when new individuals permanently move in, an act called *immigration*. In both cases, **gene flow**—a physical flow of alleles between populations—occurs. Gene flow tends to counter the effects of mutation, natural selection, and genetic drift. It keeps separated populations genetically similar.

Think of the acorns that blue jays disperse when they gather nuts for the winter. Each fall the jays visit acorn-bearing oak trees repeatedly, then bury acorns in the soil of home territories that may be as much as a mile away (Figure 12.20). Alleles flowing in with the "immigrant acorns" help reduce genetic differences among stands of oak trees.

Or think of the millions of people from politically explosive, economically bankrupt countries who seek a more stable home. The scale of their emigrations is unprecedented, but the flow of genes is not. Human history is rich with cases of gene flow that minimized many of the genetic differences among geographically separate groups. Remember Genghis Khan? His genes flowed from China to Vienna (Figure 8.21). Similarly, the armies of Alexander the Great brought the genes for green eyes from Greece all the way to India.

Gene flow is the physical movement of alleles into and out of a population, through immigration and emigration. It tends to counter the effects of mutation, natural selection, and genetic drift.

Summary

Section 12.1 Awareness of evolution, or changes in lines of descent, emerged long ago from biogeography, comparative morphology, and geology. Charles Darwin and Alfred Wallace proposed a theory that natural selection results in evolution. Here are its key points:

Populations grow in size until resources dwindle and individuals compete for them. When individuals have forms of traits that make them more competitive, they tend to have more offspring. Over generations, those forms increase in frequency. Nature "selects" variations in traits that are more effective at helping individuals survive and reproduce.

Section 12.2 Long-term, heritable adaptations are aspects of form, function, behavior, or development that improve the chance of surviving and reproducing, or at least did so under conditions that prevailed when genes for the trait first emerged.

Section 12.3 Individuals of a population generally have the same number and kinds of genes for the same traits. Mutations are the source of *new* alleles (different molecular forms of genes). Individuals who inherit different allele combinations vary in details of one or more traits. An allele at any locus may become more or less common relative to other kinds or may be lost.

Microevolution refers to changes in allele frequencies of a population brought about by mutation, natural selection, genetic drift, and gene flow (Table 12.1).

Section 12.4 Genetic equilibrium is a state in which a population is not evolving. According to the Hardy–Weinberg rule, this occurs only if there is no mutation, the population is infinitely large and isolated from all other populations of the species, there is no natural selection, mating is random, and all individuals survive and reproduce the same number of offspring. Deviations from the theoretical baseline indicate that microevolution is occurring.

Biology⊗Now
Investigate gene frequencies and genetic equilibrium with the interaction on BiologyNow.

Section 12.5 Natural selection is an outcome of differences in survival and reproduction among individuals of a population that vary in forms of a trait. Selection pressures operating over the range of phenotypic variation shift or maintain allele frequencies in the population's gene pool.

Directional selection favors the forms at one end of a phenotypic range. Stabilizing selection favors intermediate forms. With disruptive selection, forms at both ends of a range of variation are favored over intermediate forms.

Biology⊗Now
View the animation of directional selection, and the animation of disruptive and stabilizing selection. Both on BiologyNow.

Section 12.6 Sexual selection, by females or males, leads to forms of traits that favor reproductive success. Persistence in phenotypic differences between males and females (sexual dimorphism) is one outcome.

Selection may result in balanced polymorphism, with nonidentical alleles for a trait being maintained over time at frequencies greater than 1 percent.

Section 12.7 Genetic drift is a random change in allele frequencies over time due to chance alone. The random changes tend to lead to the homozygous condition and loss of genetic diversity through the generations. The effect of genetic drift is greatest in small populations, such as ones that pass through a bottleneck or arise from a small group of founders.

Biology⊗Now
Learn more about genetic drift with the interaction on BiologyNow.

Table 12.1	Summary of Microevolutionary Processes
Mutation	A heritable change in DNA
Natural selection	Changes in relative frequencies of alleles for a heritable trait in a population; outcome of differences in survival and reproduction among individuals that vary in a trait
Genetic drift	Random fluctuation in allele frequencies over generations due to chance occurrences alone
Gene flow	Individuals, and their alleles, move into and out of populations; the physical flow counters the effects of the other microevolutionary processes

Section 12.8 Gene flow shifts allele frequencies by physically moving alleles into a population (by way of immigration) and out of it (by emigration). It tends to counter mutation, natural selection, and genetic drift and keep populations of the species alike.

Self-Quiz
Answers in Appendix I

1. Individuals don't evolve, _____ do.

2. Biologists define evolution as _____ .
   a. the origin of a species
   b. heritable change in a line of descent
   c. acquiring traits during the individual's lifetime
   d. both a and b

3. _____ is the original source of new alleles.
   a. Mutation   c. Gene flow
   b. Natural selection d. All are original sources of
   c. Genetic drift   new alleles

4. Natural selection is an outcome of _____ .
   a. differences in forms of traits
   b. differences in survival and reproduction among individuals that differ in one or more traits
   c. both a and b

5. Directional selection _____ .
   a. eliminates common forms of alleles
   b. shifts allele frequencies in a consistent direction
   c. does not favor intermediate forms of a trait
   d. works against adaptive traits

6. Disruptive selection _____ .
   a. eliminates uncommon forms of alleles
   b. shifts allele frequencies in a consistent direction
   c. does not favor intermediate forms of a trait
   d. both b and c

7. _____ tends to counter changes that occur in the allele frequencies among populations of a species.
   a. Genetic drift   c. Mutation
   b. Gene flow   d. Natural selection

8. Match the evolution concepts.
   ___ gene flow   a. source of new alleles
   ___ natural   b. changes in a population's allele
       selection   frequencies due to chance alone
   ___ mutation   c. allele frequencies change owing to
   ___ genetic   immigration, emigration, or both
       drift   d. outcome of differences in survival,
                         reproduction among individuals
                         that vary in forms of shared traits

Additional questions are available on **Biology ⊜ Now™**

Critical Thinking

1. Martha is studying a population of tropical birds. The males have brightly colored tail feathers and the females do not. She suspects the difference is maintained by sexual selection. Design an experiment to test her hypothesis.

Figure 12.21 Two designer dogs: the Great Dane (*legs, left*) and the chihuahua (*right*).

2. A few families in a remote region of Kentucky show a frequency of *methemoglobinemia*, an autosomal recessive disorder. Skin of affected individuals appears bright blue. Homozygous recessives lack an enzyme that maintains hemoglobin in its normal molecular form. Without it, a blue form of hemoglobin accumulates in blood and shows through the skin. Formulate a hypothesis to explain why the blue-offspring trait recurs among a cluster of families but is rare in the human population at large.

3. About 50,000 years ago, humans began domesticating wild dogs. By 14,000 years ago, they started to favor new varieties (breeds) by way of artificial selection. Individual dogs having desirable forms of traits were selected from each new litter and, later, encouraged to breed. Those with undesired forms of traits were passed over.

   After favoring the pick of the litter for hundreds or thousands of generations, we ended up with sheep-herding border collies, badger-hunting dachshunds, bird-fetching retrievers, and sled-pulling huskies. At some point we began to delight in odd, extraordinary forms.

   In practically no time at all, evolutionarily speaking, we picked our way through the pool of variant dog alleles and came up with such extreme breeds as chihuahuas and Great Danes (Figure 12.21).

   Sometimes the canine designs exceeded the limits of biological common sense. How long would a tiny, nearly hairless, nearly defenseless, finicky-eating chihuahua last in the wild? Not long. What about English bulldogs, bred for a stubby snout and compressed face? Breeders thought these traits would let the dogs get a better grip on the nose of a bull. (Why they wanted dogs to bite bulls is a story in itself.) So now the roof of the bulldog mouth is ridiculously wide and often flabby, so bulldogs have trouble breathing. Sometimes they get so short of breath that they pass out.

   Explain what aspects of artificial selection processes are like, and unlike, natural selection processes.

Measuring Time

How do you measure time? Is your comfort level with the past limited to your own generation? Probably you can relate to a few centuries of human events, but how about a few millennia? Comprehending the distant past requires an intellectual leap from the familiar to the unknown.

Perhaps the possibility of an asteroid slamming into our planet can help you make the leap. Asteroids are chunks of rock and metal hurtling through space. Gaspra, the one in the filmstrip, is about 12 miles long. At least 6,000 asteroids orbit the sun between Mars and Jupiter, cold, stony leftovers from the formation of our solar system. Millions more frequently zip past Earth. They are hard to spot, particularly the ones heading straight for Earth, because they do not emit light. We do not see most of them until after they pass us by—and some have passed too close for comfort.

The 570-foot deep Barringer crater in Arizona is hard to miss. Geologic exploration of the site and its environs yielded tons of evidence (literally). Scientists analyzed the resulting data and figured out what had actually made the mile-wide hole: About 50,000 years ago, a 300-ton meteorite traveling at 26,800 miles per hour slammed into Earth. The impact was 150 times more powerful than an atomic bomb.

We have evidence that such gigantic meteorite impacts influenced the history of life. For instance, researchers found a thin layer of iridium around the world, and it dates to a mass extinction that wiped out the last of the dinosaurs. Iridium is an element that is rare on Earth—but not in asteroids. There are plenty of fossils of dinosaurs below this layer. Above it, there are none, anywhere. Connect the iridium layer to an asteroid big enough to coat the Earth with its debris, and you can infer plenty of reasons for the disappearance of the dinosaurs in the aftermath of a colossal collision.

Unlike today's large, globally dispersed populations of humans, early species lived in small bands. What if some of them were casualties of the twenty asteroids that collided with Earth while they were alive? What if *our* ancestors were just plain lucky?

You are about to make an intellectual leap through time. We invite you to launch yourself from this one premise: *Any aspect of the natural world, either past or present, has one or more underlying causes.* That premise is the foundation for scientific research into the history of life. It guides probes into physical and chemical aspects of Earth. This research represents a shift from experience to inference—from the known to what can only be surmised. And it has given us some astonishing glimpses into the past.

 How Would You Vote? A large asteroid impact could obliterate civilization and much of Earth's biodiversity. Should nations around the world contribute resources for locating and tracking asteroids? See BiologyNow for details, then vote online.

 ## Key Concepts

EVIDENCE OF EVOLUTION
Evidence of evolution comes from biogeography, fossils, and comparisons of body form, development, and biochemistry among major groups.

MACROEVOLUTIONARY EVENTS
Asteroid impacts, changing climates, and other global events contributed to mass extinctions and recoveries, as well as the emergence of new species.

ORGANIZING THE EVIDENCE
Naming and classifying species helps us manage data on biodiversity. Cladistics is an evolutionary classification system that groups species with respect to derived traits that evolved only once, in their last shared ancestor.

Links to Earlier Concepts

Before starting this chapter, be sure you understand how gene flow can keep populations of the same species genetically similar by countering impacts of mutation, natural selection, and genetic drift (Sections 12.7 and 12.8). You will also be applying your knowledge of radioisotopes (2.1), changes in chromosome structure and number (8.8 and 8.9), mutation (10.5), controls over gene expression (10.6), and DNA analysis techniques (11.2 and 11.3). These topics are basic to understanding the biochemical and molecular comparisons that are clarifying the evolutionary relationships among species.

You also will be taking a closer look at the three-domain system of classification (1.3).

13.1 Fossils—Evidence of Ancient Life

Turn now to fossil evidence of the connection between life's evolution and the evolution of Earth.

About 500 years ago, Leonardo da Vinci was puzzled by seashells entombed in rocks of northern Italy's high mountains, hundreds of kilometers from the sea. How did they get there? The prevailing belief was that water from a stupendous, divinely invoked flood had surged up against the mountains and deposited the shells. But many shells were thin, fragile, and intact. If they had been swept across such great distances, then wouldn't they have been battered to bits?

Leonardo also brooded about the rocks. They were stacked like cake layers. Some layers had shells, others had none. Then he remembered how large rivers swell with spring floodwaters and deposit silt in the sea. Did such depositions happen in ancient seasons? If so, then shells in the mountains could be evidence of layered communities of organisms that once lived in the seas!

By the 1700s, **fossils** were accepted as the buried remains and impressions of organisms that lived in the past. (*Fossil* comes from a Latin word for "something that was dug up.") However, fossils were still being interpreted through the prism of cultural beliefs, such as when a Swiss naturalist unveiled the remains of a giant salamander and excitedly announced they were the skeleton of a man who drowned in the great flood.

By midcentury, naturalists were questioning these interpretations. Excavations for canals, quarrying, and mining were unearthing similar rock layers as well as similar sequences of fossils in distant places, such as the cliffs on both sides of the English Channel. If those layers had been deposited over time, then a vertical series of fossils embedded in them might be a record of past life—*a fossil record.*

HOW DO FOSSILS FORM?

Most fossils are remains of bones, teeth, shells, seeds, spores, and other hard parts (Figure 13.1). Fossilized feces can contain residues of species that were eaten. Imprints of leaves, stems, tracks, burrows, and other *trace* fossils are all indirect evidence of past life.

Fossilization is a slow process that starts when an organism or traces of it become covered by sediments or volcanic ash. Water slowly infiltrates the remains. Metal ions and other inorganic compounds dissolved in the water gradually replace minerals in bones and other hardened tissues. As sediments accumulate, they exert increasing pressure on the burial site. In time, the pressure and mineralization processes transform those remains into stony hardness.

Remains that become buried quickly are less likely to be obliterated by scavengers. Preservation is also favored when a burial site stays undisturbed. Usually, however, erosion and other geologic insults deform, crush, break, or scatter the fossils. This is one reason that fossils are relatively rare.

Other factors also affect preservation. For instance, organic materials do not decompose in the absence of free oxygen. Remains are more likely to endure if sap, tar, ice, mud, or some other air-excluding substance protects them. Insects trapped in amber and frozen woolly mammoths are examples.

Figure 13.1 Two representatives of more than 250,000 ancient species known from the fossil record. *To the left*, fossilized parts of the oldest known land plant (*Cooksonia*). Its stems were about as tall as a toothpick. *To the right*, fossilized skeleton of an ichthyosaur. This marine reptile lived 200 million years ago.

Figure 13.2 A slice through time—Butterloch Canyon, Italy, once at the bottom of a sea. Its sedimentary rock layers formed over hundreds of millions of years. Here, scientists Cindy Looy and Mark Sephton climb to reach the Permian–Triassic boundary layer, where they will look for fossilized fungal spores.

Iridium is rare on Earth, not in asteroids. A thin, dark layer of it dates to the Cretaceous–Tertiary (K–T) boundary, a sign that the Earth took a huge hit. Iridium-rich dust blew skyward and then drifted all over the globe. It marks the K–T mass extinction.

FOSSILS IN SEDIMENTARY ROCK LAYERS

Stratified (stacked) layers of sedimentary rock are rich sources of fossils. They formed long ago from deposits of volcanic ash, silt, and other materials. Sand and silt piled up after rivers transported them from land to the sea, just as da Vinci had suspected. Sandstones formed from sand, and shales from silt. The deposits were not continuous. For instance, the sea level changed during ice ages. Tremendous volumes of water became locked in glaciers, rivers dried up, and depositions ended in some regions. When the climate warmed and glaciers melted, depositions resumed.

The formation of sedimentary rock layers is called **stratification**. The deepest layer was the first to form; those closest to the surface were the last. Most formed horizontally, as in Figure 13.2, because particles tend to settle in response to gravity. You may see tilted or ruptured layers, as along a road that was cut out of a mountainside. Major crustal movements or upheavals disturbed them much later in time.

Most fossils are in sedimentary rock. Understand how rock layers form, and you know that the fossils in them formed at specific times in the past. Specifically, *the older the rock layer, the older the fossils.*

INTERPRETING THE FOSSIL RECORD

We have fossils for more than 250,000 known species. Judging from the current range of biodiversity, there must have been many, many millions more. Yet the fossil record will never be complete. Why is this so?

The odds are against finding evidence of an extinct, ancient species. At least one individual had to die and be gently buried before it decomposed or something ate it. The burial site had to have escaped obliteration by erosion, lava flows, or other geologic assaults. The fossil had to end up in a place where someone could actually find it, as in a sedimentary layer exposed by a river that had slowly carved out a canyon.

Most ancient species did not lend themselves well to preservation. For example, unlike bony fishes and hard-shelled mollusks, the jellyfishes and soft worms do not show up as much in the fossil record. Yet they probably were just as common, or more so.

Also think about population density and body size. One plant population might release millions of spores in a single season. The earliest humans lived in small bands and raised few offspring. What are the odds of finding even one fossilized human bone compared to spores of plants that lived at the same time?

Finally, imagine one line of descent, a **lineage**, that vanished when its habitat on a remote volcanic island sank into the sea. Or imagine two lineages, one lasting only briefly and the other for billions of years. Which is more likely to be represented in the fossil record?

Fossils are physical evidence of organisms that lived in the distant past. Stratified layers of sedimentary rock that are rich in fossils are a historical record of life. The deepest layers generally contain the oldest fossils.

The fossil record is incomplete. Major geologic events obliterated much of it. Our sampling of the record is slanted toward species that had large bodies, hard parts, dense populations, and broad distribution. It also is slanted toward species that persisted for a long time.

Even so, the fossil record is now substantial enough to reconstruct patterns and trends in the history of life.

13.2 Dating Pieces of the Puzzle

How do we assign fossils to a place in time? In other words, how do we know how old they really are?

RADIOMETRIC DATING

People have been coming across fossils for as long as they have been digging up rocks. For a long time they could assign only *relative* ages to their fossil treasures, not absolute ones. For instance, a fossilized mollusk in a rock layer was said to be younger than a fossil below it and older than a fossil above it, and so on.

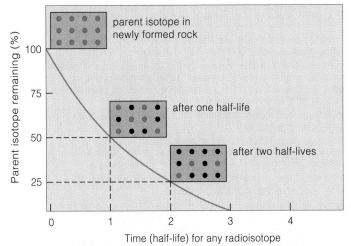

a A simple way to think about the decay of a radioisotope to a more stable form, as plotted against time.

Figure 13.3 *Animated!* (**a**) The decay of radioisotopes at a fixed rate to more stable forms. The half-life of each kind of radioisotope is the time it takes for 50 percent of a sample to decay. After two half-lives, 75 percent of the sample has decayed, and so on.

(**b–d**) Radiometric dating of a fossil. Carbon 14 (^{14}C) forms in the atmosphere. There, it combines with free oxygen, the result being carbon dioxide. Along with far greater quantities of the more stable carbon isotope (^{12}C), trace amounts of carbon 14 enter food webs by way of photosynthesis. All living organisms incorporate carbon in their body tissues.

b Long ago, trace amounts of ^{14}C carbon and a lot more ^{12}C were incorporated in tissues of a living mollusk. The carbon was part of the organic compounds making up the tissues of its prey, which it captured in the sea. As long as the mollusk was living, the proportion of ^{14}C to ^{12}C in its tissues remained the same.

c When the mollusk died, it stopped gaining carbon. Over time, the proportion of ^{14}C to ^{12}C in its remains fell because of the radioactive decay of the ^{14}C. Half of the ^{14}C had decayed in 5,370 years, half of what was left was gone after another 5,370 years, and so on.

d Fossil hunters find the fossil. They measure its $^{14}C/^{12}C$ ratio to determine half-life reductions since death. The ratio for this fossil turns out to be one-eighth of the $^{14}C/^{12}C$ ratio in living organisms. Thus the mollusk lived about 16,000 years ago.

Things changed when the nature of radioisotope decay was discovered. Remember, a radioisotope is an atom of an element with an unstable nucleus (Section 2.1). The nucleus decays spontaneously, losing energy and subatomic particles until the atom becomes a more stable element. In **radiometric dating**, the proportions of a radioisotope to its daughter elements in a sample can be used to calculate the sample's age. This works because, like the ticking of a perfect clock, the rate of decay for each isotope is constant.

We cannot predict the exact instant of decay of any one atom, but a predictable number of a radioisotope's atoms in any sample will decay in a characteristic time span. No change in pressure, temperature, or chemical state alters that rate.

The time it takes for half of a radioisotope's atoms to decay is called its **half-life** (Figure 13.3*a*). For instance, an atom of uranium 238 will decay into thorium 234, which decays into something else, and so on through intermediate isotopes until it becomes lead, the final, stable daughter element for this decay series. The half-life of uranium 238 is 4.5 billion years.

Radiometric dating can be used to reveal the age of rocks. By measuring the uranium 238–to–lead ratio in the oldest known rocks, geologists figured out that Earth formed more than 4.6 billion years ago.

Volcanic rocks can be dated, but most fossils occur in sedimentary rock. Dating a sedimentary rock yields the date its component materials formed, not the date they actually sedimented. The only way to date very old fossils in sedimentary rock is to find their position relative to volcanic rocks in the same area. Fossils that retain some carbon can be dated directly by measuring their ratio of carbon 12 to carbon 14 (Figure 13.3*b–d*).

PLACING FOSSILS IN GEOLOGIC TIME

Early geologists carefully counted backward through layer upon layer of sedimentary rock. They used their findings to construct a chronology of Earth history—a **geologic time scale**. By comparing information from around the world, they found four abrupt transitions in fossil sequences and used them as boundaries for four great intervals.

LINK TO SECTION 2.1

Eon	Era	Period	Epoch	Millions of Years Ago	Major Geologic and Biological Events That Occurred Millions of Years Ago (mya)
PHANEROZOIC	**CENOZOIC**	**QUATERNARY**	Recent	0.01	1.8 mya to present. Major glaciations. Modern humans evolve. The most recent *extinction crisis* is under way.
			Pleistocene	1.8	
		TERTIARY	Pliocene	5.3	65–1.8 mya. Major crustal movements, collisions, mountain building. Tropics, subtropics extend poleward. When climate cools, dry woodlands, grasslands emerge. *Adaptive radiations* of flowering plants, insects, birds, mammals.
			Miocene	22.8	
			Oligocene	33.7	
			Eocene	55.5	
			Paleocene	65	65 mya. Asteroid impact; *mass extinction* of all dinosaurs and many marine organisms.
	MESOZOIC	**CRETACEOUS**	Late		99–65 mya. Pangea breakup continues, inland seas form. Adaptive radiations of marine invertebrates, fishes, insects, and dinosaurs. Origin of angiosperms (flowering plants).
				99	
			Early		145–99 mya. Pangea starts to break up. Marine communities flourish. *Adaptive radiations* of dinosaurs.
				145	145 mya. Asteroid impact? Mass extinction of many species in seas, some on land. Mammals, some dinosaurs survive.
		JURASSIC		213	248–213 mya. *Adaptive radiations* of marine invertebrates, fishes, dinosaurs. Gymnosperms dominate land plants. Origin of mammals.
		TRIASSIC		248	248 mya. *Mass extinction*. Ninety percent of all known families lost.
	PALEOZOIC	**PERMIAN**			286–248 mya. Supercontinent Pangea and world ocean form. On land, *adaptive radiations* of reptiles and gymnosperms.
				286	360–286 mya. Recurring ice ages. On land, *adaptive radiations* of insects, amphibians. Spore-bearing plants dominate; cone-bearing gymnosperms present. Origin of reptiles.
		CARBONIFEROUS			
				360	360 mya. *Mass extinction* of many marine invertebrates, most fishes.
		DEVONIAN			410–360 mya. Major crustal movements. Ice ages. *Mass extinction* of many marine species. Vast swamps form. Origin of vascular plants. *Adaptive radiation* of fishes continues. Origin of amphibians.
				410	
		SILURIAN			440–410 mya. Major crustal movements. *Adaptive radiations* of marine invertebrates, early fishes.
				440	
		ORDOVICIAN			505–440 mya. All land masses near equator. Simple marine communities flourish until origin of animals with hard parts.
				505	
		CAMBRIAN			544–505 mya. Supercontinent breaks up. Ice age. *Mass extinction*.
				544	
PROTEROZOIC					2,500–544 mya. Oxygen accumulates in atmosphere. Origin of aerobic metabolism. Origin of eukaryotic cells. Divergences lead to eukaryotic cells, then protists, fungi, plants, animals.
				2,500	
ARCHEAN AND EARLIER					3,800–2,500 mya. Origin of photosynthetic prokaryotic cells.
					4,600-3,800 mya. Origin of Earth's crust, first atmosphere, first seas. Chemical, molecular evolution leads to origin of life (from proto-cells to anaerobic prokaryotic cells).

Figure 13.4 *Animated!* Geologic time scale. *Red* boundaries mark times of the greatest mass extinctions. If these spans were to the same scale, the Archean and Proterozoic portions would extend downward, spill off the page, and extend across the room. Compare Figure 13.5.

Figure 13.5 How time flies—a geologic time clock. Think of the spans as minutes on a clock that runs from midnight to noon. The recent epoch started after the very last 0.1 second before noon. Where does that put you?

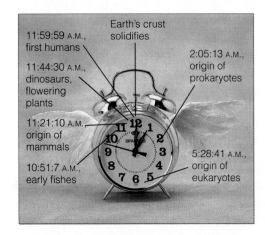

11:59:59 A.M., first humans
Earth's crust solidifies
2:05:13 A.M., origin of prokaryotes
11:44:30 A.M., dinosaurs, flowering plants
11:21:10 A.M. origin of mammals
10:51:7 A.M., early fishes
5:28:41 A.M., origin of eukaryotes

The current geologic time scale now correlates with **macroevolution**, or major patterns, trends, and rates of change among lineages. Life originated in one of the intervals, the Archean eon (Figure 13.4).

The predictable property of radioisotope decay allows scientists to accurately estimate the age of volcanic rocks and carbon-containing fossils.

A geologic time scale shows the chronology of Earth history. Intervals of life history on the scale were initially organized using evidence from the fossil record. Dates assigned to the intervals later were determined by radiometric dating.

island arc   oceanic crust   oceanic ridge   trench   continental crust

a

lithosphere   hot   athenosphere   subducting
(solid layer of mantle)   spot   (plastic layer of mantle)   plate

Figure 13.6 Some forces of geologic change.

(a) Immense, rigid parts of Earth's crust split, drift apart, and collide very slowly, at almost imperceptible rates. Huge plumes of molten material drive the movement. The plumes well up from the interior, spread laterally under the crust, and rupture it at deep mid-oceanic ridges. At these ridges, molten material seeps out, cools, and slowly forces the seafloor away from the rupture, which displaces plates from the ridges. The leading edge of one plate plows under an adjoining plate and uplifts it.

Besides these forces, superplumes ruptured the crust at what are now called "hot spots" in the mantle. The Hawaiian Archipelago has been forming this way. Continents also can rupture in their interior. Deep rifting and splitting are happening now in Missouri, at Lake Baikal in Russia, and in eastern Africa.

(b) Present configuration of Earth's crustal plates. *Red* lines indicate where spreading is occurring. *Blue* lines show where one plate is being driven under another. The Cascades, Andes, and other great mountain ranges paralleling the coasts of continents formed this way.

b

Evidence From Biogeography

By clinking their hammers against the rocks, geologists realized that Earth had not stayed put beneath our feet.

When early geologists were beginning to map vertical stacks of sedimentary rock, the theory of uniformity prevailed. The geologists knew that mountain building and erosion had repeatedly altered the surface of the Earth. Eventually, however, it became clear that those recurring geologic events were only part of a bigger picture. Like life, Earth had changed irreversibly.

AN OUTRAGEOUS HYPOTHESIS

For instance, the Atlantic coasts of South America and Africa seemed to "fit" like jigsaw puzzle pieces. Were all continents once part of a bigger one that had split into fragments and drifted apart? One model for the proposed supercontinent—Pangea—took into account the world distribution of fossils and existing species.

It also took into account glacial deposits, which held clues to ancient climate zones.

Most scientists did not accept the continental drift theory. To them, continents drifting about on their own across the Earth's mantle was an outrageous idea; they preferred to think that continents did not move.

But more evidence kept piling up. Iron-rich rocks are molten when they form. Their iron particles orient north–south in response to the Earth's magnetic poles and stay that way after the rocks cool and harden. Yet the tiny iron compasses in the 200 million year old rocks of North and South America did *not* point pole to pole. So scientists came up with a hypothetical map. They made a north–south alignment work by joining North America and western Europe—and by orienting them in a way that did not look like current maps.

Deep-sea probes also showed that the seafloor has been spreading from mid-oceanic ridges (Figure 13.6). Molten rock spewing from a ridge flows sideways in both directions, and then hardens into new crust. The spreading crust forces older crust into deep trenches elsewhere. All of the ridges and trenches are actually edges of relatively thin but enormous plates, a bit like

a 420 mya   **b** 260 mya   **c** 65 mya   **d** 10 mya

Figure 13.7 *Animated!* A series of reconstructions of drifting continents. (**a**) The early supercontinent Gondwana (*yellow*). (**b**) Later, all major land masses collided and formed the supercontinent Pangea. (**c**) Positions of fragments that drifted apart after Pangea split apart 65 million years ago, and (**d**) their positions 10 million years ago.

About 260 million years ago, seed ferns and other plants lived nowhere except on the area of Pangea that had once been Gondwana. So did mammal-like reptiles named therapsids. (**e**) Fossilized leaf of one of the seed ferns, *Glossopteris*. (**f**) *Lystrosaurus*, a therapsid about 1 meter (3 feet) long. This tusked herbivore fed on fibrous plants in dry floodplains.

pieces of a gigantic cracked eggshell. The movement is imperceptibly slow.

Together, these findings put continental drift into a broader explanation of Earth's crustal movements, the **plate tectonics theory**. Researchers soon found ways to use the new theory's predictive power.

For instance, by this theory, 420 million years ago the supercontinent **Gondwana** comprised most of the land on Earth (Figure 13.7*a*). By 260 million years ago, Gondwana had joined with all of the other major land masses to become the supercontinent **Pangea** (Figure 13.7*b*). Pangea fragmented about 65 million years ago. By 10 million years ago, the parts of Pangea that once were Gondwana had formed Antarctica, India, Africa, Australia, and South America. The rest of Pangea had become North America, Europe, Siberia, and all other land in the northern hemisphere (Figure 13.7*c,d*).

Fossils of *Glossopteris*, a seed fern, and *Lystrosaurus*, a reptile (Figure 13.7*e,f*), were found in India, Africa, Australia, and South America, but not found in North America or Europe. Scientists hypothesized that both species had evolved together on Gondwana, and they predicted that the same fossils would also be found in Antarctica. In support of the hypothesis and the plate tectonics theory, Antarctic explorers found the fossils.

A BIG CONNECTION

Let's review. In the remote past, crustal movements put huge land masses on collision courses. In time the masses converged to form supercontinents, which split at deep rifts and formed new ocean basins. Gondwana drifted south from the tropics, across the south pole, then north until it crunched into other land masses.

A new supercontinent, Pangea, now extended from pole to pole, and a single world ocean lapped against its coasts. All the while, the erosive forces of water and wind resculpted the land. Asteroids and meteorites hit the crust. The impacts and aftermaths had long-term effects on global temperature and climate.

Such changes on land and in the ocean and atmosphere influenced life's evolution. Imagine early life in shallow, warm waters along continents. Shorelines vanished as continents collided and wiped out many lineages. But even as old habitats vanished, new ones opened up for survivors—and evolution took off in new directions.

> Over the past 3.8 billion years, gradual and catastrophic events changed the Earth's crust, the atmosphere, and the ocean—and thus influenced the evolution of life.

EVIDENCE OF EVOLUTION

13.4 More Evidence From Comparative Morphology

To biologists, remember, evolution means heritable changes in lines of descent. Comparisons of body forms and structures of major groups of organisms yield clues to evolutionary trends.

Comparative morphology is the study of body forms and structures of major groups of organisms, such as vertebrates and flowering plants. Often, comparison reveals similarities in one or more body parts, which hint at inheritance from a shared ancestor. Such body parts are called **homologous structures** (*homo*– means the same). In these cases, genetic similarities persist even when the structures have different functions in different kinds of organisms.

MORPHOLOGICAL DIVERGENCE

Populations of a species diverge genetically when gene flow ends between them (Chapter 12). In time, some of the morphological traits that help define these species commonly diverge, also. Change from the body form of a common ancestor is a major macroevolutionary pattern called **morphological divergence**.

Even if the same body part of two related species become dramatically different, some other aspects of the species may remain alike. A careful look beyond unique modifications may reveal the shared heritage.

For example, all the vertebrates on land descended from the first amphibians. Divergences led to what we generally call reptiles, then to birds and mammals. A kind of "stem reptile" probably was ancestral to these groups. Fossilized, five-toed limb bones tell us that the ancestral species crouched low to the ground (Figure 13.8*a*). Its descendants diversified into many habitats on land. When environmental conditions changed, a few of the descendants that had become adapted to land returned to the seas.

The five-toed limb was evolutionary clay that was modeled into different kinds of limbs with different functions (Figure 13.8*b–g*). It was modified for flight in extinct reptiles called pterosaurs, most birds, and in bats. In lineages that led to penguins and porpoises, it became a flipper that was used for swimming. Among moles, it became stubby, useful for burrowing, and in elephants, it became strong and pillarlike. In horses, it became a long, one-toed limb suitable for fast running. The five-toed limb was also modified in humans to become the arm and hand. A thumb that was the basis of stronger and more precise movements evolved in opposition to four fingers on the hand.

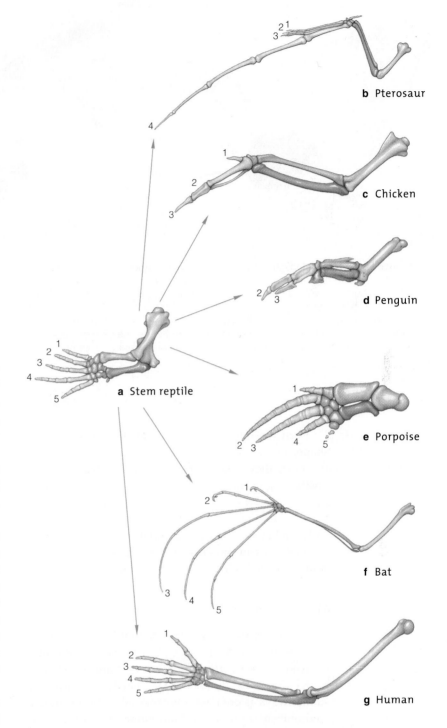

Figure 13.8 *Animated!* Morphological divergence among vertebrate forelimbs, starting with bones of a stem reptile (a cotylosaur). Similarities in the number and position of skeletal elements were preserved when diverse forms evolved. Some bones were lost over time (compare the numbers 1 through 5). The drawings are not to the same scale.

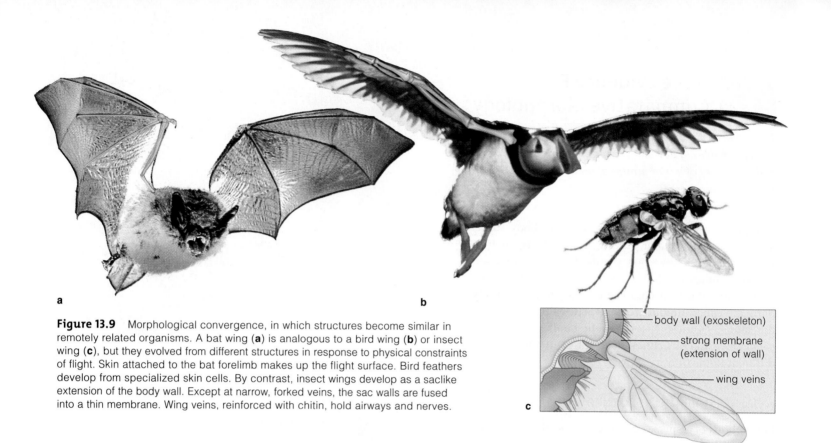

a

b

c

body wall (exoskeleton)

strong membrane
(extension of wall)

wing veins

Figure 13.9 Morphological convergence, in which structures become similar in remotely related organisms. A bat wing (**a**) is analogous to a bird wing (**b**) or insect wing (**c**), but they evolved from different structures in response to physical constraints of flight. Skin attached to the bat forelimb makes up the flight surface. Bird feathers develop from specialized skin cells. By contrast, insect wings develop as a saclike extension of the body wall. Except at narrow, forked veins, the sac walls are fused into a thin membrane. Wing veins, reinforced with chitin, hold airways and nerves.

Even though forelimbs are not the same in size, shape, or function from one group of vertebrates to the next, they clearly are alike in the structure and positioning of their bony elements. They also are alike in the patterns of nerves, blood vessels, and muscles that develop inside them. In addition, comparisons of early embryos of different vertebrates reveal strong resemblances in patterns of bone development. Such similarities point to a shared ancestor.

MORPHOLOGICAL CONVERGENCE

Body parts with similar form or function in different lineages are not always homologous. Sometimes they evolved *independently* in lineages that are not closely related. Parts that differed at first may have evolved in similar ways as organisms became subjected to similar environmental pressures. **Morphological convergence** refers to cases where dissimilar body parts evolved in similar ways in evolutionarily distant lineages.

For instance, you just read about the homologous forelimbs of birds and bats. Internal structures aside, however, the flight surfaces of bird and bat wings are *not* homologous. The flight surface of the bird wing evolved as a sweep of feathers, all derived from skin. The forelimb structurally supports the wing (Figure 13.9b). By contrast, the flight surface of the bat wing is a thin membrane, an extension of the skin itself. The wing is attached to and reinforced by the bony digits

of the bat forelimb (Figure 13.9*b*). The insect wing also resembles bird and bat wings in function—flight—but it is not homologous with either of them. The insect wing develops as an extension of the outer body wall. It has no underlying bony elements for support; it is instead reinforced with chitin (Figure 13.9*c*).

The differences between bat, bird, and insect wings are evidence that each of these animal groups adapted independently to the same physical constraints that govern how a wing can function. The flight surfaces of all three are **analogous structures**. These are structural adaptations that are the result of similar challenges in the environment, but they are derived from dissimilar body parts of distantly-related lineages. They are not modifications of the comparable body parts of closely related lineages.

With morphological divergence, comparable body parts became modified in different ways in different lines of descent from a common ancestor.

Such divergences resulted in homologous structures. Even if these body parts differ in size, shape, or function, they have an underlying similarity because of shared ancestry.

In morphological convergence, dissimilar body parts became similar in independent lineages that are not closely related.

Such body parts are analogous structures. They converged in form as an outcome of similar pressures.

Figure 13.10 How many legs? Mutations in ancestral genes may explain why animals differ in the number of appendages. A homeotic gene initiates limb development, and other genes control its expression. Expression of this gene during limb development is revealed by *green* in (**a**) a velvet walking worm, and (**b**) a sea star; and *blue* in (**c**) a mouse foot. (**d**) Cambrian legs.

13.5 Evidence From Patterns of Development

Comparing the patterns of embryonic development often yields evidence of evolutionary relationships.

Multicelled embryos of plants and animals develop step-by-step from a fertilized egg, and there are built-in constraints on how the body plan develops. Most mutations and changes in chromosome structure are selected against because they will disrupt an inherited program of development. Even so, a mutation having neutral or beneficial effects can move a lineage past one of the constraints.

Mutations in master genes can do this. Recall from Section 10.6 that homeotic genes guide the formation of tissues and organs in orderly patterns. A homeotic gene mutation can disrupt the body plan, sometimes drastically so. Such disruptions typically lead to huge problems, but occasionally an altered body plan may prove to be advantageous.

For example, a limb bud will form anywhere in an embryo that a particular homeotic gene is expressed. This gene's product is a signal for clusters of dividing embryonic cells to "stick out from the body." Where expression of this gene is suppressed, limb buds do not form. What is most remarkable is that expression of *the same gene* can initiate limb bud development in organisms as diverse as crabs, mice, beetles, sea stars, butterflies, fishes, and humans. Evolution of layers of controls over this homeotic gene may have resulted in successively fewer limbs (Figure 13.10).

Modifications in genes that influence growth rates might have caused the major proportional differences between chimpanzee and human skull bones (Figure 13.11). For humans, the facial bones and skull bones around the brain increase in size at fairly consistent rates, from infant to adult. The rate of growth is faster for chimp facial bones, so the proportions of an infant skull and an adult skull differ significantly.

In addition, transposons may have caused variation among lineages. As Section 10.5 explains, these short DNA segments spontaneously and repeatedly slip into

LINK TO SECTION 10.6

a Proportional changes in chimpanzee skull

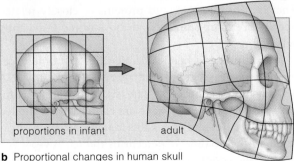

b Proportional changes in human skull

Figure 13.11
Animated!
Differences between two primates, a possible outcome of mutations that changed the timing of steps in the body's development. The skulls are depicted as paintings on a rubber sheet that is divided into a grid. Stretching both sheets deforms the grids in a way that corresponds to the differences in growth patterns between these primates.

⁺NH₃-gly asp val glu lys gly lys lys ile phe ile met lys cys ser gln cys his thr val glu lys gly gly lys his lys thr gly pro asn leu his gly leu phe gly arg lys thr gly gln ala pro gly tyr ser tyr

⁺NH₃-ala ser phe ser glu ala pro pro gly asn pro asp ala gly ala lys ile phe lys thr lys cys ala gln cys his thr val asp ala gly ala gly his lys gln gly pro asn leu his gly leu phe gly arg gln ser gly thr thr ala gly tyr ser tyr

⁺NH₃-thr glu phe lys ala gly ser ala lys lys gly ala thr leu phe lys thr arg cys leu gln cys his thr val glu lys gly gly pro his lys val gly pro asn leu his gly ile phe gly arg his ser gly gln ala glu gly tyr ser tyr

new places in genomes. For at least 30 million years, only primate DNA has carried transposons called *Alu* elements. We can expect that they were pivotal in the evolution of primates. By current estimates, 99 percent of the DNA sequences of human genes are identical to chimpanzee genes. The remaining 1 percent accounts for the species differences. Uniquely positioned *Alu* elements may be one factor.

> Similarities in patterns of development are often clues to an evolutionary relationship among plant and animal lineages.
>
> Heritable changes that alter key steps in a developmental program may be enough to bring about major differences between adult forms of related lineages. Transposons and single gene mutations can bring about such changes.

13.6 Evidence From DNA, RNA, and Proteins

All species are a mix of ancestral and novel traits, including biochemical ones. The kinds and numbers of traits they do or do not share are clues to relationships.

Each species has its own DNA base sequence. We can also expect genes to mutate in any line of descent. The more recently that two lineages diverged, the less time each will have had to accumulate unique mutations. That is why the RNA and proteins of closely related species are more similar than those of more distantly related ones.

Identifying biochemical similarities and differences among species is rapid and precise, thanks to DNA sequencing (Section 11.3). Extensive sequence data for many genomes and proteins are continually compiled into internationally accessible databases. With these data, we know that 31 percent of the 6,000 yeast cell genes have counterparts in the human genome. So do 40 percent of the 19,023 roundworm genes, and also 50 percent of the 13,601 fruit fly genes.

PROTEIN COMPARISONS

Similarities in amino acid sequences of proteins can be used to decipher connections between species. If two species have many proteins with similar or identical amino acid sequences, the species probably are closely related. When the sequences differ considerably, many mutations have accumulated, which indicates that a long time passed since the species shared an ancestor.

A few essential genes have evolved very little; they are highly *conserved* even across very diverse species. Conserved genes also are clues about ancient ancestral connections. One such gene encodes cytochrome *c*, a protein that is a crucial component of electron transfer chains in species that range from aerobic bacteria to humans. In humans, its primary structure consists of only 104 amino acids. Figure 13.12 shows the striking similarity between the entire amino acid sequence of cytochrome *c* from a yeast, a plant, and an animal.

NUCLEIC ACID COMPARISONS

Most phenotypic differences between any two species boil down to differences in the nucleotide sequences of their respective genomes. If you find the number of bases that differ between different genomes, you can estimate the relative evolutionary distance between the different species. DNA from nuclei, mitochondria, and chloroplasts, and particularly DNA regions that encode ribosomal RNA, are compared across different species for this kind of estimation.

Mitochondrial DNA (mtDNA), which mutates fast, is used to compare individuals of eukaryotic species. In sexually reproducing species, it is inherited intact from one parent—usually the mother. Thus, changes between maternally related individuals probably were caused by mutations, not by genetic recombinations.

DNA sequencing is now the most common method of comparing DNA among different species because it yields specific information, but two other methods are still in use. Nucleic acid hybridization (Section 11.2) can be used to compare DNA of two different species. DNA from both species is mixed in a way that double-thr ala ala asn lys asn lys gly ile ile trp gly glu asp thr leu met glu tyr leu glu asn pro lys lys tyr ile pro gly thr lys met ile phe val gly ile lys lys lys glu glu arg ala asp leu ile ala tyr leu lys lys ala thr asn glu-COO⁻

ser ala ala asn lys asn lys ala val glu trp glu glu asn thr leu tyr asp tyr leu leu asn pro lys lys tyr ile pro gly thr lys met val phe pro gly leu lys lys pro gln asp ala asp leu ile ala tyr leu lys lys ala thr ser ser-COO⁻

thr asp ala asn ile lys lys asn val leu trp asp glu asn asn met ser glu tyr leu thr asn pro lys lys tyr ile pro gly thr lys met ala phe gly gly leu lys lys glu lys asp arg asn asp leu ile thr tyr leu lys lys ala cys glu-COO⁻

Figure 13.12 *Animated!* Comparison of the amino acid sequence of the cytochrome *c* protein from yeast (*top row*), wheat (*middle*), and primate (*bottom*). *Gold* highlights amino acids that are identical in all three. The probability that such a resemblance resulted by chance is extremely low. Cytochrome *c* is a vital component of electron transfer chains.

stranded hybrid molecules form. The more base pairs that match, the more hydrogen bonds form between the two strands. The amount of heat that is needed to separate the strands of a hybrid DNA molecule can be used as a measure of their similarity. Why? More heat is needed to disrupt more hydrogen bonding.

Also, restriction fragments of DNA from different species are separated by gel electrophoresis. The band patterns that result can be compared just like DNA fingerprints (Section 11.4).

These comparative studies either reinforce or invite modification of ideas about evolutionary relationships that originated with morphological comparisons.

MOLECULAR CLOCKS

Some researchers estimate the timing of divergence by comparing the numbers of neutral mutations in genes that have been highly conserved in different lineages. Because the mutations have little or no effect on the individual's survival or reproduction, we can expect that neutral mutations have accumulated in conserved genes at a fairly constant rate.

The accumulation of neutral mutations to the DNA of a lineage has been likened to the predictable ticks of a **molecular clock**. Turn the hands of such a clock back, so that the total number of ticks unwind down through past geologic intervals. Where the last tick stops, that is the approximate time when molecular, ecological, and geographic events put the lineage on its own unique evolutionary road. Thus, the number of differences in DNA base sequences or amino acid sequences between species can be plotted as a series of branches on an evolutionary tree. The branches reflect relative times of divergences among species.

Biochemical similarity is greatest among the most closely related species and smallest among the most remote.

LINKS TO
SECTIONS
8.3, 12.7, 12.8

13.7 Reproductive Isolation, Maybe New Species

Speciation, a macroevolutionary event, begins with reproductive isolation and ends with a new species.

Species is a Latin word that means "kind," as in "one kind of plant." This generic definition does not help much when we are trying to figure out whether, say, a population of plants in one place belongs to the same species as a population of plants somewhere else. You may see variations in traits between the populations and variations within them, because plants can inherit diverse combinations of alleles. Also individuals of a species may vary in form because gene expression can be affected by living in different environments (Figure 13.13). In other words, we may not be able to identify a biological species on the basis of appearance alone.

Actually, individuals of *most* species vary greatly in details of form. So perhaps we should also look for a basic function that unites populations of a species and isolates them from populations of other species.

Reproduction is such a basic, defining function. It is the core of the **biological species concept**. Ernst Mayr, an evolutionary biologist, phrased it this way: Species are interbreeding natural populations that have been reproductively isolated from other such groups. In his view, it does not matter how much phenotypes vary. Populations belong to the same species as long as their members share traits that allow them to interbreed and produce fertile offspring. Being able to contribute to a shared gene pool is qualification for membership.

Mayr's species concept does not apply to asexually reproducing organisms, such as bacteria. It is a useful framework for studying the vast majority of species, which reproduce sexually. A new species arises when it achieves reproductive isolation. This isolation does not happen deliberately. *Any structural, functional, or behavioral difference that favors reproductive isolation is simply a by-product of genetic change.*

a

b

Figure 13.13 Differences in form between two plants of the same species (*Sagittaria sagittifolia*) growing (**a**) in water and (**b**) on land. Leaf shapes are responses to different conditions in the environment, not differences in genes.

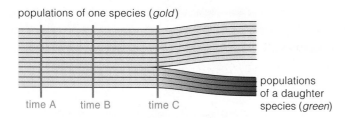

populations of one species (*gold*)

time A   time B   time C

populations
of a daughter
species (*green*)

Figure 13.14 One way to think about genetic divergence. Each horizontal band represents a different population.

LINK TO
SECTION
12.8

REPRODUCTIVE ISOLATING MECHANISMS

Gene flow alone can counter genetic changes between populations of a species. **Gene flow**, remember, is the movement of alleles into and out of a population by immigration and emigration. This microevolutionary process helps maintain diversity in a shared reservoir of alleles. **Genetic divergence** occurs after gene flow has stopped entirely (Figure 13.14). Then, mutation, natural selection, and genetic drift all begin to operate independently in each isolated population (Section 12.8). Their gene pools change. Eventually, the genetic divergence may lead to new species.

The process by which a new species arises is called **speciation**. It may happen gradually, or within a few generations, as commonly occurs among plants.

Regardless of differences in the duration or details, **reproductive isolating mechanisms** begin after gene flow has ended. These heritable aspects of body form, function, or behavior prevent interbreeding between divergent populations. Some mechanisms prevent the successful pollination or mating between individuals in the different populations. Others block fertilization or embryonic development after mating or pollination occurs. Still others cause hybrid offspring to be weak or infertile.

Mechanical Isolation. Sexual body parts of a species may not fit, mechanically, with the sexual body parts of a different species. Figure 13.15*a* shows the precise fit between a zebra orchid and its preferred pollinator, a single species of wasp.

Two sage species can keep pollen to themselves by attracting different insect pollinators. Pollen-bearing stamens of white sage extend from a nectar cup, above petals that constitute a sturdy platform. Bumblebees and large carpenter bees brush against the stamens when they land on the platform. Small honey bees that land on it are too small to reach the stamens, and so do not pick up pollen. The flower of black sage is too

Figure 13.15 *Animated!* (**a**) Mechanical isolation. Few pollinating insects fit as well as this wasp on a zebra orchid. Petals form a landing platform below the stamens. The scent of this flower stimulates male wasps to attempt mating with it. They are, of course, unsuccessful, but their actions do pollinate the flower.

(**b**) Temporal isolation. *Magicicada septendecim*, a periodical cicada that matures underground and emerges to reproduce every seventeen years. Its populations often overlap the habitats of a sibling species (*M. tredecim*), which reproduces every thirteen years. Adults live only a few weeks. Birds consider them a taste thrill and frenziedly gorge on them to the point of vomiting, which makes one suspect why periodical cicadas stay underground for so long.

(**c**) Behavioral isolation. Courtship displays precede sex among many kinds of birds, including these albatrosses. Individuals recognize tactile, visual, and acoustical signals, such as a prancing dance followed by back arching, skyward pointing bill and exposed throat, and wing spreading.

small to support the bigger bees, so this plant can be pollinated only by smaller bee species.

Temporal isolation. Diverging populations cannot interbreed when they differ in timing of reproduction. Three species of periodical cicadas often live in the same habitats in the eastern United States. They differ in size and other traits, but all mature underground. They only emerge to reproduce every 17 years (Figure 13.15*b*). Each cicada species has a *sibling* species that emerges to reproduce every 13 years. Sibling species are identical, or nearly so, but they are reproductively isolated. Each cicada species and its sibling cannot get together except once every 221 years!

Behavioral isolation. Behavioral differences bar gene flow between related species in the same vicinity. For instance, before male and female birds copulate, they often engage in courtship displays (Figure 13.15*c*). A female is genetically prewired to recognize distinctive singing, head bobbing, wing spreading, and prancing by a male of her species as an overture to sex. Females of different species usually ignore his behavior.

Ecological isolation. Populations occupying different microenvironments in a habitat might be ecologically isolated. In the open forests of seasonally dry foothills of the Sierra Nevada are two manzanita species. One lives at elevations between 600 and 1,850 meters, the other between 750 and 3,350 meters. Where the ranges overlap, these species hybridize only rarely. Like other manzanita shrubs, they depend on mechanisms that conserve water all through the dry season. One species is adapted to sheltered sites, where water stress is not intense. The other lives in drier, more exposed sites on rocky hillsides, so cross-pollination is unlikely.

Gamete mortality. Gametes of different species may have evolved incompatibilities at the molecular level. For example, when the pollen of one species lands on a plant of a different species, it usually cannot recognize the molecular signals that are supposed to trigger its growth down through the plant's tissues, to the egg.

Postzygote isolating mechanisms may act while an embryo develops. Unsuitable interactions among some genes or gene products cause early death, sterility, or weak hybrids with low survival rates. Some hybrids are sturdy but sterile. Mules, the offspring of a female horse and male donkey, are examples of such hybrids.

Figure 13.16 summarizes the points at which the different reproductive isolation mechanisms can cut a connection between species.

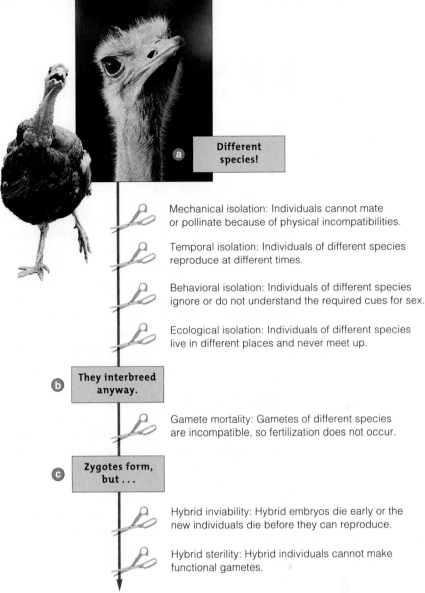

Figure 13.16 *Animated!* When certain reproductive isolating mechanisms prevent interbreeding. There are barriers to (**a**) getting together, mating, or pollination, (**b**) successful fertilization, and (**c**) survival, fitness, or fertility of hybrid embryos or offspring.

a Different species!

Mechanical isolation: Individuals cannot mate or pollinate because of physical incompatibilities.

Temporal isolation: Individuals of different species reproduce at different times.

Behavioral isolation: Individuals of different species ignore or do not understand the required cues for sex.

Ecological isolation: Individuals of different species live in different places and never meet up.

b They interbreed anyway.

Gamete mortality: Gametes of different species are incompatible, so fertilization does not occur.

c Zygotes form, but . . .

Hybrid inviability: Hybrid embryos die early or the new individuals die before they can reproduce.

Hybrid sterility: Hybrid individuals cannot make functional gametes.

No offspring, sterile offspring, weak offspring that die before reproducing, or sterility of later generations

A species is one or more populations of individuals that share a unique common ancestor, that share a gene pool under natural conditions and produce fertile offspring, and that are reproductively isolated from other species.

Speciation is the process by which a daughter species forms from a population or subpopulation of a parent species. The process varies in its details and duration.

Species originate by genetic divergence. Differences build up in the gene pools of diverging populations as mutation, natural selection, and genetic drift operate independently in each one.

Reproductive isolating mechanisms may evolve simply as by-products of genetic changes. They are heritable traits that, one way or another, prevent interbreeding.

13.8 Interpreting the Evidence: Models for Speciation

Three models for speciation differ in their basic premise of how populations become reproductively isolated.

GEOGRAPHIC ISOLATION

Genetic changes that lead to a new species often arise after a *physical separation* occurs between populations. This is called **allopatric speciation**, and it is the most common speciation route. Some physical barrier that arises stops the gene flow between two populations or subpopulations of one species. (*Allo*– means different; *patria* can be taken to mean homeland.) Reproductive isolating mechanisms evolve in the two populations. In time, speciation is complete. Interbreeding will no longer be possible even if the daughter species come into contact with one another.

How effective any geographic barrier is at blocking gene flow depends on the organisms' means of travel, how fast they travel, and also whether they are forced or behaviorally inclined to disperse. Some measurable distance often separates populations of a species, so gene flow among them is an intermittent trickle, not a steady stream. Barriers can arise abruptly and shut off the trickles. In the 1800s, a strong earthquake buckled part of the Midwest. The Mississippi River changed its course, cutting through habitats of insects that could not swim or fly. Gene flow across the river stopped in an instant of geologic time.

The fossil record suggests that geographic isolation also occurs over great spans of time. This happened after glaciers advanced down into North America and Europe during the ice ages and cut off populations of plants and animals from one another. After the glaciers retreated, descendants of the separate populations met up. Although related by descent, some were no longer reproductively compatible. They had evolved into new species. Genetic divergence was not as great between other populations; their descendants still interbreed. Reproductive isolation in their case was incomplete, so speciation did not follow.

Also, remember how Earth's crust is fractured into gigantic plates? Slow but colossal movements of the plates have altered configurations of land masses. As Central America was forming, part of an ancient ocean basin was uplifted and became a land bridge—called the Isthmus of Panama. Some camelids crossed that bridge into South America. Geographic separation led to new species, llamas and vicunas (Figure 13.17).

THE INVITING ARCHIPELAGOS

Archipelagos are island chains at some distance away from a continent. Some are so close to the mainland that gene flow is more or less unimpeded, and there is little if any speciation. The Florida Keys are a case in point. Other archipelagos are so isolated that they can

Figure 13.17 Allopatric speciations. The earliest camelids, no bigger than a jackrabbit, evolved in the Eocene grasslands and deserts of North America. By the end of the Miocene, they included the now-extinct *Procamelus*. The fossil record and comparative studies indicate that *Procamelus* may have been the common ancestral stock for llamas (**a**), vicunas (**b**), and camels (**c**). One descendant lineage that dispersed into Africa and Asia evolved into modern camels. A different lineage, ancestral to the llamas and vicunas, dispersed into South America after gradual movements of Earth's crust formed a land bridge between the two continents.

Late Eocene paleomap, before a land bridge formed between North and South America. At that time, North America and Eurasia were connected by a land bridge.

a A few individuals of a species on the mainland reach isolated island 1. Speciation follows genetic divergence in a new habitat.

b Later in time, a few individuals of the new species colonize nearby island 2. In this new habitat, speciation follows genetic divergence.

c Speciation may also follow colonization of islands 3 and 4. And it may follow invasion of island 1 by genetically different descendants of the ancestral species.

Akepa (*Loxops coccineus*)

Akekee (*L. caeruleirostris*)

Nihoa finch (*Telespiza ultima*)

Palila (*Loxioides bailleui*)

Maui parrotbill (*Pseudonestor xanthrophrys*)

Apapane (*Himatione sanguinea*)

Po' ouli (*Melamprosops phaeosoma*)

Alauahio (*Paroreomyza montana*)

Kauai Amakihi (*Hemignathus kauaiensis*)

Akiapolaau (*H. munroi*)

Akohekohe (*Palmeria dolei*)

Iiwi (*Vestiaria coccinea*)

d

Figure 13.18 *Animated!* (**a–c**) Sketch of allopatric speciation on an isolated archipelago. (**d**) Representatives of more than twenty species of Hawaiian honeycreepers, an example of a burst of speciations on an isolated archipelago. Honeycreeper bills are adapted to diverse food sources, such as insects, seeds, fruits, and floral nectar cups.

The shared ancestor of all of Hawaii's honeycreepers probably looked like this housefinch (*Carpodacus*)

serve as model laboratories for the study of evolution. Foremost among them are the Hawaiian Archipelago, nearly 4,000 kilometers west of the California coast, and the Galápagos Islands, 900 kilometers or so from the Ecuadoran coast. These islands are the very tops of volcanoes that, in time, broke the surface of the sea. We can therefore expect that their fiery surfaces were initially barren, devoid of life.

In one view, winds or ocean currents carry a few individuals of some mainland species to such islands (Figure 13.18a–c). Descendants colonize other islands in the chain. Habitats and selection pressures differ within and between islands, so allopatric speciation proceeds by way of divergences. Later, new species may even invade islands that were colonized by their ancestors. Distances between islands in archipelagos are enough to favor divergence but not enough to stop the occasional colonizers.

Flurries of speciation in the Hawaiian Archipelago appear to be the result of a few colonizers in a species-poor place. The youngest island, Hawaii, formed less than a million years ago. Here alone we see diverse habitats, ranging from cooled lava beds, rain forests, and alpine grasslands to high, snow-capped volcanoes. When ancestors of Hawaiian honeycreepers arrived, they found a buffet of fruits, seeds, nectars, and tasty insects, and not many competitors for them. The near absence of competition from other species triggered rapid allopatric speciations into the vacant adaptive zones. Figure 13.18d hints at the variation that arose among Hawaiian honeycreepers. Like thousands of other animal and plant species, they are unique to this island. As another example of speciation potential, the Hawaiian Islands combined make up less than 2 percent of the world's land masses. Yet they are home to 40 percent of all fruit fly (*Drosophila*) species.

Figure 13.19 A small, isolated crater lake in Cameroon, West Africa, where different species of cichlids may have originated by way of sympatric speciation.

LINKS TO
SECTIONS
8.8, 8.9

SYMPATRIC SPECIATION

By the model for **sympatric speciation**, a species may form *within* the home range of an existing species, in the absence of a physical barrier. (*Sym–* means together with, as in "together with others in the homeland.")

EVIDENCE FROM CICHLIDS IN AFRICA In Cameroon, West Africa, many species of freshwater fishes called cichlids may have arisen by sympatric speciation. The fish live in lakes that formed in the collapsed cones of small volcanoes (Figure 13.19). The cichlids probably colonized the lakes before volcanic action severed the inflow from a nearby river system.

Nuclear DNA and mitochondrial DNA from eleven cichlid species in Barombi Mbo, one of the small crater lakes, were compared to the DNA from cichlid species in nearby lakes and rivers. Comparisons of the DNA sequences showed that cichlid species in Barombi Mbo are more closely related to one another than to any neighboring species. All of the Barombi Mbo cichlids are descended from the same ancestral species—and speciation must have occurred *within* this lake.

What could have caused the divergences that led to speciation? The lake is only 2.5 kilometers across, with no obvious barriers. Water condition throughout the lake is uniform. Even the shoreline is uniform. Also, cichlids are very good swimmers, and so individuals of different species often meet.

However, the Barombi Mbo cichlid species do show some *ecological* separation, in that feeding preferences localize the different species in different regions of the lake. Yet all the species breed close to the lake bottom, in sympatry. Perhaps small-scale ecological separation was enough to foster sexual selection among potential mates. Sexual selection may have led to reproductive isolation, then speciation.

POLYPLOIDY'S IMPACT Reproductive isolation might happen within a few generations through **polyploidy**, in which individuals inherit three or even more sets of chromosomes characteristic of their species. Either a somatic cell fails to divide mitotically after its DNA is duplicated, or nondisjunction at meiosis results in an unreduced chromosome number (Section 8.9).

Plant speciation is rapid when polyploids produce fertile offspring by self-fertilizing or cross-fertilizing with an identical polyploid. The ancestor of common bread wheat may have been a species of wild wheat that spontaneously hybridized about 11,000 years ago with another wild species (Figure 13.20). Much later, a spontaneous chromosome doubling and an additional hybridization resulted in *Triticum aestivum*, a common bread wheat with a chromosome number of 42.

Triticum monococcum (einkorn)   Unknown species of wild wheat   Hybridization was followed by spontaneous chromosome doubling.   *T. turgidum* (wild emmer)   *T. tauschii* (a wild relative)   *T. aestivum* (one of the common bread wheats)

14AA   x   14BB ⟶ 14AB ⟶ 28AABB   x   14DD ⟶ 42AABBDD

a Einkorn has a diploid chromosome number of 14 (two sets of 7). It probably hybridized with another wild wheat species that had the same number of chromosomes.

b About 8,000 years ago, wild emmer originated from a hybrid wheat plant in which the chromosome number had doubled. Wild emmer is tetraploid; it has two sets of 14 chromosomes.

c A wild relative of wheat that had a diploid chromosome number of 14 probably hybridized with wild emmer. Common bread wheats have a chromosome number of 42.

Figure 13.20 *Animated!* Presumed sympatric speciation in wheat. Wheat grains 11,000 years old and diploid wild wheats have been found in the Near East.

MACROEVOLUTIONARY TRENDS

Figure 13.21 Parapatric speciation. (**a**) Giant velvet worm and (**b**) blind velvet worm. (**c**) These rare species of velvet walking worms live in overlapping regions of northeastern Tasmania. Hybrid offspring are sterile, which may be the main reason these two species maintain separate identities in the absence of an obvious physical barrier between their habitats.

About 95 percent of fern species and 30–70 percent of flowering plants are polyploid species. So are a few conifers, mollusks, insects, and other arthropods, as well as fishes, amphibians, and reptiles.

PARAPATRIC SPECIATION

By the model for **parapatric speciation**, populations that are maintaining contact along a common border become distinct species. (*Para–* means near, as in "near another homeland.") This happens when populations face different selection pressures across a single area. Hybrids form in a contact zone between populations. They are restricted to this zone because they are less fit than nonhybrid individuals (Figure 13.21).

By the allopatric speciation model, some type of physical barrier intervenes between populations or subpopulations of a species and prevents gene flow among them. Gene flow ends, and genetic divergences give rise to daughter species.

By the sympatric speciation model, daughter species arise from a group of individuals within an existing population. Polyploid flowering plants probably formed this way.

By the parapatric speciation model, populations in contact along a common border evolve into distinct species.

13.9   Patterns of Speciation and Extinctions

All species, past and present, are related by descent. They share genetic connections through lineages that extend back in time to the molecular origin of the first prototypic cells. Later chapters focus on evidence that supports this view. Here, simply start thinking about ways to interpret the large-scale histories of species.

BRANCHING AND UNBRANCHED EVOLUTION

The fossil record reveals two patterns of evolutionary change, one branching, the other unbranched. The first is called **cladogenesis** (from *klados*, branch; and *genesis*, origin). In this pattern, a lineage splits, the populations become genetically isolated, and they diverge. This is the pattern of speciation described earlier.

In **anagenesis**, one species can evolve into another species without branching. Mutations and changes in allele frequencies accumulate in the entire lineage as gene flow continues among its populations. In time, the species becomes so different from its ancestors that it may be classified as a new species, and the ancestral species is considered extinct.

EVOLUTIONARY TREES AND RATES OF CHANGE

Evolutionary trees summarize information about the continuity of relationships among groups. The sketch below is a simple way to start thinking about how tree diagrams are constructed. A *branch* in a tree represents a single line of descent from a common ancestor. Each *branch point* represents a time of genetic divergence:

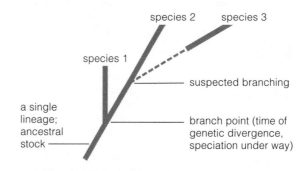

When plotted against time, a branch that ends before the present (the "treetop") signifies that the lineage has become extinct. A dashed line signifies that we know something about a lineage but not exactly where it fits in the tree.

The **gradual model of speciation** holds that species originate by small changes over long time spans. It fits with many fossil sequences. For example, sedimentary rock layers often have vertical sequences of fossilized foraminiferan shells that show incremental changes in structure (Figure 13.22).

The **punctuation model of speciation** offers an explanation for different patterns of speciation. Most morphological changes are said to occur only during a brief period when populations start to diverge—within hundreds or thousands of years. Bottlenecks, founder effects, directional selection, or a combination of these three foster rapid speciation (Sections 12.5 and 12.7). Daughter species recover quickly from the adaptive wrenching, then do not change much over the next 2 to 6 million years.

Stability has indeed prevailed for all but 1 percent of the history of most lineages but the fossil record also has episodes of abrupt change. Both models help explain the speciation patterns because change has been gradual, abrupt, or both. Species have originated at different times, and have differed in how long they last. Some did not change much over millions of years; others were the start of adaptive radiations.

Figure 13.22 Fossilized foraminiferan shells from a vertical sequence of sedimentary rock layers. The first shell (*bottom*) is 64.5 million years old. The most recent (*top*) is 58 million years old. Microscopic analysis of shell patterns confirmed that their evolutionary order matches their geological sequence.

Figure 13.23 Adaptive radiations of mammals. Branch widths show relative biodiversity. Mammals arose 220 million years ago, but did not radiate until after the K–T impact removed the dinosaurs. Not all lineages are shown. *Above*, 125-million-year-old fossil of *Eomaia scansoria* (Greek for ancient mother climber). It is the earliest known placental mammal.

ADAPTIVE RADIATIONS

An **adaptive radiation** is a burst of divergences from a single lineage. It gives rise to many new species, such as the Hawaiian honeycreepers shown in Figure 13.18. The burst requires **adaptive zones**, or a set of different niches that become filled typically by a related group of species. Think of a *niche* as a particular way of life, such as "burrowing in ocean sediments" or "catching winged insects in the air at night." A lineage can enter a vacant adaptive zone or it competes well enough to displace the resident species.

First, a species must have physical access to a niche when it opens up. For example, mammals were once distributed in the tropical regions of the supercontinent Pangea. When Pangea broke up into land masses that drifted to different parts of the globe, its habitats and resources changed. This set the stage for independent adaptive radiations of mammals (Figure 13.23).

Second, a species may enter an adaptive zone by **key innovation**: A chance modification in some body structure or function gives it the opportunity to exploit the environment more efficiently or in a novel way.

Once a species enters an adaptive zone, genetic divergences give rise to other species that can fill a variety of niches within the zone. For example, when the forelimbs of certain vertebrates evolved into wings, novel niches opened up for the ancestors of modern birds and bats.

EXTINCTIONS—END OF THE LINE

An **extinction** is an irrevocable loss of a species. By current estimates, more than 99 percent of all species that ever lived are now extinct. In addition to ongoing, small-scale extinctions, the fossil record indicates that there were at least twenty or more **mass extinctions**. These are catastrophic losses of entire families or other major groups. For example, 95 percent of all species were abruptly lost 250 million years ago. Afterward, biodiversity slowly recovered as new species filled the vacant adaptive zones.

Lineages have changed. Species originated at different times and have differed in how long they have persisted.

An adaptive radiation is the rapid origin of many species from a single lineage. It happens when an adaptive zone, a new set of niches, opens up and the lineage has physical, evolutionary, and ecological access to it.

Repeated and often large extinctions happened in the past. After times of reduced biodiversity, new species formed and occupied new or vacated adaptive zones.

MACROEVOLUTIONARY TRENDS

Bacteria	Protista	Plantae	Plantae	Animalia	Animalia	KINGDOM
Proteobacteria	Sarcomastigophora	Coniferophyta	Anthophyta	Arthropoda	Chordata	PHYLUM
Epsilonproteobacteria	Zoomastigophorea	Coniferopsida	Monocotyledonae	Insecta	Mammalia	CLASS
Campylobacterales	Trichomonadida	Coniferales	Asparagales	Diptera	Primates	ORDER
Helicobacteraceae	Trichomonadidae	Cupressaceae	Orchidaceae	Muscidae	Hominidae	FAMILY
Helicobacter	*Trichomonas*	*Juniperus*	*Vanilla*	*Musca*	*Homo*	GENUS
H. felis	*T. vaginalis*	*J. occidentalis*	*V. planifolia*	*M. domestica*	*H. sapiens*	SPECIES
none	none	western juniper	vanilla orchid	housefly	human	COMMON NAME

Figure 13.24 Taxonomic classification of six organisms. Each has been assigned to ever more inclusive categories (higher taxa)—in this case, from species to kingdom.

13.10 Organizing Information About Species

So far, you have been thinking about what species are and how they arise and vanish. Turn now to how taxonomists organize this information in ways that help us understand relationships between species.

NAMING, IDENTIFYING, AND CLASSIFYING SPECIES

Taxonomists identify, name, and classify species. They assign each a two-part name. Junipers, vanilla orchids, houseflies, humans—each has a scientific name that gives people all over the world a way to know they are talking about the same species (Figure 13.24). The name's first part is descriptive of similar species of the same kind of organism that we group together. As you saw in Chapter 1, a group of species is a **genus** (plural, genera). The second part of the name is the specific epithet. Together with the genus name, it designates one species alone.

All **classification systems** are organized systems of retrieving information about how species fit in the big picture of living things. Ever more inclusive groupings of species are **higher taxa** (singular, taxon). Family, order, class, phylum, and kingdom are examples. Now, most classification systems are phylogenic, meaning that they reflect perceived evolutionary connections within and between higher taxa in addition to patterns of evolutionary change.

A **six-kingdom classification system** prevailed for some time. All prokaryotes were assigned to kingdoms

Bacteria and Archaea, and all single-celled eukaryotes (and some multicelled forms) to kingdom Protista. Like the animals, nearly all plants and fungi are multicelled, and were each ranked as a separate kingdom. Figure 13.25*a* shows these groupings.

New fossil finds, and new insights from geology, morphological studies, and biochemical comparisons, caused many researchers to reconsider the six-kingdom system. Most of them decided to subsume the groups into a **three-domain system**, in which the three highest taxa are Bacteria, Archaea, and Eukarya (Figure 13.25*b*).

LINK TO SECTION 1.3

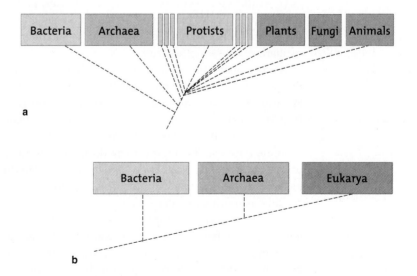

Figure 13.25 *Animated!* (**a**) Six-kingdom system of classification. The kingdom Protists is now being divided into more kingdoms. (**b**) The three-domain system of classification. Protists, plants, fungi, and animals share features that unite them in domain Eukarya.

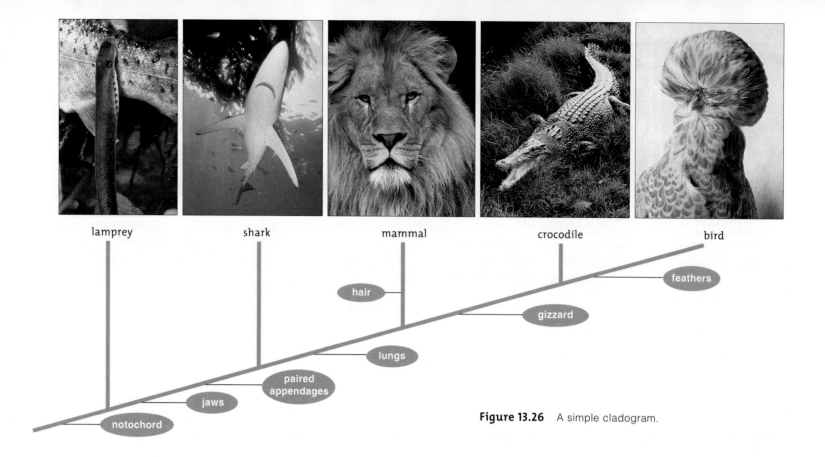

lamprey   shark   mammal   crocodile   bird

feathers

hair

gizzard

lungs

paired appendages

jaws

Figure 13.26   A simple cladogram.

notochord

WHAT'S IN A NAME? A CLADISTIC VIEW

Each set of species descended from just one ancestral species is called a **monophyletic group**, or **clade** (from *klados*, the Greek word for branch or twig). A cladistic classification system defines clades in terms of their history of divergences, or branch points in time. Only species that share traits derived from the last common ancestor are in the same clade. A **derived trait** is a novel feature that evolved only in one species, and is present only in its descendants. This emphasis means that descendants within a clade can differ—sometimes exuberantly so—in other traits.

Evolutionary tree diagrams called **cladograms** use the position of the branch point from the last shared ancestor to convey inferred evolutionary relationships (phylogenies) among taxa. A cladogram is an estimate of "who came from whom." It has no time bar with absolute dates, so it cannot convey differences in rates of evolution among taxa. Even so, a cladistic approach has already reinforced part of the fossil record, and it is making us reevaluate interpretations of the past.

As a very simple example, sharks, crocodiles, birds, and mammals have paired appendages. Sharks do not have lungs, but crocodiles, birds, and mammals do. All crocodiles and birds have a gizzard, but no mammal does. Birds alone have feathers, and mammals alone have fur. You can use these traits to put together a cladogram, as in Figure 13.26. From it, you can see that

birds and crocodiles share a more recent ancestor than the one they share with mammals, so they are more closely related to each other than they are to mammals. Crocodiles are not ancestors of birds (both are modern clades), but birds, crocodiles, and mammals all share a more recent ancestor than they do with, say, sharks.

Cladograms reflect morphological *and* biochemical comparisons. Their simple branchings do not convey the thousands of traits that are analyzed to make the connections. Cladograms such as the one that is shown in Figure 13.27 are designed only to help us visualize monophyletic groups as sets within sets, but not as absolute rankings. As such they change as quickly as phylogenies become clarified. This is not a problem if we accept that the rankings are arbitrary.

Taxonomists identify, name, and classify sets of organisms into ever more inclusive categories, the higher taxa.

Classification systems organize information about species. Phylogenetic systems show our understanding of the evolutionary relationships among species.

Reconstructing the history of a given lineage is based on the fossil record, and morphological and biochemical comparisons.

Recent evidence, especially from comparative biochemistry, favors grouping organisms into a three-domain system of classification—the archaea, bacteria, and eukarya (protists, plants, fungi, and animals).

ORGANIZING THE EVIDENCE

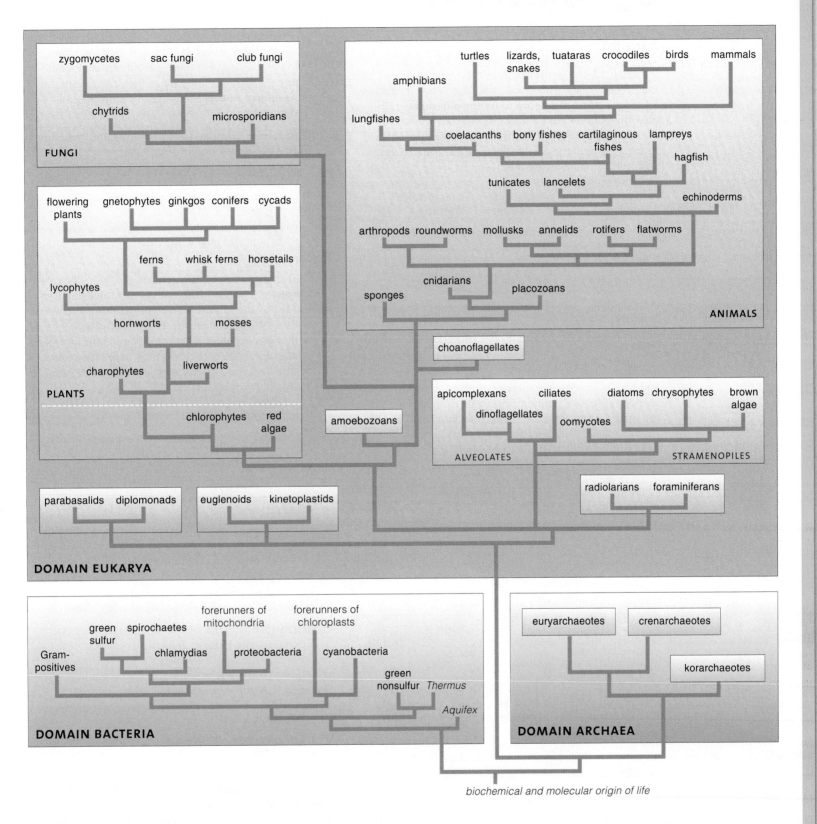

Figure 13.27 Animated! Evolutionary tree based on morphological and molecular comparisons of most major groups. Each set of organisms (taxon) has living representatives. Each branch point represents the last common ancestor of the set above it.

Summary

Section 13.1 Fossils are physical evidence of species that lived in the distant past. They are found in stacked layers of sedimentary rock, with the oldest generally near the bottom and the most recent near the top.

Although the fossil record is incomplete and biased toward certain types of species, it does reveal many patterns in the history of life.

Biology⊛Now
Learn more about fossil formation and the geologic time scale with the animations on BiologyNow.

Section 13.2 Examining fossil layers provides a relative time scale that helps explain macroevolution: patterns, rates, and trends in evolution (Table 13.1). The geologic time scale uses abrupt transitions in the fossil record as boundaries. Radiometric dating has allowed us to assign absolute dates to the scale.

Biology⊛Now
Learn more about the half-life of a radioisotope's atoms with the animated interaction on BiologyNow.

Section 13.3 The global distribution of land masses and fossils, magnetic patterns in volcanic rocks, and seafloor spreading from mid-oceanic ridges support the plate tectonics theory. Over time, movements of the Earth's crustal plates shifted continents in ways that profoundly influenced life's evolution.

Biology⊛Now
Learn more about drifting continents with the interaction on BiologyNow.

Section 13.4 Comparative morphology reveals evidence of evolution. Homologous structures are one of the clues. These body parts recur in different lineages, but they became modified in different ways after the lineages diverged from a shared ancestor. Researchers distinguish these parts from analogous structures: dissimilar body parts that became similar in independent lineages and are not indicative of shared ancestry.

Section 13.5 Similarities in patterns and structures of embryonic development suggest common ancestry. Even minor genetic changes can alter the onset, rate, and completion time of developmental stages. They can have major impact on the adult form.

Biology⊛Now
Explore proportional changes in embryonic development with the animated interaction on BiologyNow.

Section 13.6 We are now clarifying evolutionary relationships through comparisons of DNA, RNA, and proteins between different species. Investigative methods include nucleic acid hybridization and, more recently, DNA sequencing and fingerprinting.

Mutations accumulate at predictable rates in DNA; like ticks of a molecular clock, they help researchers calculate the times of divergences between lineages.

Biology⊛Now
Learn more about amino acid comparisons with the interaction on BiologyNow.

Section 13.7 All populations of a species share a unique common ancestor and a gene pool, and they can interbreed and produce fertile offspring under natural conditions. Speciation may occur after gene flow ceases between populations. Then, reproductive isolation may develop gradually, as mutation, natural selection, and genetic drift operate independently in each population. In other cases, reproductive isolation happens within a few generations.

Biology⊛Now
Use the animation and interaction on BiologyNow to explore how species become reproductively isolated.

Section 13.8 By the allopatric speciation model, a physical barrier cuts off gene flow between two or more populations. By sympatric speciation, genetic divergence starts while populations are still in physical contact. By parapatric speciation, divergences start between populations that share a common border.

Biology⊛Now
Learn more about speciation on an archipelago, and the effects of sympatric speciation in wheat, with animations on BiologyNow.

Section 13.9 Speciation has differed in its timing, rate, and direction among lineages. Adaptive radiation occurs when a new habitat spurs rapid divergence from a single lineage.

Most species have by now become extinct. Mass extinctions and recoveries have occurred in the past.

Section 13.10 Each species has a unique, two-part scientific name. Taxonomy deals with identifying, naming, and classifying species. In classification

---

Table 13.1 Summary of Processes and Patterns of Evolution

Microevolutionary Processes

Mutation	Original source of alleles	Stability or change in a species is the outcome of balances or imbalances among all of these processes, the effects of which are influenced by population size and by the prevailing environmental conditions.
Gene flow	Preserves species cohesion	
Genetic drift	Erodes species cohesion	
Natural selection	Preserves or erodes species cohesion, depending on environmental pressures	

Macroevolutionary Processes

Genetic persistence	Basis of the unity of life. The biochemical and molecular basis of inheritance extends from the origin of first cells through all subsequent lines of descent.
Genetic divergence	Basis of life's diversity, as brought about by adaptive shifts, branchings, and radiations. Rates and times of change varied within and between lineages.
Genetic disconnect	Extinction. End of the line for a species. Mass extinctions are catastrophic events in which major groups are lost abruptly and simultaneously.

systems, sets of organisms (taxa) are organized into ever more inclusive categories.

A current three-domain classification system is based largely on phylogenetic evidence. It recognizes three domains: Bacteria, Archaea, and Eukarya. The Eukarya includes diverse lineages known informally as protists, as well as plants, fungi, and animals.

Biology⊗Now
Review biological classification systems with the animation on BiologyNow.

Figure 13.28
Rama the cama, a llama–camel hybrid, displays his short temper.

Self-Quiz

Answers in Appendix I

1. Life originated in the _____ eon.
   a. Archean
   b. Proterozoic
   c. Phanerozoic
   d. Cambrian

2. Which of these supercontinents formed first: Pangea or Gondwana?

3. Morphological convergences may lead to _____ .
   a. analogous structures
   b. homologous structures
   c. divergent structures
   d. both a and c

4. Reproductive isolating mechanisms _____ .
   a. stop interbreeding
   b. stop gene flow
   c. reinforce genetic divergence
   d. all of the above

5. Most species originate by a(n) _____ route.
   a. allopatric
   b. sympatric
   c. parapatric
   d. parametric

6. In evolutionary trees, each branch point represents a(n) _____ .
   a. single species
   b. extinction
   c. time of divergence
   d. adaptive radiation

7. *Pinus banksiana, Pinus strobus,* and *Pinus radiata* are _____ .
   a. three families of pine trees
   b. three different names for the same organism
   c. three species belonging to the same genus
   d. both a and c

8. Individuals of a monophyletic group _____ .
   a. are all descended from an ancestral species
   b. demonstrate morphological convergence
   c. have a derived trait that first evolved in their last shared ancestor
   d. both a and c

9. Match these terms suitably.
   _____ phylogeny
   _____ stratification
   _____ fossils
   _____ homologous structures
   _____ cladogram
   _____ analogous structures
   _____ adaptive radiation

   a. burst of divergences from one lineage into an adaptive zone
   b. tree of branching lineages
   c. similar body parts in different lineages with shared ancestor
   d. insect wing and bird wing
   e. evolutionary relationships among species, from ancestors through descendants
   f. layers of sedimentary rock
   g. evidence of life in distant past

Additional questions are available on **Biology⊗Now**™

Critical Thinking

1. At the end of your backbone is a coccyx, a few small bones that are fused together. Could the human coccyx be a vestigial structure—all that is left of a tail of some distant vertebrate ancestors? Or is it the start of a newly evolving structure? Formulate a hypothesis, then design a way to test predictions based on the hypothesis.

2. Speculate on what might have been a key innovation in human evolution. Describe how that innovation might be the basis of an adaptive radiation in environments of the distant future.

3. Shannon thinks there are too many kingdoms and sees no reason to make another one for something as small as archaea. "Keep them with the other prokaryotes!" she says. Taxonomists would call her a "lumper." But Ann is a "splitter." She sees no reason to withhold kingdom status from archaea simply because they are part of a microscopic world that not many people know about. Which may be the most useful: more or fewer boundaries between groups? Explain your answer.

4. *Rama the cama,* a llama–camel hybrid, was born in 1997 (Figure 13.28). Camels and llamas have a shared ancestor but have been separated for 30 million years. Veterinarians collected semen from a male camel weighing close to 1,000 pounds, then used it to artificially inseminate a female llama one-sixth his weight. The idea was to breed an animal having a camel's strength and endurance and a llama's gentle disposition.

Instead of being large, strong, and sweet, Rama is smaller than expected and has a camel's short temper. Rama resembles both parents, with a camel's long tail and short ears but no hump, and llama-like hooves rather than camel footpads. Now old enough to mate, he is too short to get together with a female camel and too heavy to mount a female llama. He has his eye on Kamilah, a female cama born in early 2002, but will have to wait several years for her to mature. The question is, will any offspring from such a match be fertile?

What does Rama's story tell you about the genetic changes required for irreversible reproductive isolation in nature? Explain why a biologist might not view Rama as evidence that llamas and camels are the same species.

Looking for Life in All the Odd Places

In the 1960s, microbiologist Thomas Brock looked into Yellowstone National Park's hot springs and pools. He found a simple ecosystem of microscopically small cells, including *Thermus aquaticus*. This prokaryote lives in *really* hot water, on the order of 80°C (176°F)!

Brock's work had two unexpected results. First, it pointed the way to discovery of the Archaea, one of the three domains of life. Second, it led to a faster way to amplify DNA to useful quantities. *T. aquaticus* happens to make a heat-resistant enzyme. And that enzyme can catalyze the polymerase chain reaction—PCR—which became the premier tool for making multiple copies of a desired piece of DNA (Section 11.2).

So *bioprospecting* became the new game in town. Many companies started looking closely at thermal pools and other harsh environments for species that might yield more valuable products. What did they find? Forms of life that can withstand extraordinary levels of temperature, pH, salinity, and pressure.

Extreme thermophiles would find Yellowstone's hot springs too cool. They prefer life at hydrothermal vents on the seafloor, where water temperatures reach 110°C (230°F). Other species live in acidic springs, where pH approaches zero, and in highly alkaline soda lakes. In polar regions, some bacteria cling to life in salt ponds that never freeze and glacial ice that never melts.

It's not just prokaryotes that survive at extremes. Populations of snow algae tint mountain glaciers red. Another red alga, *Cyanidium caldarium,* grows in hot acidic springs. Free-living cells called diatoms inhabit extremely salty lakes, where the hypertonicity would make cells of most forms of life shrivel up and die. Some euglenoids thrive in acidic, metal-tainted lakes.

Life has taken hold in almost every place where there is a source of energy. Bacteria have even been filtered from the atmosphere 41 kilometers (25 miles) above the Earth. The nanobes in the filmstrip at right were found 3.8 kilometers (3 miles) below the Earth's surface in truly hot rocks—170°C (338°F).

Being one-tenth the size of most bacteria, nanobes can't be observed without electron microscopes. They look like the simplest fungi. Some argue that they are too small to be alive, to hold the metabolic machinery that runs life processes. But nanobes do have DNA and they seem to grow. Are they more like the proto-cells that preceded the origin of life? Maybe.

The rest of this unit presents a slice through time. It invites you to look at life in new ways and to trace the many lines of descent that led to the present range of biodiversity. The story is incomplete. Yet evidence from many avenues points to a principle that helps us organize information about its immense journey: *Life is a magnificent continuation of the physical and chemical evolution of the universe and the planet Earth.*

 How Would You Vote? *Private companies make millions of dollars selling an enzyme first isolated from cells at Yellowstone National Park. Should the government let companies bioprospect in national parks, as long as it shares in profits from any discoveries? See BiologyNow for details, then vote online.*

 Key Concepts

ORIGIN OF LIFE
Experiments show that organic molecules that serve as the building blocks of life can form spontaneously, as can cell-like structures. There were the first steps toward the origin of life.

PROKARYOTES
The first cells were prokaryotic. Existing prokaryotes are the most widespread, numerous, and metabolically diverse organisms. We recognize two distinct prokaryotic lineages.

THE SIMPLEST EUKARYOTES
Protists, a diverse collection of producers and consumers, are now being sorted into smaller groups of relatives. Fungi are single-celled and multicelled consumers.

THE BAD BUNCH
Viruses are infectious particles that use a host cell's metabolic machinery to replicate. Like other disease-causing organisms, they coevolve with specific hosts. Viroids and prions are even simpler infectious agents.

Links to Earlier Concepts

Sections 1.2 and 1.3 introduced the characteristics of life and the major groups of organisms. Here we begin to fill in the details. This chapter describes how major kinds of organic molecules (2.5) could have originated and how cell membranes (3.3) and metabolic pathways (4.2) arose. It expands on the earlier introductions to prokaryotic and eukaryotic cell structure (3.4, 3.5), endosymbiosis (3.6), and classification (13.10).

Section 5.1 explained how evolution of the non-cyclic pathway of photosynthesis changed the Earth's atmosphere. Here you will meet the cells that brought about that change. You will also learn more about the metabolic diversity of life (5.4).

14.1 Origin of the First Living Cells

Life originated when Earth was a thin-crusted inferno, so we probably will never find evidence of the first cells. Yet we can arrive at a plausible explanation of how they emerged by considering three questions. What were the prevailing conditions? Do known principles of physics, chemistry, and evolution support the hypothesis that the first cells could have arisen through known chemical and molecular processes? And can we design experiments to test that hypothesis? Let's take a look.

CONDITIONS ON THE EARLY EARTH

Figure 14.1*a* shows part of one of the many vast clouds in the universe. It is mostly hydrogen gas, with water, iron, silicates, hydrogen cyanide, ammonia, methane, formaldehyde, and other small inorganic and organic substances. Between 4.6 billion and 4.5 billion years ago, the cloud that would become our solar system most likely had a similar composition. Clumps of minerals and ice at that cloud's edges grew more massive as they swept up asteroids and other objects. Over time, one of these clumps condensed into the early Earth.

Four billion years ago, hot gases blanketed the first patches of Earth's thin, fiery crust. Most likely, this first atmosphere was a mixture of gaseous hydrogen, nitrogen, carbon monoxide, and carbon dioxide. There was little free oxygen. How can we tell? When oxygen is present, some binds to iron in rocks and forms iron oxides. But geologists have found that such oxides did not appear until fairly recently in Earth's history.

The relatively low oxygen levels on the early Earth probably made the origin of life possible. Free oxygen is highly reactive. If it had been present, the organic compounds characteristic of life would not have been able to form and persist. Oxygen atoms would have attacked and destroyed compounds as they formed.

What about water? At first, all of the water that fell on the molten surface would have evaporated at once. After Earth's crust cooled and became solid, however, rainfall and runoff eroded mineral salts from rocks. Over millions of years, salty waters collected in crustal depressions and formed early seas. If liquid water had not accumulated, membranes could not have formed, because they take on their bilayer structure in water. No membrane, no cell, and no life.

Cells appeared less than 200 million years after the crust solidified, so carbohydrates, lipids, proteins, and nucleic acids must have formed by then. The synthesis of organic molecules always requires an energy input. What energy sources powered the assembly of the first

organic molecules? Lightning, sunlight, or heat from hydrothermal vents may have fueled reactions.

Stanley Miller was the first to test the hypothesis that the simple organic molecules that now serve as the building blocks of life can form by chemical processes. He put water, methane, hydrogen, and ammonia in a reaction chamber (Figure 14.1*b*) and zapped the mix with sparks to simulate lightning. In less than a week, amino acids and other organic molecules had formed. Later, other researchers carried out experiments using mixes that were more like the early atmosphere. These tests also yielded organic compounds, including some that can act as building blocks of nucleic acids.

By another hypothesis, simple organic compounds formed in outer space. Scientists have detected amino acids in interstellar clouds and some meteorites. One meteorite that fell to Earth in Australia contained eight amino acids identical to those in living organisms.

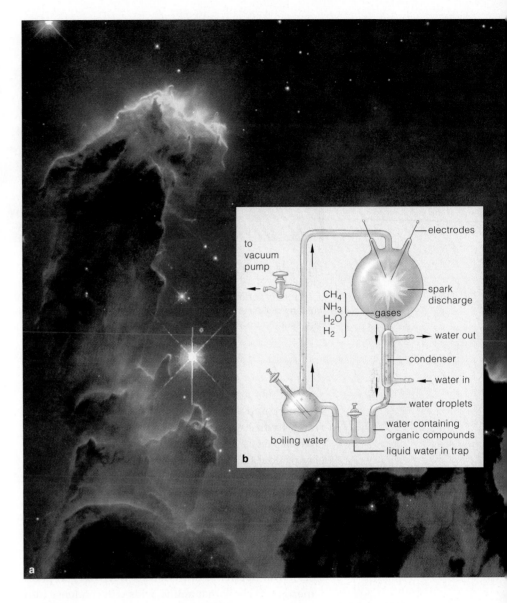

Figure 14.1 *Animated!*
(**a**) Part of the Eagle nebula, a hotbed of star formation. Each of these pillars is wider than our solar system. New stars shine brightly on the tips of gaseous streamers.

(**b**) A diagram of the reaction chamber Stanley Miller used to test whether simple organic molecules, like those used to build living cells, can form spontaneously through chemical processes.

Figure 14.2 Where did the first complex organic compounds form? Two candidates: (**a**) clay templates in tidal flats, and (**b**) iron-sulfide rocks at hydrothermal vents on the deep ocean floor. Laboratory simulations of conditions near these vents produce rocks riddled with tiny cell-sized chambers (**c**). By one hypothesis, such chambers could have served as protected environments in which early membranes formed and reactions took place.

20 µm

What about proteins, DNA, and the other complex organic compounds? Where did they form? In open water, hydrolysis reactions would have broken them apart as fast as they assembled. By one hypothesis, the clay of tidal flats bound and protected newly forming polymers (Figure 14.2a). Clays are rich in mineral ions that can attract amino acids or nucleotides. Once a few of these molecules bind to clay, others bond to them, forming chains that resemble proteins or nucleic acids.

By another hypothesis, biological molecules could have formed at deep-sea hydrothermal vents. (Figure 14.2b). The depths of ancient seas were oxygen-poor. Experiments show that amino acids will condense into proteinlike structures when heated in water.

ORIGIN OF AGENTS OF METABOLISM

Metabolism is a key defining characteristic of life. The word refers to all the reactions by which cells harness and use energy. In the first 600 million years or so of Earth history, enzymes, ATP, and other crucial organic compounds probably assembled spontaneously. If they did so in the same places, their close association might have promoted the start of metabolic pathways.

Imagine an ancient estuary, where seawater mixes with mineral-rich water being drained from the land. Beneath the sun's rays, organic molecules stick to clay. At first there are quantities of an amino acid; call it **D**. Molecules of **D** are added to proteins assembling on the clay—until **D** starts to run out. Imagine an enzyme-like protein has already been assembled. It makes **D** by acting on a plentiful, simpler substance **C**.

By chance, clumps of organic molecules include the enzyme-like protein. Now suppose **C** molecules start to run out. Molecular clumps that can produce **C** from simpler substances **B** and **A** will continue to persist and enlarge, while those that lack this ability will not. Suppose that **B** and **A** are carbon dioxide and water. The air and seas contain unlimited amounts of both.

In this way, a synthetic pathway could have arisen by chance:

$$A + B \longrightarrow C \longrightarrow D$$

This sort of "chemical evolution" may have been at work before cells appeared, increasing the frequency of certain enzymes and complex organic compounds.

ORIGIN OF SELF-REPLICATING SYSTEMS

Another defining characteristic of life is a capacity for reproduction, which now starts with protein-building instructions in DNA. As you learned in Section 10.4, using the DNA instructions to build proteins requires RNA, enzymes, and other molecules.

Did an **RNA world** in which RNA was a template for protein synthesis *precede* the origin of DNA? It's possible. As you know, RNA and DNA are a lot alike. Their sugars (ribose and deoxyribose) differ by only a hydroxyl group. Three of four bases are identical, and RNA's uracil differs only a bit from DNA's thymine.

Strands of RNA form spontaneously if nucleotides are mixed with phosphates and heated. Many existing enzymes get help from coenzymes, some of which are identical to RNA subunits.

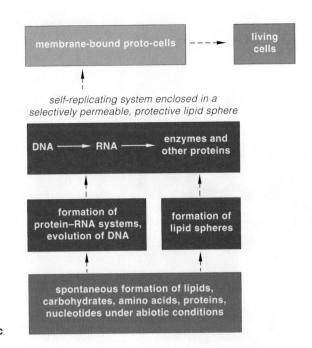

Figure 14.3 Laboratory-formed proto-cells. (**a**) Selectively permeable vesicles with a protein outer membrane were formed by heating amino acids, then wetting the resulting protein chains. (**b**) RNA-coated clay (*red*) surrounded by a membrane of fatty acids and alcohols (*green*). The mineral-rich clay catalyzes RNA polymerization and promotes the formation of membranous vesicles.

(**c**) Model for the steps that may have led to living cells.

Simple self-replicating systems of RNA, enzymes, and coenzymes have been synthesized in laboratories. So we know RNA can indeed serve as an information-storing template for making simple proteins. Also, the rRNA in ribosomes catalyzes peptide bond formation during protein synthesis (Figure 10.4). Ribosomes, you will recall, are highly conserved structures. The ones in eukaryotic cells are not that much different from those in prokaryotic cells. This suggests that the ability of rRNA to speed such reactions could have emerged early in Earth history.

If we assume that the first self-replicating systems were RNA-based, the question arises, why was there a switch to DNA? The structure of DNA may hold the answer. DNA molecules are more stable than RNAs and hold more information. As Section 9.2 describes, DNA is a double-stranded molecule that coils into helices. Helical coiling of a molecule maximizes the number of subunits that can be packed into a small space. Being single-stranded, RNA cannot form a helix.

The story of life's origin will always be incomplete because no human was there to witness it. But many details are filling in. For instance, researchers fed data on inorganic compounds and energy sources into an advanced supercomputer program. They asked the computer to put selected compounds through random chemical competition and natural selection—as might have happened untold billions of times in the distant past. They ran the program again and again. And they always got the same result: *Simple precursors evolved, the precursors spontaneously organized themselves as large, complex molecules, and they started interacting as systems.*

ORIGIN OF CELL MEMBRANES

The outermost membrane of every living cell is a lipid bilayer with diverse proteins, a plasma membrane. It controls which substances move into and out of a cell (Section 3.1). Without its control, cells can neither exist nor reproduce. Therefore, a step in life's origin had to be the formation of **proto-cells**, simple membrane sacs that could surround and protect information-storing templates and metabolic events.

In one experiment, amino acids formed protein-like chains. These were placed in heated water. After being cooled, the chains spontaneously assembled into tiny, stable, selectively permeable spheres (Figure 14.3a). In other experiments, fatty acids and alcohols formed into membranous vesicles around bits of clay (Figure 14.3b). These vesicles "grew," by adding more fatty acids, and the clay inside catalyzed the formation of RNA.

Also, researchers found out something intriguing about hydrothermal vents. Smokestacks at these vents are honeycombed with cell-sized chambers, as shown in Figure 14.2c. Mineral ions that projected into such chambers could have been templates on which RNA, proteins, DNA, and lipids could replicate themselves. If molecules became concentrated and protected from their surroundings inside such chambers, this could have favored the spontaneous formation of cells.

In short, there are gaps in the story of life's origin. But there also is experimental evidence that chemical evolution led to the organic molecules and structures characteristic of life. Figure 14.3c shows likely steps in the chemical evolution that preceded the first cells.

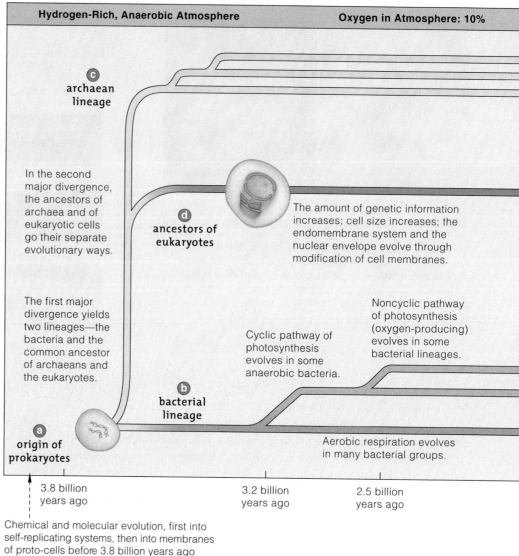

c archaean lineage

In the second major divergence, the ancestors of archaea and of eukaryotic cells go their separate evolutionary ways.

The first major divergence yields two lineages—the bacteria and the common ancestor of archaeans and the eukaryotes.

d ancestors of eukaryotes

The amount of genetic information increases; cell size increases; the endomembrane system and the nuclear envelope evolve through modification of cell membranes.

Cyclic pathway of photosynthesis evolves in some anaerobic bacteria.

Noncyclic pathway of photosynthesis (oxygen-producing) evolves in some bacterial lineages.

b bacterial lineage

Aerobic respiration evolves in many bacterial groups.

a origin of prokaryotes

3.8 billion years ago

3.2 billion years ago

2.5 billion years ago

Chemical and molecular evolution, first into self-replicating systems, then into membranes of proto-cells before 3.8 billion years ago

LINKS TO SECTIONS 3.6, 5.1, 5.2, 6.1

THE FIRST CELLS AND BEYOND

The first living cells originated in the **Archean** eon, which lasted from 3.8 billion to 2.5 billion years ago. Fossils show they were like existing **prokaryotic cells**, with no nucleus. Maybe they were membrane-bound, self-replicating sacs of RNA, DNA, and other organic molecules. Free oxygen was absent, so they got energy by anaerobic pathways such as fermentation. Geologic processes had enriched the seas naturally with organic compounds. So "food" was available, there were no predators, and the cells were free from oxygen attacks.

Some populations diverged in two major directions shortly after the time of origin. One lineage gave rise to **bacteria**. The other gave rise to the shared ancestor of **archaea** and **eukaryotic cells** (Figure 14.4a–d).

By 3.2 billion years ago, light-trapping pigments, electron transfer chains, and other innovations led to the cyclic pathway of photosynthesis in some bacteria. These new cells tapped into sunlight, an unlimited energy source. Some formed flattened mats in which sediments collected. The mats accumulated, one atop the other. Minerals hardened and preserved the mats, which are now called **stromatolites** (Figure 14.4).

By 2.5 billion years ago, photosynthetic machinery had become altered in some bacterial species, and the noncyclic pathway of photosynthesis emerged. Free oxygen—a by-product of this pathway—started to accumulate. Here we pick up the threads of a story that we began in Chapters 3 and 5.

An atmosphere enriched with free oxygen had two irreversible effects. First, *it stopped the further chemical*

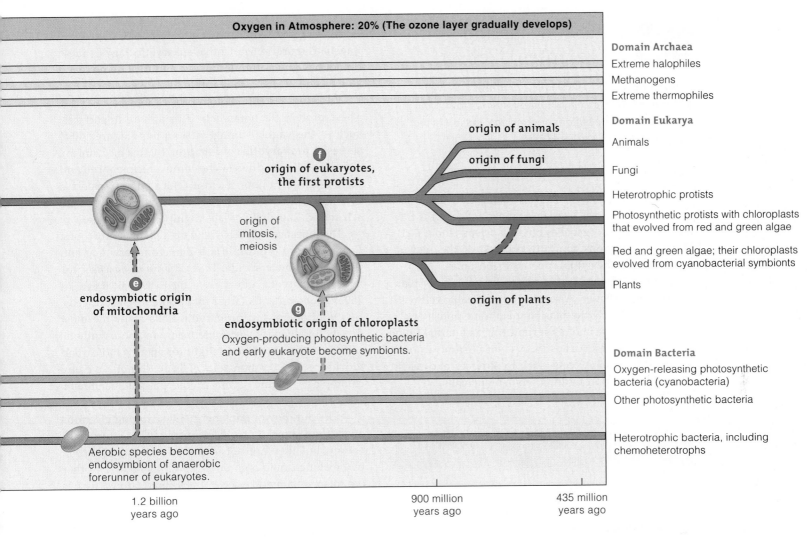

Oxygen in Atmosphere: 20% (The ozone layer gradually develops)

Domain Archaea
Extreme halophiles
Methanogens
Extreme thermophiles

Domain Eukarya
Animals
Fungi
Heterotrophic protists
Photosynthetic protists with chloroplasts that evolved from red and green algae
Red and green algae; their chloroplasts evolved from cyanobacterial symbionts
Plants

origin of animals

origin of fungi

f origin of eukaryotes, the first protists

origin of mitosis, meiosis

e endosymbiotic origin of mitochondria

g endosymbiotic origin of chloroplasts
Oxygen-producing photosynthetic bacteria and early eukaryote become symbionts.

origin of plants

Domain Bacteria
Oxygen-releasing photosynthetic bacteria (cyanobacteria)
Other photosynthetic bacteria
Heterotrophic bacteria, including chemoheterotrophs

Aerobic species becomes endosymbiont of anaerobic forerunner of eukaryotes.

1.2 billion years ago   900 million years ago   435 million years ago

Figure 14.4 *Animated!*   Milestones in the history of life. The photograph at *far left* shows stromatolites at low tide in Australia's Shark Bay. These mounds, 2,000 years old, are structurally similar to fossils more than 3 billion years old; the painting at the top shows how they may have looked then. The photo of the cut stromatolite shows layers of fossilized cyanobacterial mats.

origin of living cells. Except in a few anaerobic habitats, complex organic compounds could no longer assemble spontaneously without being oxidized. Second, *aerobic respiration became the dominant energy-releasing pathway.* It was a key innovation in the evolution of eukaryotes. As described in Section 3.6, some aerobic bacteria gave rise to mitochondria by endosymbiosis (Figure 14.4*e*). Photosynthetic bacteria that carried out the noncyclic pathway gave rise to the chloroplasts (Figure 14.4*g*).

By 900 million years ago, the major lineages that would give rise to protists, plants, fungi, and animals had diverged. All of these groups arose in the seas.

By 600 million years ago, tiny multicelled animals barely visible to the naked eye were scooting around the ocean floor. Some even had tissues and a digestive system, like most modern animals. Later, during the

Cambrian, some of their descendants would start the first adaptive radiation of animals.

Experiments provide indirect evidence that the complex organic molecules characteristic of life could have formed under conditions that probably prevailed on the early Earth.

The first cells evolved by about 3.8 billion years ago. They were prokaryotic and made ATP by anaerobic pathways.

Early on, the lineage that led to the archaea and eukaryotes diverged from the lineage that led to modern bacteria.

Oxygen-releasing photosynthetic bacteria evolved. In time, the oxygen-enriched atmosphere put an end to the further spontaneous chemical origin of life. That atmosphere was a key selection pressure in the evolution of eukaryotic cells.

14.2 What Are Existing Prokaryotes Like?

Prokaryotic cells are the smallest organisms, as well as the most widespread, abundant, and metabolically diverse. They arose billions of years ago, and all living organisms are their descendants.

GENERAL CHARACTERISTICS

Prokaryotic cells arose before nucleated cells did. (*Pro–* means before; *karyon* is taken to mean nucleus.) A rare few have simple organelles but no nucleus (Table 14.1). Metabolic reactions typically proceed at the plasma membrane, in the cytoplasm, or at ribosomes.

Figure 14.5 shows a typical prokaryotic body plan and common shapes. A semirigid, permeable cell wall usually surrounds the plasma membrane and helps a cell hold its shape and resist rupturing as internal fluid pressure increases. A capsule or layer of slime around the wall helps many cells stick to surfaces and resist attacks by infection-fighting white blood cells.

coccus

bacillus

spirillum

a

Table 14.1   Characteristics of Prokaryotic Cells

1. No membrane-bound nucleus.
2. Generally a single chromosome (a circular DNA molecule) with no associated proteins; many species also contain plasmids.
3. Cell wall present in most species.
4. Reproduction mainly by prokaryotic fission.
5. Collectively, great metabolic diversity among species.

Many prokaryotes move by means of one or more flagella (Figure 14.6*a*). Unlike eukaryotic flagella, these do not have microtubules inside, and they do not bend side to side. Instead, they rotate like a propeller.

Pili (singular, pilus) often project above a cell wall. These protein filaments help cells adhere to surfaces, such as the vaginal lining. A sex pilus (Figure 14.6*b*) serves in **prokaryotic conjugation**. By this mechanism, a donor cell's sex pilus latches onto a recipient cell and then retracts, pulling it close. The donor transfers a plasmid to the recipient. A **plasmid**, recall, is a small, self-replicating circle of DNA with just a few genes.

The tiniest cells barely show up in light microscopes. *E. coli* is more typical; it is 1–2 micrometers long and 0.5–1.0 micrometers wide. *Thiomargarita namibiensis*, the largest prokaryote known, is visible to the naked eye. At 750 micrometers, it's larger than some eukaryotic cells.

Of all organisms, prokaryotic cells show the most metabolic diversity. *Photoautotrophs* are photosynthetic self-feeders that utilize sunlight for energy and carbon dioxide as their carbon source. They do not have any chloroplasts; pigments are in the plasma membrane. *Chemoautotrophs* are self-feeders that also use carbon dioxide, but they obtain energy by stripping electrons away from compounds. *Photoheterotrophs* capture light energy, but they are not self-feeders; they get carbon by breaking down organic compounds made by other organisms. Some *chemoheterotrophs* break down tissues of a living host to obtain necessary energy and carbon. Others survive by breaking down organic products, wastes, or remains.

Among the chemoheterotrophs we find **pathogens**. These infectious, disease-causing agents invade target species and multiply in or on them. Disease follows if a pathogen's metabolic activities damage host tissues.

LINK TO SECTION 3.4

Figure 14.5 *Animated!* **(a)** The three most common shapes among the prokaryotic cells: spheres, rods, and spirals. **(b)** Generalized body plan of a prokaryotic cell.

cytoplasm, with ribosomes

DNA, in nucleoid

pilus

bacterial flagellum

outer capsule

cell wall

plasma membrane

b

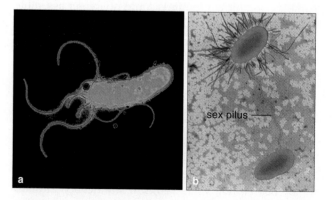

sex pilus

a

b

Figure 14.6 **(a)** *Helicobacter pylori*, a flagellated bacterial pathogen that is the most common cause of stomach ulcers. **(b)** A sex pilus connects these two *Escherichia coli* cells.

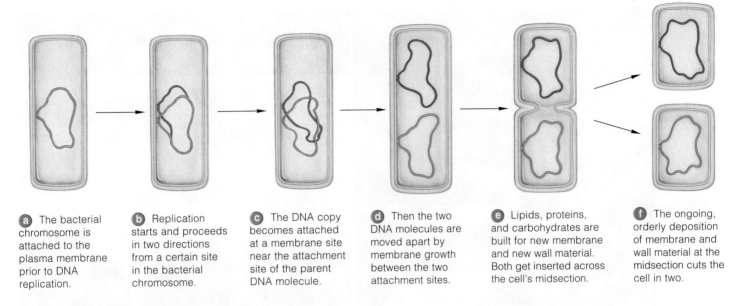

a The bacterial chromosome is attached to the plasma membrane prior to DNA replication.

b Replication starts and proceeds in two directions from a certain site in the bacterial chromosome.

c The DNA copy becomes attached at a membrane site near the attachment site of the parent DNA molecule.

d Then the two DNA molecules are moved apart by membrane growth between the two attachment sites.

e Lipids, proteins, and carbohydrates are built for new membrane and new wall material. Both get inserted across the cell's midsection.

f The ongoing, orderly deposition of membrane and wall material at the midsection cuts the cell in two.

Figure 14.7 *Animated!* Reproduction by way of prokaryotic fission. Only bacteria and archaea reproduce by this cell division mechanism.

When we describe bacterial pathogens in this and later chapters keep the following in mind: *Bacteria that cause human diseases are a tiny minority among the prokaryotes.* Many more prokaryotic species are harmless or even beneficial. Among other services, they put oxygen into the atmosphere, make essential nutrients available to plants, and break down wastes and remains of nearly all organisms. Without the prokaryotes, organic debris would pile up and the nutrient cycles that sustain all ecosystems would grind to a halt. Prokaryotes may be mostly microscopic, but their collective actions have enormous impacts on life on Earth.

GROWTH AND REPRODUCTION

The reproductive potential of prokaryotic cells can be staggering. Some types divide every twenty minutes. A cell divides in two, two become four, four become eight, and so forth. Before the cell divides, it nearly doubles in size. After division, each daughter cell has a single bacterial chromosome—one circular double-stranded molecule of DNA with few proteins.

In some species, a daughter cell buds from a parent cell. More commonly, a cell reproduces by prokaryotic fission (Figure 14.7). Briefly, a parent cell replicates its single chromosome, and this DNA replica attaches to the plasma membrane next to the parent molecule. As more membrane is added, the two DNA molecules are moved apart. Eventually, the membrane and cell wall extend into the cell's midsection and divide it into two genetically equivalent daughter cells.

PROKARYOTIC CLASSIFICATION

Once, reconstruction of the evolutionary history of the prokaryotes seemed an impossible task. Most ancient prokaryotic groups are either poorly represented in the fossil record or missing entirely.

Today, comparisons of gene sequences can clarify relationships. For example, studies of genes for rRNA opened the way for the division of all organisms into three domains: Archaea, Bacteria, and Eukarya:

The two prokaryotic domains show some marked differences in genes, membranes, and metabolism. In some ways, archaeans are more similar to eukaryotes than to the bacteria. For example, only eukaryotes and archaeans wrap their DNA around histone proteins.

A recent discovery that most prokaryotic genomes contain hereditary material from more than one source complicates classification efforts a bit. Mixed genomes arise by **lateral gene transfer**, the transfer of genetic information between cells (often of different species) by conjugation or some other process.

LINKS TO
SECTIONS
5.2, 11.2

5 μm

Figure 14.8 Life in extreme environments. (**a**) In saltwater evaporation ponds in Utah's Great Salt Lake, extreme halophiles (certain archaea and red algae) tint the water pink. (**b**) Extreme thermophiles live in Yellowstone National Park's hot springs. (**c**) The parasitic *Nanoarchaeum equitans* (smaller *blue* spheres) survives only when attached to *Ignicoccus* (larger spheres). Both archaeans were isolated from 100°C water near a hydrothermal vent.

EXAMPLES OF ARCHAEAN DIVERSITY

Many of the archaea withstand conditions as harsh as those on the early Earth, so they may resemble the first cells. Hence their group name; *archae–* means "beginning." Physiologically, many of the archaea are extreme halophiles (salt lovers), methanogens (cells that make methane), or extreme thermophiles (heat lovers). Species that live in more hospitable habitats tend to have an ancestor in one of these groups.

Extreme halophiles live in the Dead Sea, Great Salt Lake, seawater evaporation ponds, and other highly salty habitats (Figure 14.8*a*). Most can produce ATP by aerobic reactions, but they switch to photosynthesis if oxygen gets scarce. Their plasma membrane includes bacteriorhodopsin, a unique light-absorbing pigment.

Methanogens live in oxygen-free habitats, such as wetland mud, the gut of termites and mammals, and hydrothermal vents. They produce ATP by stripping electrons from hydrogen gas or from alcohols. Most use carbon dioxide as their carbon source and as the final electron acceptor for reactions that yield methane. All are anaerobes; free oxygen kills them. As a group, the methanogens release about 2 billion tons of methane per year. Their activity affects atmospheric levels of carbon dioxide and the global carbon cycle.

Sulfur-rich hot springs typify the homes of **extreme thermophiles** (Figure 14.8*b*). Like the methanogens,

they are strict anaerobes. Unlike them, they use sulfur as the electron acceptor or donor for making ATP. All thrive at temperatures above 80°C. The first one to be discovered, *Solfolobus*, flourishes in acidic hot springs. Some bacteria, such as *Thermus aquaticus*, the source of the enzyme used in PCR (Section 11.2) also survive in hot spring environments.

Some archaean extreme thermophiles are the basis for food webs in the sediments around hydrothermal vents, where water temperatures exceed 110°C. These cells obtain electrons from hydrogen sulfide released from the vents. The existence of archaeans that can tap into this energy source suggests that life could have originated near such vents on the ancient seafloor.

Nanoarchaeum equitans, the smallest known cell, was discovered beside a hydrothermal vent off the coast of Iceland (Figure 14.8*c*). It also has the tiniest genome, with fewer than 500,000 base pairs in its chromosome.

EXAMPLES OF BACTERIAL DIVERSITY

By far, most of the 400 named genera of prokaryotes are bacteria. Among them, the cyanobacteria have the longest fossil record. As described in Section 14.1, they were already forming stromatolites more than three billion years ago. Cyanobacteria contain chlorophyll *a*, carry out photosynthesis by the noncyclic pathway, and release oxygen as a by-product. As you learned in

PROKARYOTES

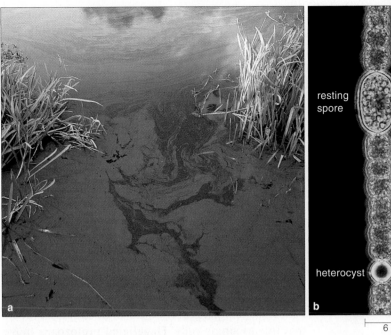

Figure 14.10
Endospore forming in *Clostridium tetani*.

— DNA

spore coat

Figure 14.11
Thiomargarita namibiensis, the largest bacterium. Cells can be seen with the naked eye. This marine species concentrates nitrate from seawater and has roles in nitrogen and sulfur cycles.

resting spore

heterocyst

b

|— 6 µm —|

Figure 14.9 Cyanobacteria. (**a**) Cyanobacterial populations often form dense mats near the surface of nutrient-enriched ponds and lakes. (**b**) Some cells develop into resting spores and nitrogen-fixing heterocysts when conditions restrict growth.

Section 3.6, cyanobacteria gave rise to chloroplasts by endosymbiosis. We can trace almost all free oxygen in the atmosphere back to free-living cyanobacteria or to chloroplasts that are descended from them.

Modern cyanobacteria are common in ponds. They often remain together as mucus-sheathed chains that form slimy mats in nutrient-enriched habitats (Figure 14.9*a,b*). *Anabaena* and other nitrogen-fixing kinds can convert nitrogen gas (N_2) to ammonia, which is used in biosynthesis. Some cells become heterocysts, which make and share nitrogen compounds with other cells in the chain. In return, they receive carbohydrates.

Gram-positive bacteria are named for the stain that colors their thick, multilayered walls purple. Most are chemoheterotrophs. *Lactobacillus* species form lactate by fermentation reactions. Some are used to make yogurt, pickles, and other foods. *L. acidophilus* lowers the pH of skin and of vaginal and intestinal linings. This helps keep disease-causing bacteria from taking hold. Use of antibiotics can inhibit *L. acidophilus* and other beneficial bacteria. The result is often vaginitis or diarrhea.

Clostridium and *Bacillus* are Gram-positive bacteria that survive hostile conditions by forming endospores. This resistant structure forms inside the cell, around the chromosome and a bit of cytoplasm (Figure 14.10). It survives boiling, irradiation, and drying out. When conditions improve, the endospore will germinate. A bacterium emerges, and normal function resumes.

Endospores that germinate in the human body can be lethal. In 2001, *Bacillus anthracis* endospores sent in the mail caused fatal cases of *anthrax* in five people. *Clostridium botulinum* often taints fermented grain and food in improperly sterilized or sealed cans and jars. Toxins produced by the bacteria that emerge after the endopores are eaten cause *botulism*, a potentially lethal food poisoning. A related species, *C. tetani*, causes the disease *tetanus*, which is described in Section 21.4.

Proteobacteria—the most diverse bacterial group—include photosynthetic cells that don't release oxygen, free-living heterotrophs with roles in nutrient cycles, and many pathogens. For example, *Helicobacter pylori* can cause stomach ulcers (Figure 14.6*a*), *Vibrio cholerae* infection leads to cholera (Section 31.5), and *Neisseria gonorrhoea* is the agent of gonorrhea (Section 27.7).

The best studied prokaryote is *E. coli* (Figure 14.6*b*). Grown in laboratories around the world, its natural habitat is the mammalian gut. Here it makes vitamin K and helps keep harmful bacteria at bay. But mutant strains can cause a deadly foodborne illness.

Thiomargarita namibiensis is the largest prokaryote ever discovered (Figure 14.11). Nitrate in its vacuole acts as the acceptor for electrons it strips from sulfur for energy. *Rhizobium* lives in nodules on the roots of peas and some other plants and converts nitrogen gas to nitrates, which plants take up and use. *Nitrobacter* lives in soil and converts nitrites to nitrates.

Figure 14.12 *Borrelia burgdorferi*, causes Lyme disease. This spirochete is spread from host to host by ticks, such as the deer tick, *Ixodes*.

BACTERIAL BEHAVIOR

Bacteria are small. Their insides are not elaborate. But some display remarkably complex behavior. Consider the spirochete *Borrelia burgdorferi*. Like all spirochetes, it resembles an overstretched spring (Figure 14.12). Ticks spread it from one mammal to another, where it causes Lyme disease. After being taken up in a blood meal, spirochetes find their way from a tick's gut to its salivary glands. Once there, they await the bite that will carry them to their next mammalian host.

> Archaea and bacteria, the only prokaryote cells, make up two domains of life. Nearly all are microscopically small, and they alone reproduce by prokaryotic fission.
>
> Prokaryotic cells inhabit almost all environments in and on Earth and its waters. Collectively, they are the most metabolically diverse and far-flung organisms.

14.3 The Curiously Classified Protists

Protists comprise a diverse group of eukaryote lineages. They include producers ranging from microscopic cells to seaweeds. Others are consumers: decomposers, parasites, or predators. The vast majority are single-celled species, but many lineages also have multicelled members.

Of all existing species, **protists** are the most like the first eukaryotic cells. Unlike prokaryotes, protist cells have a nucleus. Most also have mitochondria, ER, and Golgi bodies. They have more than one chromosome, each consisting of DNA with many proteins attached. They have a cytoskeleton that includes microtubules. Protist cells divide by way of mitosis, meiosis, or both.

Most protists are single-celled species. But there are colonial forms and lots of multicelled species. Many are photoautotrophs that have chloroplasts. Others are predators, parasites, and decomposers. A number of them can form spores.

Until recently, protists were defined largely by what they are *not*—not prokaryotes, not plants, not fungi, not animals. They were lumped together in a kingdom between prokaryotes and the "higher" forms of life. Thanks to improved comparative methods, protists are now being assigned to groups that reflect evolutionary relationships. Figure 14.13 shows these groupings.

FLAGELLATED PROTOZOANS AND EUGLENOIDS

Protozoans is a general term for heterotrophic protists that live as single cells. **Flagellated protozoans** move using one or more flagella.

Among the most ancient are the diplomonads and parabasalids. Both have multiple flagella at one end of the cell and a single flagellum at the other end. *Giardia lamblia* is a diplomonad. It has no mitochondria, Golgi bodies, or lysosomes. *G. lamblia* lives as an intestinal parasite inside humans, cattle, and some wild animals. The cells can survive outside the host body, in cysts. Among microbes, a **cyst** is a protective covering that formed from cell secretions. Drink water from a stream contaminated with *G. lamblia* cysts, and you could end up with *giardiasis*. Symptoms range from mild cramps to very severe diarrhea that can persist for weeks.

As an example of a parabasalid, consider *Trichomonas vaginalis*, a sexually transmitted parasite of humans (Figure 14.14). It causes *trichomoniasis*. Infection can permanently damage urinary and reproductive tracts.

Figure 14.13 Evolutionary tree for major living protist groups based on morphological and genetic comparisons.

THE SIMPLEST EUKARYOTES

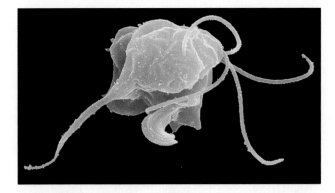

Figure 14.14 Scanning electron micrograph of *Trichomonas vaginalis*, which causes trichomoniasis, a common sexually transmitted disease.

Cells of *T. vaginalis* do not have mitochondria, but a similar organelle produces ATP by anaerobic means. Like mitochondria, this organelle is most likely the descendant of a symbiotic proteobacteria.

Trypanosomes are flagellated cells that live inside a diverse array of hosts, including plants, insects, and most vertebrates. Many infect humans. *Trypanosoma brucei* infection causes *African sleeping sickness*, which damages the central nervous system (Figure 14.15*a*). *T. cruzi* infection causes *Chagas disease*, which harms the liver, brain, and heart. Both diseases can be fatal.

Euglenoids are flagellated cells that live in ponds and lakes. They are related to trypanosomes, and the early ones were heterotrophs. Some still are, but many more have acquired chloroplasts by endosymbiosis. Like other freshwater protozoans, the euglenoids have a contractile vacuole that expels any excess water that diffuses inward (Figure 14.15*b*). You read about this organelle in Chapter 3.

As in many other protozoans, a **pellicle** holds the euglenoid body in and gives it shape. This is a flexible body covering constructed of parallel, spiraling strips of a translucent, protein-rich material. Pigments form an eyespot, which shields a light-sensitive receptor. This receptor helps photosynthetic euglenoids locate and stay in the places most suitable for photosynthesis.

RADIOLARIANS AND FORAMINIFERANS

Radiolarians and foraminiferans are related groups of mostly marine protists that secrete shells (Figure 14.16). Both groups use pseudopods to trap and engulf tiny prey, such as bacteria. Most foraminiferans live on the seafloor. The shell is hardened with calcium carbonate. Radiolarians have a silica-strengthened shell. Most are part of the marine **plankton**, microscopic organisms that drift or swim weakly in water.

red blood cell
undulating membrane attached to flagellum

Figure 14.15 *Animated!*
Two related flagellates.
(**a**) Light micrograph of *Trypanosoma brucei*, which causes African sleeping sickness. (**b**) Light micrograph of *Euglena* in a drop of pond water and a sketch of the body plan.

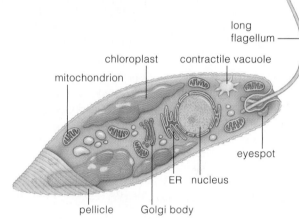

long flagellum

chloroplast · contractile vacuole

mitochondrion

eyespot

ER · nucleus

pellicle · Golgi body

Figure 14.16 (**a**) England's white chalk cliffs at Dover are the calcium carbonate-rich remains of countless foraminiferans and other shelled protists. Remains were laid down over many millions of years, then uplifted. The cliffs now stand 300 meters high. (**b**) One living foraminiferan. The many tiny yellow spheres on its thin pseudopods are algal cells that live inside it. (**c**) The silica-rich shells of two radiolarians.

LINK TO
SECTION
4.8

THE CILIATES

Molecular studies reveal that the ciliated protozoans, or **ciliates,** the dinoflagellates, and the apicomplexans are close relatives. Collectively, they are known as the *alveolates,* because all have alveoli—tiny membranous sacs—just beneath their outer membranes.

Ciliates live in fresh water, saltwater, and the soil. Most are free-living cells. Others attach to a surface or form colonies. About 30 percent live inside animals. A cell's cilia beat in synchrony to direct food toward an oral cavity and to move the cell. Ciliates eat bacteria, algae, and one another.

Paramecium is a freshwater species (Figure 14.17). A mature cell is about 200 micrometers long. Outer membranes form a pellicle. An oral cavity that opens on the surface leads to the gullet. Ingested food gets enclosed in enzyme-filled digestive vesicles. Excess water is expelled by one or more contractile vacuoles.

Most ciliates are able to reproduce both sexually and asexually. Each typically has two types of nuclei that it distributes to the daughter cells. Some ciliates engage in binary fission—they reproduce asexually by dividing in half near the midpoint of the cell.

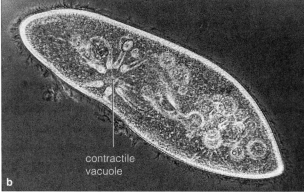

Figure 14.17 (**a**) Cilia at the surface of *Paramecium*, a single-celled predator. (**b**) A look at its organelles.

Figure 14.18 Effects of dinoflagellates. (**a**) *Karenia brevis* produces toxins that can cause fish kills (**b**) if populations reach high density during algal blooms. (**c**) *Noctiluca scintillans*, a bioluminescent dinoflagellate. (**d**) Near Vieques Island, Puerto Rico, each liter of water may hold 5,000 of these cells. Each flashes with blue light when agitated.

DINOFLAGELLATES

Dinoflagellates are single cells, usually with cellulose plates beneath their plasma membrane. Most species have a flagellum at the posterior end and another in a groove that runs around the cell. About half of the species are aquatic producers in freshwater or marine habitats. Their chloroplasts are descendants of red or green algae. A few photosynthetic species live inside corals and supply them with sugars. Most predatory dinoflagellates eat bacteria. Still others are parasites of marine or freshwater fish and invertebrates.

At high densities, certain dinoflagellates can have dramatic effects (Figure 14.18). Disturb bioluminescent species in tropical seas and the water will shimmer. Less pretty are the huge fish kills that result from algal blooms (also known as red tides). An **algal bloom** is a population explosion of a dinoflagellate or another aquatic protist. Algal blooms tend to occur in warm, shallow, nutrient-enriched waters. Increasingly, such nutrient enrichment is caused by pollutants.

APICOMPLEXANS AND MALARIA

Apicomplexans are parasitic alveolates equipped with a unique microtubular device that can attach to and penetrate a host cell. More than 4,000 apicomplexan species are known to parasitize animals ranging from insects to humans. Here we will focus on four species

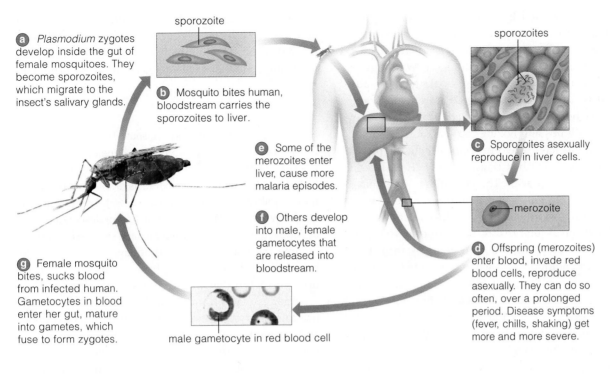

a *Plasmodium* zygotes develop inside the gut of female mosquitoes. They become sporozoites, which migrate to the insect's salivary glands.

sporozoite

b Mosquito bites human, bloodstream carries the sporozoites to liver.

e Some of the merozoites enter liver, cause more malaria episodes.

f Others develop into male, female gametocytes that are released into bloodstream.

g Female mosquito bites, sucks blood from infected human. Gametocytes in blood enter her gut, mature into gametes, which fuse to form zygotes.

male gametocyte in red blood cell

sporozoites

c Sporozoites asexually reproduce in liver cells.

merozoite

d Offspring (merozoites) enter blood, invade red blood cells, reproduce asexually. They can do so often, over a prolonged period. Disease symptoms (fever, chills, shaking) get more and more severe.

sporozoites

merozoites about to rupture a red blood cell

Figure 14.19 *Animated!* Life cycle of one of the *Plasmodium* species that cause malaria.

of *Plasmodium* that cause the disease *malaria*. These species have infected more than 300 million people in tropical parts of the world, especially Africa.

The bite of an infected female *Anopheles* mosquito puts the parasite in a host. Sporozoites, a motile stage, travel in the bloodstream to the liver. They reproduce asexually in liver cells until the cells burst and release another stage—the merozoites. Some merozoites will reinfect the liver. Others enter red blood cells, where they develop into immature gametes known as male and female gametocytes (Figure 14.19).

Plasmodium gametocytes cannot mature in humans because the parasites are sensitive to the temperature and oxygen levels inside their hosts' body. Humans are warm-bodied and their blood has little free oxygen; most of it is bound to hemoglobin. Gametocytes can mature only in a female mosquito, which has a lower body temperature and higher internal oxygen levels. In this insect's gut, zygotes form, divide, and give rise to sporozoites. This infective stage then migrates to the mosquito's salivary glands to await a new bite.

Malaria symptoms start when infected liver cells rupture and release merozoites, metabolic wastes, and cellular debris into the bloodstream. Shaking, chills, a burning fever, and drenching sweats follow. After one episode, symptoms usually subside for a few weeks or months. Infected individuals might feel healthy, but they should expect relapses. Jaundice, kidney failure, convulsions, and coma are outcomes of the disease.

Many *Plasmodium* strains are now resistant to older antimalarial drugs. Artemisinin, a synthetic version of a chemical in the wormwood plant (*Artemisia*) is a new drug. It lowers fever and reduces parasite numbers in the blood, so it can slow the course of infection.

Efforts to design a vaccine are ongoing. If things go smoothly and a vaccine seems effective, widespread clinical trials will probably get under way in 2006 or 2007. The sooner the better—every thirty seconds, one African child dies of malaria.

DIATOMS, BROWN ALGAE, AND RELATIVES

Colorless flagellates, water molds, and photosynthetic species—the diatoms, yellow-green, golden, and brown algae—are now grouped as stramenopiles. The shared trait of this diverse group is a shaggy-looking flagella.

Water molds, downy mildews, and white rusts are heterotrophs known as oomycotes. The name means egg fungus and refers to how the egg cells form. They are not actually fungi. Water molds help decompose organic debris and dead organisms in aquatic habitats. The downy mildews are important plant pathogens.

Another oomycote pathogen, *Phytophthora infestans*, causes *late blight*, a rotting of plant parts. During the

LINK TO
SECTION
5.1

mid-1800s, it destroyed Irish potato crops, causing the famine that killed and displaced millions. *P. ramorum,* a related species, kills trees. It is the cause of an ongoing epidemic of *sudden oak death* in Oregon, Washington, and California. Tens of thousands of oaks have already died, and the pathogen has jumped to redwoods and other novel hosts. Signs of *P. ramorum* infection can include dead limbs and oozing cankers on tree trunks.

The diatoms have a two-part silica shell; the pieces fit together like a hat box (Figure 14.20). They live in freshwater and cool seas. The coccolithophores have calcium carbonate plates and live in warm seas. Both groups are photosynthetic and together release about as much oxygen as land plants. Both have survived for many millions of years and their remains form vast deposits. Coccolithophore shells contributed to those high cliffs at Dover, England. Silica-rich diatomaceous earth is quarried for use in filters and abrasives.

Most brown algae inhabit temperate or cool seas. Accessory pigments tint them olive, golden, or dark brown. In size, they range from microscopic filaments to giant kelps, such as *Macrocystis,* that reach thirty meters in height. Brown algae reproduce sexually and asexually, and a typical life cycle includes both gamete-producing and spore-producing stages.

The giant kelps are of great ecological importance. The sporophyte (spore-producing stage) of a giant kelp is the familiar seaweed (Figure 14.21*a*). It has stipes (stemlike parts), blades (leaflike parts), and holdfasts (structures that hold the alga in place). Hollow, gas-filled bladders impart buoyancy to stipes and blades, and keep them upright, near the sunlit surface waters. Diverse species feed on and take shelter in kelp beds.

Sargassum, Macrocystis, and other brown algae have commercial uses. Extracts are used to manufacture ice cream, pudding, jelly beans, toothpaste, cosmetics, and other products. Alginic acid from algal cell walls is used to produce algins, which serve as thickeners, emulsifiers, and suspension agents in many products.

RED ALGAE

Nearly all red algae live in the seas. They are abundant in tropical seas, as far 265 meters beneath the surface. Encrusting types help form coral reefs. Red algae are tinted red to black by phycobilins, accessory pigments that can absorb blue-green wavelengths that penetrate deep into water. Their chloroplasts are descended from cyanobacteria. In turn, some ancient single-celled types became the chloroplasts of dinoflagellates. Life cycles have sexual and asexual phases. Multicelled stages often form, but they lack true tissues (Figure 14.21*b*).

Agar is made from the walls of a few species. We use this inert gel to preserve moisture in baked goods and cosmetics, as a setting agent for jellies, and to package drugs in soft capsules. We use carrageenan extracts to stabilize paints, dairy products, and other emulsions. Some species of *Porphyra* are used in sushi as *nori,* a wrapping for rice and fish.

GREEN ALGAE

Of all protists, green algae are most plantlike in their structure and biochemistry, and they share a common ancestor with the land plants. Like land plants they are photosynthetic, with the same chlorophylls and starch grains inside their chloroplasts. Some have cell walls that include polysaccharides typical of plants.

Single-celled, sheetlike, tubular, filamentous, and colonial forms abound. Figures 14.22 and 14.23 show

Figure 14.20 Shells of two important aquatic producers. *Above,* silica shell of *Stephanodiscus niagarae,* a diatom that is common in temperate zone lakes. *Below,* the calcium carbonate shell from a coccolithophore.

bladder

blade

stipe

holdfast

Figure 14.21 Examples of algae. (**a**) Body plan and photograph of the giant kelp, *Macrocystis,* one of the brown algae. (**b**) A red alga, *Antihamnion plumula.*

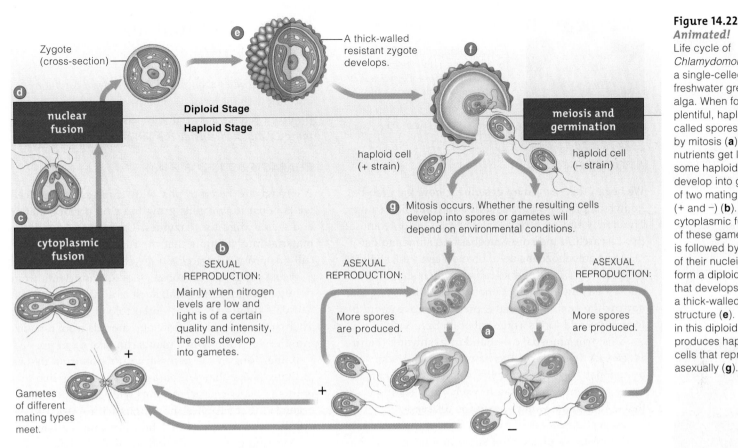

Zygote (cross-section)

Zygote (cross-section)

A thick-walled resistant zygote develops.

Figure 14.22
Animated!
Life cycle of *Chlamydomonas*, a single-celled freshwater green alga. When food is plentiful, haploid cells called spores multiply by mitosis (**a**). When nutrients get low, some haploid cells develop into gametes of two mating strains, (+ and −) (**b**). The cytoplasmic fusion of these gametes (**c**) is followed by fusion of their nuclei (**d**) to form a diploid zygote that develops into a thick-walled resting structure (**e**). Meoisis in this diploid cell (**f**) produces haploid cells that reproduce asexually (**g**).

d nuclear fusion

Diploid Stage
Haploid Stage

meiosis and germination

haploid cell (+ strain)

haploid cell (− strain)

c cytoplasmic fusion

g Mitosis occurs. Whether the resulting cells develop into spores or gametes will depend on environmental conditions.

b
SEXUAL REPRODUCTION:
Mainly when nitrogen levels are low and light is of a certain quality and intensity, the cells develop into gametes.

ASEXUAL REPRODUCTION:
More spores are produced.

a

ASEXUAL REPRODUCTION:
More spores are produced.

Gametes of different mating types meet.

−   +

+

−

just a few. Many cannot be seen without the aid of a microscope. Most live in freshwater. Green algae also grow in seawater, in soil, and on rocks, bark, and snow. Some are symbionts with fungi in lichens.

Green algae have diverse modes of reproduction. Consider *Chlamydomonas* (Figure 14.22), a single-celled alga about twenty-five micrometers wide. It sexually reproduces. It also forms sixteen or so daughter cells asexually, by mitotic cell divisions that take place within the parent cell wall. Daughter cells live in the parent for a while. Then they get away by secreting enzymes that digest what's left of the parent cell.

Figure 14.23 Green algae. (**a**) Sea lettuce (*Ulva*) grows in estuaries and attaches to kelps in the seas. (**b**) *Volvox*, a colony of cells that resemble free-living *Chlamydomonas* cells. This colony is rupturing. Released cells will found new colonies. (**a**) *Chara*, a charophyte. Members of this group are the closest relatives of plants. This one is commonly known as muskweed for its skunky odor.

b

aggregating cells

c

thousands of cells
migrating as a "slug"

d

forming a fruiting body

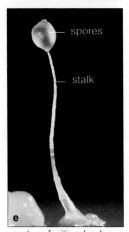

spores

stalk

e

mature fruiting body

Figure 14.24 (**a**) *Amoeba proteus*, a true amoeba that lives in freshwater ponds and streams. (**b–e**) Stages in the life cycle of a cellular slime mold, *Dictyostelium discoideum*. Some slugs include as many as 100,000 cells.

AMOEBOZOANS

We began this section by describing how the grab-bag known as protists is being sorted into smaller groups. The newly united group Amoeboza is an outcome of this research. It includes amoebas and slime molds.

Most amoebozoans do not have a cell wall, shell, or constrictive pellicle and so can alter their shape. A cell may be a compact blob at one moment, then long and narrow the next. Cells engulf prey and move about by forming pseudopods as described in Section 3.7.

Most true amoebas are found in freshwater (Figure 14.24*a*). A few live in the animal gut and some are human pathogens. Worldwide, *amoebic dysentery* is the cause of about 100,000 deaths each year. Resting cysts that contaminate water spread the disease.

Most slime molds live on the floor of the temperate zone forests. In plasmodial slime molds, an amoeba-like cell goes through multiple rounds of mitosis but does not divide its cytoplasm. A multinucleated mass called a plasmodium is formed by these divisions. It streams along the forest floor, engulfing bacteria. If conditions get bad, some cells become resting spores.

Cellular slime molds spend most of the life cycle as free-living amoeba-like cells that engulf and devour bacteria. When this food runs out, the cells secrete a chemical signal and come together to form a tiny sluglike mass that can migrate as a unit. The "slug" responds to light and to gradients in moisture. A fruiting body forms as cells differentiate into spores or part of the stalk that holds them. Air currents disperse the spores. Figure 14.24*b-e* shows stages in the life cycle of one cellular slime mold.

Diverse lineages of single-celled eukaryotic organisms were traditionally grouped in a single kingdom as protists. Spore-forming parasitic and predatory molds, and free-living single-celled decomposers, predators, grazers, and parasites, were placed in this kingdom. So were free-living, colonial, and multicelled photoautotrophs of aquatic habitats.

Molecular and biochemical comparisons are now clarifying relationships. The revised classification systems being constructed are far better reflections of the origin and evolution of these lineages.

14.4 The Fabulous Fungi

Molecular studies suggest the fungi share a common ancestor with the amoebozoans and the animals. The fungi are heterotrophs that absorb nutrients from material digested outside the body.

CHARACTERISTICS OF FUNGI

All fungi are heterotrophs with chitin-reinforced cell walls. Most fungal cells grow in or on organic matter and secrete digestive enzymes. The enzymes break the material into subunits that the fungus can absorb. We call this process *extracellular digestion and absorption.*

Most fungi are **saprobes**; they get nutrients from nonliving organic material, such as a fallen tree trunk. Saprobic fungi play an essential role in nutrient cyles. Their actions speed the flow of materials from remains into the soil, from which plants can take them up.

Other fungi draw necessary nutrients from tissues of living hosts; they are parasites. Still others live in a mutually beneficial partnership with a photosynthetic organism that supplies the fungus with nutrients. We call this type of win-win relationship a **mutualism**.

Most fungi reproduce both sexually and asexually. Both forms of reproduction may involve formation of resting spores. A fungal spore consists of a cell or cells surrounded by a protective wall.

In multicelled species, germination of a spore gives rise to a **mycelium** (plural, mycelia). This network of branched filaments has a high surface-to-volume ratio and grows over or into organic matter. A filament in a mycelium is a **hypha** (plural, hyphae). Cytoplasm of cells in a hypha often interconnects, so nutrients flow easily throughout a mycelium.

FUNGAL DIVERSITY

About 56,000 fungal species have been named, and there may be a million more. Fungi have been around

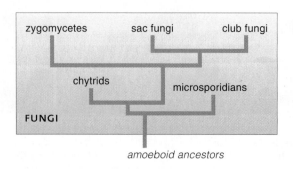

Figure 14.25 Fungal family tree.

THE SIMPLEST EUKARYOTES

Figure 14.26 *Animated!*

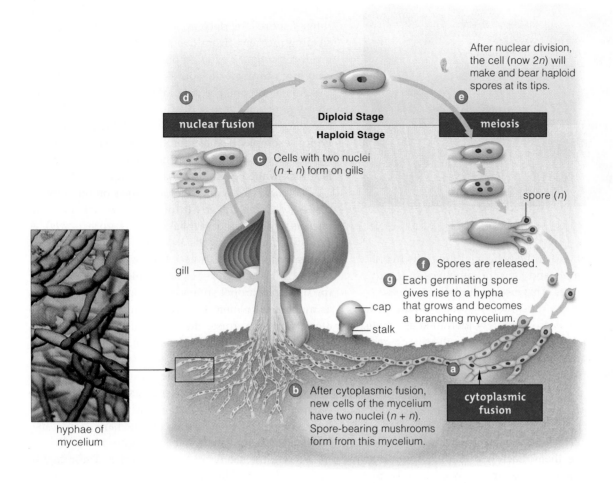

After nuclear division, the cell (now 2n) will make and bear haploid spores at its tips.

d **nuclear fusion** — Diploid Stage / Haploid Stage — **meiosis** **e**

c Cells with two nuclei (n + n) form on gills

spore (n)

gill

f Spores are released.

g Each germinating spore gives rise to a hypha that grows and becomes a branching mycelium.

cap

stalk

b After cytoplasmic fusion, new cells of the mycelium have two nuclei (n + n). Spore-bearing mushrooms form from this mycelium.

a cytoplasmic fusion

hyphae of mycelium

Figure 14.26 *Animated!* Generalized life cycle of many club fungi. Hyphae of different mating strains often grow through the same patch of soil. (**a**) When hyphal cells of two compatible strains contact each other, their cytoplasm fuses but their nuclei do not. (**b**) Mitotic cell divisions yield a mycelium in which each cell contains two nuclei. When conditions are favorable, many hyphae of this mycelium start intertwining into the type of fruiting body called a mushroom. (**c**) Club-shaped structures form on the mushroom's gills. (**d**) Inside the terminal cell of each structure, two nuclei fuse and form a diploid zygote. (**e**) Four haploid spores form by meiosis. They migrate into the cytoplasmic extensions at the tip of the club-shaped structure. After spores are released (**f**), each may germinate and give rise to a new haploid mycelium (**g**).

and diversifying for a very long time. Some fossilized forms date to 900 million years ago. Microsporidians and chytrids are considered the most evolutionarily ancient fungal lineages (Figure 14.25). The spread of the parasitic chytrid *Batrachochytrium dendrobatidis* is believed to be contributing to a decline in amphibian populations (Section 29.5). Some microsporidians, such as *Enterocytozoon bieneusi*, can cause chronic diarrhea in AIDS patients.

Besides these two basal groups, three major fungal lineages became established by 300 million years ago. They are the **club fungi**, **sac fungi**, and **zygomycetes**. These groups are the focus of the rest of the section.

THE CLUB FUNGI

Most familiar mushrooms are reproductive parts of club fungi, and they usually appear only very briefly. Most of the fungus is a mycelium buried in soil or decaying wood. Each mushroom has a stalk and a cap with fine tissue sheets (gills) lining its underside. Spores form on gills, and air currents disperse them.

When they alight and conditions favor germination, spores give rise to a haploid mycelium.

A mycelium may grow extensively, and then give rise to mushrooms when nutrients and moisture favor reproduction. Figure 14.26 shows how two different "mating" strains of hyphae start the process of sexual reproduction when they come together in the soil.

Mushrooms, shelf fungi, coral fungi, puffballs, and stinkhorns are among the thousands of species of club fungi. Figure 14.27 shows examples. Some saprobes are decomposers of litter on and in soil. Other species live as symbionts with young tree roots. Fungal rusts and smuts often ruin crops of wheat, corn, and other plants meant to feed the human population. Cultivation of *Agaricus bisporus*, the mushroom most commonly sold in grocery stores, is a multimillion-dollar business.

One honey mushroom, *Armillaria ostoyae*, is among the oldest and largest organisms alive. This individual has been spreading in an Oregon forest for 2,400 years. It now extends through 2,200 acres to an average depth of 3 feet. Imagine 1,665 football fields side by side and you get an idea of how big that is. *A. ostoyae*

Figure 14.27 Club fungi. (**a**) Purple coral fungus (*Clavaria*) and (**b**) sulfur shelf fungus (*Polyporus*).

Figure 14.28 (a) The fly agaric mushroom (*Amanita muscaria*), a hallucinogenic species. In India, Russia, and Central America, it was used in ancient rituals to induce trances. (b) *A. phalloides*, the death cap mushroom. Nausea and diarrhea occur 5 to 48 hours after eating it. Initial symptoms pass, but without treatment, the liver is severely damaged, kidneys fail, and coma and death can follow.

definitely is not one of the beneficial fungi. Its hyphae penetrate tree roots and clog the pipelines that move water and nutrients. This eventually kills the tree.

Before eating any mushrooms gathered in the wild, be sure they have been accurately identified as edible. Many can be deadly (Figure 14.28). As the saying goes, there are old mushroom hunters and bold mushroom hunters, but no old, bold mushroom hunters.

SPORES AND MORE SPORES

Fungal spores are usually miniscule and dry, and air currents disperse them. Each germinating spore may start a hypha, then a mycelium. Stalked reproductive structures may form on many hyphae, then produce asexual spores. After the spores germinate, each may be the start of *another* mycelium. In no time at all, the staggering numbers of descendants from that single fungal spore are decomposing organic materials or stealing nutrients from hosts. Each fungal group also produces *unique* sexual spores. As you just learned, club fungi form sexual spores in club-shaped cells.

Only about one percent of the named fungal species are zygomycetes. Most live in soils, dung, and other decaying matter. *R. stolonifer*, the black bread mold, is a typical example. Sexual reproduction occurs when the hyphae of two mating types meet and fuse. The result is a thick-walled diploid zygospore (Figure 14.29*a*).

Pilobolus is a zygomycete that grows on the dung of cows and other grazers. It forms its asexual spores in packets at the tips of hyphal stalks (Figure 14.29*b*). A fluid-filled region swells the stalk below the spores. Exposure to sunlight causes the fluid pressure in this region to rise until the hypha explodes. The force of the blast can propel spores onto a fresh patch of grass as much as two meters away. When a cow eats spores along with the grass, then deposits them in a fresh pile of dung, the cycle begins again.

The sac fungi are the most diverse fungal group. Multicelled species form their sexual spores in sac-shaped cells called asci (singular, ascus). The cells are often enclosed in reproductive structures composed of tightly woven hyphae. Different kinds are shaped like flasks, globes, and shallow cups (Figure 14.29*c*,*d*).

Among the sac fungi are the truffles. These highly prized culinary delicacies grow as symbionts with the roots of oak trees. Some sell for hundreds of dollars an ounce. Most food-spoiling molds are multicelled sac fungi. So are many of the single-celled fungi that we

asexual spores

Figure 14.29 (a) *Rhizopus stolonifer*, the black bread mold, produces zygospores. (b) Hyphae of *Pilobolus*, another zygomycete. When the fluid-filled part of a hypha bursts, spores on top are propelled into the air. (c,d) *Sarcoscypha coccinia*, the scarlet cup fungus. On the cup's inner surface, spores form in saclike structures. (e) Another sac fungus: *Candida albicans* causes "yeast" infections of the vagina, mouth, and skin. When yeast cells fuse, they become a spore-producing sac.

THE SIMPLEST EUKARYOTES

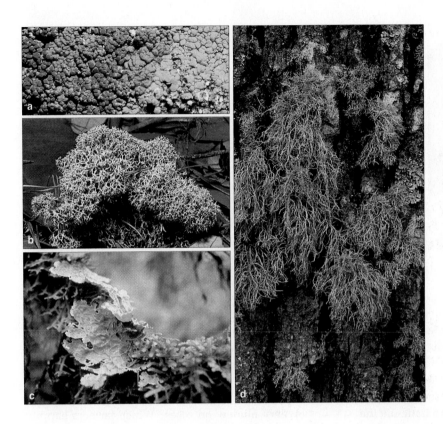

dispersal fragment (cells of fungus and its photosynthetic partner)

cortex (outer layer of fungal cells)

photosynthetic cells

inner layer of loosely woven hyphae

cortex

e

Figure 14.30 (**a**) One of the encrusting lichens, growing on a sunlit rock. (**b**) *Cladonia rangiferina*, an erect, branching lichen. (**c**) *Lobaria oregana*, a leaflike lichen that converts nitrogen gas to forms that trees in the Pacific Northwest's old-growth forests can use for growth. (**d**) *Usnea*, a pendant lichen commonly called old man's beard. (**e**) Body plan of a stratified lichen, cross-section.

call yeasts (others are club fungi). A packet of baker's yeast holds thousands of tiny spores of *Saccharomyces cerevisiae*. In a warm, moist spot, such as bread dough set out to rise, spores germinate. The cells that emerge will reproduce asexually by budding. Carbon dioxide released by their aerobic respiration causes the bread dough to rise. Fermentation by other yeast species is used to make alcoholic beverages. Still other yeasts are human pathogens (Figure 14.29*e*).

FUNGAL SYMBIONTS

Ever since plants invaded the land, fungi were there, many as symbionts. **Symbiosis**, recall, refers to species that interact closely during their life cycles. In cases of mutualism, their interaction benefits both the partners. Lichens and mycorrhizae generally are like this.

LICHENS Each **lichen** is a vegetative body in which a fungus intertwines with one or more photosynthetic organisms—usually green algae and cyanobacteria. It forms after the tip of a fungal hypha binds with a host cell. Both lose their wall, and their cytoplasm fuses or the hypha induces the host to cup around it. The two species multiply together into a multicelled form that usually has distinct layers. The growth pattern may be flattened, erect, leaflike, or pendulous (Figure 14.30).

Lichens typically colonize sites that are too hostile for most organisms, such as sunbaked or frozen rocks,

fences, gravestones, and plants, even the tops of giant Douglas firs. Almost always, the fungus is the larger component. Cyanobacteria typically live in a separate structure inside or outside the main body. The fungus gets a fine long-term source of nutrients. Its nutrient withdrawals do hurt its food-producing partner a bit, but that partner may benefit by getting shelter in return.

Lichens absorb minerals. They produce antibiotics against bacteria that can decompose them and toxins that can kill invertebrate larvae that graze upon them. Coincidentally, their metabolic products enrich soil or help form new soil from bedrock. This is what happens when lichens colonize barren sites, including bedrock exposed by retreating glaciers. Conditions improve, so other species move in and replace the pioneers. This is probably what happened when plants invaded land.

Lichens that include cyanobacteria are important in the cycling of nitrogen. They capture nitrogen gas and convert it to a form that plants use (Figure 14.30*c*).

Lichens also provide early warning of deteriorating environmental conditions. How? They all absorb toxins but cannot get rid of them. From studies in New York City and in England, we know that when lichens die around human habitats, air pollution is getting bad.

MYCORRHIZAE Some fungi and young tree roots also are mutualists. Their interaction is a **mycorrhiza** (plural, mycorrhizae), which means "fungus-root." Underground parts of a fungus grow through the soil and functionally

Figure 14.31 (a) Mycorrhiza on a hemlock root. (b) Effects of the presence or absence of mycorrhizae on plant growth in sterilized, phosphorus-poor soil. The juniper seedlings at left were controls; they grew without a mycorrhizal fungus. The seedlings at right are the experimental group. They were grown with a partner fungus. All of the seedlings are the same age.

increases the tree's absorptive surface area. Young roots start out with hairlike absorptive structures, but fungal hyphae are thinner and better at growing around soil particles. When phosphorus and other nutrients are in good supply, hyphae take them up from the soil. When these nutrients get scarce, the hyphae release some of them to their plant partner. The fungus benefits because the plant gives it some sugars. The loss is a trade-off; the plant does not grow as efficiently without the fungus.

Figure 14.31 shows an example of a mycorrhiza and its effects on plant growth. We'll take a closer look at this mutualistic interaction in Chapter 18.

AS FUNGI GO, SO GO THE FORESTS Since the early 1900s, collectors have recorded data on wild mushroom populations in many European forests. These records show that the number and kinds of fungi are declining. The mushroom gatherers are not to blame; inedible as well as edible species are disappearing. But the decline does correlate with rising levels of air pollution.

Under natural conditions, just about all trees have fungal symbionts on or in their roots. Compared to tree roots, fungal hyphae are smaller and better able to grow into small crevices where they can absorb inorganic minerals that they share with the tree.

As a tree ages, one mycorrhizal fungus gives way to another, in predictable patterns. When fungi die, trees lose this essential support system, and become more vulnerable to damage by severe frost, drought, and acid rain. Are North American forests at risk too? Conditions there are changing in similar ways.

THE UNLOVED FEW

You know that you are a serious student of biology when you view organisms in terms of their roles in nature, rather than their effects on humans in general and you in particular. You salute the saprobic fungi as essential decomposers and praise parasitic ones that help keep populations in check. The true test is when you open the refrigerator to get your strawberries and a fungus has beaten you to them. Or when a fungus starts feeding on the warm, damp tissues between your toes (Figure 14.32a).

Which home gardeners wax poetic about black spot or powdery mildew on roses? Which farmers happily hand over millions of dollars a year to sac fungi that infect corn, wheat, peaches, and apples? Who rejoices that a certain sac fungus, *Cryphonectria parasitica*, killed off the chestnut trees in eastern North America?

Who willingly inhales airborne spores of *Ajellomyces capsulatus*? In soil this fungus forms a mycelium. In lung tissues it causes *histoplasmosis*, a potentially fatal respiratory disease.

And household molds! Allergies to mold promote sinus, ear, and lung infections, cause rashes, and can increase the frequency of asthma attacks.

Some fungi have even altered human history. One species, *Claviceps purpurea*, is a parasite of rye and other cereal grains (Figure 14.32b). Eating rye bread that was made from contaminated flour can lead to *ergotism*. Symptoms include vomiting, diarrhea, hallucinations, hysteria, and convulsions. Untreated, it can be fatal.

Ergotism outbreaks were common in Europe in the Middle Ages. They thwarted Peter the Great, a Russian czar who became obsessed with conquering ports along the Black Sea for his nearly landlocked empire. Soldiers laying siege to the ports ate mostly rye bread and fed rye to their horses. Soldiers went into convulsions and horses into "blind staggers."

Figure 14.32 Love those fungi! (a) On one young individual, athlete's foot, caused by *Epidermophyton floccosum*. (b) Rye plant infected by the historically notable fungus *Claviceps purpurea*.

Ever since their time of origin, fungal decomposers have been cyclers of nutrients that help sustain webs of life just about everywhere. It might help to keep this global view in mind when the few unloved ones make life miserable.

14.5 Viruses, Viroids, and Prions

The noncelluar particles we call viruses are like stripped-down versions of life. There's no fossil record for viruses and biologists are still debating whether they arose before cellular life forms did or are derived from the genes of cellular organisms. Either way, molecular comparisons can be used to group them into related lineages and to study their ongoing evolution.

CHARACTERISTICS OF VIRUSES

In ancient Rome, *virus* meant "poison" or "venomous secretion." In the late 1800s, this rather nasty word was bestowed on newly discovered pathogens, smaller than bacteria. Many viruses deserve the name. They attack animals, plants, fungi, protists, bacteria—you name it, there are viruses that infect it.

A **virus** is a noncellular infectious agent with two characteristics. Each viral particle consists of a protein coat wrapped around genetic material and a few viral enzymes. It can't reproduce itself; a host cell is tricked into making copies of the virus.

Viral genetic material is DNA *or* RNA. The coat has one or more types of protein subunits organized into a rodlike or many-sided shape, as in Figure 14.33. The coat protects the genetic material during the journey to a new host cell. It also holds proteins that can bind to specific receptors on host cells. An envelope of mostly membrane remnants from the previously infected cell encloses some types of coats. It bristles with spikes of glycoproteins. Coats of complex viruses have sheaths, tail fibers, and other accessory structures attached.

The vertebrate immune system detects certain viral proteins. But genes for many of them mutate often, so a virus can slip past the immune fighters. For instance, people with respiratory problems might get new "flu shots" each year because spikes on the enveloped coat of influenza viruses keep changing.

Each kind of virus multiplies only in certain hosts. It cannot be studied easily except in living cells. This is why researchers study interactions between certain bacterial cells and **bacteriophages**, the class of viruses that infect them. Unlike the cells of humans and other complex, multicelled organisms, both can be cultured easily and fast. Both were used in early experiments to determine the function of DNA. They are still used as research tools and in genetic engineering.

Animal viruses contain double- or single-stranded DNA or RNA. They range in size from parvoviruses (a tiny 18 nanometers) to brick-shaped poxviruses (350 nanometers). Many kinds cause diseases, including the

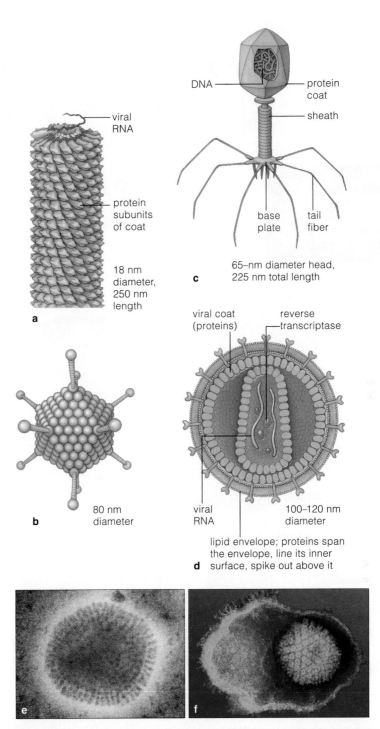

Figure 14.33 Body plans of viruses. (**a**) Protein subunits bound to nucleic acid form a spiral in rod-shaped viruses, including this tobacco mosaic virus. (**b**) Complex viruses, such as this bacteriophage, have structures attached to the coat. (**c**) Polyhedral (many-sided) viruses include this adenovirus. (**d**) A lipid envelope derived from membrane fragments of a lysed host cell surrounds HIV and other enveloped viruses. A capsid of regularly arrayed protein subunits encloses the viral RNA and a few enzymes.

common cold, warts, herpes, influenza, and certain cancers. One virus, HIV, causes *AIDS* by infecting white blood cells. It weakens the immune system's ability to fight infections that normally are not life threatening.

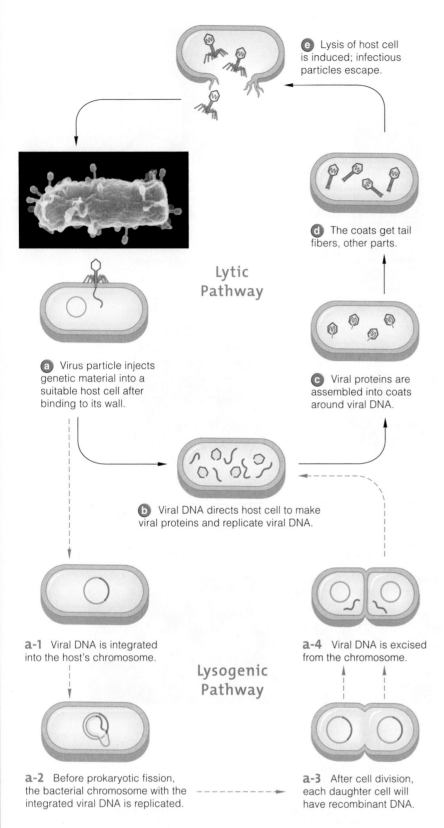

e Lysis of host cell is induced; infectious particles escape.

Lytic Pathway

d The coats get tail fibers, other parts.

a Virus particle injects genetic material into a suitable host cell after binding to its wall.

c Viral proteins are assembled into coats around viral DNA.

b Viral DNA directs host cell to make viral proteins and replicate viral DNA.

a-1 Viral DNA is integrated into the host's chromosome.

Lysogenic Pathway

a-4 Viral DNA is excised from the chromosome.

a-2 Before prokaryotic fission, the bacterial chromosome with the integrated viral DNA is replicated.

a-3 After cell division, each daughter cell will have recombinant DNA.

Figure 14.34 Generalized multiplication cycle for some bacteriophages. Virus particles may enter a lytic pathway. The lytic pathway may expand to include a lysogenic pathway for some types of viruses.

VIRAL MULTIPLICATION CYCLES

What happens during viral infections? Once host cells are infected, different viruses multiply in a variety of ways. Even so, nearly all multiplication cycles proceed through five basic steps, as outlined here:

1. *Attachment.* A virus attaches to a host cell through molecular groups that chemically recognize and lock on to specific molecular groups at the cell surface.

2. *Penetration.* Either the complete virus or its genetic material alone penetrates the cell's cytoplasm.

3. *Replication and protein synthesis.* In an act of molecular piracy, the viral DNA or RNA directs the host cell into producing many copies of the virus's nucleic acids and proteins, including enzymes.

4. *Assembly.* Viral nucleic acids and proteins become organized as new infectious particles.

5. *Release.* New virus particles are released from the cell by one mechanism or another.

Figure 14.34 shows two pathways that are common among bacteriophages. Steps 1 through 4 occur rapidly in a *lytic* pathway, and new particles are released when a host cell undergoes **lysis**. Lysis means that damage to a host cell's plasma membrane, wall, or both lets the cytoplasm dribble out. New virus particles dribble out also as the cell dies. The host itself built a viral enzyme that caused the damage and swift destruction.

In a *lysogenic* pathway, a latent period extends the cycle. The virus doesn't kill a host cell outright. A viral enzyme cuts the host chromosome, then integrates the viral genes into it. Before the cell divides, it replicates the recombinant molecule. Miniature time bombs thus await all of its descendants. Later, a molecular signal or some other stimulus may reactivate the cycle.

Latency is typical of many viruses, such as Type I *Herpes simplex*, which causes *cold sores* (fever blisters). This virus hides in sensory nerves. When stress or a sunburn reactivates it, the virus moves to epithelial cells in skin and causes painful eruptions. Like other enveloped viruses, it enters a host cell by a version of endocytosis, then leaves by budding from the surface.

RNA viruses have a different multiplication cycle. Viral RNA functions as a template for making mRNA or DNA. One of the enzymes that the retrovirus HIV carries into host cells catalyzes the formation of DNA by reverse transcription (Sections 11.1 and 23.8).

VIROIDS AND PRIONS

Stripped-down as viruses are, **viroids** are simpler still. They are circles of RNA with no protein coat and no protein-coding genes. Many viroids are pathogens that kill millions of dollars' worth of crops annually. So far, only one type has been found in humans. It helps to cause hepatitis D, a deadly liver disease.

A **prion** is an infectious protein. The best studied ones cause "mad cow disease," or bovine spongiform encephalopathy (BSE), and human variant Creutzfeldt-Jacob disease (vCJD). Normal versions of these prion proteins occur in the brain and spinal cord, where they are folded in a particular way. Infectious prions fold up wrong, and they induce normal versions to misfold too. The result is a degenerative nervous disease that is invariably fatal.

A virus consists of nucleic acid enclosed in a protein coat and sometimes an outer membranous envelope. It can replicate only in specific types of host cells. Viroids and prions are even simpler infectious particles.

14.6 Evolution and Infectious Diseases

Just by being human, you are a potential host for the kinds of diverse pathogens and parasites you have just read about. Before leaving the chapter, reflect on what this means from an evolutionary perspective.

Infection is the invasion of a cell or multicelled body by a pathogen (or by the pathogen's genetic material). **Disease** follows when the pathogen multiplies and the activities of its descendants disrupt body activities. Table 14.2 and Figure 14.35 give some examples.

It takes direct contact with an infected person or their secretions for a contagious disease to spread. The *sporadic* diseases, such as whooping cough, break out irregularly among very few people. *Endemic* diseases, such as tuberculosis, are more or less always present in a population but are confined to a small part of it.

During an *epidemic*, some disease quickly spreads through part of a population for a limited time, then subsides. *Pandemic* refers to many epidemics of the same disease breaking out in several countries during the same interval.

We discuss the ongoing AIDS pandemic in detail in Chapter 23. The 2003 outbreak of SARS (*Severe Acute Respiratory Syndrome*) was a brief pandemic caused by a coronavirus. Before quarantines halted its spread, 774 died. A new threat is H5N1, a bird influenza virus that has infected and killed people in Asia. Bird flu is not yet easily transmitted from person to person, but a mutation could change this, with deadly results.

Table 14.2	The Eight Deadliest Infectious Diseases		
Disease	Main Agents	Estimated New Cases per Year	Estimated Deaths per Year
Acute respiratory infection*	Bacteria, viruses	1 billion	4.7 million
Diarrhea**	Bacteria, viruses, protozoans	1.8 billion	3.1 million
Tuberculosis	Bacteria	9 million	3.1 million
Malaria	Sporozoans	110 million	2.5–2.7 million
AIDS	Virus (HIV)	5.6 million	2.6 million
Measles	Viruses	200 million	1 million
Hepatitis B	Virus	200 million	1 million
Tetanus	Bacteria	1 million	500,000

* Includes pneumonia, influenza, and whooping cough.
** Includes amoebic dysentery, cryptosporidiosis, and gastroenteritis.

Now consider disease in terms of the pathogen's prospects for survival. A pathogen stays around only for as long as it has access to outside sources of energy and raw materials. To a pathogen, the human body is a jackpot. Inside it, a pathogen may multiply to huge population sizes. The point? Evolutionarily speaking, the ones that leave the most descendants win.

Two barriers prevent pathogens from taking over the world. First, any species with a history of being attacked by a specific pathogen has coevolved with it and has built-in defenses against it. The immune system of vertebrates is one example. Second, if some pathogen kills its host too quickly, it might vanish along with that individual. This is one reason most pathogens are not fatal. Infected individuals who live will spread the pathogen and thus contribute to its reproductive or replicative success.

An individual is most likely to die if it becomes host to overwhelming numbers of a pathogen, if it is a novel host that lacks coevolved defenses, or if a novel pathogen emerges and breaches current defenses.

Antibiotics are compounds synthesized by one organism that can kill another, and we use them as weapons against bacterial pathogens. Penicillin and other antibiotics have saved millions of human lives. However, these drugs act as agents of selection, and antibiotic-resistant forms of bacteria are increasing.

Lateral gene transfers among bacterial cells of the same or different species contribute to the problem. Many bacterial pathogens swap genes like teens swap music files. The genes that confer antibiotic resistance move about in plasmids or as a free piece of DNA.

Figure 14.35 *Above*, HIV, the agent of AIDS. *Below*, two particles of the coronavirus that causes SARS.

In evolutionary terms, those pathogens that leave the most descendants win. They are at a selective advantage.

Summary

Section 14.1   Experiments demonstrate that complex organic compounds could have self-assembled under conditions on early Earth. Chemical and molecular processes may have given rise to proto-cells. The first cells were prokaryotic. An early divergence gave rise to bacteria and to the shared ancestors of archaean and eukaryotic cells.

Biology ⊛ Now
Learn how organic compounds can spontaneously assemble with the animation on BiologyNow.

Section 14.2   Archaea and bacteria are the only prokaryotes. They are metabolically diverse. They alone can reproduce by prokaryotic fission. Many archaea live in places too hostile for most organisms. Bacteria are the most common prokaryotes. Many function as decomposers and photosynthesizers; some can cause disease. Table 14.3 compares features of prokaryotes and eukaryotes.

Biology ⊛ Now
Review the components of a typical prokaryotic cell with the animated interaction on BiologyNow.

Section 14.3   The "protist kingdom" is now being revamped to reflect new understanding of evolutionary relationships. Most protists are single-celled, but many lineages also have multicelled species. Nearly all types live in water or moist habitats, including host tissues.

Ancient lineages include euglenoids and related flagellated protozoans. The amoeboid protozoans use pseudopods to move about and capture prey. Ciliates, dinoflagellates, and apicomplexans are close relatives. Algal blooms can kill fish and other aquatic organisms; they are population explosions of dinoflagellates or other protists. Apicomplexans are parasites that spend part of their life cycle inside cells of their host.

Diatoms and coccolithophores are single-celled, shelled, aquatic producers. They are close relatives of the oomycotes—which include many plant pathogens— and the brown algae. Brown algae include the giant kelps, which are the largest protists. Green algae are the closest relatives of plants. Phycobilins mask chlorophylls in red algae. Slime molds spend part of their life cycle as a single cell and part in a larger cohesive group that can migrate and differentiate to form a spore-bearing body.

Biology ⊛ Now
View the body plan of a euglena and the life cyle of a green alga with the animation on BiologyNow.

Section 14.4   All fungi are heterotrophs. Most are decomposers. A few are parasites or pathogens. All engage in extracellular digestion. Sexual and asexual spores form during the life cycle of the three major lineages: zygomycetes, sac fungi, and club fungi. In multicelled species, germinating spores undergo cell divisions that produce a mycelium.

Lichens and mycorrhizae are cases of mutualism. In lichens, a fungal symbiont lives with a photosynthetic species, most often a green alga or cyanobacterium. In mycorrhizae, a fungus and young roots are symbionts.

Biology ⊛ Now
Observe the life cycle of a club fungus with the animation on BiologyNow.

Section 14.5   A virus is a noncellular infectious particle that can multiply only by pirating metabolic machinery of a host cell. It is DNA or RNA packaged in a protein coat. Some replication cycles of viruses include a latent phase, when viral genetic material is integrated into the host genome. Viroids are infectious RNA particles. Prions are infectious proteins.

Biology ⊛ Now
Use the animation on BiologyNow to explore viral structure and multiplication pathways.

Table 14.3   Comparison of Prokaryotes With Eukaryotes

	Prokaryotes	Eukaryotes
Organisms represented:	Archaea, bacteria	Protists, fungi, plants, and animals
Ancestry:	Two major lineages that evolved more than 3.5 billion years ago	Ancient prokaryotic ancestors gave rise to forerunners of eukaryotes, which evolved more than 1.2 billion years ago
Level of organization:	Single-celled	Protists, single-celled or multicelled. Nearly all others multicelled; division of labor among differentiated cells, tissues, and often organs
Typical cell size:	Small (1–10 micrometers)	Large (10–100 micrometers)
Cell wall:	Most with distinctive wall	Cellulose or chitin; none in animal cells
Membrane-enclosed organelles:	Rarely; no nucleus, no mitochondria	Typically profuse; nucleus present; most with mitochondria, many with chloroplasts
Modes of metabolism:	Both anaerobic and aerobic	Aerobic modes predominate
Genetic material:	One chromosome; plasmids in some	Chromosomes of DNA plus many associated proteins in a nucleus
Mode of cell division:	Prokaryotic fission, mostly; some reproduce by budding	Nuclear division (mitosis, meiosis, or both) associated with one of various modes of cytoplasmic division

Section 14.6 Evolution shapes the characteristics of pathogens, as it does for all organisms. Pathogens and their hosts have coevolved. Antibiotics select for antibiotic-resistant pathogens.

Self-Quiz
Answers in Appendix I

1. The _____ are major decomposers.
   a. fungi   b. viruses   c. ciliates   d. dinoflagellates

2. All of the _____ are prokaryotic.
   a. archaea   c. protists   e. all are correct
   b. bacteria   d. a and b

3. Chloroplasts are descended from _____ .
   a. mitochondria   c. archaeans
   b. cyanobacteria   d. viroids

4. The _____ are parasitic eukaryotes that spend part of their life cycle inside specific cells of host organisms.
   a. viruses   c. euglenoids   e. both a and b
   b. apicomplexans d. slime molds f. all are correct

5. Chitin reinforces the cell walls of _____ .
   a. coccolithophores   c. foraminiferans
   b. diatoms   d. fungi

6. Some of the _____ are human pathogens.
   a. slime molds   c. flagellated protozoans
   b. archaea   d. both a and c

7. When a spore of a multicelled fungus germinates, it gives rise to a _____ .
   a. viroid   c. mycelium
   b. fruiting body   d. mycorrhiza

8. The genetic material of a _____ may be DNA or RNA.
   a. zygomycete   b. ciliate   c. dinoflagellate   d. virus

9. Match these terms suitably.
____ green algae	a. protist population explosion
____ virus	b. silica-shelled producer
____ bacteria	c. most diverse prokaryotes
____ brown algae	d. noncellular, not alive
____ mycorrhiza	e. include the largest protists
____ lichen	f. fungus plus producer cells
____ algal bloom	g. closest relative of plants
____ mushroom	h. fossilized cyanobacteria
____ diatom	i. a "fungus-root"
____ stromatolite	j. spore-bearing fruiting body of a club fungus

Additional questions are available on **Biology ⒺNow™**

Critical Thinking

1. The cellular slime mold *Dictyostelium discoideum* is now being studied by more than 660 biologists. Each year, hundreds of papers about its biology appear in scientific journals. It is one of the "model organisms" that help us learn about cell differentiation and other developmental processes. A single cell can be allowed to divide again and again to form a huge population. When such a population is starved, cells aggregate and form fruiting bodies, as shown in Figure 14.24. Although genetically identical, some cells will go on to become part of the stalk, while others become spores. How is fruiting body formation of genetically identical slime mold cells similar to the process by which a multicelled organism develops?

Figure 14.36 A radura, international symbol for irradiated food. It means that a product has been exposed to high-energy radiation to kill pathogens. Irradiation cannot make food itself radioactive.

2. Runoff from heavily fertilized fields, animal waste, and raw sewage promotes algal blooms that can result in massive kills of fish and other aquatic species. If you find this environmental cost unacceptable, how would you stop the pollution? Bear in mind that we now depend on high-yield (and heavily fertilized) crops. How do you suggest we dispose of the waste from farms and cities?

3. The fungus *Fusarium oxysporum* is a plant pathogen. Some view it as a potential weapon in the war on drugs. Why? Strains of the fungus attack and kill only specific plants, such as the coca plants used to produce cocaine. Proposals to spray *F. oxysporum* to kill marijuana plants in Florida were abandoned after public outcries. What concerns, if any, would you have about the widespread spraying of natural mycoherbicides to kill off plants that are sources of illegal drugs?

4. Curtis Suttle studies microbial communities in ocean water. In one experiment, he selectively removed viruses from seawater and found that algae stopped growing. Further studies revealed that the algae required nutrients released by dying bacteria that lysed after viral infection. Make a list of some of the other ways in which viruses might help sustain natural ecosystems or be beneficial to human health.

5. One way to prevent bacterial food poisoning is by *food irradiation*, exposing food to high-energy rays that kill pathogens. The process also slows spoilage and can prolong shelf life. Some think this is a safe way to protect consumers. Others worry that irradiation could produce harmful chemicals. They point out that this treatment does not kill endospores that can cause botulism. In their view, a better approach would be to tighten and enforce food safety standards.

Irradiated meat, poultry, spices, and fruits are now in supermarkets. By law, they must be marked with a special symbol (Figure 14.36). Would noticing this symbol on a package influence your likelihood of purchasing that product? Explain your answer.

Beginnings, and Endings

Change is the way of life. About 300 million years ago, in the Carboniferous, swamp forests carpeted the wet, warm lowlands of continents. Tree-high ancestors of today's tiny club mosses and horsetails were dominant. Then things changed. As the global climate became cooler and drier, moisture-loving plants declined. Hardier plants—cycads, ginkgos, and conifers—rose to dominance. They were the gymnosperms, and they flourished for millions of years.

Things changed again. Other plants had evolved, and they held far more potential to radiate into diverse environments. The angiosperms—flowering plants—started to take over. They still dominate most regions. But in the far north, at high elevations, and in parts of the Southern Hemisphere, some conifers retained the competitive edge and held on as great forests.

Then small bands of humans learned to grow some flowering plants as crops, which became the foundation for population growth. Human populations did grow, spectacularly, and they required far more resources than crops. Conifers had the misfortune to become premier sources of lumber, paper, resins, and other goods. They became vulnerable to **deforestation**, the removal of all trees from large tracts of land.

In California, only 4 percent of the original coastal redwood forest is still undisturbed. In Maine, an area the size of Delaware has been logged over in just the past fifteen years. Many logs end up as paper or pulp. The United States also exports logs to lumber mills overseas. At the same time, it imports timber from tropical rain forests as far away as New Zealand.

Clearing the world's forests has cascading ecological effects. Nutrients wash away from exposed soils, and sediments clog streams. Herbicides applied to stop nontimber species from taking over interfere with ecosystem recovery. Species that depend on the forest for food and shelter become threatened or extinct.

Almost 750 species of plants in the United States alone are now on the endangered list. And with this bit of perspective on change, we turn to the beginning—and end of the line—of some ancient lineages.

 How Would You Vote? Demand for paper is a big factor in deforestation. But using recyled paper can add to the cost of a product. Would you be willing to pay more for papers, books, and magazines that are printed on recycled paper? See BiologyNow for details, then vote online.

Key Concepts

EVOLUTIONARY TRENDS
Plants arose from a group of green algae. Their evolutionary story is one of increasing adaptation to life in drier environments.

BRYOPHYTES
All bryophytes are small plants, such as mosses. The gamete-producing stage is the longer lived and more conspicuous part of their life cycle.

THE RISE OF VASCULAR PLANTS.
The life cycle of vascular plants is dominated by a spore-producing body with specialized tissues that serve as pipelines for moving water and dissolved sugars.

THE RISE OF SEED PLANTS
Pollen grains and seeds were key innovations that allowed seed plants (gymnosperms and angiosperms) to radiate into diverse habitats. Only angiosperms produce flowers and fruits.

Links to Earlier Concepts

In Section 14.3, you learned about green algae, which include the closest relatives of the plants. You studied gamete formation in Section 7.7, and a brief look back may help you to understand how plant life cycles changed as lineages moved to increasingly drier habitats.

Much of what we know about plant evolution is based on the fossil record (13.1). This is a story of adaptive radiations and extinctions (13.9). A working knowledge of the overall history of life and the geologic time scale (13.2) may help you put the events we discuss in this chapter into a larger perspective.

Pioneers in a New World

We share the world with at least 295,000 kinds of plants. Within their tremendously diverse kingdom, we find recurring structures that correlate with present and past functions.

About five hundred million years ago, the invasion of land was getting under way. Why then? As described in the previous chapter, the noncyclic photosynthetic pathway had evolved in cyanobacteria and the oxygen released as a by-product had changed the atmosphere. Far above Earth, the sun's energy had converted some oxygen into the ozone layer—a shield against deadly doses of ultraviolet radiation. Until that time, life had survived only beneath the surface of water and mud.

Green algae were evolving at the water's edge, and one group gave rise to plants. By 430 million years ago, *Cooksonia*, a simple branching plant a few centimeters high, was common (Figure 15.1*a*). It took another 160 million years for the much taller *Psilophyton* to evolve (Figure 15.1*b,c*). Then a great adaptive radiation began. It took only 60 million more years for plants to radiate from lowland swamps to mountain peaks and nearly everywhere in between. Changes in structure, function, and reproductive modes made this radiation possible.

ROOTS, STEMS, AND LEAVES

Underground absorptive structures evolved as plants colonized the land, and in some lineages they became systems of roots. A root system helped plants absorb water and dissolved mineral ions from soil. As they do today, young roots formed a close association with mycorrhizal fungi. Roots also helped anchor the plant.

Aboveground, shoot systems evolved. Stems and leaves intercepted sunlight energy and took in carbon dioxide from the air. A glue-like polymer called lignin appeared in the cell walls of some lineages. It allowed stems to grow taller and plants to stand more erect.

For many lineages, internal pipelines contributed to the evolution of roots and shoots. They evolved as components of vascular tissues—xylem and phloem. Xylem carries water and mineral ions throughout the plant. Phloem distributes sugars and other products of photosynthesis to all tissues.

Having enough water for metabolism was not a problem in most aquatic habitats. It became a greater challenge on the land. A waxy outer layer—a **cuticle**—evolved and helped plants to conserve water inside shoots on hot, dry days. Water and gases could enter or leave shoots only at **stomata**, gaps that spanned the

Figure 15.1 Early plants. (**a**) Scanning electron micrograph of a *Cooksonia* fossil shows a leafless branching structure. Spores formed at the branch tips. (**b**) A fossil of *Psilophyton*, which was common in swamps during the Devonian period (**c**).

cuticle. Evolution of stomata allowed control of water loss. A stomata-pierced cuticle helps to regulate water balance in nearly all existing land plants (Section 6.6).

FROM HAPLOID TO DIPLOID DOMINANCE

Life cycles were altered as plants evolved. We find multicelled haploid and diploid body forms in the life cycles of all plants, but the relative size of the two stages varies. The **gametophyte** is a haploid gamete-making body. When two gametes fuse at fertilization, a diploid zygote forms. It grows and develops into a **sporophyte**, in which haploid spores are produced by meiosis. A **spore** is a resting structure that withstands unfavorable environmental conditions. A plant spore gives rise to a new gametophyte (Figure 15.2).

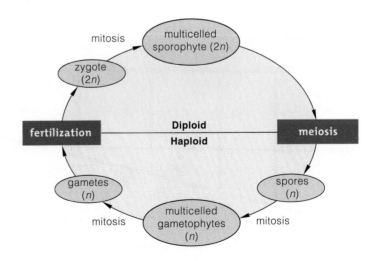

Figure 15.2 Generalized life cycle for all plants.

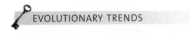

LINKS TO
SECTIONS
7.7, 14.3

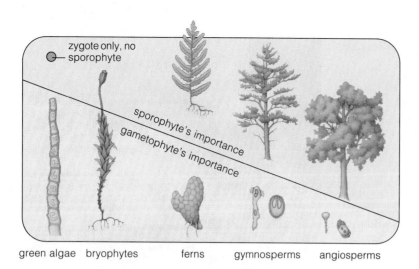

Figure 15.3 *Animated!* One major evolutionary trend in plant life cycles. Each plant's gametophyte (haploid) stage is on the *gold* background. Its sporophyte stage (diploid) is on the *green* background. For each group, the relative size of the green and gold areas indicates relative size and life span of the stages. Most algae, and all bryophytes, put more of their energy into gamete-producing structures. The other groups evolved in seasonally dry habitats on land. They put the most energy into complex structures that produce spores and also retain, nourish, and protect gametes through harsh times.

In the charophyte algae that gave rise to plants, the diploid phase of the life cycle is unicellular; a haploid, gamete-producing stage dominates the life cycle. As plants radiated onto land, and into higher and drier habitats, production and dispersal of spores took on greater importance. The fossil record reveals a trend toward sporophyte dominance.

In most plant lineages, the sporophyte became the larger, longer lived portion of the life cycle. Shrubs, trees, and other large forms having internal transport pipelines evolved. The spores became enclosed within hard capsules. Larger-bodied plants were better able to retain, nourish, and protect gametophytes, as well as the developing sporophytes. Figure 15.3 shows this evolutionary trend from haploid to diploid dominance.

SEEDS AND POLLEN GRAINS

Figure 15.4 is an evolutionary tree diagram for major groups of plants. Only 24,000 of the 295,000 existing species are *nonvascular* plants, or bryophytes. *Vascular*

types have internal tissue systems that conduct water and solutes. They became the success stories on land. The lycophytes, horsetails, and ferns are among the *seedless* vascular plants. Cycads, ginkgos, conifers, and gnetophytes are gymnosperms, a group of *seed-bearing* vascular plants. The other group of seed plants is the angiosperms. They alone produce flowers.

Seed-bearing plants produce two types of spores. The larger ones are megaspores, which develop into egg-producing female gametophytes. Microspores are smaller. They develop into the **pollen grains**. Sperm-bearing male gametophytes form inside pollen grains.

Male gametes of nonvascular plants and seedless vascular plants must swim to eggs for fertilization to occur. Most of the seed-bearing plants have sperm that do not have flagella. Air currents, insects, birds, and so on deliver their pollen grains to eggs. Pollen grains are a major reason why the seed-bearing plants were able to radiate into higher and drier environments.

Seeds were also useful in diverse habitats. Female gametophytes of seed plants form within the tissues of

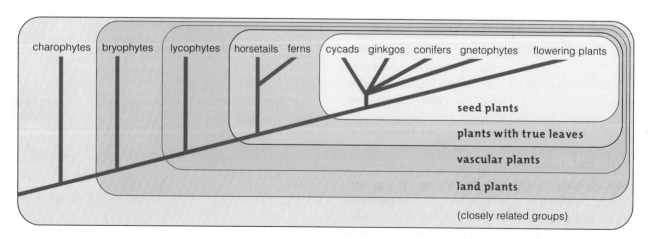

Figure 15.4 An evolutionary tree for plants, showing nested monophyletic groups.

sporophytes. Each **seed** is an embryo sporophyte, the tissues that nourish it, and an outer coat. Inside its coat, the embryo waits out seasons that do not favor its growth. When conditions become more favorable, the seed germinates and a new sporophyte develops.

The plant kingdom includes multicelled, photosynthetic species called bryophytes, seedless vascular plants, and seed-bearing vascular plants.

Most plant lineages became structurally adapted to life on land. They have root and shoot systems, a waxy cuticle, stomata, vascular tissues, and lignin-reinforced tissues.

Sporophytes with well-developed roots, stems, and leaves came to dominate the life cycles of most land plants. Parts of these complex sporophytes nourish and protect the new generation until conditions favor dispersal and growth.

The most recently evolved groups produce pollen grains and seeds. These were key innovations that allowed the seed plants to spread widely into diverse habitats.

Figure 15.5 Representative bryophytes. (**a**) Moss plants growing on damp rocks. (**b**) Peat moss (*Sphagnum*). A few sporophytes—brown, jacketed structures on white stalks—are attached to a pale gametophyte. (**c,d**) *Marchantia*. Like other liverworts, this one reproduces sexually as well as asexually. Unlike the other types, it has (**c**) female and (**d**) male reproductive parts on different plants.

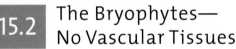

15.2 The Bryophytes— No Vascular Tissues

The first plants were bryophytes. A gamete-forming body dominates their life cycle. There are no specialized tissues to distribute water and sugars, so plants remain small.

Modern **bryophytes** include 24,000 species of mosses, liverworts, and hornworts. None of these nonvascular plants is taller than twenty centimeters (eight inches). Despite their small size, they have leaflike, stemlike, and usually rootlike parts. Their **rhizoids** are thread-like cells or structures that absorb water and dissolved mineral ions. They also attach gametophytes to soil.

Bryophytes do not have specialized vascular tissues and most grow in moist habitats. Some can dry out, then revive after absorbing water. This is one way that some hardy species survive in deserts and windswept plateaus in Antarctica.

Mosses are the bryophytes that most people know about (Figure 15.5*a*). Like others of this group, they are highly sensitive to air pollution. Where air quality is poor, mosses are few. Most moss gametophytes have a low mounded or feathery form. Gametes develop at their shoot tips, in vessels enclosed in a tissue jacket. Rain causes these jackets to open, and sperm swim to the eggs through tiny water droplets that cling to the plants. The moss sporophyte consists of a stalk and a jacketed structure in which haploid spores develop.

Sphagnum, a peat moss, is the dominant plant in bogs that cover more than 350 million acres in cooler regions of Europe, North Asia, and North America (Figure 15.5*b*). Peat bogs can be as acidic as vinegar. Mosses share them with acid-tolerant plants, such as cranberries. In some places, accumulated, compressed remains of bog plants can be meters thick and are an important store of carbon. In Ireland especially, these remains have traditionally been cut and dried, then burned as fuel. Peat is even burned in power plants to generate electricity. Compared to coal, burning peat produces far fewer air pollutants. But, like fossil fuels, peat is a finite resource, and industrial-scale harvests have depleted most of Ireland's bogs. Conservationists are now calling for the preservation and protection of those that still remain.

Acidic metabolic products of bryophytes hamper the growth of bacterial and fungal decomposers. Well-preserved human bodies about 2,000 to 3,000 years old have been found in peat bogs in Europe. These bogs may have been the sites for ritual human sacrifices. The high acidity of the bogs slowed bacterial growth and kept the victims' bodies from decomposing.

Compared to cotton, *Sphagnum* soaks up five times more water, which enters into dead cells in the plant's leaflike parts. Because of the antiseptic properties and absorbency of peat moss, it was used during World War I as a substitute for unavailable cotton bandages.

Early land plants may have resembled hornworts and liverworts (Figure 15.5*c,d*). Even today, these are

Zygote grows, develops into a sporophyte while still attached to gametophyte.

zygote

a fertilization

Mature sporophyte (spore-producing structure and stalk), still dependent on gametophyte.

Diploid Stage

Haploid Stage

b meiosis

Spores form by way of meiosis and are released.

Sperm reach eggs by moving through raindrops or film of water on the plant surface.

rhizoid

Spores germinate. Some grow and develop into male gametophytes.

sperm-producing structure at shoot tip of male gametophyte

c

egg-producing structure at shoot tip of female gametophyte

Other germinating spores grow and develop into female gametophytes.

Figure 15.6
Animated! Life cycle of a moss (*Polytrichum*), one of the bryophytes. (**a**) Fertilization will occur after sperm swim to the eggs. It produces a zygote, which develops into a nonphotosynthetic sporophyte that will remain attached to and dependent on the gametophyte. (**b**) The sporophyte produces spores by meiosis. (**c**) These spores develop into new gametophytes, which will produce eggs or sperm.

The northern spotted owl and about 450 moss and 170 liverwort species live in the Pacific Northwest's old-growth forests. Logging of this region threatens both the owl and some bryophytes with extinction.

among the hardy pioneer species that colonize barren habitats. Gametophytes are most often small ribbon-like or "leafy" forms.

All bryophytes share three traits that evolved in early land plants. *First*, a waxy cuticle helps conserve water in aboveground parts. *Second*, a cellular jacket around the gamete-producing parts keeps in moisture. *Third*, the large gametophytes do not draw nutrients from sporophytes as they do in other plants. It is the other way around. In bryophytes, a sporophyte grows from and stays attached to the gametophyte. It draws nutrients from the gametophyte for part or all of the life cycle (Figure 15.6).

Bryophytes are nonvascular plants with flagellated sperm that require liquid water to reach and fertilize the eggs.

In these plants, a sporophyte develops within gametophyte tissues. It remains attached to the gametophyte and receives some nutritional support from it.

15.3 Seedless Vascular Plants

Seedless vascular plants differ from bryophytes in three respects. The sporophyte does not remain attached to a gametophyte, it has true vascular tissues, and it is the larger, longer lived phase of the life cycle.

About 350 million years ago, during the Carboniferous, mild climates prevailed. Swamp forests carpeted the warm, wet lowlands of continents. Consistently balmy temperatures allowed plants to grow through much of the year. Sporophytes that had lignin-braced tissues and a vascular system were at a selective advantage. These traits arose in the first seedless vascular plants. Lineages that survived to the present are **lycophytes**, **horsetails**, and **ferns**.

Lycophytes are known informally as club mosses. Some Carboniferous ones, such as *Lepidodendron*, were fifty-meter-tall giants (Figure 15.17). Some relatives of

Lepidodendron

modern-day horsetails reached twenty meters high. Their aboveground stems grew from rapidly spreading **rhizomes,** or underground stems.

Most existing seedless vascular plants survive only in damp habitats. Gametophytes do not have vascular tissue and flagellated sperm require water to reach the egg. Species that live in drier places become dormant when rainfall is scarce and reproduce sexually during seasonal brief but heavy rains.

CLUB MOSSES

About 1,100 far smaller lycophyte species made it to the present. Club mosses are the most familiar group. Their sporophytes have tiny leaves and a branching rhizome that gives rise to the vascularized roots and stems (Figure 15.8*a*).

On stem tips, tight clusters of modified leaves form spore-producing chambers. Such conelike reproductive structures made of modified leaves are found in many plant groups. Each is a **strobilus** (plural, strobili).

HORSETAILS

The tree-sized sphenophytes flourished in the ancient swamp forests. Twenty-five smaller species of a single genus (*Equisetum*) still survive today (Figure 15.8*b,c*). You may be familiar with them as "horsetails." They thrive along streams and roadsides, and in disturbed areas. Their sporophytes have rhizomes and hollow, jointed, silica-reinforced stems with a ringlike array of xylem and phloem. Some species produce spores at the tips of their photosynthetic stems. In others, spore-bearing stems with little chlorophyll appear early in the season, and photosynthetic ones come along later. Free-living gametophytes about the size of a pin head develop from the released spores.

Figure 15.7 Reconstruction of a Carboniferous swamp forest. The sea level rose and fell fifty times during this geologic interval. Each time the sea receded, swamp forests flourished. When the sea moved back in, forest trees became submerged and buried in sediments, which protected them from decay.

Sediments slowly compressed the saturated, undecayed remains into peat. Each time more sediments accumulated, the increased heat and pressure made the peat more and more compact. In time, the compressed organic remains were transformed into great seams of coal. With its high percentage of carbon, coal is energy rich and one of our premier "fossil fuels."

It took a fantastic amount of photosynthesis, burial, and compaction to form each major coal seam. It has taken us only a few centuries to deplete much of the world's known coal deposits. Often you will hear about annual production rates for coal or some other fossil fuel. How much do we really produce in a year? None. We simply *extract* some portion. Fossil fuels are a nonrenewable source of energy.

vegetative stem

strobilus on fertile stem

Figure 15.8 (**a**) Sporophyte of a lycophyte (*Lycopodium*). (**b**) Mature vegetative stem of a horsetail (*Equisetum*). (**c**) Strobili at the tips of the nonphotosynthetic fertile stems produce and release spores. The young vegetative stem will later grow into the characteristic bushy form.

THE RISE OF VASCULAR PLANTS

Figure 15.9
Animated!

Life cycle of a chain fern (*Woodwardia*). (**a**) Fertilization occurs when sperm swim to eggs. It produces a diploid sporophyte, which grows into the familiar large, leafy photosynthetic form. (**b**) Meiosis in cells on the underside of fronds produces spores. (**c**) After their release, the spores germinate and grow into tiny heart-shaped gametophytes that produce eggs and sperm.

The sporophyte (still attached to the gametophyte) grows, develops.

zygote

rhizome

one sorus (a cluster of spore-producing structures)

Diploid Stage

a fertilization

Haploid Stage

meiosis b

Spores develop.

Spores are released.

egg

egg-producing structure

mature gametophyte (underside)

sperm

sperm-producing structure

c

A spore germinates, grows into a gametophyte.

Figure 15.10
Forest of tree ferns (*Cyathea*) in Australia's Tarra-Bulga National Park.

Before the invention of modern abrasive cleansers, silica-rich stems of some *Equisetum* species were used to scrub pots and to polish metals. These species are commonly referred to as scouring rushes. Their high silica content may help them resist insect predators.

FERNS

With 12,000 or so species, the ferns are the largest and most diverse group of seedless vascular plants. All but about 380 species are native to the tropics, but we find them all over the world (Figures 15.9 and 15.10). There is an enormous variation in size. The fronds of certain floating ferns can be less than 1 centimeter wide. Some tree ferns grow 25 meters tall.

In most fern species, roots and leaves grow from vascularized rhizomes. Exceptions include the tropical tree ferns and species growing as epiphytes. *Epiphyte* refers to any aerial plant that grows attached to tree trunks or branches. The young fern leaves, or fronds, develop in a coiled pattern that looks somewhat like a fiddlehead, and are uncoiled by maturity. Fronds of many species, such as the chain ferns, are divided into leaflets (Figure 15.9).

Rust-colored patches appear on the lower surface of most fern fronds. Each patch, one tiny cluster of spore-forming chambers, is called a sorus (plural, sori). Most chamber walls are only one cell thick, and when they pop open, spores are catapulted through the air. After a spore germinates, it will develop into a heart-shaped gametophyte, only centimeters across.

Seedless vascular plants include lycophytes, horsetails, and ferns. Their sporophytes are adapted to conditions on land, yet they have not entirely escaped their aquatic ancestry. Films of water must be available at the time of sexual reproduction if their flagellated sperm are to reach eggs.

THE RISE OF VASCULAR PLANTS

15.4 The Rise of Seed-Bearing Plants

Seed-bearing plants arose about 360 million years ago, as the Devonian gave way to the Carboniferous. In diversity, numbers, and distribution, they became the most successful groups of the plant kingdom.

Seed-bearing plants made their entrance in the late Devonian. The first kinds were close relatives of ferns, but seeds, in some cases as big as walnuts, formed on their fernlike fronds. Those "seed ferns" flourished through the Carboniferous, and a few made it through a mass extinction that ended the Permian. Long before the last seed ferns went extinct, cycads, conifers, and other gymnosperms started adaptive radiations. They became the dominant plants of Mesozoic times. (What some folks call the Age of Dinosaurs, botanists call the Age of Cycads.) However, flowering plants originated during the late Jurassic or early Cretaceous. By 120 million years ago, the angiosperms were diversifying as gymnosperms were on the decline. They have now assumed supremacy in nearly all land habitats.

Think back on the factors that promoted the rise of land plants. Structural modifications, such as a water-conserving cuticle, were one reason they endured in seasonally dry, often cold habitats. Just as important were the extraordinarily adaptive ways in which two kinds of spores evolved in the seed-bearers.

Seed plants form two types of spores. The smaller **microspores** develop into pollen grains. Pollen grains are walled structures that enclose the immature male gametophytes. The male gametophyte is carried inside a pollen grain to female plant parts that contain the eggs. Pollen grains are tiny, and are easily spread by air currents or by animals. Regardless of how pollen grains are dispersed, **pollination** refers to the arrival of pollen on female reproductive parts of a seed plant. The evolution of pollen grains meant that sperm could travel to eggs without water. Keep in mind, most seed plants have sperm that cannot move by themselves.

The larger **megaspores** form on a sporophyte in the specialized structures called ovules. Each megaspore gives rise to a female gametophyte, which produces egg cells. Other gametophyte parts form a nutritious tissue and a seed coat. Sperm fertilize the egg and an embryo develops inside an ovule. *What we call a "seed" is a fully mature ovule.* The seed's outer coat protects an embryo sporophyte from harsh conditions. Inside each seed, there is a supply of nutrient-rich tissue that the embryo can draw upon to power its renewed growth once conditions favor germination.

A footnote to this story: Evolution of some seed plants has been shaped by interactions with humans. By half a million years ago, *Homo erectus* was stashing nuts and rose hips in caves, and roasting seeds. By 11,000 years ago, modern humans were domesticating seed plants as reliable sources of food. We now know 3,000 or so species as edible, and we plant staggering numbers of about 200 species to supply food (Figure 15.11). In this sense, modern agricultural practices are still contributing to the ongoing evolutionary success of at least some seed-bearing plants.

pine pollen grains

> *Seed-bearing plants are known for pollen grains that can be dispersed without water, ovules that mature into seeds, and tissue adaptations to dry conditions.*

Figure 15.11 Edible treasures from flowering plants. (**a**) A sampling of fruits, which function in seed dispersal. (**b**) Mechanized harvesting of bread wheat, *Triticum*. (**c**) Indonesians picking shoots of tea plants. Only the terminal bud and two or three of the youngest leaves are picked for the finest teas. (**d**) From Hawaii, a field of sugarcane, *Saccharum officinarum*. Sap extracted from its stems is boiled to make sucrose crystals (table sugar).

15.5 Gymnosperms—Plants With "Naked" Seeds

The modern gymnosperms include diverse types from vines to shrubs to trees (Figure 15.12). Seeds form in ovules that are not enclosed within a flower or a fruit. Gymnos means naked; sperma is taken to mean seed.

The cycads and ginkgos were diverse in dinosaur times. Today, they are the only existing seed plants in which pollen releases ciliated sperm that can swim to eggs.

About 130 species of **cycads** made it to the present. Their pollen-bearing and seed-bearing cones form on separate plants (Figure 15.12b). Pollen is carried from male to female plants mainly by beetles. Most cycads live in tropical and subtropical regions, where people grind their seeds and trunks into a flour. Gardeners employ many cycads as ornamental plants, and some cycad species are disappearing from the wild because of over-collection of plants for this market.

Only a single species of ginkgo still survives, the maidenhair tree, *Ginkgo biloba* (Figure 15.12c-f). Like a few other gymnosperms, ginkgos are deciduous; they shed all their leaves in the fall. Several thousand years ago, ginkgos were widely planted around temples in China. Natural populations nearly became extinct; they may have been logged for firewood. Today, male trees are often planted along city streets. They have pretty fan-shaped leaves, are resistant to insects and disease, and tolerate air pollutants well. Few female trees are planted. Their fleshy seeds are the size of plums and give off a sharp, offensive smell as they decay.

Gnetophytes include tropical trees and leathery leafed vines as well as desert shrubs. They resemble flowering plants in some respects and are thought to be their closest living relatives. *Ephedra* (Figure 15.12g) forms photosynthetic stems. It is the source of the now-banned herbal supplement ephedrine. *Welwitschia* lives in hot African deserts. The sporophyte has a long taproot and a woody stem with two strap-shaped leaves that split lengthwise as a plant ages (Figure 15.12h).

Conifers are the most diverse gymnosperms. Most are woody trees or shrubs with needlelike or scalelike leaves. Most conifers shed some leaves all year long yet still stay leafy, or evergreen. A few are deciduous. Only the conifers have true **cones**. These reproductive structures—clusters of papery or woody scales—bear exposed ovules on their upper surface. The conifers include the tallest trees in the Northern Hemisphere (redwoods), as well as the most abundant (pines), and the oldest (bristlecone pines) (Figure 15.12a).

Figure 15.13 shows the life cycle of a conifer. The familiar tree is the sporophyte. Gametophytes form in cones (Figure 15.13a,b). In female cones, megaspores develop into female gametophytes that produce eggs. In male cones, microspores become pollen grains. Each spring, pollen grains are dispersed from the tree by the

Figure 15.12 Representative gymnosperms. (**a**) Bristlecone pine (*Pinus longaeva*) on a windswept mountain in the Sierra Nevada. (**b**) Male strobilus of *Dioon*, a cycad. *Ginkgo biloba*: (**c**) Fleshy seeds, (**d**) new leaves, (**e**) fossilized leaf, and (**f**) fall foliage. Two gnetophytes: (**g**) *Ephedra viridins* and (**h**) *Welwitschia mirabilis*, with strappy leaves and seed-bearing strobili.

Figure 15.13 *Animated!* Life cycle of a conifer, the ponderosa pine.

Labels within figure:

section through one ovule (the red "cut" in the diagram to the left):

surface view of one cone scale (houses two ovules)

ovule

a

b

section through a pollen-producing sac (red cut):

surface view of one cone scale (houses a pollen-producing sac)

h mature sporophyte

seed coat

embryo

nutritive tissue

zygote

seedling

g seed formation

f fertilization

Diploid Stage

Haploid Stage

c meiosis

d meiosis

pollen tube

(view inside an ovule)

pollination (wind deposits pollen grain near ovule)

Microspores form, develop into pollen grains.

Megaspores form; one develops into the female gametophyte.

sperm-producing cell

eggs

female gametophyte

e Germinating pollen grain (the male gametophyte). Sperm nuclei form as the pollen tube grows toward the egg.

wind. Pollination occurs when a pollen grain lands on an exposed ovule of a female cone scale. The pollen grain germinates and a tubular structure grows from it—the sperm-bearing male gametophyte. After a year, growth of the tube delivers nonmotile sperm to an egg, and fertilization takes place. The resulting zygote develops into an embryo sporophyte, which—along with female gametophyte tissues—becomes a seed.

After fertilization occurs, it takes yet another year for the seed to mature and be released from the parent pine. After seed release, the embryo sporophyte inside is nourished by tissues derived from the gametophyte. Other gametophyte tissues form a hardened coat that encloses and protects the seed.

Like their Permian ancestors, gymnosperms are adapted to seasonally dry climates. They bear seeds on the exposed surfaces of cones or other spore-producing structures.

LINK TO
SECTION
13.9

15.6 Angiosperms—The Flowering Plants

Only angiosperms produce the specialized reproductive structures called flowers. Angeion, which means vessel, refers to the female reproductive parts at the center of a flower. The enlarged base of the "vessel" is the floral ovary, where ovules and seeds develop.

We find flowering plants almost everywhere, from icy tundra to deserts to oceanic islands. What accounts for their distribution and diversity? Consider the **flower**, a specialized reproductive shoot (Figures 15.14 and 15.15). When plants invaded land about 435 million years ago, insects that ate decaying plant parts and spores were not far behind. After pollen-producing

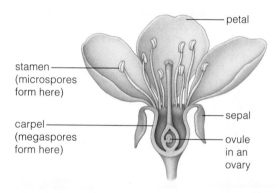

Figure 15.14 Floral structures typical of angiosperms.

- petal
- stamen (microspores form here)
- carpel (megaspores form here)
- sepal
- ovule in an ovary

plants evolved, insects made the connection between "plant parts with pollen" and "food." Plants lost some pollen to the insects but gained a reproductive edge: The insects delivered pollen to the plants' ovules. So plants evolved in ways that made them more enticing to insects. Insects became specialized in locating and gathering the pollen from particular plants, instead of searching at random for sources of food.

In time, plants with flowers, fragrances, and nectar coevolved with diverse pollinators. A **pollinator** is any agent that transfers pollen to female floral parts of the same species. In addition to insects, pollinators include bats, birds, and other animals. **Coevolution** is a term for two or more species evolving jointly as a result of some close ecological interaction. Because of this interaction, a heritable change in one species can select for changes in the other, which evolves also.

Recruiting pollinators as assistants in reproduction contributed to the dominance of flowering plants for the past 100 million years. We know of at least 260,000 species. They range from tiny floating duckweeds one millimeter long to towering *Eucalyptus* trees over 100 meters tall. A few species, including the mistletoes, are no longer photosynthetic; they steal needed nutrients from other plants (Figure 15.15*d*).

The three major groups of flowering plants are the **magnoliids**, **eudicots**—true dicots—and the **monocots**. More ancient groups include the water lilies.

The 9,200 magnoliids include magnolias, avocados, nutmeg, and peppers. Among the 170,000 eudicots are most of the herbaceous (nonwoody) plants such as

Figure 15.15 (**a**) Examples of flowers. Their colors, patterns, and shapes attract pollinators. (**b**) Sacred lotus (*Nelumbo nucifera*), an aquatic species. The flower's radial pattern is typical of ancient lineages. (**c**) More recently evolved lineages, including the pansies (*Viola*), display a bilateral pattern, with symmetrical left and right parts. (**d**) Dwarf mistletoe (*Arceuthobium*). It draws nutrients from trees and stunts their growth.

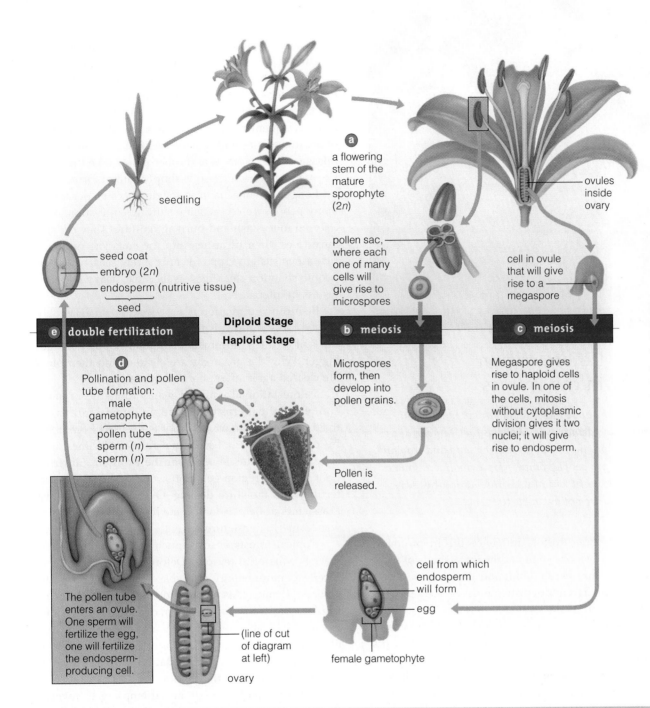

seedling

a flowering stem of the mature sporophyte (2n)

ovules inside ovary

seed coat
embryo (2n)
endosperm (nutritive tissue)
seed

pollen sac, where each one of many cells will give rise to microspores

cell in ovule that will give rise to a megaspore

e double fertilization | **Diploid Stage** | **b** meiosis | **c** meiosis
Haploid Stage

d
Pollination and pollen tube formation: male gametophyte
pollen tube
sperm (n)
sperm (n)

Microspores form, then develop into pollen grains.

Megaspore gives rise to haploid cells in ovule. In one of the cells, mitosis without cytoplasmic division gives it two nuclei; it will give rise to endosperm.

Pollen is released.

The pollen tube enters an ovule. One sperm will fertilize the egg, one will fertilize the endosperm-producing cell.

(line of cut of diagram at left)

ovary

cell from which endosperm will form

egg

female gametophyte

Figure 15.16 *Animated!*
Example of a flowering plant life cycle. This lily (*Lilium*) is one of the monocots.

(a) The sporophyte is the dominant part of the life cycle. Gametophytes form inside flowers. Pollen grains develop from microspores in pollen sacs (**b**) and contain male gametophytes. Female gametophytes develop from megaspores inside an ovary (**c**). After pollination, a tube grows from the pollen grain into the ovule and delivers two sperm (**d**). (**e**) "Double" fertilization occurs in all flowering plant life cycles. One sperm fertilizes the egg. The other fertilizes a cell that gives rise to endosperm, a tissue that will nourish the embryo. Sections 19.1 and 19.2 detail the flowering plant life cycles, using a dicot as the example.

tomatoes, cabbages, roses, and daisies, most flowering shrubs and trees, and cacti. The 80,000 or so monocots include orchids, palms, lilies, and grasses, such as rye, wheat, corn, rice, sugarcane, and other valued plants.

The next unit takes a close look at the structure and function of flowering plants. As in the other vascular plants, the sporophyte dominates the life cycle. The flowering plants are unique in that they nourish new generations with a nutritious tissue called endosperm. It surrounds an embryo sporophyte in a seed. As the seed develops, ovary tissues mature into fruit, which functions in seed dispersal. Figure 15.16 illustrates a monocot life cycle. We will show a eudicot life cycle in Figure 19.4.

Angiosperms are the most successful plants in terms of diversity, numbers, and distribution. They alone produce flowers. Most species coevolved with animal pollinators.

Monocots and eudicots are the two most diverse lineages.

Figure 15.17
Tropical rain forest
in Southeast Asia.

15.7 Deforestation in the Tropics

*Paralleling the huge increases in human population
are increasing demands for wood as fuel, lumber, and
other forest products, and for grazing land and cropland.
More and more people compete for dwindling resources
for economic profit, but also because alternative ways
of life simply are not available to them.*

The world's great forests influence life in profound
ways. Like enormous sponges, the watersheds of these
forested regions absorb, hold, and then release water
slowly. By mediating downstream flows, they prevent
erosion, flooding, and sedimentation that can disrupt
rivers, lakes, and reservoirs. Strip forests away, and
the exposed soil is vulnerable to leaching of nutrients
and to erosion, especially on steep slopes.

Again, **deforestation** means the removal of all trees
from large tracts for logging, agriculture, and grazing.
This chapter started with a look at what is going on in
once-sweeping temperate forests of the United States,
Canada, Europe, Siberia, and elsewhere. But tropical
forests are in even more peril (Figure 15.15). For 10,000
years or more, these forests have been habitat for 50 to
90 percent of all land-dwelling species. In less than
four decades, humans have destroyed more than half
of them. Logging over millions of acres per year is like
leveling about thirty-four city blocks every minute.

Deforestation is now greatest in Brazil, Indonesia,
Colombia, and Mexico. If the clearing and destruction
continue at present rates, then by 2010, only Brazil and
Zaire will still have large tropical forests. By 2035, most
of their forests also will be gone.

In tropical regions, clearing forests for agriculture
sets the stage for long-term losses in productivity. The
irony is that tropical forests are one of the worst places
to grow crops and raise animals. In intact forests, litter
can't accumulate because the high temperatures and
heavy, frequent rainfall promote rapid decomposition
of organic wastes and remains. As fast as decomposers
release nutrients, trees and other plants take them up.
Deep, nutrient-rich topsoils simply cannot form.

Long before mechanized logging became pervasive,
many people were practicing shifting cultivation, once
referred to as slash-and-burn agriculture. They cut and
burn trees, then till ashes into the soil. The nutrient-
rich ashes sustain crops for only a few seasons. Later,
people abandon cleared plots, because heavy leaching
results in infertile soil.

If shifting cultivation is practiced on small, widely
spaced plots, a forest isn't necessarily damaged much.
But soil fertility plummets as human populations get
bigger. Why? Larger areas are cleared, and then plots
are cleared again after shorter intervals.

Deforestation also affects evaporation rates, runoff,
and regional patterns of rainfall. In tropical forests, 50
to 80 percent of the water vapor is released from trees.
In logged-over regions, annual rainfall declines, and
rain swiftly drains away from the exposed, nutrient-
poor soil. The region gets hotter and drier, and its soil
fertility and moisture decline. Over time, sparse, dry
grasslands or deserts may come to prevail where there
was formerly dense forest.

Tropical forests absorb much of the sunlight that
reaches equatorial regions. Deforested land is shinier;
it reflects more energy into space, possibly adding to
global warming. Also, the photosynthetic activity of
tropical forests affects the global cycling of carbon and
oxygen. Extensive tree harvesting and burning of tree
biomass release stored carbon as carbon dioxide. Some
researchers think that deforestation will amplify the
greenhouse effect for that reason.

Conservation biologists are attempting to reverse
the trend. As three examples, a coalition of 500 groups
is dedicated to preserving Brazil's remaining tropical
forests. In India, women have already constructed and
installed 300,000 inexpensive, smokeless wood stoves.
In the past decade, the stoves saved more than 182,000
metric tons of trees by cutting demands for fuelwood.
In Kenya, women have planted millions of trees as a
source of wood and to counter erosion.

*Once-vast forests helped sustain rapid increases in human
population sizes. But lineages that began hundreds of
millions of years ago are ending because of deforestation.*

THE RISE OF SEED PLANTS

Summary

Section 15.1 Plants evolved from green algae about 500 million years ago. Nearly all existing plants are multicelled photoautotrophs. Table 15.1 summarizes the major groups.

Plants arose near the water's edge, then radiated into drier habitats. Key innovations included a waterproof cuticle with stomata, and internal pipelines of vascular tissue (xylem and phloem).

Complex diploid sporophytes evolved; they hold, nourish, and protect spores and gametophytes. Male gametes that can be dispersed without water (pollen grains) and seeds were key innovations that allowed the radiation of gymnosperms and flowering plants. Table 15.2 summarizes these trends.

Biology⊗Now
Compare the relative contributions of haploid and diploid stages to different plant life cycles with the interaction on BiologyNow.

Section 15.2 Mosses, liverworts, and hornworts are bryophytes. These nonvascular plants lack complex vascular tissues. Their flagellated sperm can reach the eggs only by swimming through films or droplets of water. The sporophyte begins development inside the gametophyte tissues. It remains attached to and often dependent upon the gametophyte even when mature.

Biology⊗Now
Observe the life cycle of a typical moss with the animation on BiologyNow.

Section 15.3 Lycophytes, horsetails, and ferns are among the seedless vascular plants. Their sporophytes have vascular tissues and are the larger, longer lived phase of the life cycle. Like bryophytes, they produce flagellated sperm and require free water for fertilization.

Biology⊗Now
Observe the life cycle of a typical fern with the animation on BiologyNow.

Section 15.4 Gymnosperms and flowering plants (angiosperms) are seed-bearing vascular plants. They produce two types of spores. Microspores give rise to pollen grains, the sperm-bearing male gametophytes. Even in the absence of water, pollen grains can deliver non-flagellated sperm to egg cells. Megaspores form in ovules, and consist of a female gametophyte (with egg) and tissues that will become the seed coat.

Section 15.5 The conifers, cycads, ginkgos, and gnetophytes are gymnosperms. They are adapted to dry climates and bear seeds on exposed surfaces of spore-bearing structures. In conifers, these spore-bearing structures are distinctive cones.

Biology⊗Now
Learn about the stages in the life cycle of a typical gymnosperm with the animation on BiologyNow.

Table 15.1 Comparison of Major Plant Groups

Bryophytes — 24,000 species. Moist, humid habitats.

Nonvascular land plants. Fertilization requires free water. Haploid dominance. Cuticle and stomata present in some.

Club mosses — 1,100 species with simple leaves. Mostly wet or shady habitats.

Horsetails — 25 species of single genus. Swamps and disturbed habitats.

Ferns — 12,000 species. Wet, humid habitats in mostly tropical and temperate regions.

Seedless vascular plants. Fertilization requires free water. Diploid dominance. Cuticle and stomata present.

Conifers — 550 species, mostly evergreen, woody trees and shrubs with pollen- and seed-bearing cones. Widespread distribution.

Cycads — 130 slow-growing tropical, subtropical species.

Ginkgo — 1 species, a tree with fleshy-coated seeds.

Gnetophytes — 70 species. Limited to some deserts, tropics.

Gymnosperms—vascular plants with "naked seeds." Free water not required for fertilization. Diploid dominance. Cuticle and stomata present.

Flowering plants

Monocots — 80,000 species. Floral parts often arranged in threes or in multiples of three; one seed leaf; parallel leaf veins common; pollen with one furrow.

Eudicots (true dicots) — At least 170,000 species. Floral parts often arranged in fours, fives, or multiples of these; two seed leaves; net-veined leaves common; pollen with three or more pores and/or furrows.

Magnoliids and basal groups — 9,200 species. Many spirally arranged floral parts or fewer, in threes; two seed leaves, net-veined leaves common; pollen with one furrow.

Angiosperms—vascular plants with flowers, protected seeds. Free water not required for fertilization. Diploid dominance. Cuticle and stomata present.

Table 15.2 Evolutionary Trends Among Plants

Bryophytes	Ferns	Gymnosperms	Angiosperms
Nonvascular	Vascular		
Haploid dominance	Diploid dominance		
Spores of one type	Spores of two types		
Seedless		Seeds	
Flagellated sperm			Nonmotile sperm*

* Require pollination by wind, insects, animals, etc.

Section 15.6 Angiosperms are the dominant land plants. They alone produce flowers inside of ovaries. Most kinds coevolved with pollinators. Endosperm, a nutritious tissue, surrounds the embryo sporophyte inside seeds, which have a hard coat. Fruit encases most seeds and aids in dispersal.

Biology⊗Now
Explore the angiosperm life cycle with the animation on BiologyNow.

Section 15.7 Deforestation threatens ecosystems worldwide. Forests play a role in the global cycling of water, carbon dioxide, and oxygen. Their destruction may be amplifying the greenhouse effect.

Figure 15.18 Kenyan Wangari Maathai, who in 2004 won a Nobel Peace Prize for her work in support of women's rights and sustainable development.

Self-Quiz

Answers in Appendix I

1. Which of the following statements is *not* correct?
   a. Gymnosperms are the simplest vascular plants.
   b. Bryophytes are nonvascular plants.
   c. Lycophytes and angiosperms are vascular plants.
   d. Only angiosperms produce flowers.

2. Which does *not* apply to gymnosperms or angiosperms?
   a. vascular tissues   c. single spore type
   b. diploid dominance   d. all of the above

3. Bryophytes have independent _____ and attached, dependent _____ .
   a. sporophytes;   c. rhizoids; zygotes
      gametophytes   d. rhizoids; stalked
   b. gametophytes;   sporangia
      sporophytes

4. Lycophytes, horsetails, and ferns are classified as _____ plants.
   a. multicelled aquatic   c. seedless vascular
   b. nonvascular seed   d. seed-bearing vascular

5. The _____ produce flagellated sperm.
   a. ferns   c. monocots
   b. conifers   d. a and c

6. The _____ produced in the male cones of a conifer develop into pollen grains.
   a. sori   c. megaspores
   b. microspores   d. a and c

7. A seed is _____ .
   a. a female gametophyte   c. a mature pollen tube
   b. a mature ovule   d. an immature spore

8. The majority of _____ rely on animal pollinators.
   a. lycophytes   c. angiosperms
   b. gymnosperms   d. bryophytes

9. Match the terms appropriately.
   ____ gymnosperm   a. gamete-producing body
   ____ sporophyte   b. help control water loss
   ____ lycophyte   c. "naked" seeds
   ____ bryophyte   d. spore-producing body
   ____ gametophyte   e. nonvascular land plant
   ____ stomata   f. seedless vascular plant
   ____ angiosperm   g. flowering plant

Additional questions are available on **Biology⊗Now™**

Critical Thinking

1. The 2004 Nobel Peace Prize was awarded to Wangari Maathai, who founded the Green Belt Movement in 1977 (Figure 15.18). Since then, group members—mostly poor rural women —have planted more than 25 million trees. Reforestation efforts help halt soil erosion, provide sources of fruits, food, and firewood, and empower women to work together for the long-term good of their communities. Find out if any nonprofit groups are working to protect forests in your area. Do you support their goals and methods?

2. Early botanists admired ferns but found their life cycle perplexing. In the 1700s, they learned to propagate them by sowing what appeared to be tiny dustlike "seeds"from the undersides of fronds. Despite many attempts, the scientists could not find the pollen source, which they assumed must stimulate the "seeds" to develop. Imagine you could write to one of these botanists. Compose a note that would clear up their confusion.

3. Today, the tallest bryophytes reach a maximum height of 20 centimeters or so. So far as we know from fossils, there were no giants among their ancestors. Lignin and vascular tissue first evolved in the lycophytes, and some extinct species stood 40 meters high. Among seed plants, *Sequoia* (a gymnosperm) and *Eucalyptus* (an angiosperm) can be more than 100 meters high. Explain why evolution of vascular tissues and lignin would have allowed such a dramatic increase in maximum plant height. How might an increase in maximum height give one plant species a competitive advantage over another?

4. In cycads and ginkgos, after a pollen grain lands on a receptive female cone, a pollen tube grows toward the ovary. Once the tube gets to the ovary, it releases flagellated sperm that swim a short distance through a fluid-filled egg chamber to reach the egg. In all other gymnosperms and all angiosperms, the sperm do not have flagella. The pollen tube delivers them directly to the egg. Speculate about how the transition from flagellated to nonflagellated sperm came about. How might it be selectively advantageous?

Interpreting and Misinterpreting the Past

In Darwin's time, the presumed absence of transitional forms was an obstacle to acceptance of his new theory of evolution. Skeptics wondered, if new species evolve from older ones, then where are the transitional forms in the fossil record, the "missing links"? Where are the fossils that bridge major groups? Ironically, one such form was unearthed by workers at a limestone quarry in Germany just one year after Darwin's *On the Origin of Species* was published.

The fossil, about the size of a large crow, looked like a small meat-eating dinosaur. It had a long, bony tail, three long, clawed fingers on each forelimb, and a heavy jaw fringed with short, spiky teeth. Later, diggers found another fossil of the same type. Later still, someone noticed the feathers. If these were dinosaur fossils, what were they doing with feathers? The feathers looked like those of modern birds. The specimen type for this new fossil was named *Archaeopteryx* (ancient winged one).

Between 1860 and 1988, six *Archaeopteryx* specimens and a fossilized feather were found. Anti-evolutionists tried to dismiss them as forgeries. Someone, they said, had pressed modern bird bones and feathers against wet plaster. The dried, imprinted plaster casts merely looked like fossils. But microscopic examination confirmed the fossils are real. Further confirmation of their age came from the remains of obviously ancient jellyfishes, worms, and other organisms in the same limestone layers.

Today, radiometric dating shows *Archaeopteryx* lived 150 million years ago. Why are its remains preserved so well after so much time? *Archaeopteryx* lived in forests near a shallow lagoon. The lagoon was large, warm, and stagnant; coral reefs barred inputs of oxygenated water from the sea. So it was not a particularly inviting place for scavenging animals that might have feasted on—and obliterated the remains of—*Archaeopteryx* and other organisms that fell from the sky or drifted offshore.

Each storm surge drove fine sediments over the reefs. They gently buried the carcasses littered at the bottom of the lagoon. In time, the soft, muddy sediments were slowly compacted and hardened. They became a lasting limestone tomb that preserved more than 600 species, including *Archaeopteryx*.

No one was around to witness such transitions in the history of life. But fossils are real, just as the morphology, biochemistry, and molecular makeup of living organisms are real. Radiometric dating assigns fossils to their place in time. *Life's evolution is not "just a theory."* Remember, a scientific theory differs from speculation because its predictive power has been tested in nature many times, in many different ways that might disprove it. If a theory stands, there is a high probability it is not wrong.

In this chapter, you will learn that Darwin's theory of evolution by natural selection goes a long way toward explaining the patterns of animal evolution, including the evolution of our own lineage.

 How Would You Vote? *Private fossil hunters have discovered important fossils, but some poach fossils from public lands for sale. Should sale of vertebrate fossils be regulated? See BiologyNow for details, then vote online.*

 ## Key Concepts

ANIMAL CHARACTERISTICS
Animals are multicelled heterotrophs that move about during some part of the life cycle. Most have tissues and organs and undergo development.

INVERTEBRATES
The first animals were invertebrates and, in terms of both species and number of individuals, most still are. Animals arose in the seas, but many groups now have representatives on land.

VERTEBRATES
A backbone, jaws, paired fins, and lungs were key innovations in the evolution of fishes, amphibians, reptiles, birds, and mammals.

HUMAN EVOLUTION
Human traits emerged through modification of traits that evolved in earlier mammalian, primate, and then early hominid lineages.

Links to Earlier Concepts

In this chapter you will investigate the diversity of animals, a group you met in Section 1.3. Much information about animal evolution comes from fossils (13.1, 13.2). You will learn how asteroid impacts (Chapter 13 Introduction) and changes in the distribution of continents (13.3) affected animal diversity, and how we use the evidence derived from comparative methods (13.4–13.6).

You may wish to review Sections 13.9 and 13.10. They introduced the patterns of speciation and extinction, and the main methods of organizing information about biological diversity.

16.1 Overview of the Animal Kingdom

We already know of more than 2 million animal species, and no simple definition can cover that much diversity.

We define **animals** by a list of their characteristics:

1. Animals are multicelled, and most (not all) species have tissues, organs, and organ systems.

2. All animals are aerobic and heterotrophic. They eat other organisms or withdraw nutrients from them.

3. Animals are sexual reproducers. Many species also reproduce asexually. Life cycles usually include stages of development. In embryos, cell divisions give rise to primary tissues—**ectoderm**, **endoderm**, and in most cases **mesoderm**—from which adult tissues develop.

4. Animals are motile; they actively move about during some or all of the life cycle.

All animal lineages are a monophyletic group; they all share a common ancestor, as shown in Figure 16.1. The ones traditionally known as mammals, birds, reptiles, amphibians, and fishes are **vertebrates**—animals that have a backbone. Yet all but 50,000 of the 2 million or so named species are backboneless **invertebrates**.

ANIMAL ORIGINS

How did animals arise? Most likely, they evolved from colonies of flagellated cells, much like the ones formed today by choanoflagellates. These single-celled protists resemble the body cells of sponges, the simplest living animals, and DNA sequence comparisons point to an evolutionary connection. In some ancient colonies, cells could have mutated in ways that led to specializations in feeding and other tasks that now define "animals."

However the animals arose, tiny, soft-bodied species shaped like disks, fronds, and blobs were living on or in seafloor sediments by 610 million years ago. We call them the **Ediarcarans**. Some resembled existing animal groups and are probably their earliest representatives.

During the Cambrian (543 to 490 million years ago), animals underwent an enormous adaptive radiation. Before it was over, all major groups were represented by diverse forms in the seas. What caused this great explosion of diversity? Perhaps changes in the climate, sea level, and distribution of land masses triggered it. Or maybe mutations in genes that act in development kicked it off. Possibly predators and prey exerted new selective pressures on one another. Or, these and other factors may have collectively promoted speciations.

A LOOK AT BODY PLANS

How might we get a conceptual handle on animals as different as flatworms and dinosaurs, hummingbirds and humans? Comparing similarities and differences with respect to some basic features is one way to start.

BODY SYMMETRY AND CEPHALIZATION Animals can be described in part by their symmetry (Figure 16.2). With **radial symmetry**, body parts are arrayed around a central point, like the spokes of a wheel. Usually the mouth is at the center. Radial animals live in the water and respond to stimuli that arrive from all directions. In contrast, animals with **bilateral symmetry** have two halves that mirror each other. Their main body axis runs from *anterior* (front) to *posterior*. This axis divides a body into right and left halves, with a *ventral* surface (underside) and *dorsal* surface (the back). Most of the bilateral groups show **cephalization**—a concentration of nerve and sensory cells in the anterior end, the end that is the first to encounter stimuli.

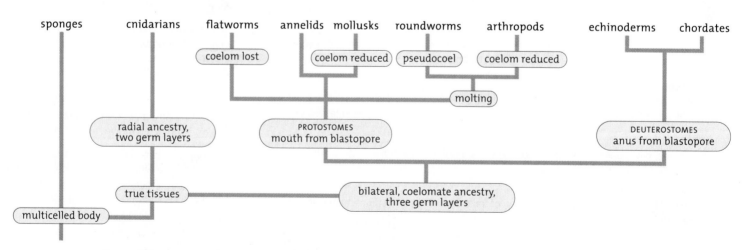

Figure 16.1 An evolutionary tree for animal groups covered in this text.

ANIMAL CHARACTERISTICS

TYPE OF GUT The animal **gut** digests food and then absorbs the released nutrients. An *incomplete* digestive system, in which the gut is like a sac, evolved in early animals. Cnidarians, such as sea anemones and jellyfish, still have this type of gut. So do the flatworms.

A *complete* digestive system has a tubelike gut that starts with a mouth and ends at an anus. How this gut develops is one way that we define two major lineages of bilateral animals. In both lineages, the first opening that appears on the embryo's surface develops into one of the gut's openings. In the **protostomes**, which include flatworms, mollusks, annelids, roundworms, and arthropods, the initial embryonic opening (called the blastopore) becomes the mouth. In echinoderms, chordates, and other **deuterostomes**, this first opening becomes the anus.

A tubular gut has distinct advantages. The mouth can be modified with teeth or other structures instead of being a single, generic opening for taking in food and expelling residues. Also, regions of the tube can become specialized for food breakdown, absorption, storage, and elimination of undigestible wastes.

BODY CAVITIES A body cavity called a **coelom** lies between the body wall and the gut of most bilateral animals (Figure 16.3). Peritoneum, a tissue that arises from mesoderm, lines the coelom. This lining helps to hold the soft internal organs in place.

The coelom was a key innovation in the evolution of larger, active animals from smaller ancestral forms. Why? It cushions and protects internal organs, yet lets them move independently of the body wall. Some of the less complex invertebrates do not have any body cavity; tissues fill the region between the gut and the body wall. Others have a pseudocoel, a "false coelom." This is a body cavity that is not fully lined.

SEGMENTATION Many of the bilateral groups show some degree of **segmentation**, the division of a body into interconnecting units that are repeated one after the next along the main body axis. We clearly see body segments in annelids, such as earthworms. We also find clues to our own segmented origins in the segmented muscles that appear in human embryos.

Animals are multicelled, aerobically respiring heterotrophs. Nearly all animals have tissues, organs, and organ systems.

Animals reproduce sexually and, in many cases, asexually. They go through a period of embryonic development, and most are motile during at least part of the life cycle.

Animal bodies differ in their symmetry, cephalization, type of gut, type of body cavity, and degree of segmentation.

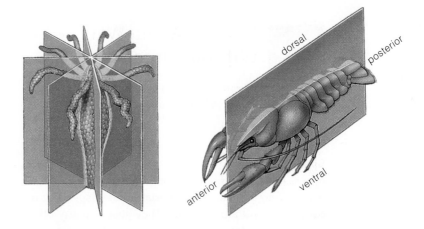

Figure 16.2 *Left*, radial body symmetry of *Hydra*, a cnidarian. *Right*, bilateral symmetry of a lobster, an arthropod. Like other bilateral animals, a lobster shows cephalization. Nervous and sensory organs—a brain, eyes, and antennae—are concentrated at the anterior or head end.

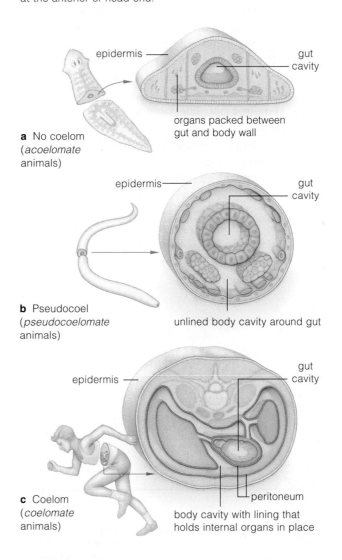

a No coelom (*acoelomate* animals)

epidermis — gut cavity

organs packed between gut and body wall

b Pseudocoel (*pseudocoelomate* animals)

epidermis — gut cavity

unlined body cavity around gut

c Coelom (*coelomate* animals)

epidermis — gut cavity

peritoneum

body cavity with lining that holds internal organs in place

Figure 16.3 Types of body cavity. (**a**) A flatworm is acoelomate; its gut is surrounded by tissues. (**b**) Pseudocoelomates, such as roundworms, have an unlined cavity around the gut. (**c**) Like all vertebrates, humans have a lined body cavity, or coelom.

16.2 Getting Along Well Without Organs

*Simple animals can endure—as demonstrated
by the sponges and the cnidarians.*

SPONGES

Sponges—animals that have no symmetry, tissues, or organs—are one of nature's success stories. They have been abundant since Precambrian times, especially on coral reefs and along coasts. They shelter many other animals, including worms and shrimps. Some are large enough to sit in; others are tiny. They can be flattened, lobed, tubelike, or vaselike (Figure 16.4).

Tough protein fibers and sharp, glassy spicules of calcium carbonate or silica often stiffen a sponge body. These protective components may be one reason why sponges have surived so long. Predators discover that biting a sponge is like eating glass splinters in fibrous gelatin. Many of the sponges also produce chemicals that smell or taste bad to potential predators.

Sponges have no tissues. There are three main cell types. Flattened cells line the outer body surface and surround the pores in the body wall. Water flows into a sponge through these pores, then out through one or more larger openings. Flagella of many thousands of phagocytic collar cells of the inner lining drive the flow. The "collars" of these cells trap food, which is then engulfed. Amoeba-like cells prowl between inner and outer cell layers; they process food, distribute it, cart away wastes, and secrete spicules.

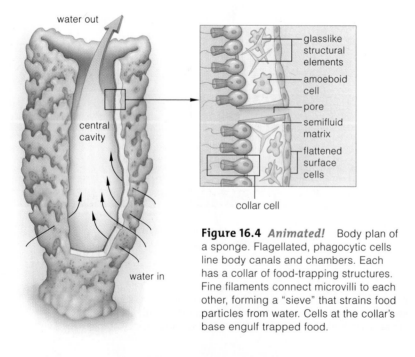

Figure 16.4 Animated! Body plan of a sponge. Flagellated, phagocytic cells line body canals and chambers. Each has a collar of food-trapping structures. Fine filaments connect microvilli to each other, forming a "sieve" that strains food particles from water. Cells at the collar's base engulf trapped food.

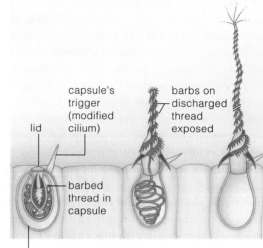

nematocyst (capsule at free surface of epidermal cell)

Figure 16.5 Animated! Action of a type of nematocyst. When prey brushes against the trigger, the capsule gets "leakier." As water diffuses in, internal pressure rises, and the thread is forced outward. Its barbed tip pierces the prey.

Most sponges reproduce sexually. They release their sperm into the water but retain fertilized eggs, which develop into a microscopic, swimming larva. A **larva** (plural, larvae) is a sexually immature stage that has a different body form than the adult. Many sponges also reproduce asexually by budding. The new individual develops on the parent body, then drops or is broken off and lives on its own.

CNIDARIANS

Cnidarians are radial, tentacled animals that include jellyfishes, sea anemones, corals, and hydras. Nearly all are marine. All have cells organized as tissues but no organs. Tentacles have distinctive capsules called nematocysts. If something brushes against a tentacle, the capsules snap open and substances pierce, poison, or entangle prey (Figure 16.5).

Cnidarian bodies take two main forms (Figure 16.6). The sea anemones, corals, and hydras are tube-shaped polyps. Usually one end of the polyp attaches to some surface. Tentacles surround a mouth at the other end. A jellyfish is a medusa, which appears like a saucer or bell. Medusae swim or drift through the water.

Like all animals except the sponges, cnidarians have layers of **epithelium**, a tissue that lines internal and external body surfaces. In cnidarians, epithelial tissue called *epidermis* covers the outer surface. *Gastrodermis* lines the saclike gut. Running through the epidermal layers are nerve cells that function as a *nerve net*. This simple nervous system interacts with contractile cells to alter body shape and move the body or its parts.

The "jelly" inside a jellyfish is mesoglea, a secreted gelatinous material found between the epidermis and the gastrodermis. Mesoglea makes a jellyfish buoyant and contractile cells exert force against it. As a jellyfish bell contracts, water streams out from beneath the bell, and the jellyfish is propelled forward. Any fluid-filled cavity or cell mass that contractile cells can act against is known as a *hydrostatic* skeleton.

Cnidarians reproduce both asexually and sexually. Sexual reproduction yields a planula, a bilateral ciliated larva. The larva moves about, and then settles down and develops into a polyp. In jellyfish, medusae develop on and are released by specialized polyps. The life cycle of some other cnidarians does not include a medusa stage.

Many cnidarians live in colonies. The reef-building coral polyps are cemented side by side in a structure built up from their own calcium carbonate secretions. Photosynthetic dinoflagellates in their tissues provide sugars that supplement feeding activities.

> *Sponges have no symmetry, tissues, or organs; they are at the cellular level of construction.*
>
> *Cnidarians are the simplest living animals with cells that are structurally and functionally organized in tissue layers.*

Figure 16.6 *Animated!* Cnidarian body plans: (**a**) medusa and (**b**) polyp. The *Obelia* life cycle includes both free-swimming medusae (**c**) and polyps. Sea anemones (**d**) and reef-building corals (**e**) live only as polyps.

16.3 Flatworms—Introducing Organ Systems

Beyond the cnidarians are animals ranging from simple worms to humans. Most are bilateral; all have organs.

With this section, we begin our survey of protostomes, one of the two lineages of animals that develop from a three-layered embryo. Embryonic tissues—exoderm, endoderm, and mesoderm—develop into organs. An **organ** is two or more tissues that interact in a common task, such as maintaining fluid balance. One or more organs interact to maintain conditions in the internal environment. Protostomes include flatworms, annelids, mollusks, roundworms, and arthropods.

Flatworms, the simplest animals to have any organ systems, include turbellarians, flukes, and tapeworms. These cephalized, bilateral animals all have a flattened body with longitudinal and circular muscles. As you saw in Figure 16.3*a*, there is no body cavity; the area between their gut and body wall is filled with tissues. Their digestive system is a saclike branching gut with a pharynx, a muscular tube that sucks in food or lets out waste (Figure 16.7). Most are hermaphrodites with female *and* male gonads and a penis. During mating, each gives some sperm to the other. The turbellarians

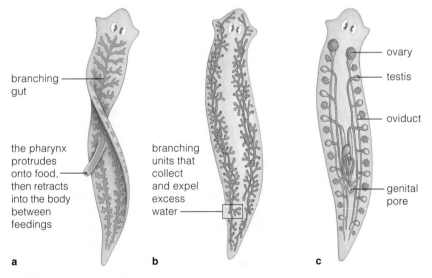

Figure 16.7 *Animated!* (**a**) Digestive, (**b**) water-regulating, and (**c**) reproductive systems of a planarian. Section 17.2 explains the flatworm mode of gas exchange. Section 25.3 includes a sketch of a flatworm nervous system.

called planarians also reproduce by transverse fission. The body splits in two crosswise and each piece grows the missing parts.

Planarians live in freshwater, but most of the other turbellarians are marine. All prey on or scavenge tiny animals. The flukes and tapeworms are parasites that

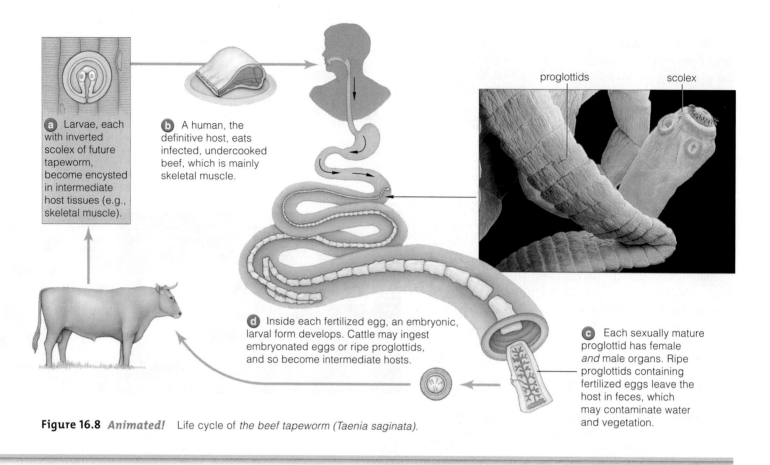

a. Larvae, each with inverted scolex of future tapeworm, become encysted in intermediate host tissues (e.g., skeletal muscle).

b. A human, the definitive host, eats infected, undercooked beef, which is mainly skeletal muscle.

proglottids   scolex

d. Inside each fertilized egg, an embryonic, larval form develops. Cattle may ingest embryonated eggs or ripe proglottids, and so become intermediate hosts.

c. Each sexually mature proglottid has female *and* male organs. Ripe proglottids containing fertilized eggs leave the host in feces, which may contaminate water and vegetation.

Figure 16.8 *Animated!* Life cycle of *the beef tapeworm (Taenia saginata)*.

live in or on one to four different hosts during their life cycle. Like most parasites, tapeworms usually do not kill their host, at least not until after they reproduce. A *definitive* host harbors the mature parasites. Immature stages live in one or more *intermediate* hosts. Humans get infected when they ingest tapeworm larvae in raw, undercooked, or poorly preserved pork, beef, or fish. Adult tapeworms live in the host's gut, where they are constantly bathed in nutrients. They hold on to the intestinal lining with a barbed scolex (Figure 16.8).

In tropical regions, blood flukes commonly cause *schistosomiasis*. Freshwater larvae bore into the skin of humans wading in the water. Inside this host, flukes mature, mate, and lay eggs. Immune responses to the fluke eggs cause rash, fever, and muscle aches and can damage the liver, spleen, and other organs. Fertilized eggs pass in the feces and hatch into ciliated larvae that enter freshwater snails. Flukes multiply asexually in the snails, then develop into swimming forms that can attack humans and start the cycle once again.

Flatworms are simple bilateral, cephalized, acoelomate animals with organ systems. They include free-living turbellarians, as well as parasitic flukes and tapeworms.

16.4 Annelids—Segments Galore

Annelids have so many segments, it makes you wonder about controls over how the animal body plan develops.

Have you ever seen an earthworm wriggling out of its flooded burrow after a heavy rain (Figure 16.9)? It is a familiar example of 15,000 or so species of bilateral, segmented animals called **annelids**. Other members of this group are leeches and the less well-known, mostly marine polychaetes (Figure 16.10).

Except in the leeches, most annelid segments have two or more chitin-reinforced bristles, or setae, on each side. Bristles extended into sediments provide the traction for crawling and burrowing. Earthworms, with only a few bristles per segment, are oligochaetes. Polychaete worms typically have numerous bristles per segment. (*Oligo–* means few; *poly–* means many).

Earthworms have a complete gut and they eat their way through the soil. They draw in dirt through the muscular pharynx, digest any organic bits, and excrete wastes through an anus. Their activities help to aerate the soil. Like other annelids, they have a hydrostatic skeleton. Muscles work against their fluid-cushioned

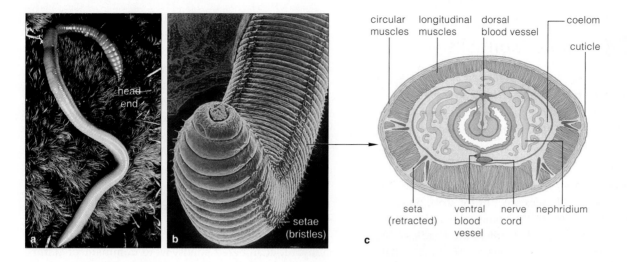

circular muscles · longitudinal muscles · dorsal blood vessel · coelom · cuticle

seta (retracted) · ventral blood vessel · nerve cord · nephridium

a · b · c

Figure 16.9 (**a**) A familiar annelid—one of the earthworms. (**b**) Externally, nearly all of the body segments of this bilateral worm look alike. Inside are specialized regions along the main body axis. (**c**) Cross-section through one segment. An outer cuticle surrounds two muscle layers, which surround internal organs. A nerve cord runs along the lower side of the body, connecting a cluster of nerves in each segment. Blood vessels running the length of the body supply small vessels that service each segment. Each segment also has an excretory organ or nephridium.

coelomic chambers. Inner and outer layers of muscles contract and relax in ways that shorten and fatten the segments, one after the other. The bristles alternately plunge into soil to gain traction, then retract, moving the worm forward.

In oligochaetes and polychaetes, heartlike pumps push blood through a continuous system of vessels in a closed circulatory system. The rudimentary **brain**, a cluster of nerve cell bodies, integrates sensory input and outputs calling for muscle responses. Also, a pair of **nerve cords** extend out from the brain and run the length of the body. They allow rapid communication among body regions and coordinated movements.

Annelids provide a glimpse into the evolutionary potential of a segmented body plan. Segmentation sets the stage for the evolution of regional specializations. Some segments can become modified and take on new functions, while others carry out their original tasks. In an earthworm, most segments look alike on the outside, but there are internal specializations. The leeches have specialized suckers at both ends. Polychaetes often have an elaborate head and many segments have fleshy-lobed parapods ("closely resembling feet").

Annelids are bilateral, coelomate, segmented worms that have complex organ systems. Some show the great degree of specialization possible with a segmented body plan.

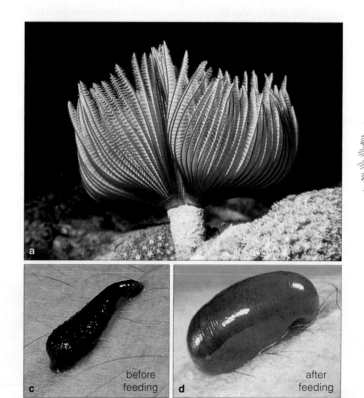

a · b · before feeding · c · after feeding · d

Figure 16.10 Annelid diversity. (**a**) A tube-dwelling polychaete with mucus-coated, featherlike structures at its head end. (**b**) As a group, the polychaetes have a dizzying variety of modifed segments. They are one of the most common animals along coasts. Many species are predators or scavengers; others dine on algae. Similarly, most species of leeches are scavengers or predators of small invertebrates, but *Hirundo medicinalis* has a taste for mammalian blood (**c**,**d**). It is used in medicine to draw off pooled blood and to slow clotting.

16.5 The Evolutionarily Pliable Mollusks

From our explorations in gardens and books, most of us would have no trouble recognizing a land snail when we see one. Yet few know much about its relatives in one of the largest of all animal groups, the mollusks. There are more than 110,000 known species.

Mollusks have a reduced coelom and a soft, fleshy, bilateral body (*molluscus* is Latin for soft). Body plans vary widely (Figure 16.11). Some have a distinct head, with eyes and a brain, but others are headless. Many have a shell or a reduced version of one. All have a mantle, which drapes like a skirt over the main body mass. This fleshy covering is unique to mollusks.

Mollusks range from tiny snails in treetops to huge predators of the open oceans. Most are gastropods, or "belly foots." The name refers to the underside of their body mass, which many use like a giant foot to glide across surfaces. Snails, slugs, and nudibranchs are the familiar members of this spectacularly diverse group.

Maybe it was their fleshy, soft bodies, so forgiving of chance evolutionary alterations, that gave mollusks the potential to radiate in so many ways. Let's explore this idea by focusing on a few representatives.

HIDING OUT, OR NOT

If you were small, soft of body, and *tasty*, a shell might discourage predators. Scallops, oysters, and mussels are bivalves. Members of this molluscan group have a hinged two-part shell that can snap shut when there is a threat. Scallops can also "swim" away by clapping the two shell halves together. Water forced out propels them backward, ideally far enough to get away from a predator that would like to make a meal of them.

If you were small, soft of body, and *toxic*, a shell would not be needed if there were a way to advertise your toxicity to predators. Some nudibranchs do just that. Gland cells of some species secrete sulfuric acid and other noxious substances. Other species graze on cnidarians, then incorporate nematocysts in their own tissues. They advertise the toxicity with vivid warning colors. Any predators that nibble at a colorful Spanish shawl nudibranch get stung by the stored nematocysts and quickly learn to avoid the species (Figure 16.11*c*).

ON THE CEPHALOPOD NEED FOR SPEED

About 500 million years ago cephalopods, with their buoyant, chambered shells, were dominant predators in Ordovician seas. Their living descendants include the fastest invertebrates (squids), the largest (the giant squid), and the smartest (octopuses) (Figure 16.11*d*).

In all but one existing descendant of thousands of ancestral species, the shell is reduced or entirely gone. Fossils show that disappearance of the shell coincided with an adaptive radiation of fishes that preyed upon cephalopods or competed with them for prey. In what might have been a long-term battle of speed and wits, most cephalopods lost their shell and became highly streamlined and—as invertebrates go—smart. Of all mollusks, cephalopods have the largest brain relative

Figure 16.11 A medley of mollusks. (**a**) Body plan of a snail, one of the gastropods. (**b**) Scallop, a bivalve, with light sensitive "eyes" (blue dots) fringing its shell. (**c**) Spanish shawl nudibranchs (*Flabellina iodinea*) store nematocysts in their red respiratory organs. (**d**) An octopus. Like other cephalopods, it has a complex nervous system.

to their body size, and demonstrate the most complex and varied behavior.

For cephalopods, *jet propulsion* became the name of the game. Each forces a jet of water from the mantle cavity through the funnel-shaped siphon. Water is drawn into the cavity as mantle muscles relax. Then muscles contract while the mantle's free edge closes over the head. These contractions shoot a jet of water out through the siphon. By guiding siphon action, the brain controls the direction of escape or pursuit.

Cephalopods are the only mollusks with a closed circulatory system. Blood remains inside vessels, as it is pumped over gills and to and from body tissues by three hearts. This system delivers plenty of oxygen to the highly active tissues and supports the metabolic activities that fuel bursts of speed.

Mollusks are bilateral, soft-bodied, coelomate animals that differ tremendously in body size, body details, and life-styles. They include the largest, fastest, and smartest invertebrates.

16.6 Amazingly Abundant Roundworms

Roundworms are among the most abundant living animals. Sediments in shallow water often hold a million per square meter; a cupful of topsoil teems with them.

Roundworms, more formally known as nematodes, are more closely related to the insects than they are to other worms. At least 12,000 species are known. Most are microscopic or only millimeters long. The body is cylindrical and bilateral. There is a complete gut and a false coelom, which is filled by reproductive organs.

Roundworms have a flexible cuticle, which is molted repeatedly as the animal grows. **Molting** is the name for a periodic casting off of an outer covering.

Most roundworms are free-living species that are decomposers. Many play vital roles in the cycling of nutrients through ecosystems. *Caenorhabditis elegans* is a favorite experimental organism among biologists. It has the same tissue types as more complex animals, but it is tiny and transparent Researchers can monitor the developmental fate of each of its 950 or so somatic cells. Also, this worm has a very small genome, and hermaphroditic forms can be self-bred to yield worms that are homozygous for specific desired traits.

Many soil-dwelling roundworms infect plant roots and are agricultural pests. Others parasitize animals including humans, dogs, and insects. More than one billion people are infected by *Ascaris lumbricoides*, an extremely large roundworm (Figure 16.12*a*). Infection is most common in Latin America and Asia.

Pigs and game animals can carry *Trichinella spiralis*. Adults of this parasite live attached to the intestinal lining. Their eggs develop into juveniles, which move through blood vessels to muscles, where they become encysted (Figure 16.12*b*). *Trichinosis*, the disease that results from infection, can be fatal. It is extremely rare in the United States, but common worldwide.

Repeated infections by *Wuchereria bancrofti* causes *elephantiasis*, a severe case of edema. Worms obstruct flow through lymph nodes, so fluid backs up in tissue spaces and enlarges lower body parts (Figure 16.12*c*). Mosquitoes are intermediate hosts for this parasite.

Roundworms are mostly tiny worms that have a false coelom. They shed their cuticle as they grow. Many are abundant in soil and some are plant pests. Others live as parasites of animals, including humans.

Figure 16.12 (**a**) Living roundworms (*Ascaris lumbricoides*). Worms can block ducts that empty into the intestinal tract, or the intestine itself. Larvae can migrate into and damage the lungs. (**b**) Juveniles of *Trichinella spiralis* in the muscle tissue of a host animal. (**c**) A case of elephantiasis. This type of severe edema results from an infection by the parasitic roundworm *Wuchereria bancrofti*.

16.7 Arthropods—The Most Successful Animals

Evolutionarily speaking, "success" means having the most offspring, species, and habitats; fending off threats and competition efficiently; and having the capacity to exploit the greatest amounts and kinds of food. Arthropods own these features.

Researchers have identified more than a million kinds of arthropods, mostly insects, and they are discovering new ones weekly. Of five lineages, trilobites are extinct. Chelicerates (spiders and mites), crustaceans, insects, and myriopods (centipedes and millipedes) survive.

KEYS TO SUCCESS

The success of arthropods—insects especially—can be traced to six key features: a hard exoskeleton, jointed appendages, fused and modified segments, specialized respiratory structures, efficient sensory systems, and often a division of labor in the life cycle.

In the next section, we will survey the diversity of arthropod groups. But first, let us consider the six key adaptations that contributed to the great success of the lineage as a whole.

A hardened exoskeleton. Arthropod cuticle acts as an external skeleton, called an **exoskeleton**. Exoskeletons might have evolved as defenses against predation. They took on added functions when some of the arthropods invaded the land. They support a body no longer made buoyant by water, and their waxes ⌐ water loss. Arthopods' muscles attach to and act against the exoskeleton.

An arthropod's cuticle would limit growth if it were not periodically molted. Hormones control molting. During certain stages of the life cycle, hormones induce a soft new cuticle to form under the old one, which is then shed (Figure 13). Body size is expanded by the uptake f air or water and by increased rapid cell divisions before the new cuticle hardens. Molting of the cuticle is a trait that the arthropods share with roundworms.

Jointed appendages. If cuticle was rigid everywhere, it would prevent all movement. Arthropod cuticle, however, thins in places where it spans *joints*, or regions where two body parts abut. Muscle contraction makes cuticle at joints bend, and moves body parts. The jointed exoskeleton was a key innovation that allowed the evolution of specialized appendages, including wings, antennae, and diverse types of legs. ("Arthropod" means jointed leg.)

Figure 16.14 Two marine arthropods. (**a**) Fossil of a trilobite, a now extinct group, and (**b**) a horseshoe crab (*Limulus*). Horseshoe crabs are the closest living relatives of trilobites. They have survived basically unchanged for more than 250 million years, but are now threatened by human activities.

Specialized segments and fused-together segments. The first arthropods were segmented, and their segments were all more or less similar. In most existing lineages, some of segments became fused together or modified in specialized ways. For example, compare the highly similar body segments of a centipede with the winged and otherwise modified segments of a butterfly.

Respiratory structures. Many freshwater and marine arthropods use gills as respiratory organs. Among the insects and other arthropods on land, air-conducting tubes called tracheae evolved. The tubes start at pores on the body surface. Their extensive branching ends in fluid near tissues. Diffusion of oxygen into this fluid supports the high rates of aerobic respiration essential for flight and other energy-demanding activities.

Specialized sensory structures. For many arthropods, diverse sensory structures also contributed to success. For instance, some evolved into eyes that can sample a big part of the visual field and provide the brain with information from many directions.

Specialized developmental stages. Many arthropods divide the task of surviving and reproducing among different stages of development. The division of labor is most common among insects. For some species, the new individual is a *juvenile*—a miniaturized version of the adult that simply grows in size until it is sexually mature. Other species undergo **metamorphosis**. Their body changes—often in dramatic ways—between the

Figure 16.13 Molting by a red-orange centipede wriggling out of its old exoskeleton.

Figure 16.15 Arachnids. (**a**) Tarantula with one fearless female. Like all spiders, it helps keep insects in check. Its bite has not harmed humans. (**b**) Brown recluse, with a violin-shaped mark on its forebody. Its bite can be severe to fatal. (**c**) Female black widow, with a red hourglass-marking on the underside of a shiny black abdomen. The males are smaller and don't bite. (**d**) Dust mite. Dust mite feces and corpses cause allergies in susceptible people.

embryo and adult forms. Hormones help regulate this tissue reorganization and remodeling.

Eating to sustain rapid growth is the specialty of the immature stages; adults specialize in reproduction and in dispersal. (Think of how flightless caterpillars devour leaves, then turn into butterflies that fly away, mate, and deposit eggs.) The different developmental stages are adapted to different conditions, especially seasonal shifts in food sources, water supplies, and availability of potential mates.

SPIDERS AND THEIR RELATIVES

Like the extinct trilobites, the chelicerates originated in shallow seas during the early Paleozoic. The only surviving marine species are a few mites, sea spiders, and horseshoe crabs (Figures 16.14). Familiar kinds on land—spiders, scorpions, ticks, and chigger mites—are all arachnids (Figure 16.15). And we might say this: Never have so many been loved by so few.

Spiders and scorpions are predators that sting, bite, or subdue prey with venom. Spiders have appendages that spin silk threads for webs and egg cases. Most webs are netlike. The bola spider spins a thread with a ball of sticky material at the end, releases a scent that mimics a female moth's sex attractant, then uses a leg to swing the ball at male moths lured by the scent!

All spiders of the genus *Loxosceles* are poisonous to humans and other mammals, but we seldom see most of them. One, the brown recluse (*L. reclusa*), often hides in warm places in houses, especially near refrigerator motors and clothes dryers, and in closets. Its body has a distinctive violin-shaped marking (Figure 16.15*b*). Its bite ulcerates the skin and the wound is slow to heal.

About five people die each year in the United States as the result of black widow spider (*Latrodectus*) bites (Figure 16.15*c*). This spider's venom contains a toxin that interferes with nervous control of muscle tissue. A bite can paralyze muscles involved in breathing, so medical attention should be prompt.

Of all arachnids, mites are among the smallest, most diverse, and most widely distributed. Of an estimated 500,000 mites, the few parasites—ticks especially—are harmful; many transmit pathogens that cause serious diseases in humans. As you learned earlier, one tick is the vector for *Borrelia burgdorferi*, the bacterium that causes Lyme disease (Figure 14.12). Others transmit the agents of Rocky Mountain spotted fever, typhus, tularemia, encephalitis, and other diseases. Ticks do not fly or jump. They cling to grasses or shrubs, then crawl onto animals brushing past. Look for them after a walk in the wild. Finally, house dust mites are tiny but their remains and feces can cause allergic reactions and trigger asthma in some people (Figure 16.15*d*).

CRUSTACEANS

Shrimps, lobsters, crabs, barnacles, pillbugs, and other crustaceans got their name because they have a tough yet flexible "crust"—a cuticle hardened with calcium. The vast majority live in marine habitats, where they are so abundant they are sometimes called the insects of the seas. Humans harvest many edible types.

All crustaceans have two pairs of antennae that they use to feel and smell. There are many body segments, more than sixty in some groups. Usually, the cuticle extends back from the head as a shield above some or all segments. The head has jawlike appendages, and appendages used to handle food. "Decapods," such as the lobsters, shrimps, and crabs, have ten walking legs (five pairs). Figure 16.16 shows a typical crab life cycle.

Barnacles are the only arthropods that secrete an external "shell." This structure consists of plates that can close up tightly to protect them from predators, drying winds, and surf. Adults glue themselves by their antennae to rocks, piers, and even whales, then filter food from the seawater using their feet (Figure

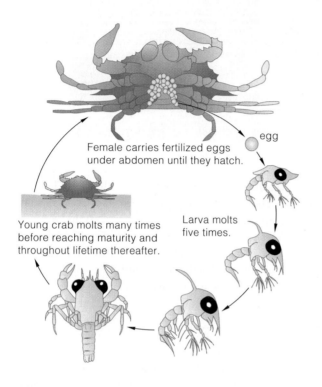

Female carries fertilized eggs under abdomen until they hatch.

egg

Young crab molts many times before reaching maturity and throughout lifetime thereafter.

Larva molts five times.

Figure 16.16 *Animated!* Life cycle of a crab.

16.17*b*). Given that a barnacle is glued in place at its head end, you might think sex could be challenging. But barnacles tend to cluster in groups and most are hermaphrodites. Each extends its penis—which can be several times the body length—out to its neighbors.

Copepods are less than two millimeters long. These crustaceans are the most numerous animals in aquatic habitats and are abundant on land (Figure 16.17*c*). A number parasitize different invertebrates and fishes. Most ingest phytoplankton. Copepods are themselves eaten by other invertebrates, fishes, and whales.

A LOOK AT INSECT DIVERSITY

All insects share the six adaptations listed earlier. Their segmented body is divided into a head, an abdomen, and a thorax with three pairs of legs and usually two pairs of wings. They have paired mouthparts for biting, chewing, and other tasks, and one pair of antennae. The only winged invertebrates are in this group.

An insect body has a foregut, a midgut where food is digested, and a hindgut where water is reabsorbed. A system of small tubes disposes of waste materials. After proteins are digested, the residues diffuse from blood into the tubes. There, enzymes convert them to uric acid, which is eliminated with feces. This system lets land-dwelling insects get rid of potentially toxic wastes without losing precious water.

Figure 16.18 shows a few of the more than 800,000 species. If numbers and distribution are the measure, the most successful ones are small and reproduce often. By one estimate, if all the offspring of one female fly survived and reproduced for six more generations, she would have more than 5 trillion descendants!

Figure 16.17 (**a**) Lobster in the Caribbean Sea, off the coast of Belize. (**b**) Goose barnacles. Adults cement themselves to one spot. You might mistake them for mollusks, but when they open their hinged shell to filter-feed, you see jointed appendages—the hallmark feature of arthropods. (**c**) One of the copepods. Different kinds are free-living filter feeders, predators, and parasites.

The most successful insects are winged. They move among food sources that are too widely scattered to be exploited by other kinds of animals. The capacity for flight contributed enormously to their success on land.

Insect life cycles often include stages that permit exploitation of different resources at different times. In some groups, such as the springtails, eggs hatch into young that look like minature adults. In others, such as butterflies, a larva that is specialized for feeding turns into a pupa, which undergoes metamorphosis. This is resumption of growth and the transformation of the post-embryonic stages into the sexually mature, adult form. Tissue reorganization and a total remodeling of many body parts occur during this process.

The very features that contribute to insect success also make them our most aggressive competitors for resources. Insects destroy much of what we value, such as crops, stored food, wool, paper, and timber. Some transmit pathogens that make us sick.

Yet bees and other insects are necessary to pollinate flowering plants, including highly valued crop plants. And many "good" insects attack or parasitize the ones we would rather do without.

Figure 16.18 Examples of insects. (**a**) Newly hatched stinkbugs. (**b**) In the center of this group of honeybee workers, one honeybee is signaling the position of a food source as described in Section 32.2.

With more than 300,000 species of beetles, Coleoptera is the largest order of animals. Ladybird beetles (**c**) and a staghorn beetle from New Guinea (**d**) are two examples.

(**e**) A swallowtail butterfly and (**f**) a luna moth. (**g**) A damselfly. (**h**) Mediterranean fruit fly. Its larvae destroy citrus and other crops.

A duck louse (**i**) eats bits of feathers and skin. A flea (**j**) has legs adapted for jumping onto and off of its animal hosts. Fleas transmit diseases, including bubonic plague.

Arachnids, crustaceans, insects, and the millipedes and centipedes are living arthropods. As a group, they show enormously diverse modifications on a basic body plan.

All arthropods have an exoskeleton of hardened cuticle, jointed appendages, fused and modified body segments, respiratory structures, specialized sensory organs, and often a division of labor during the life cycle.

In terms of distribution, number of species, population sizes, competitive adaptations, and exploitation of diverse foods, insects are the most successful animals.

Figure 16.19 A few echinoderms. (**a**) Sea star, with a close-up of its little tube feet. (**b**) Sea urchins, which can move about on spines and a few tube feet. (**c**) Brittle star. Its slender arms (rays) make rapid, snakelike movements. (**d**) Sea cucumber, with rows of tube feet along its elongated body.

LINKS TO
SECTIONS
13.4, 13.5, 13.10

16.8 The Puzzling Echinoderms

Reflect, for a moment, on Figure 16.1. Echinoderms and chordates show similar patterns of early development, which reflect their shared ancestry. Together, they are classified as deuterostomes. But unlike chordates, the adult echinoderms have radial features.

Echinoderm means spiny skinned. The body wall of these coelomate animals has spines, spicules, or plates of calcium carbonate. Included in this marine group are sea stars, brittle stars, sea urchins, sea cucumbers, and sea lilies (crinoids). Nearly all echinoderms have an internal skeleton of calcium carbonate and other substances. Adults have a generally radial body with some bilateral features, but the larval body is bilateral. Along with DNA data, this suggests that echinoderms arose from bilateral ancestors.

Adult echinoderms have no brain. A decentralized nervous system receives and responds to signals about touch, light, and chemicals. For instance, any arm of a sea star that touches the shell of a mussel can become the leader, directing the rest of the body to move in a direction suitable for capturing prey.

Echinoderms have tube feet. These are fluid-filled, muscular structures that have suckerlike adhesive disks (Figure 16.19). Sea stars use their feet for burrowing, walking, clinging to rocks, or gripping a clam or snail at feeding time. Eager sea stars swallow prey whole. Others push part of their stomach outside the mouth and start digesting prey before swallowing. Sea stars eliminate any coarse, undigested residues through the mouth. Their have an anus, but it is too small to expel chunks of shell.

Echinoderm sexes are usually separate. Eggs and sperm are released into the water. After fertilization, a bilateral free-swimming larva develops. The larva will grow and develop into the mostly radial adult form.

The radial echinoderms have bilateral larvae and probably had bilateral ancestors. Though brainless and spiny-skinned, they belong to the same lineage as the chordates.

INVERTEBRATES

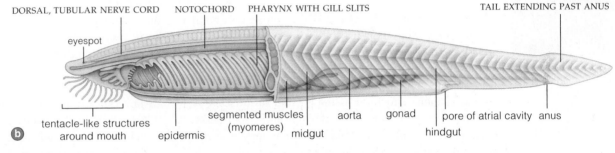

Figure 16.20
Animated! (**a**) The chordate family tree. All groups are multicelled, bilateral, coelomate animals. (**b**) Photo and body plan of a lancelet. This small filter-feeder shows the four defining features of chordates.

In the tree (a): tunicates, lancelets, hagfishes, lampreys, cartilaginous fishes, ray-finned fishes, lobe-finned fishes, lungfishes, amphibians, "reptiles," birds, mammals. Nodes: ancestral chordates, craniates, vertebrates, jawed vertebrates, lungs or swim bladder, tetrapods, amniotes.

In the body plan (b): DORSAL, TUBULAR NERVE CORD, NOTOCHORD, PHARYNX WITH GILL SLITS, TAIL EXTENDING PAST ANUS, eyespot, tentacle-like structures around mouth, epidermis, segmented muscles (myomeres), midgut, aorta, gonad, hindgut, pore of atrial cavity, anus.

16.9 Evolutionary Trends Among Vertebrates

We start our look at vertebrates and their chordate origins with a reminder: Each kind of organism—including humans—is a mosaic of traits. Many traits are conserved from remote ancestors, and others are unique to its branch on the family tree.

Figure 16.20*a* is an evolutionary tree for the **chordates**. These coelomate, bilateral animals have four unique features that appear in their embryos and often persist into adulthood:

First, a notochord—a strong but flexible rod—runs lengthwise through the body and helps to support it. *Second*, a nerve cord parallels the notochord and gut; its anterior end develops into a brain. In contrast to the ventral nerve cords of many invertebrates, this one is dorsal. *Third*, embryos have gill slits in the wall of their pharynx, a type of tube that functions in feeding, respiration, or both. *Fourth*, a tail develops in embryos and extends past the anus.

The two groups of invertebrate chordates, tunicates and lancelets, filter food from currents of water that pass through gill slits in their pharynx. In lancelets, the characteristic chordate traits are visible in the adult, as shown in Figure 16.20*b*.

In the tunicates, the notochord, tail, and nerve cord occur only in a tadpole-shaped larval stage. The adults retain only the pharynx with gill slits (Figure 16.21). Water flows into an adult's saclike body through one siphon, is filtered through gill slits in the pharynx, then exits the body by way of the other siphon. Tunicates known as sea squirts are commonly found attached to piers and boats.

EARLY CRANIATES

Fishes, amphibians, reptiles, birds, and mammals are **craniates**. The distinguishing feature of this group is a cranium, a chamber of cartilage or bone that encloses the brain. The first craniates arose before 530 million years ago. They had fins, a notochord, and segmented muscles. These active swimmers resembled larvae of lampreys, one group of modern jawless fishes.

Figure 16.21 *Above*, A free-swimming tunicate larva. It will attach by its head to some surface and then take on the adult form (*below*). Like lancelets, both stages are filter feeders.

LINKS TO
SECTIONS
13.4, 13.5, 13.10

The earliest jawless fishes, including ostracoderms, had armor-like plates composed of bone and dentin, a hard tissue still found in teeth. The armor plating may have protected them against the pincers of giant sea scorpions. But it was far less effective against **jaws**, the hinged, bony feeding structures that evolved in other lineages (Figure 16.22). When jawed craniates began their radiation, ostracoderms vanished.

Placoderms were among the earliest jawed fishes. Armor plates protected their head, but not much else. Before, chordate feeding strategies had been limited to filtering, sucking, and rasping food. Some placoderms evolved into huge predators. The changes sparked the evolution of offensive and defensive body structures that has continued to the present.

THE KEY INNOVATIONS

Vertebrates profoundly changed the course of animal evolution. Unlike their nearly brainless, filter-feeding relatives, they have an internal skeleton composed of mineral-hardened secretions. Segments made of bone or cartilage and called **vertebrae** (singular, vertebra) replaced their notochord as the partner of muscles. A vertebral column was more flexible and less clunky than external armor plates. The union of vertebrae and muscles promoted maneuverability, and allowed more forceful contractions. The outcome? *Agile, fast-moving fishes began their dominance of the seas.*

Jaws were part of a related trend. They had evolved from the first of a series of structural elements that had supported pharyngeal gill slits (Figure 16.22b). Their

appearance started a race for better ways to overcome and escape them. For example, selection now favored better vision for detecting predators and bigger brains to guide escapes. *A trend toward complex sensory organs and nervous systems started among the ancient fishes and continued among vertebrates on land.*

Another trend began with the evolution of paired fins. **Fins** are appendages that help propel, stabilize, and guide the fish body through water. Some groups had fleshy ventral fins with internal skeletal supports that became the forerunners of limbs. *Paired, fleshy fins were a starting point for all legs, arms, and wings that evolved among amphibians, reptiles, birds, and mammals.*

A change in respiration started another trend. In lancelets, oxygen and carbon dioxide simply diffuse across the skin. In most early vertebrates, gills evolved. **Gills** are respiratory organs with moist, thin folds and adjacent blood vessels. They have a large surface area that exchanges gases between the environment and the body. Oxygen diffuses *from* water in a fish mouth into gills, as carbon dioxide diffuses out, *into* the water.

Among the larger, active fishes, gills became more efficient. But gills do not work out of water; their thin surfaces stick together without a flow of water to keep them moist. In the fishes that were ancestral to land vertebrates, two small outpouchings formed on the gut wall. The pouches evolved into **lungs**—internally moistened sacs for gas exchange. In a related trend, modifications to the heart and blood vessels enhanced the pumping of oxygen and carbon dioxide through their body. *The ancestors of land vertebrates relied more on paired lungs than on gills—and more efficient circulatory systems accompanied the evolution of lungs.*

Placoderms and many of the jawless fishes became extinct during the Carboniferous. Through adaptation, key innovations, and plain luck, some descendants of some jawless and jawed groups made it to the present.

supporting structures

a

b — gill slit

c — jaw

d

Figure 16.22 (**a**) A whale shark, with a mouth used in capturing zooplankton, squids, and sometimes fishes, including tuna. (**b**) In early jawless fishes, supporting elements reinforced a series of gill slits on both sides of the body. (**c**) In placoderms, the first elements in the series had become modified and functioned as jaws. (**d**) The human at left gives an idea of the size of one ancient jawed fish, the placoderm *Dunkleosteus*.

Chordate embryos alone have this combination of traits: a notochord, a tubular dorsal nerve cord, a pharynx with fine slits in its wall, and a tail extending past the anus. These traits are legacies from a shared ancestor, an early invertebrate chordate.

Tunicates and lancelets are living invertebrate chordates. They are aquatic filter feeders that strain food from water that flows through gill slits in the wall of their pharynx.

Craniates are chordates with a brain chamber of cartilage or bone. Vertebrates are craniates with a vertebral column (cartilaginous or bony segments).

Jaws, paired fins, and later, lungs, were key innovations that led to the adaptive radiation of vertebrates.

16.10 Major Groups of Jawed Fishes

Unless you study underwater life, you may not know that fishes are the world's dominant vertebrates. Their numbers exceed those of all other vertebrate groups combined. They are also the most diverse; they include about half of all vertebrates.

There are two major groups of living fishes, one with a skeleton of cartilage, the other with a skeleton of bone. Their form and behavior hint at the challenges of life in water. Water is 800 times denser than air and resists motion through it. In response to this constraint, many fish species became streamlined for pursuit or escape. The predatory sharks are an example (Figure 16.23*a*). A long, trim body reduces friction; their tail muscles provide the propulsive force for rapid movement. By contrast, a manta ray has a flattened body and glides through the water by flapping fins attached along the length of the body (Figure 16.23*b*).

Sharks and rays are the most well-known kinds of **cartilaginous fishes**. All have a skeleton of cartilage, and five to seven pairs of gill slits. Sharks shed and replace their teeth on an ongoing basis. These teeth are modified scales that are most often used to grab and tear chunks of flesh. Human surfers and swimmers can resemble typical prey, and rare attacks by a few species have given all of the sharks a bad reputation. Worldwide, shark attacks kill about 25 people a year.

Bony fishes have a skeleton made of bone, which is heavier than cartilage. A gas-filled swim bladder or lung helps these fish maintain buoyancy. By the early Devonian—about 400 million years ago—three main lineages had emerged: ray-finned fishes, lung fishes, and lobe-finned fishes.

In **ray-finned fishes** flexible fins are supported by rays that are derived from skin rather than cartilage or bones (Figure 16.24). With more than 21,000 species, these fishes make up nearly half of all the vertebrates. Ray-fined fishes exhibit a diverse variety of colors and body shapes that adapt them to a wide array of aquatic habitats. Elongated eels squeeze into crevices, flattened flounders lie on sandy bottoms, and box-shaped fishes can wedge themselves among corals.

Most of the fish we eat are ray-finned fish, such as salmon, sardines, sea bass, swordfish, snapper, trout, tuna, halibut, carp, and cod. Many species have been over-fished. Fish have been harvested faster than they can reproduce, so stocks have declined. For example, it is estimated that the population of Atlantic cod (*Gadus morhua*) off the coast of Nova Scotia has declined by 96 percent or more since the 1850s.

Figure 16.23 Cartilaginous fishes. (**a**) Galápagos shark and (**b**) a manta ray. Two projections on the ray's head unfurl and waft plankton to a mouth between them. Notice the pairs of gill slits.

Figure 16.24 Ray-finned bony fishes. (**a**) A perch showing its ray-supported fins. (**b**) Coral grouper. (**c**) Flying fish. Like bird wings, the modified pectoral fins help these fishes fly out of the water when they are being pursued by predators.

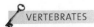

LINKS TO
SECTIONS
13.4,13.5, 13.10

a

b

c

inside lobed fins, bony or cartilaginous
structures undergoing modification

d

e   limb bones of an early tetrapod

Coelacanths are the only living lobe-finned fishes. Their ventral fins include fleshy body extensions with bony elements inside them (Figure 16.25a). Lungfishes have similar fins (Figure 16.25b). As the name suggests, they have gills *and* one or two lunglike sacs. The fish inflates these sacs by gulping air. Like the air in our lungs, air in the sacs exchanges gases with the blood. Some lungfishes are totally dependent on this type of gas exchange; they drown if kept underwater!

Morphological and genetic evidence suggests that lungfishes are probably the sister group to **tetrapods**: amphibians, reptiles, birds, and mammals. Both have a similar arrangement of skull bones and the same type of circulation, with a separate circuit that carries blood to and from lungs. The "limbs" of lobe-finned fishes, lungfishes, and tetrapods are about the same in their size, position, and structure (Figure 16.25d,e).

Limb joints, digits, and other traits of early tetrapod fossils indicate that walking began in the water, rather than on land. By the late Devonian, some species were using their limbs for crawling, as well as swimming.

Ray-finned fishes are now the most diverse and abundant vertebrates. The lobe-finned fishes have fleshy ventral fins reinforced with skeletal parts. The lungfishes have simple lunglike sacs that supplement respiration by gills.

Walking probably originated in water during the Devonian, among the aquatic forerunners of land tetrapods.

16.11 Early Amphibious Tetrapods

Even a small genetic change could have transformed lobed fins into limbs. Remember those enhancers that control gene transcription? One of them influences how digits form on limb bones. Even a single mutation in one master gene can lead to a big change in morphology.

Amphibians are vertebrates with four legs (or a four-legged aquatic ancestor); they were the first tetrapods on land. Their body plan and reproductive mode are somewhere between fishes and reptiles.

Figure 16.25 (a) Living coelacanth (*Latimeria*), of the only lineage of existing lobe-finned fishes. (b) Australian lungfish. (c) Painting of Devonian tetrapods. *Acanthostega*, submerged, and *Ichthyostega*. Their skull, caudal tail, and fins were fishlike. Unlike fishes, they both had a short neck and four limbs with digits. Compare Figure 13.10. (d,e) Proposed evolution of skeletal elements inside fins into the limb bones of early amphibians.

Figure 16.26 Some familiar amphibians. (a) Red-spotted salamander (*Notophthalamus*). (b) A frog jumping and displaying its tetrapod heritage. (c) The American toad.

The earliest tetrapods probably spent most or all of their time in water. But by the end of the Devonian, some types were semiaquatic and spending time on land. Life in this new habitat was both dangerous and promising. Temperatures shifted much more on land than in the water, air did not support the body as well as water did, and water was not always plentiful. But air is richer in oxygen. Amphibian lungs continued to be modified in ways that enhanced oxygen uptake. The heart, newly divided into three chambers, pumped more oxygen-rich blood to cells.

New sensory information also challenged the early amphibians. Swamp forests supported vast numbers of insects and other invertebrate prey. On land, vision, hearing, and balance became the most advantageous senses. Brain regions concerned with interpreting and responding to sensory input expanded.

All frogs, toads, and salamanders alive today are descended from those first amphibians (Figure 16.26). Most still live in water or in moist habitats. Even when species use gills or lungs, they also exchange oxygen and carbon dioxide across their thin skin. Respiratory surfaces must be kept moist to function, however, and skin dries easily in air.

Some existing amphibians spend their entire life in water. Others release eggs in water or produce aquatic larvae. The species adapted to dry habitats either lay eggs in moist places or, in a few cases, protect embryos during development inside the moist adult body.

Many amphibians are now in trouble (Figure 16.27) Six frog, four toad, and eleven salamander species are now listed as threatened or endangered in the United

Figure 16.27 Frog deformity. A parasitic fluke (*Ribeiroia*) burrows into tadpole limb buds and alters cells in their embryos. The infected tadpoles grow extra legs or no legs at all. Fluke cysts have been found in weirdly legged frogs and salamanders across North America.

States and Puerto Rico. One, the California red-legged frog, is the largest frog native to the western United States. This was the species that inspired Mark Twain's famous short story "The Celebrated Jumping Frog of Calaveras County."

Researchers correlate these declines with shrinking habitats. Developers and farmers commonly fill in low-lying ground that once filled up seasonally and provided pools of standing water. Nearly all species of amphibians cannot breed unless they can deposit their eggs in water, in which the larvae will develop.

In body form and behavior, amphibians show resemblances to aquatic tetrapods and reptiles. Most species have not fully escaped dependency on aquatic or moist habitats to complete their life cycle.

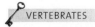

LINKS TO
SECTIONS
13.4, 13.9

Figure 16.28 (**a**) Jurassic scene, with *Archaeopteryx* gliding in the foreground. Left to right: Plant eaters *Stegosaurus* and *Apatosaurus*, the carnivore *Saurophaganax* ("king of reptile eaters"), and *Camptosaurus*, a horny-beaked plant eater. A climbing mammal looks out from the shrubbery. *Apatosaurus* was one of the largest land animals ever. This tiny-brained plant eater grew as long as 27 meters (90 feet) and weighed 33 to 38 tons. (**b**) Galápagos tortoise. (**c**) Spectacled caiman, a crocodilian. Its peglike upper and lower teeth do not match up. (**d**) Two hognosed snakes emerge from leathery amniote eggs. (**e**) *Sphaerodactylus ariasae*, the smallest lizard known.

16.12 The Rise of Amniotes

Amniotes were the first vertebrates to adapt to dry land habitats, through modifications in their organ systems and eggs. The eggs have four membranes that conserve water and support the embryo's development.

THE "REPTILES"

Late in the Carboniferous, some amphibians gave rise to **amniotes**, which were better adapted to life on dry land. The membranes in an amniote egg help conserve water and protect a developing embryo. Amniote skin is rich in the protein keratin, which helps waterproof it. A pair of well-developed kidneys help to conserve water, and fertilization is always internal.

Synapsids and sauropsids are two major groups of amniotes. Existing mammals and extinct mammal-like groups are synapsids. Sauropsids are "**reptiles**," and birds. We put "reptile" in quotes because this is not a formal taxonomic group. It is not monophyletic; the name refers to several lineages that all share the basic amniote features but not the derived traits that define birds or mammals.

The first **dinosaurs** descended from Triassic reptiles. They were turkey-sized and possibly warm-blooded. Many sprinted about on two legs. Adaptive zones opened up for dinosaurs about 213 million years ago, when huge fragments from an asteroid or comet fell to Earth. Craters in France, Canada, and North Dakota mark where they hit. Nearly all of the animals that survived the impacts were small, had high rates of metabolism, and tolerated big temperature changes.

VERTEBRATES

yolk sac embryo amnion chorion allantois

hardened shell   albumin ("egg white")

a

b

Figure 16.29 (**a**) Sketch of a typical amniote egg. (**b**) A Laysan albatross with a wingspan of more than two meters weighs less than 10 kilograms (22 pounds). Its bones are lightweight and attach to powerful muscles.

Descendants of the surviving dinosaurs became the ruling reptiles—which endured for 140 million years. Some reached monstrous proportions (Figure 16.28*a*). Many dinosaurs died in a mass extinction at the end of the Jurassic, others during a pulse of extinctions in the Cretaceous. Some lineages recovered, and new lineages evolved in the forests, swamps, and open habitats.

The last of the reptiles traditionally called dinosaurs vanished 65 million years ago, with a mass extinction. According to the **K–T asteroid impact theory**, collision with an asteroid as big as Mount Everest caused this extinction. A crater wider than Connecticut, in what is now Mexico's Yucatán peninsula, supports the theory. So does iridium in sediments from this time. Iridium is an element common on asteroids, but rare on Earth.

Many reptiles survived the asteriod impact. Figure 16.28*b–e* shows some of their descendants. The turtles and tortoises have a shell attached to their skeleton. If threatened, many can pull their head and limbs inside. All lay eggs that develop outside the body.

Lizards are the most diverse of the living reptiles. The largest, the monitor lizard known as the komodo dragon, can grow up to ten feet long. Most lizards lay eggs that develop outside the body. Some species give birth to live young; although eggs develop inside the mother, they are not nourished by her tissues.

Snakes arose from ancestral lizards. All the modern snakes are carnivores. Some retain bony remnants of ancestral hindlimbs. Like lizards, most lay eggs, but some give birth to live young.

Crocodilians are the most intelligent of the reptiles. Crocodiles, alligators, and caimans all live in or near water and are predators with powerful jaws, a long snout, and sharp teeth. The crocodilians are the only reptiles with a four-chambered heart, like mammals and birds. They lay eggs, and like many of the birds,

they show parental behavior, such as guarding nests and assisting the hatchlings.

BIRDS

Birds are vertebrates with feathers. They originated when Mesozoic reptiles began an adaptive radiation. They diverged from a lineage of theropod dinosaurs, a type of small carnivore that ran about on two legs. Like existing reptiles, birds have a cloaca, an opening that serves in both excretion and reproduction. Birds make amniote eggs that are fertilized internally and develop outside the body (Figure 16.29*a*).

The evolution of bird flight involved more than feathers. It involved modification of the entire body, from internal bone structure to highly efficient modes of respiration and circulation. Elastic sacs that connect to lungs greatly enhance the uptake of oxygen. Like mammals, birds have a large, four-chambered heart. The heart pumps oxygen-rich blood to the lungs and to the rest of the body in separate circuits.

Each bird wing, a modified forelimb, consists of feathers and lightweight bones attached to powerful muscles. Air cavities in bones make them lightweight. The flight muscles attach to an enlarged breastbone (sternum) and to the upper limb bones attached to it. When the muscles contract, they produce a powerful downstroke (Figure 16.29*b*).

Modern birds differ in size, proportions, colors, and their capacity for flight. The tiny bee hummingbird weighs 1.6 grams (0.6 ounces). The ostrich, the largest living bird, can weigh 150 kilograms (330 pounds). It cannot fly, but it can sprint on powerful legs. Many of the birds show impressive capacities for learning and for social behavior. Bird song and other behaviors are topics in Chapter 32.

Figure 16.30 Representatives of the three mammalian groups: (**a**) Humans are placental mammals. (**b**) An egg-laying mammal, the platypus. (**c**) Koala, a pouched mammal.

The smallest and the largest living placental mammals: (**d**) Kitti's hognosed bat, at 1.5 grams; and (**e**) blue whale, up to 200 tons.

MAMMALS

How do we define **mammals**? They alone have hair, although they differ in its distribution, amount, and type. Only mammalian females can secrete milk from mammary glands. Jaws and teeth are also distinctive. Mammals have a single lower jawbone and their four types of upper and lower teeth match up. The reptiles have a hinged two-part jaw, all teeth are one type only, and the teeth do not match up.

Where did mammals come from? Remember, the extinct mammal-like reptiles and all extinct and living species of mammals are *synapsids*. The defining trait of this group is an opening in a particular location on each side of the skull that functions in the attachment and alignment of jaw muscles. The trait arose in the Carboniferous. Later on, in the Permian, the synapsids underwent adaptive radiations and became the major land vertebrates. At the close of the Permian, many groups died out in a mass extinction.

Fortunately for humans, the group called *therapsids* endured. Over time, limbs became positioned upright under their trunk. Their skeletal arrangement made it easier to walk erect, and brain regions that function in balance and spatial positioning increased in size.

The earliest groups of modern mammals evolved early in the Jurassic. These were mouselike forms that coexisted with dinosaurs. By 130 million years ago,

three lineages had evolved: **monotremes** (egg-laying mammals), **marsupials** (the pouched mammals), and **eutherians** (placental mammals). Figure 16.30 shows examples of each group.

Three species of monotremes survive and—with the exception of opossums—most pouched mammals are in Australia or New Zealand. In contrast, the placental mammals are widely distributed. What gave them the competitive edge? The placental mammals had higher metabolic rates, better body temperature control, and a more efficient method to nourish embryos. You will read more about this in later chapters. For now, note that the success of placental mammals was achieved at the expense of the other mammal lineages. When the placental mammals entered areas where monotremes or marsupials had evolved in their absence, placental mammals drove most nonplacental ones to extinction.

Amniotes, the first vertebrates to escape reproductive dependence on outside water, have water-conserving skin and kidneys, and amniote eggs. Synapsids (mammals and extinct mammal-like forms) are one major group. Sauropsids ("reptiles" and birds) are another.

Much of amniote history was a matter of luck, of species with particular traits being in the right or wrong places on a changing geologic stage at particular times.

Figure 16.31 A few primates. (**a**) A tarsier, small enough to fit on a human hand, is a climber and leaper. (**b**) A gibbon has arms and legs adapted to swinging between trees. (**c**) Male gorilla, a knuckle-walker. (**d**) Comparison of skeletal organization and stances. These images are not at the same scale. A monkey has long, thin limbs relative to its torso, and its long, thin digits are good at gripping, not at supporting the body's weight. Gorillas use their forelimbs to climb and to support body weight when walking on knuckles and two legs. Humans are two-legged walkers.

16.13 From Early Primates to Humans

So far, you've traveled 570 million years through time, from tiny flattened animals to craniates, then on to the jawed fishes with a backbone. You met some early tetrapods, then amniotes—reptiles, birds, and mammals. Now you are about to access what we know about an evolutionary road that led to modern humans.

Tarsiers, prosimians, and anthropoids are the living **primates**. Prosimians include lemurs and lorises. The anthropoids are monkeys, apes, and humans. Apes are gibbons, siamangs, orangutans, gorillas, chimpanzees, and bonobos (Figure 16.31).

The oldest primate fossils are 55 million years old, but molecular evidence suggests the lineage arose 30 or so million years earlier. Molecular comparisons also reveal that apes and humans are close relatives. They are grouped together as hominoids. All extinct human-like species and modern humans are **hominids**.

Most primate species live in tropical or subtropical forests, woodlands, or grasslands with scattered trees. Most are tree dwellers. Yet no one feature sets them apart from other mammals, and each lineage has its defining traits. Five trends define the lineage that led to hominids, and then to humans.

First, there was less reliance on the sense of smell and more on daytime vision. *Second*, skeletal changes promoted **bipedalism**—upright walking—which freed hands for novel tasks. *Third*, bone and muscle changes led to diverse hand motions. *Fourth*, teeth became less specialized. *Fifth*, the evolution of the brain, behavior, and culture became interlinked. **Culture** is the sum of all behavioral patterns of a social group, passed on through generations by way of learning and symbolic behavior. In brief, *our uniquely human traits emerged by modification of traits that arose earlier, in ancestral species.*

Enhanced daytime vision. Early primates had an eye on each side of their head. Later species had forward-directed eyes, which were better at sampling shapes and movements in three dimensions. Other changes in vision helped individuals respond to variations in light intensity (dim to bright) and color.

Upright walking. How a primate walks depends on arm length and the shape and position of the shoulder blades, pelvic girdle, and backbone. With arm and leg bones about the same length, a monkey can climb and run on four legs. A gorilla walks on two legs and the knuckles of its long arms. Humans and bonobos stride on two legs. Humans have the shortest, most flexible backbone, which has an S-shaped curve. Bipedalism contributed to such structural differences.

Power grip and precision grip. So how did we get our amazing hands? Early mammals spread their toes

LINKS TO
SECTIONS
13.3, 13.4

apart to support the body as they walked or ran on four legs. Among the ancient tree-dwelling primates, modifications to hand bones allowed them to curl their fingers around an object (*prehensile* movements) and to touch their thumb to fingertips (*opposable* movements). Later, power and precision grips evolved among other lineages of primates:

power grip

precision grip

These hand positions gave human ancestors a capacity to make and to use tools. They were a basis for unique technologies and cultural development.

Teeth for all occasions. Modifications to the jaws and teeth accompanied a shift from eating insects to fruits and leaves, then to a mixed diet. Rectangular jaws and lengthy canines evolved in monkeys and apes. A bow-shaped jaw evolved in early hominid lineages. So did teeth that were smaller, and all about the same length.

Brains, behavior, and culture. Shifts in reproductive and social behavior accompanied the shift to life in the trees. In many lineages, parents invested greater effort in fewer offspring. Maternal care became more intense, and the young required longer periods of dependency before setting out on their own.

New forms of behavior promoted new connections in brain regions that deal with sensory inputs. A brain having more intricate wiring favored new behavior. As the brain coevolved with behavior, culture also evolved over time. It did so through enrichment in learning and symbolic behaviors, especially language. The capacity for language arose among hominids. It required more volume in the bony chamber holding the brain, as well as greater brain complexity.

ORIGINS AND EARLY DIVERGENCES

Fossils suggest that primates arose in tropical forests. Like the tiny rodents and tree shrews they resembled (Figure 16.32), they had voracious appetites and spent their nights foraging for eggs, insects, seeds, and buds beneath trees or scrambling about in lower branches. They had a long snout and an acute sense of smell.

During the Eocene (about 55 to 37 million years ago), some primates began to radiate into a new part of the habitat. They began to climb higher up in the trees. Compared with the earlier primates, these high climbers had a shorter snout, a larger brain, far better

Figure 16.32 Tree shrew (*Tupaia*) of Southeast Asia.

daytime vision, and much better ways to grasp and hold objects. *How did these traits evolve?*

Consider the trees. Trees offered food and safety from predators on the ground. They also were adaptive zones of uncompromising selection. Picture an Eocene morning with dappled leaves swaying in the breeze, colorful fruit, and predatory birds. An odor-sensitive, long snout would not have been of much use; breezes disperse odors. But a brain that could assess motion, depth, shape, and color was favored. So were skeletal changes that increased fitness. For instance, eye sockets facing forward instead of on the sides of the skull are better for depth perception.

By 36 million years ago, tree-dwelling anthropoids had evolved. They were on or close to the lineage that led to monkeys and apes. One had forward-directed eyes, a snoutless, flattened face, and an upper jaw that sported shovel-shaped front teeth. It probably used its hands to grab food. Some early anthropoids lived in swamps infested with predatory reptiles. Was that one reason why it became so imperative to think fast, grip strongly, and avoid adventures on the ground? Maybe.

Between 23 and 5 million years ago, in the Miocene, apelike forms—*the first hominoids*—evolved and spread through Africa, Asia, and Europe. Shifts in land masses and ocean circulation were causing a long-term change in climate. Africa became cooler and drier, with more seasonal change. Tropical forests, with their edible soft fruits, leaves, and abundant insects, gave way to open woodlands and later to grasslands. Food had become drier, harder, and more difficult to find.

Sahelanthropus tchadensis
7–6 million years

Australopithecus afarensis
3.6–2.9 million years

A. africanus
3.2–2.3 million years

A. garhi
2.5 million years
(first tool user?)

Paranthropus boisei
2.3–1.4 million years
(huge molars)

P. robustus
1.9–1.5 million years

a

Figure 16.33 (**a**) Gallery of fossils of African hominids that lived between 7 and 1.4 million years ago. (**b**) Fossilized bones of "Lucy" (*Australopithecus afarensis*), a two-legged walker who lived 3.2 million years ago.

(**c**) At Laetoli in Tanzania, Mary Leakey found these footprints, made in soft, damp, volcanic ash 3.7 million years ago. The arch, big toe, and heel marks of these footprints are signs of bipedal hominids. Unlike apes, the early hominids didn't have a splayed-out big toe, as the chimpanzee in (**d**) obligingly demonstrates.

THE FIRST HOMINIDS

Hominoids that had evolved in the lush forests had two options: Move into new adaptive zones or die out. Most died out, but not the common ancestor of apes and, much later in time, modern humans.

When did the hominid lineage arise? *Sahelanthropus tchadensis* evolved in Central Africa as early as 6 or 7 million years ago (Figure 16.33*a*). Is it an ape or early hominid? Opinions vary. The braincase of the single fossil discovered thus far is no bigger than that of a chimpanzee. Yet, like the bigger brained hominids that evolved later in time, it has a shorter, flattened face, a pronounced brow ridge, and smaller canines.

The Miocene on through the Pliocene was a "bushy" time of evolution for hominids in central, eastern, and southern Africa. By this we mean that many diverse forms were evolving. *Ardipithecus ramidus* had evolved by 4.4 million years ago. It is the earliest of the fossil hominids now informally known as **australopiths** (for "southern apes"). *Australopithecus* and *Paranthropus* are others (Figure 16.33*a*,*b*).

We don't know exactly how australopiths are related to one another or to humans. But one or more species of *Australopithecus* are most likely among our ancestors.

Like apes, australopiths had a large face relative to their braincase, a relatively small brain, and jaws that protruded. But they differed in critical respects from earlier hominoids. For one thing, their thick-enameled molars could grind up harder food. For another, they walked upright. We know this from fossil pelvis and limb bones—and from footprints. About 3.7 million years ago, a group of *A. afarensis* walked across some newly fallen volcanic ash. A light rain transformed the ash to a stone-like hardness and preserved their tracks for millions of years (Figure 16.33*c*).

As bipedal hominids evolved in the late Miocene, they were still adapted to forest life. Their hands were adapted for strong gripping, not precision movement. Over time, descendants grew more fully upright and the dexterity of their hands increased. Some may have made tools. In an area of Ethiopian desert, researchers discovered antelope bones that appear to have been cut using stone tools. Fossils of *A. garhi* were found in the same sediments and date back 2.5 million years.

EMERGENCE OF EARLY HUMANS

What do fossilized fragments of early hominids tell us about human origins? The record is still too sketchy to reveal how all the diverse forms were related, let alone which were our ancestors. Besides, exactly which traits characterize **humans**—members of the genus *Homo*?

What about brains? Our brain provides the basis of analytical and verbal skills, complex social behavior, and technological innovation. It sets us apart from apes that have a far smaller skull volume and brain size. Yet this feature alone can't tell us when certain hominids made the evolutionary leap to becoming human. Their brain size fell within the range of apes. They did make simple tools—but so do chimps and some birds. We have few clues to their social behavior.

We are left to speculate on physical traits—skeletal evidence of bipedalism, manual dexterity, larger brain volume, small face, and small, more thickly enameled teeth. These traits emerged late in the Miocene. They are features of what might be the first humans—*Homo habilis*. The name means "handy man."

Between 2.4 and 1.6 million years ago, australopiths and early forms of *Homo* lived together in woodlands bordering the savannas of eastern and southern Africa (Figure 16.34). Fossilized teeth suggest that they ate hard-shelled nuts, dry seeds, soft fruits and leaves, and insects—all seasonal foods. Most likely, *H. habilis* had to think ahead, to plan when to gather and store foods for cool, dry seasons ahead.

H. habilis
2.4?–1.6 million years

a

H. habilis shared its habitat with carnivores such as saber-tooth cats. A big cat's teeth could impale prey and tear flesh but could not break open bones to get at the marrow. Marrow and bits of meat still clinging to bones may have provided an important protein source to early hominids. *H. habilis* and other forms may have enriched their diet by scavenging carcasses, perhaps breaking open bones by striking them with stones.

We have many stone tools dating to the time of *H. habilis*, although we cannot be sure if they made them. Paleontologist Mary Leakey was the first to discover evidence of stone toolmaking. In Tanzania's Olduvai Gorge, she showed that the deepest sediments include crudely chipped stones that might have been used to smash food, dig up roots, and probe for insects in bark (Figure 16.35). The shallower, more recent layers yield more complex tools.

Production of even simple stone tools required a fair amount of brainpower and dexterity. Tool makers did not simply slam rocks together, but rather chose as raw materials rocks that were suited for sharpening, then shaped them with repeated angled blows from another rock.

EMERGENCE OF MODERN HUMANS

Between 1.9 and 1.6 million years ago, strikingly new forms evolved on the plains of Africa. Compared to *H. habilis*, they had bigger brains, stood taller, and were better adapted for walking upright for long distances. The most impressive fossil is the nearly complete skeleton of a boy, estimated to be nine to twelve years old, who died 1.6 million years ago. It is thought that had he lived, he would have stood nearly six feet tall.

Some researchers place all of these more modern-looking hominids into one species, *H. erectus*. Others distribute them into *H. erectus* and *H. ergaster* (Figure 16.36). All now agree that the new hominids were more creative toolmakers, and that social organization and communication skills must have been well developed.

H. erectus was the first hominid to leave Africa. By 1.8 million years ago, this species had expanded its range into Soviet Georgia. It eventually spread as far as Indonesia. *H. ergaster* stayed put in Africa and gave rise to *H. heidelbergensis*. Some members of this new species stayed in Africa; they are considered the most likely ancestors of *H. sapiens*. Other populations made their way into Europe, where they probably gave rise to Neandertals (*H. neanderthalensis*) (Figure 16.37).

b

Figure 16.34 (**a**) Fossil of an early human form. (**b**) Reconstruction of *Homo habilis* in an East African woodland. Two australopiths are in the distance.

Figure 16.35 Olduvai Gorge in Africa, and a sampling of stone tools. *Left to right:* crude chopper, more refined chopper, hand ax, and cleaver.

Figure 16.36 A timeline for the appearance and extinction of major hominid species (currently placed in four genera). The lines at the upper right depict one view of the evolutionary relationships among the most recent species. The number of hominid species, which fossils should be assigned to each species, and how the species are connected are all matters of heated debate among paleontologists.

At this writing, analysis of fossils from Ethiopia has pushed the earliest appearance of *Homo sapiens* back to 190,000 years ago. This new species had smaller teeth, facial bones, and jawbones. Many individuals also had a novel feature—a prominent chin. They had a higher, rounder skull and a bigger brain.

Like *H. sapiens*, the Neandertals had big brains, but they were heavier built, probably as an adaptation to life in far colder climates. They became extinct about 30,000 years ago, about the time that modern humans entered their range. There is no evidence that matings between the two species produced offspring or had any significant effect on the human gene pool. DNA from Neandertal fossils includes many sequences that are not represented in any present-day humans.

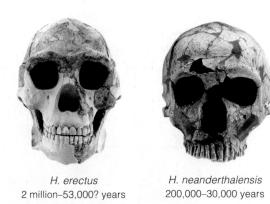

H. erectus
2 million–53,000? years

H. neanderthalensis
200,000–30,000 years

Figure 16.37 Fossils of two species of early humans. Evidence indicates that both species coexisted for a time with fully modern humans, *Homo sapiens.*

HUMAN EVOLUTION

LINK TO
SECTION
13.7

WHERE DID MODERN HUMANS ORIGINATE?

If we are interpreting the fossil record correctly, then Africa was the cradle for us all. At this writing, every human fossil older than 2 million years is from Africa. Between 2 million and 500,000 years ago, *H. erectus* populations left Africa in waves. They dispersed into the grasslands, forests, and mountains of Europe and then continued along into Asia.

Did the geographically separated populations of *H. erectus* evolve in regionally distinct ways in response to local selection pressures? Probably. In 2003, a team working on the Indonesian island of Flores found the 18,000-year-old remains of an adult female hominid who stood only a meter tall. Like *H. erectus*, she had a small brain (even relative to her size) and heavy brow ridges—but *H. erectus* disappeared 200,000 years ago. They named the fossil as a new species, *H. floresiensis*, and postulated that it is descended from *Homo erectus*. The species' small stature may be a trait that arose as an adaptation to life on the island. Island species often evolve to be smaller than their continental ancestors.

Could other regionally specialized populations of *H. erectus* have similarly given rise to subpopulations of *H. sapiens*? This is the idea behind the *multiregional* model. By this model, *H. sapiens* emerged in numerous locales, as a result of selective pressure on *H. erectus* populations in differing habitats. This model suggests that some differences among human groups arose in ancestral *H. erectus* populations.

The *African emergence* model has modern humans arising in sub-Saharan Africa between 200,000 and 100,000 years ago, then moving out of Africa (Figure 16.38). In each region where they settled, they replaced *H. erectus* populations that had preceded them. Then regional phenotypic differences were added on top of a basic *H. sapiens* body plan. The multiregional model emphasizes the differences among subpopulations of humans; this model emphasizes similarities.

In support of this model, the oldest *H. sapiens* fossils are from Africa. Also, in Zaire, finely wrought barbed-bone tools indicate that African populations were as skilled at producing tools as any *Homo* populations in Europe. Gene patterns from forty-three ethnic groups in Asia suggest that modern humans migrated from Central Asia, along India's coast, then into Southeast Asia and southern China. They later dispersed north and west across China and Russia into Siberia, where they crossed the Bering Strait into the Americas.

For the past 40,000 years, our cultural evolution has been outpacing macroevolution of the human species—and so we complete our story. One point to remember: Humans spread rapidly through the world by devising *cultural* means to deal with their diverse range of environments. Hunters and gatherers persist in certain parts of the world even as other groups have moved from "stone-age" technology to the age of high tech. Such differences highlight the great behavioral plasticity and depth of human adaptations.

Primates evolved from small, rodentlike mammals that moved into the trees 60 million years ago. By 7 million years ago, divergences had given rise to the first hominids.

These key adaptations evolved along the pathway from arboreal primates to modern humans: complex, forward-directed vision; bipedalism; refined hand movements; more generalized teeth; and an interlocked elaboration of brain regions, behavior, and culture.

Cultural evolution has outpaced the biological evolution of the only remaining human species, H. sapiens. *Today, humans everywhere rely on cultural innovation to adapt rapidly to a broad range of environmental challenges.*

H. sapiens fossil from Ethiopia, 160,000 years old

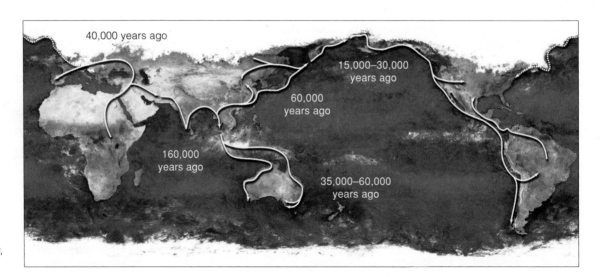

Figure 16.38 The estimated times when early *H. sapiens* reached different parts of the world, based on the radiometric dating of fossils. White lines show the presumed dispersal routes. Molecular data show that very few populations left Africa. Genetically, all living humans are extremely similar.

Summary

Section 16.1 Animals are multicelled heterotrophs that ingest or parasitize other organisms. Most have diploid body cells organized into tissues, organs, and organ systems. Animals reproduce sexually; many also reproduce asexually. They develop through embryonic stages. Most are motile during at least part of the life cycle. Table 16.1 summarizes the major groups.

Section 16.2 Sponges are simple filter feeders with no body symmetry or tissues. The radially symmetrical cnidarians, such as jellyfishes, sea anemones, corals, and hydras, are at the tissue level of organization. They are the only animals that make nematocysts: capsules with tubular, dischargeable threads that help capture prey.

Biology Now
See the water flow through a sponge and the action of a cndiaran nematocyst with animation on BiologyNow.

Sections 16.3, 16.4 Flatworms have the simplest organ systems. They have a saclike gut and no coelom. Free-living and parasitic turbellarians, tapeworms, and flukes are in this group. Annelids, highly segmented worms, have bristled segments. They are oligochaetes (such as earthworms), polychaetes, and leeches.

Biology Now
Learn about the organ systems of a flatworm and the life cycle of a tapeworm with animation on BiologyNow.

Section 16.5 Only mollusks form a mantle, which is an extension of the body mass that drapes back on itself like a skirt. Bivalves, gastropods, and cephalopods (that move by water-jet propulsion) are representatives.

Sections 16.6, 16.7 Roundworms and arthropods have a cuticle that they molt as they grow. Roundworms are free-living or parasitic, have a complete gut, and have a false coelom. Arthropods have a coelom and a jointed exoskeleton. The body has modified segments and specialized appendages. Living arthropods are the arachnids, crustaceans, centipedes, millipedes, and insects. Insects include the only winged invertebrates and are the most numerous and diverse animals.

Section 16.8 Echinoderms such as sea stars have an exoskeleton of spines, spicules, or plates of calcium carbonate. Adults are radial, but bilateral ancestry is evident in their larval stages and other features.

Section 16.9 Four features help define the chordates. The embryos develop a notochord, a dorsal hollow nerve cord (which becomes a brain and spinal cord), a pharynx with gill slits, and a tail extending past the anus. Some or all of these features persist in adults.

Lancelets and tunicates are invertebrate filter-feeding chordates. Craniates are chordates with a brain chamber of cartilage or bone. Most craniates are vertebrates; they have a vertebral column (backbone) of cartilage or bone. Tetrapods are vertebrates with four limbs or with four-limbed ancestors. Amniotes are the only vertebrates that make amniote eggs. Jaws, paired fins, and later, lungs, were key innovations that led to the adaptive radiation of vertebrates.

Biology Now
Explore the body plan and chordate features of a lancelet with animation on BiologyNow.

Section 16.10 Cartilaginous fishes and bony fishes are aquatic vertebrates. Bony fishes are the most diverse vertebrates, and one group gave rise to the tetrapods.

Section 16.11 Amphibians, the first tetrapods on land, share traits with aquatic tetrapods and reptiles. Existing groups include salamanders, frogs, and toads. Many species are now threatened or endangered.

Section 16.12 Amniotes were the first vertebrates that did not require external water for reproduction. They have water-conserving skin and kidneys, and produce amniote eggs. Mammals and their ancestors are one amniote lineage. "Reptiles" and birds (the only animals that have feathers) are another.

There are three lineages of mammals: egg-laying mammals, pouched mammals, and placental mammals. Today, placental mammals are the most diverse group and have displaced others in most habitats.

Table 16.1	Major Animal Groups	
Group	Some Representatives	Existing Species
Poriferans	Sponges	8,000
Cnidarians	Hydras, jellyfishes	11,000
Flatworms	Turbellarians, flukes, tapeworms	15,000
Annelids	Leeches, earthworms, polychaetes	15,000
Mollusks	Snails, slugs, squids, octopuses	110,000
Arthropods	Crustaceans, spiders, insects	1,000,000+
Roundworms	Pinworms	20,000
Echinoderms	Sea stars, sea urchins, sea cucumbers	6,000
Chordates	Invertebrate chordates: Tunicates, lancelets	2,100
	Vertebrate chordates:	
	Jawless fishes	84
	Jawed fishes	21,000
	Amphibians	4,900
	Reptiles	7,000
	Birds	8,600
	Mammals	4,500

Section 16.13 The hominoid branch of the primate family tree evolved in Africa as the climate became cooler and drier. It includes apes and the hominids: humans and their most recent ancestors (Table 16.2).

Australopiths were early hominids, and they walked upright. The relationships among the diverse forms are sketchy. The first known members of the genus *Homo* lived in Africa between 2.4 million and 1.6 million years ago. *H. erectus* was the first hominid to migrate out of Africa. *H. sapiens* evolved by 190,000 years ago.

Self-Quiz

Answers in Appendix I

1. Only _____ have a notochord, a tubular dorsal nerve cord, a pharynx with gill slits, and a tail past the anus.
   - a. earthworms
   - b. echinoderms
   - c. chordates
   - d. both b and c

2. The _____ have a cuticle and molt as they grow.
   - a. sponges
   - b. cnidarians
   - c. annelids
   - d. arthropods
   - e. both c and d

3. Reptiles moved fully onto land owing to _____ .
   - a. tough skin
   - b. internal fertilization
   - c. good kidneys
   - d. amniote eggs
   - e. a and c
   - f. all of the above

4. _____ preceded the emergence of modern humans.
   - a. Craniates
   - b. Jawed vertebrates
   - c. Tetrapods
   - d. Amniotes
   - e. Hominoids
   - f. all of the above

5. A coelom is a _____ .
   - a. lined body cavity
   - b. resting stage
   - c. sensory organ
   - d. bristle

6. The _____ are filter feeders.
   - a. cephalopods
   - b. flatworms
   - c. lancelets
   - d. sea stars

7. Earthworms are most closely related to _____ .
   - a. leeches
   - b. flukes
   - c. tapeworms
   - d. roundworms

8. All animals _____ .
   - a. are motile for at least some stage in the life cycle
   - b. consist of tissues arranged as organs
   - c. can reproduce asexually as well as sexually
   - d. both a and b

9. Match the organisms with the appropriate description.
   - ____ sponges
   - ____ cnidarians
   - ____ flatworms
   - ____ roundworms
   - ____ annelids
   - ____ arthropods
   - ____ mollusks
   - ____ echinoderms
   - ____ amphibians
   - ____ fishes
   - ____ "reptiles"
   - ____ mammals

   - a. most diverse vertebrates
   - b. no true tissues, no organs
   - c. jointed exoskeleton
   - d. mantle over body mass
   - e. segmented worms
   - f. tube feet, spiny skin
   - g. nematocyst producers
   - h. first with amniote eggs
   - i. hair or fur
   - j. complete gut, false coelom
   - k. first terrestrial tetrapods
   - l. saclike gut, no coelom

Additional questions are available on Biology ⑧ Now™

Critical Thinking

1. In the summer of 2000, only 10 percent of the lobster population off Long Island Sound survived after a massive die-off. Many lobstermen in New York and Connecticut lost small businesses that their families had owned for generations. Some believe the die-off followed heavier sprays of pesticides, used to control mosquitoes that carry the West Nile virus. Explain how a chemical substance that targets mosquitoes might also harm or kill lobsters.

2. During the Triassic, hundreds of species of reptiles that resemble living tuataras roamed Pangea. Now only two species survive on a few small cold islands off the coast of New Zealand (Figure 16.39). Speculate on why these relic populations have survived there, despite disappearing from the rest of their once extensive range.

3. In 1798, a stuffed platypus specimen was delivered to the British Museum. At the time, many biologists were sure it was a fake, assembled by a clever taxidermist. Soft brown fur and beaverlike tail put the animal firmly in the mammalian camp. But a ducklike bill and webbed feet suggested an affinity with birds. Reports that the animal laid eggs only added to the confusion.

We now know that platypuses burrow in riverbanks and forage for prey under water. Webbing on their feet can be retracted to reveal claws. The highly sensitive bill allows the animal to detect prey even with its eyes and ears tightly shut.

To modern biologists, a platypus is clearly a mammal. Like other mammals, it has fur and the females produce milk. Young animals have more typical mammalian teeth that are replaced by hardened pads as the animal matures. Why do you think modern biologists can more easily accept the idea that a mammal can have some reptilelike traits, such as laying eggs? What do they know that gives them an advantage over scientists living in 1798?

4. Some paleontologists assign fossils to only four species in the genus *Homo*. Others split the genus into a many as seven species. Why is it more difficult to classify fossils than it is to classify existing species? What kinds of evidence would be useful in classifying new fossils?

Figure 16.39 Where did everybody go? One of the two species of tuatara (*Sphenodon*) that survived to the present.

Too Hot To Handle

Heat can kill. In the summer of 2001 Korey Stringer, a football player for the Minnesota Vikings, collapsed after morning practice. His team had been working out in full uniform on a field where temperatures were in the high 90s. High humidity put the heat index above 100.

With an internal body temperature of 108.8°F and blood pressure too low to record, Stringer was rushed to the hospital. Doctors immersed him in an ice-water bath, then wrapped him in cold, wet towels. It was too late.

Stringer's blood-clotting mechanism shut down and he started to bleed internally. Then his kidneys faltered and he was placed on dialysis. He stopped breathing on his own and was put on a respirator. His heart gave up. Less than twenty-four hours after football practice had started, Stringer was pronounced dead. He was twenty-seven years old.

All organisms function best within a limited range of internal operating conditions. For us, "best" is when the body's internal temperature remains between about 97°F and 100°F. Past 104°F, blood is transporting a great deal of metabolically generated heat. Transport controls divert that heat away from the brain and other internal organs to the skin. A person's skin transfers heat to the outside environment—as long as it isn't too hot outside. Profuse sweating can dissipate more heat—but only if humidity is low enough for sweat to evaporate quickly.

As happened in Stringer's case, these normal cooling mechanisms fail above 105°F. The body's production of sweat falters, and its temperature starts to climb rapidly. The heart beats faster and often the individual becomes confused or faints. These are signs of *heat stroke*, which

denatures the enzymes and other proteins that keep us alive. When heat stroke is not countered fast enough, brain damage or death is the expected outcome.

We use this sobering example as our passport to the world of anatomy and physiology. **Anatomy** is the study of body form—that is, its morphology. **Physiology** is the study of patterns and processes by which an individual survives and reproduces in the environment. It deals with how structures are put to use and how metabolism and behavior are adjusted when conditions change. It also deals with how, and to what extent, physiological processes can be controlled.

You can use a lot of this information to interpret what goes on in your own body. More broadly, you can use it to perceive commonalities among all animals and plants. Regardless of the species, *the structure of a given body part almost always has something to do with a present or past function*. Most aspects of form and function are long-standing adaptations that evolved as responses to environmental challenges.

That said, nothing in the evolutionary history of the human species suggests that the organ systems making up our body are finely tuned to handle intense football practice on hot, humid days.

 How Would You Vote? *Should a coach or team doctor who does not stop team practice or a game when the heat index soars be held liable if a player dies from heat stroke? See BiologyNow for details, then vote online.*

 Key Concepts

MULTICELLULAR ORGANIZATION
Each cell of a multicelled organism engages in activities that help assure its own survival. The structural and functional organization of all cells collectively sustains the whole body.

RECURRING CHALLENGES
Plants and animals both require mechanisms of internal transport, gas exchange, nutrition, ion-water balance with the external environment, and defense.

HOMEOSTASIS
Homeostasis is the maintenance of internal conditions within limits that allow the body to function normally. Both plants and animals have mechanisms of homeostasis, including systems of negative feedback control.

CELL-TO-CELL COMMUNICATION
Signaling mechanisms control and integrate body activities. Cells send messages that are received and interpreted by other cells, which alter their activities in response.

Links to Earlier Concepts

This chapter emphasizes the common processes that take place in cells and bodies of plants and animals. You learned about limitations on cell size in Section 3.2 and about the cellular effects of diffusion and osmosis in Sections 4.4 to 4.6. Keep in mind the trends in plant evolution (15.1) and animal evolution (16.1) that were introduced earlier. We will again touch on some examples of eukaryotic controls over gene expression (10.6) and how loss of controls can cause cancer (7.8).

LINKS TO
SECTIONS
15.1, 16.1

17.1 Levels of Structural Organization

This chapter introduces some important concepts that will be applied in the next two units of the book. They can help deepen your understanding of how plants and animals function under stressful as well as favorable environmental conditions.

FROM CELLS TO MULTICELLED ORGANISMS

In most plants and animals, cells, tissues, organs, and organ systems split up the work of keeping the body alive. A separate cell lineage gives rise to body parts that will function in reproduction. Said another way, the plant or animal body shows a *division of labor*.

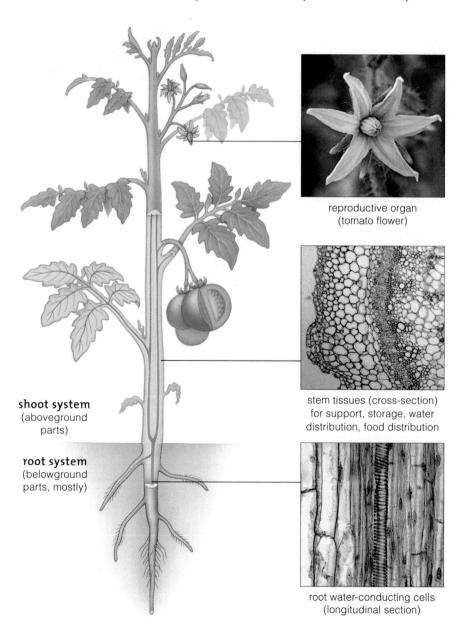

reproductive organ
(tomato flower)

stem tissues (cross-section)
for support, storage, water
distribution, food distribution

shoot system
(aboveground
parts)

root system
(belowground
parts, mostly)

root water-conducting cells
(longitudinal section)

A **tissue** is a collection of particular kinds of cells and intercellular substances that interact in a task or tasks. Wood tissue and bone tissue support the body. All wood is composed of the same types of cells, just as the composition of all bone is similar.

An **organ** consists of two or more tissues, organized in specific proportions and patterns, that perform a task or tasks. A leaf that carries out photosynthesis and an eye that responds to light are examples of organs. An **organ system** is two or more organs that interact physically, chemically, or both. A plant's shoot system, with organs of photosynthesis and reproduction, is an example (Figure 17.1). Another is an animal's digestive system, which takes in food, breaks it down, absorbs breakdown products, and expels the residues.

GROWTH VERSUS DEVELOPMENT

A plant or animal body becomes structurally organized during growth and development. Generally, **growth** of a multicelled organism means its cells increase in number, size, and volume. **Development** is a succession of stages, during which specialized tissues, organs, and organ systems form. In other words, we measure growth in *quantitative* terms, and development in *qualitative* terms.

STRUCTURAL ORGANIZATION HAS A HISTORY

The structural organization of each tissue, organ, and organ system has an evolutionary history. Think back on how plants invaded the land and you have some idea of how their structure relates to function. Plants that left aquatic habitats found plenty of sunlight and carbon dioxide for photosynthesis, as well as oxygen for aerobic respiration. As they dispersed farther away from their aquatic cradle, however, they faced a new challenge: how to keep from drying out in air.

Consider that challenge when you see micrographs that reveal the internal structure of stems, and leaves (Figure 17.1). These pipelines carry water from soil to leaves. Stomata, small gaps across a leaf's epidermis, open and close in ways that conserve water (Section 5.3). Cells that have lignin-reinforced walls collectively support the upright growth of stems. Remember that

Figure 17.1 Morphology of a tomato plant (*Lycopersicon esculentum*). Different cell types make up vascular tissues that conduct either water and dissolved minerals or organic substances. Vascular tissues thread through other tissues making up most of the plant. A different tissue covers the surfaces of root and shoot systems.

Figure 17.2 Some of the structures that function in human respiration. Their cells carry out specialized tasks. Airways to a pair of organs called lungs are lined with ciliated cells that whisk away bacteria and other airborne particles that might cause infections. Inside the lungs are tubes (blood capillaries), a fluid connective tissue called blood, and thin air sacs (epithelial tissue), all with roles in gas exchange.

organs (lungs), part of an organ system (the respiratory tract) of a whole organism

ciliated cells and mucus-secreting cells that line respiratory airways

lung tissue (tiny air sacs) laced with blood capillaries—one-cell-thick tubular structures that hold blood, which is a fluid connective tissue

challenge also when you examine root systems. A soil region that has a greater concentration of water and dissolved mineral ions will stimulate root growth in its direction; hence the branching roots.

Similarly, respiratory systems of land animals are adaptations to life in air. Gases can move into and out of the body only by diffusing across a moist surface. This is not a problem for organisms that live in water, but moist surfaces dry out in air. Land animals have moist sacs for gas exchange *inside* their body (Figure 17.2). This brings us to the body's own environment.

THE BODY'S INTERNAL ENVIRONMENT

To stay alive, plant and animal cells must be bathed in a fluid that provides nutrients and carries away metabolic wastes. In this they are no different from the free-living single cells. But a plant or animal consists of thousands to many trillions of living cells. All must draw nutrients from the fluid that bathes them and dump wastes into it.

Fluid not inside body cells—extracellular fluid—is an **internal environment**. Changes in its composition and volume affect cell activities. The concentration and number of solutes in the internal environment must be maintained within the range that supports metabolism. *Survival of any plant or animal, simple or complex, requires a stable fluid environment for all of its living cells.* As you will see, this concept is central to understanding how plants and animals work.

HOW DO PARTS CONTRIBUTE TO THE WHOLE?

The next two units describe how a plant or an animal functions. The body provides its cells with a stable fluid environment. It acquires energy and essential nutrients and other raw materials, distributes them, and disposes of wastes. It has ways to protect itself against injury and attack. It also has the capacity to reproduce, and often it helps nourish and protect new individuals during their early growth and development.

The big picture is this: Each living plant or animal cell engages in metabolic activities that ensure its own survival. But the structural organization and collective activities of cells in tissues, organs, and organ systems sustain the whole body. They work to keep operating conditions in the internal environment within tolerable limits for individual cells—a state called **homeostasis**.

The structural organization of plants and animals emerges during growth and development.

Cells, tissues, and organs all require a favorable internal environment, and they collectively maintain it. All body fluids not contained in cells make up that environment.

Basic body functions include maintaining a stable internal environment, securing substances and distributing them to cells, disposing of wastes, protecting cells and tissues, reproducing, and often nurturing offspring.

17.2 Recurring Challenges to Survival

Plants and animals have such spectacularly diverse body plans that we sometimes forget how much they have in common.

Figure 17.3 What do these organisms have in common besides their good looks?

How often do you think about the similarities between Tina Turner and a tulip (Figure 17.3)? Connections are there. They begin with what individual cells are doing inside both of those multicelled bodies. Nearly all are working in the best interest of the whole body. At the same time, the whole body is working in the best interest of individual cells.

REQUIREMENTS FOR GAS EXCHANGE

Cells in Tina and the tulip would not last long if they could not continually exchange certain gases across their plasma membrane. Like most other heterotrophs, for instance, Tina must supply her aerobically respiring cells with oxygen and dispose of carbon dioxide end products. Like most other autotrophs, the plant must supply its photosynthetic cells with carbon dioxide. In addition, it must keep enough oxygen around for *its* aerobically respiring cells and dispose of the excess.

Certain aspects of the structure and function of all multicelled bodies are responses to this challenge: *How can gaseous molecules move to and from individual cells?*

Remember **diffusion**? This is a net movement of molecules or ions of a substance from a region where they are concentrated to an adjacent region where they are not as concentrated. Animals and plants have ways to keep gases diffusing in the directions most suitable for metabolism and cell survival. How? That question will lead you to the stomata at leaf surfaces (Chapter 18) and to the respiratory and circulatory systems of animals (Chapter 22).

REQUIREMENTS FOR INTERNAL TRANSPORT

For cells in a plant or animal to stay alive, they need a constant supply of oxygen and nutrients. If it takes too long for these substances to diffuse through the body, or from its surface, the body will shut down. This is one reason that cells and multicelled organisms have the sizes and shapes that they do.

Recall from Section 3.2 that with every increase in size, the ratio of a cell or body's surface to its volume decreases. That is the essence of the **surface-to-volume ratio**. If a cell or body were to get too big, it would not have enough surface area, relative to volume, for fast, efficient exchanges with the environment.

When a body or some body part is thin, as it is for the flatworm and lily pads in Figure 17.4, substances can easily diffuse between its individual cells and the environment. In massive bodies, however, individual cells far from an exchange point with the environment depend on systems of rapid internal transport.

In most plants and animals, substances move to and from cells through vascular tissues that act as internal pipelines. A vein in a plant leaf is a bundle of vascular tissue (Figure 17.5a). One tissue in the vein distributes soil water and mineral ions. Another distributes sugars.

In large animals, vascular tissues are usually blood vessels, which function as part of a circulatory system. Each time Tina takes a breath, she moves oxygen into her lungs and carbon dioxide out of them. Her blood vessels weave through lung tissues where gas exchange takes place. Blood vessels also supply her body tissues, where gases diffuse to and from the interstitial fluid and cells (Figure 17.5b).

As in plants, the animal vascular system transports nutrients and water. It also connects with other organ systems. Cells of the immune system use the blood vessels as their highway to move about the body. A urinary system adjusts the concentration and volume of the internal environment.

Figure 17.4 A flatworm gliding along in water, and lily leaves floating on water. These body plans are responses to the same constraint. What is it?

MAINTAINING A SOLUTE—WATER BALANCE

Plants and animals continually take in and lose water and solutes. They regularly produce metabolic wastes. Given these inputs and outputs, how do they keep the volume and composition of their internal environment within a tolerable range? Plants and animals deal with this challenge differently. Yet we can find similarities in their responses at the molecular level.

Substances tend to follow concentration gradients as they move into and out of the body, or from one of its compartments into another. At such interfaces, we see sheetlike tissues where cells passively and actively transport substances (Section 4.5). **Active transport**, recall, is a pumping of substances *against* gradients.

Active transport mechanisms in roots help control which solutes move into the plant. In leaves, they help control water loss and gas exchange by opening the stomata only at certain times. In animals, we find such mechanisms in kidneys and other organs. The active transport of ions across the membranes of cells in the kidney sets up osmotic gradients that cause water to move in ways that concentrate the urine. As you will see again and again in the next two units, *active and passive transport help maintain homeostasis in the internal environment by controlling the movements of substances into and out of all body cells.*

REQUIREMENTS FOR INTEGRATION AND CONTROL

We could go on about the structural and functional similarities between plants and animals. But we will end this discussion with a final point. In both groups, cells make signals that influence other cells and thus coordinate the body as a whole. These intercellular signaling mechanisms control how an animal or plant grows, develops, maintains, and even protects itself. The remaining sections in this chapter provide some examples of these signaling mechanisms.

ON VARIATIONS IN RESOURCES AND THREATS

Now move beyond the common challenges. Resources and dangers differ greatly among habitats. A **habitat** is the place where individuals of a species normally live.

What are the physical and chemical characteristics of the habitat? Is water plentiful, with suitable kinds and amounts of solutes? Is the habitat rich or poor in nutrients? Is it sunlit, shady, or dark? Is it warm or icy, windy or calm? How much do outside temperatures change from day to night? Are seasonal changes slight or highly pronounced?

Consider also the biotic (living) components of the habitat. What are the numbers and types of producers,

predators, prey, pathogens, and parasites? Are there other species that compete for sunlight, nesting sites, or some other resource? These sorts of interactions can shape a species' anatomy and physiology.

Even in all that diversity, we often observe similar responses to similar challenges. Sharp cactus spines or sharp porcupine quills are effective deterrents to most animals that might eat the cactus or porcupine. Both arise from specialized epidermal cells. Vascular tissues and other body parts helped nurture those cells. And by contributing to building defensive structures at the body surface, epidermal cells help protect the body as a whole against an environmental threat.

Figure 17.5 Transport lines to and from cells. (**a**) Veins visible in a decaying dicot leaf, and (**b**) some human veins and capillaries.

Plant and animal cells function in ways that help ensure survival of the body as a whole. At the same time, tissues and organs that make up the body function in ways that help ensure survival of individual living cells.

The connection between each cell and the body as a whole is evident in the requirements for—and contributions to— gas exchange, nutrition, internal transport, stability in the internal environment, and defense.

17.3 Homeostasis in Animals

In preparation for your trek through the next two units, take a moment to get a firm grasp on what homeostasis means to survival. Animal physiologists were the first to identify this state and describe how it is maintained, so we will start with animals.

Like other adult humans, your body has many trillions of living cells. Each cell must draw nutrients from and dump wastes into the same fifteen liters of fluid—less than sixteen quarts. Remember, fluid not inside cells is the extracellular fluid. Much is **interstitial fluid**; it fills spaces between cells and tissues. The rest is **plasma**, the fluid portion of blood. Interstitial fluid exchanges many substances with cells and with blood.

Homeostasis, again, is a state in which the body's internal environment is being maintained within the range that its cells can tolerate. In nearly all animals, three components interact and maintain homeostasis. They are sensory receptors, integrators, and effectors.

Sensory receptors are cells or cell parts that detect stimuli, which are specific forms of energy. The brain is one example of an **integrator**, a central command

STIMULUS input into the system

```
receptor          integrator        effector
(such as a        (such as          (a muscle
free nerve        the brain         or gland)
ending in         or spinal
the skin)         cord)
```

RESPONSE to stimulus causes change. The change is "fed back" to receptor. In *negative* feedback, the system's response cancels or counters the effect of the original stimulus.

Figure 17.6 *Animated!* Three components of negative feedback systems in multicelled organisms.

post that pulls together information about stimuli and issues signals to the muscles, glands, or both. Muscles and glands are **effectors**. In reaction to signals from an integrator, they bring about a suitable response.

Consider how levels of carbon dioxide and oxygen are controlled in the blood. In this example, the brain is the integrator. It receives information from receptors that sense how much carbon dioxide is dissolved in blood. The brain compares this information against a set point for what the carbon dioxide level should be, and sends signals to effectors—the muscles involved in breathing. In response, breathing is altered.

Now think about what happens when a climber scales a mountain. As altitude increases, the density of the atmosphere declines, so the climber takes in fewer oxygen molecules in every breath. At altitudes above 3,300 meters, air is about half as dense as it is at sea level, and about half of climbers suffer *altitude sickness*. Receptors alert the brain, which calls on the body to hyperventilate (breathe deeper and more rapidly). If oxygen levels do not increase, headache, nausea, and vomiting will follow. All are warnings that cells are starved for oxygen.

NEGATIVE FEEDBACK

Feedback mechanisms are key homeostatic controls. In a **negative feedback mechanism**, an activity changes some condition, and when that condition shifts past a certain level, it triggers a response that reverses the change. Figure 17.6 shows one model for such control.

Think of a furnace with a thermostat. A thermostat senses the surrounding air's temperature relative to a preset point on a thermometer built into the furnace's control system. When the temperature falls below the preset point, it sends signals to a switching mechanism that turns on the furnace. When the air becomes heated enough to match the prescribed level, the thermostat signals the switch, which shuts off the furnace.

Similarly, feedback mechanisms work to keep the internal body temperature of humans and many other

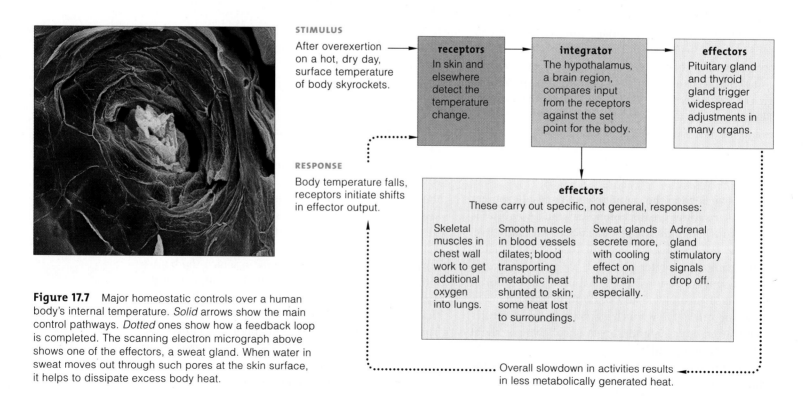

STIMULUS

After overexertion on a hot, dry day, surface temperature of body skyrockets.

receptors
In skin and elsewhere detect the temperature change.

integrator
The hypothalamus, a brain region, compares input from the receptors against the set point for the body.

effectors
Pituitary gland and thyroid gland trigger widespread adjustments in many organs.

RESPONSE

Body temperature falls, receptors initiate shifts in effector output.

effectors
These carry out specific, not general, responses:

Skeletal muscles in chest wall work to get additional oxygen into lungs.

Smooth muscle in blood vessels dilates; blood transporting metabolic heat shunted to skin; some heat lost to surroundings.

Sweat glands secrete more, with cooling effect on the brain especially.

Adrenal gland stimulatory signals drop off.

Overall slowdown in activities results in less metabolically generated heat.

Figure 17.7 Major homeostatic controls over a human body's internal temperature. *Solid* arrows show the main control pathways. *Dotted* ones show how a feedback loop is completed. The scanning electron micrograph above shows one of the effectors, a sweat gland. When water in sweat moves out through such pores at the skin surface, it helps to dissipate excess body heat.

mammals near 98.6°F (37°C) even during hot or cold weather. When a football player runs around on a hot summer day, the body becomes hot. Receptors trigger changes that should slow down the entire body *and* its cells (Figure 17.7). Under typical conditions, these and other control mechanisms avert overheating. They curb activities that naturally generate metabolic heat, and increase heat transfer to the surrounding air.

POSITIVE FEEDBACK

With a **positive feedback mechanism**, some change in conditions causes the body to respond in a manner that increases the intensity of the change. As a result, the system moves farther and farther from its initial state. Positive feedback often culminates in some event that returns the system to its starting state.

For example, the flow of sodium across the plasma membrane of a neuron allows the body to send and receive signals quickly. A stimulus causes the opening of a few sodium channels in the neuron membrane. As sodium ions flow in through the open channels, they increase the sodium levels inside the cell. This leads to the opening of additional channels and the flow of still more sodium inward, as you will learn in Chapter 25.

As another example, during labor, the fetus exerts pressure on the wall of the chamber enclosing it, the uterus. Pressure induces the production and secretion of a hormone, oxytocin, that makes muscle cells in the wall contract. Contractions exert pressure on the fetus, which puts more pressure on the wall, and so on until the fetus is expelled from the mother's body.

We have been introducing a pattern of detecting, evaluating, and responding to a flow of information about the internal and external environment. During this activity, organ systems work together. In time you will be asking these questions about their operation:

1. Which physical or chemical aspects of the internal environment are organ systems working to maintain?

2. How are organ systems kept informed of change?

3. By what means do they process the information?

4. What mechanisms are set in motion in response?

As you will read in Unit Five, organ systems of nearly all animals are under nervous and endocrine control.

Homeostatic controls maintain the body's internal environment within ranges that favor cell activities.

With negative feedback a change triggers a response that reverses the change. With positive feedback a response intensifies the change that stimulated it.

17.4 Does Homeostasis Occur in Plants?

Plants differ from animals in important respects. Even so, they, too, have controls over the internal environment.

Direct comparisons between plants and animals are not always possible. In young plants, new tissues arise only at the tips of actively growing roots and shoots. In animal embryos, tissues form all through the body. Plants do not have a centralized integrator, such as a brain. Yet plants have decentralized mechanisms that protect the internal environment and help ensure the survival for the body as a whole. It may be a bit of a leap, but we also can apply the concept of homeostasis to plants. Two examples will make the point.

RESISTING PATHOGENS

Like animals, plants are subject to infection by diverse pathogens—viruses, bacteria, fungi, and other agents that cause disease. In animals, white blood cells and blood proteins act against pathogens by mechanisms you will learn about in Chapter 23. But plants do not have any blood or blood cells. Instead, a wound or infection sets in motion a mechanism we call **systemic acquired resistance**. Signaling molecules are released at a damaged site. Here, the signals call for production of defensive compounds. Signals are also distributed to other undamaged tissues in more distant regions. Here, the signals induce cells to synthesize and release compounds that make tissues less vulnerable to future attacks for days to months. Synthetic forms of some of the compounds that act in systemic defenses are now being used to enhance the disease resistance of crops and ornamental plants.

Many woody plants can also protect their internal environment by building a fortress of thickened cell walls around wounds. They exude phenols and other toxic compounds. Many also secrete resins. The heavy flow of these gooey compounds saturates and protects bark and wood at an attack site. It also can seep into soil around roots. This type of plant response to attack is called **compartmentalization** (Figure 17.8).

Some toxins are so potent that they also kill cells of the tree itself. As compartments form around injured or infected or poisoned sites, the tree lays down new tissues over them. This works well when the tree acts

strong

moderate

weak

Figure 17.8 *Animated!* Patterns of drilling into stems for an experiment to test compartmentalization responses. From top to bottom, the decay patterns (*green*) for three species that make strong, moderate, or weak compartmentalizing responses.

1 A.M.

6 A.M.

NOON

3 P.M.

10 P.M.

MIDNIGHT

c

Figure 17.9 *Animated!* (**a,b**) Yellow bush lupine, *Lupinus arboreus*, in a sandy shore habitat. Introduced into northern California in the early 1900s, this exotic species is taking over the sand dunes. It outcompetes native species.

(**c**) Observational test of the rhythmic movements of the leaves of a young bean plant (*Phaseolus*). The investigator, Frank Salisbury, kept this plant in complete darkness for twenty-four hours. Its leaves continued to fold and unfold rhythmically even without the cues from changes in sunlight.

strongly or is not under massive attack. Drill holes in a tree species that makes a strong response and it walls off the wound rapidly. Drill holes in a tree species that makes a moderate response, and it decays lengthwise and somewhat laterally. Drill holes in a species that is a weak compartmentalizer, though, and spread of fungi and bacteria from the wound will cause massive decay. Even the strong compartmentalizers live only so long. When they form too many walls, they shut off the vital flow of water and solutes to living cells.

SAND, WIND, AND THE YELLOW BUSH LUPINE

Anyone who has tiptoed barefoot across sand near the coast on a hot, dry day can understand why very few plants grow in it. One of the exceptions is the yellow bush lupine, *Lupinus arboreus* (Figure 17.9*a*).

L. arboreus is a colonizer of soil exposed by fires or abandoned after being cleared for agriculture. It also has become common along windswept, sandy shores of the Pacific Northwest. Nitrogen-fixing bacteria live in association with the lupine's roots. This gives these plants a competitive edge in nitrogen-deficient soils.

A big environmental challenge near the beach is the scarcity of fresh water. Leaves of a yellow bush lupine are structurally adapted for water conservation. Each leaf has a surprisingly thin cuticle, but a dense array of fine epidermal hairs projects above it, especially on the lower leaf surface. Collectively, the hairs trap moisture escaping from stomata. The trapped, moist air slows evaporation and helps keep water inside the leaf.

These leaves also maintain homeostasis by folding up lengthwise. This helps them to resist the drying force of the wind. The folded leaf is better at stopping moisture loss from stomata than a broad unfolded leaf would be (Figure 17.9*b*).

Leaf folding by *L. arboreus* is a controlled response to changing conditions. When strong winds blow and the potential for water loss is great, these leaves fold tightly. The least-folded leaves are near the center of the plant or on the side most sheltered from the wind. Folding is a response to heat as well as to wind. When air temperature is highest during the day, leaves fold at an angle that helps reflect the sun's rays from their surface. The response minimizes heat absorption.

ABOUT RHYTHMIC LEAF FOLDING

For another example of leaf folding as a coordinated homeostatic response, look at Figure 17.9*c*. This plant holds its leaves out horizontally in the day and folds them closer to its stem at night. Even if you keep the plant in full darkness for a few days, it will continue to move its leaves into and out of the "sleep" position, at the appropriate times. This folding response might be a way to reduce heat losses during the night, when the air becomes cooler. It may help to maintain the plant's internal temperature within a tolerable range.

Rhythmic leaf movements are just one example of **circadian rhythm**, a biological activity that is repeated in cycles, each lasting for close to twenty-four hours. Circadian means "about a day." A pigment molecule called phytochrome, described in Chapter 19, may be part of homeostatic controls over leaf folding.

> *Homeostatic mechanisms are at work in plants, although they are not governed from central command posts as they are in most animals. Compartmentalization and rhythmic leaf movements are two examples of responses to specific environmental challenges.*

HOMEOSTASIS

Signal to die docks at receptor.

Signal leads to activation of enzymes that break down proteins and others that cut up DNA.

Figure 17.10 Artist's depiction of apoptosis. Body cells self-destruct when they finish their functions or if they become infected or cancerous and could threaten the body as a whole.

17.5 How Cells Receive and Respond to Signals

Signal reception, transduction, and response—this is a fancy way of saying cells chatter among themselves and make things happen.

Reflect on the earlier overview of how adjoining cells communicate, as by plasmodesmata in plants and gap junctions in animals (Section 3.8). Also think back on how free-living *Dictyostelium* cells issue signals to get together and make a spore-bearing structure. They do so in response to dwindling supplies of food, which is an environmental cue for change (Figure 14.28).

In all large multicelled organisms, one cell signals others in response to cues from the internal or external environment. Assorted local and long-distance signals lead to local and regional changes in metabolism, gene expression, growth, and development.

The molecular mechanisms by which cells "talk" to one another evolved early in the history of life. They involve three events: 1) activation of a specific receptor as by reversible binding of a signaling molecule, 2) the transduction of the signal into a molecular form that can operate in the cell, and 3) a functional response.

Most receptors are membrane proteins of the sort shown in Section 3.3. When activated, many change shape and start signal transduction. One enzyme might activate many molecules of a different enzyme, which then activates many molecules of another kind, and so on in cascading reactions that amplify the signal.

In the next two units, you will come across diverse cases of signal reception, transduction, and response. For now, a simple example will give you a sense of the kinds of events the signals set in motion.

The very first cell of a new multicelled individual holds the blueprints that will guide its descendants through growth, development, and reproduction, and often death. As part of that program, many cells heed calls to self-destruct when they finish their prescribed function. They also can execute themselves when they become infected, cancerous, or otherwise pose a threat to the body as a whole.

Apoptosis (app-uh-TOE-sis) is the formal name for programmed cell death. It starts up with signals that unleash enzymes of self-destruction, which each body cell synthesizes and stores in case they are needed.

Proteases are among the self-destruction enzymes. They can break apart structural proteins, including the protein subunits of the cytoskeleton and the histones that keep DNA structurally organized (Figure 17.10). Other enzymes called nucleases chop up the DNA.

A cell in the act of apoptosis shrinks away from its neighbors. Its surface bubbles outward and inward. No longer are its chromosomes extended through the nucleoplasm; they are bunched up near the nuclear envelope. The nucleus and then the cell break apart. Phagocytic cells that patrol and protect body tissues swiftly engulf the suicidal cells and remnants of them.

Cells were busily committing suicide when your own hands were developing. The human hand, recall, starts out as a paddlelike structure. If all goes well, apoptosis makes the paddle split up into individual

Figure 17.11 *Animated!* Human finger development. (**a**) Forty-eight days after fertilization, webs still hold embryonic fingers together. (**b**) Three days later, after apoptosis, no more webs. (**c**) Fingers stay stuck together if cells do not die on cue.

fingers over the course of just a few days. Figure 17.11 shows the outcome.

Often the timing of cell death can be predicted. For instance, keratinocytes are the most common cells in the skin; they usually survive for about three weeks. Dead ones are continually sloughed off and replaced at the skin surface. But these and other cells can be induced to die well ahead of schedule. All it takes is a sensitivity to signals that can call up the enzymes that carry out apoptosis.

Control genes suppress or trigger programmed cell death. One, *bcl-2*, helps keep normal body cells from dying before their time. Self-destruction fails in cancer cells, which are supposed to respond to signals to die, but do not. Maybe they lose the normal receptors that keep them in contact with signaling neighbors. Maybe they get abnormal signals about when to grow, divide, or stop dividing. In some kinds of cancer, for instance, the suppressor gene *bcl-2* has been tampered with or shut down. Apoptosis is no longer a control option.

What about plants? The controlled death of xylem cells creates pipelines that carry water throughout the plant. Also, controlled cell death helps plants to slow the spread of pathogens as part of acquired systemic resistance and compartmentalization.

Cells within a plant or animal body communicate with one another by secreting signaling molecules into extracellular fluid and selectively responding to signals from other cells.

Communication involves receiving signals, transducing them, and bringing about change in a target cell's activity.

Signal transduction may involve shape changes in receptors and other membrane proteins. It also involves activation of enzymes and other molecules in the cytoplasm.

Summary

Section 17.1 Anatomy is the study of body form. Physiology is the study of how the body functions in the environment. The structure of most body parts correlates with current or past functions.

A plant or animal's structural organization emerges during growth and development. Each cell in the body performs metabolic functions that ensure its survival. Cells are organized in tissues, organs, and often organ systems. These structural units function in coordinated ways in the performance of specific tasks.

Section 17.2 Plants and animals have responded in similar ways to certain environmental challenges. All have the means for exchanging gases, transporting substances to and from cells, maintaining the body's water and solute concentrations within tolerable limits, and integrating and controlling body parts in ways that ensure survival of the whole organism. Both plant and animal cells have mechanisms for sensing and responding to environmental changes.

Section 17.3 Homeostasis is a state in which the body's internal environment is being maintained within a range that keeps cells alive. In animal cells, sensory receptors, integrators, and effectors interact to maintain homeostasis, often by feedback controls.

With negative feedback mechanisms, a change in some condition, such as body temperature, triggers a response that results in reversal of the change. With positive feedback mechanisms, a change leads to a response that intensifies the condition that caused it.

Biology Now
Learn about negative feedback and the control of body temperature with animation on BiologyNow.

Section 17.4 Plants have decentralized mechanisms that help maintain the internal environment and ensure their survival. Plants respond to pathogens and wounds with protective responses, such as systemic acquired resistance and compartmentalization. Some respond to changes in environmental conditions by folding leaves. Rhythmic leaf folding is a circadian rhythm that may help a plant maintain its temperature at night.

Biology Now
Observe compartmentalization and rhythmic leaf folding with animation on BiologyNow.

Section 17.5 Cell communication often involves signal reception and transduction, then a response by the target cell. Many signals are transduced through shape changes in membrane proteins that trigger reactions in the cell. Reactions may alter the activity of genes or enzymes that take part in cellular events. For example, after a signal unleashes protein-cleaving enzymes during apoptosis, a target cell self-destructs.

Biology Now
See how apoptosis helps shape a human hand with video on BiologyNow.

b   c   d

Figure 17.12 (**a**) Biologist Consuelo De Moraes, who studies plant stress responses. (**b**) Caterpillar (*Heliothis virescens*) chewing a tobacco plant leaf. In response, the plant secretes chemicals. (**c**) The airborne chemicals attract a parasitic wasp, *Cardiochiles nigriceps*. (**d**) The wasp grabs the caterpillar, then inserts one egg inside it. The egg will grow and develop into a larva that will feed on the caterpillar's tissues and kill it.

Self-Quiz

Answers in Appendix I

1. An increase in the number, size, and volume of plant cells or animal cells is called _____ .
   - a. growth
   - b. development
   - c. differentiation
   - d. all of the above

2. The internal environment consists of _____ .
   - a. all body fluids
   - b. all fluids in cells
   - c. all body fluids outside cells
   - d. interstitial fluid

3. _____ influences the concentrations of water and solutes in the internal environment.
   - a. Diffusion
   - b. Active transport
   - c. Passive transport
   - d. All are correct

4. As basic functions, a plant or animal must _____ .
   - a. maintain a stable internal environment
   - b. obtain and distribute water and solutes
   - c. dispose of wastes
   - d. defend the body against threats
   - e. all of the above

5. Interstitial fluid is found _____ .
   - a. inside cells
   - b. in blood vessels
   - c. in spaces between cells
   - d. all of the above

6. Cell communication typically involves signal _____ .
   - a. reception
   - b. transduction
   - c. response
   - d. all of the above

7. A sweat gland is an example of _____ .
   - a. an integrator
   - b. an effector
   - c. a receptor
   - d. none of the above

8. Match the terms with their most suitable description.
   - ____ circadian rhythm
   - ____ homeostasis
   - ____ integrator
   - ____ apoptosis
   - ____ negative feedback
   - ____ effector
   - ____ receptor
   - a. programmed cell death
   - b. 24-hour or so cyclic activity
   - c. stable internal environment
   - d. central command center
   - e. an activity changes some condition, then the change causes its own reversal
   - f. detects some stimulus
   - g. carries out some response

Additional questions are available on **Biology ⊜ Now**™

Critical Thinking

1. Consuelo De Moraes studies the interactions among plants, caterpillars, and parasitoid wasps. A *parasitoid* is a kind of parasite that lays its eggs in its host. After eggs hatch, they develop into larvae that feed on the host's tissues, killing it by devouring it from the inside out.

   A caterpillar chewing on a tobacco plant leaf secretes saliva. Some chemicals in the saliva trigger the leaves to release molecules that diffuse through the air. A parasitoid wasp follows the concentration gradient to the leaves, grabs the caterpillar, and deposits an egg in it. Eggs will develop into caterpillar-munching larvae (Figure 17.12).

   Plant responses are highly specific. Leaf cells release different chemicals in response to different caterpillar species. Each chemical attracts only the wasps that attack the kind of caterpillar that triggers the chemical's release.

   This interaction benefits the wasps and the plants. But how could it evolve? Give a possible explanation for this plant–wasp interaction in terms of benefits to each species and natural selection theory.

2. The Arabian oryx (*Oryx leucoryx*), an endangered antelope, evolved in the harsh deserts of the Middle East. Most of the year there is no free water, and temperatures routinely reach 117°F (47°C). The most common tree in the region is the umbrella thorn tree (*Acacia tortilis*). List common challenges that the oryx and the acacia face. Also research the morphological, physiological, and behavioral responses of both organisms to the challenges.

3. High humidity increases the risk of heat stroke; it reduces the rate of evaporative cooling. So does wearing tight clothing that doesn't "breathe," like the uniform Korey Stringer wore during his last practice.

   The elderly and the very young also are susceptible to heat stroke. Other factors that increase the risk are obesity, poor circulation, dehydration, and alcohol intake. Using Figure 17.7 as a reference, suggest how these risk factors might interfere with the homeostatic controls over the body's internal temperature.

The Human Touch

In 1722, on Easter morning, a European explorer landed on a small volcanic island and found a few hundred skittish, hungry Polynesians living in caves. He saw dry grasses and scorched shrubs, but no trees. He noticed about 200 massive stone statues near the coast and 700 unfinished, abandoned ones in inland quarries. Some weighed fifty tons. What did they represent?

Two years later James Cook visited and saw only four canoes on the whole island. Nearly all of the statues had been tipped over, often onto face-shattering spikes.

Later, researchers solved the mystery of the statues. Easter Island, as it came to be called, is only 165 square kilometers (64 square miles) in size. Voyagers from the Marquesas discovered this eastern outpost of Polynesia around A.D. 350. The place was a paradise. Its fertile soil supported dense forests and lush grasses. New arrivals built canoes from long, straight palms strengthened with rope made of fibers from hauhau trees. They used wood as fuel to cook fish; they cleared forests to plant crops. They also had many children.

By 1400, possibly as many as 15,000 people were living on the island. Crop yields declined; erosion and harvesting had depleted the soil's nutrients. In time, fish vanished from nearshore waters around the island. All the native birds had been eaten and people were raising rats as food.

Survival was at stake; those in power appealed to the gods. They directed the people to carve divine images of unprecedented size, and to use a system of greased logs to move them over miles of rough terrain to the coast. By about 1550, no one ventured offshore to fish. No one

could build canoes because people had cut down all the palms and used all the hauhau trees for firewood. Easter Islanders turned to their last source of protein. They started to hunt and eat one another.

Central authority crumbled and gang wars raged. Those on the rampage burned the remaining grasses to destroy hideouts. The dwindling population retreated to caves and launched raids against perceived enemies. Winners ate losers and tipped over the statues. There was nowhere to go. What could they have been thinking when they chopped down the last palm?

From archeological and historical records, we know Easter Island sustained a society that flourished amid abundant resources, and then abruptly fell apart as those resources were exhausted. What happened on this small bit of land in the vastness of the Pacific is a chilling reminder that all resources are finite and no population can soar forever.

With this example in mind, we turn to the principles that govern the growth of all populations. Later in the chapter, we'll consider how they apply to the 6.4 billion humans now inhabiting Earth, our own island in space.

 How Would You Vote? With few predators, deer populations are soaring so high that they are harming many forests. Deer hunting is the least expensive way to keep deer populations in check. Do you support encouraging hunting in areas where the presence of too many deer is degrading the habitat and threatening other species? See BiologyNow for details, then vote online.

Key Concepts

DESCRIBING POPULATIONS
Beyond its shared phenotypic traits, every population has a characteristic size, density, and age structure. It also shows patterns of distribution and growth.

LIFE HISTORY PATTERNS
Longevity, fertility, and the age when reproduction begins are examples of life history traits. Like other heritable traits, they are subject to natural selection.

HUMAN POPULATIONS
Humans have sidestepped certain limits on population growth that affect other species. Our population size has soared as a result.

Links to Earlier Concepts

As described in Section 1.1, a population is one level of organization in nature. Populations grow when individuals reproduce sexually or asexually (7.1, 14.2). Environmental challenges (17.2) can limit population growth, and so can infectious disease (14.6). Think back on how humans evolved (16.13), as you read about how they side-stepped many controls on population growth.

Recall too how natural selection can influence the frequency of traits that characterize a population (12.3, 12.5). This will help you understand how life history traits can be altered.

18.1 Characteristics of Populations

*Certain principles govern the growth and sustainability of populations. They are the bedrock of **ecology**, the systematic study of how organisms interact with one another and their physical and chemical environment. The interactions start within and between populations, and extend through communities, ecosystems, and the biosphere. These are topics of this last unit of the book.*

From earlier discussions, you know that a population consists of all the members of a species living in some defined area at a particular time. The gene pool of a population, along with environmental factors, is the basis for morphological, physiological, and behavioral traits. When studying a population, ecologists collect information about the genes, modes of reproduction, and behavior of its component individuals. They also consider **demographics**. These are vital statistics that describe the population, such as its size, age structure, density, and distribution.

OVERVIEW OF THE DEMOGRAPHICS

Population size refers to the number of individuals that make up a population's gene pool. A population's **age structure** is the number of individuals in each age category. For instance, members may be grouped into *pre-reproductive*, *reproductive*, and *post-reproductive* ages. Pre-reproductive individuals will have the capacity to produce offspring later, when they mature. Together with the individuals that are already able to reproduce, they make up the population's **reproductive base**.

Population density is the number of individuals in some specified area or volume of a habitat, such as the number of frogs per acre of rainforest. A *habitat*, recall, is the type of place where a species normally lives. We characterize a habitat by physical features, chemical features, and other species sharing the same habitat.

Crude density is a measured number of individuals in some specified area. You'll see how ecologists make counts that can help them track changes in population density over time. The counts do not reveal how much of a habitat is utilized as living space. Even areas that seem uniform, such as a long, sandy beach, are more like tapestries of light, moisture, temperature, mineral composition, and other variables. Also, environmental conditions may change from day to night or with the seasons, so part of a habitat might be more suitable than other parts some or all of the time.

Different species that share the same area typically compete for energy, nutrients, living space, and other resources. As later chapters explain, species interact in ways that benefit or harm one another. For now, it's enough to know that species interactions influence a population's density and dispersion through a habitat.

Population distribution describes how individuals are dispersed in a specified area. There are three types of distribution patterns: clumped, nearly uniform, or random (Figure 18.1). In nature, individuals of most species tend to clump together, rather than spread out randomly or uniformly.

Why is a clumped distribution so common? First, each species is adapted to a limited set of ecological conditions that usually occur in patches. For example, animals often cluster in an area that provides access to food, water, or nesting sites. Plants survive only where soil, water, and sunlight are adequate. Second, many

LINK TO SECTION 1.1

clumped

nearly uniform

random

Figure 18.1 Three patterns of population distribution: clumped, as in squirrelfish schools; more or less uniform, as a penguin breeding colony; and random, as when wolf spiders prowl on a forest floor.

animal species form social groups. These groups can promote survival and reproduction, as by providing more opportunities for mating and mutual defense against predators. Third, the offspring of many species do not disperse far from their parents. As the saying goes, the nut doesn't fall very far from the tree. Many larvae and immature forms of animals also have little mobility. For example, sponges are usually clumped because the larvae settle down close to their parents. Asexual reproduction also results in clumping.

When individuals are more evenly spaced than we would expect them to be by chance, they show a near uniform dispersal pattern. This pattern often arises as a result of territoriality or stiff competition for some essential resource. Many nesting colonies of seabirds show this distribution because each pair actively keeps other individuals from settling too close to them.

We observe random dispersion only when habitat conditions are nearly uniform, resource availability is fairly constant, and members of a population neither attract nor avoid one another. Wolf spiders, which are solitary hunters on forest floors, are an example. Each generation may be randomly spaced.

The regional scale we use to consider distribution can affect what we observe. For example, although sea birds will display a nearly uniform distribution within their colony, their colony sites are usually clumped. In addition, seabirds that show a clumped distribution at mating time are randomly distributed during other times of the year.

ELUSIVE HEADS TO COUNT

From a practical standpoint, measuring a population's size or density, or determining how it is distributed, can can be challenging. Making an absolute count of all individuals within the population's range is nearly always impossible. Many species are difficult to locate, with members distributed over a wide area. With few exceptions, biologists study population characteristics by making their counts in some portion or portions of a population's range, and then using that data to make estimates about the population as a whole.

For example, you could get a map of some region and divide it into small plots, or quadrats. **Quadrats** are sampling areas of the same size and shape. You could then count and collect data about individuals in these plots and use that to make estimates about the general population. Ecologists often use this method when they are studying plants and other species that do not move around much.

Suppose you want to study a population of animals that roam from place to place. How can you determine if the individuals you count in a given plot are the same ones you counted earlier in a different plot?

The population density of mobile animals is very often estimated by **capture–mark–recapture methods**. The idea is to capture individuals and mark them in some way. Deer get collars, squirrels get dye, salmon get fin tags, birds get leg rings, butterflies get wing markers, and so on (Figure 18.2). Marked animals are released at time 1. Some time later, at time 2, the traps are reset. The proportion of marked individuals in the second group of captives is taken to be representative of the proportion marked in the whole population:

$$\frac{\text{Marked individuals in sampling at time 2}}{\text{Total captured in sampling 2}} = \frac{\text{Marked individuals in sampling at time 1}}{\text{Total population size}}$$

Ideally, the marked and unmarked individuals of the population are captured at random, none of the marked animals die during the course of the study, and none migrate out of the study area. In the real world, recapturing marked individuals might *not* be random. Squirrels marked after being attracted to bait in boxes might now avoid the boxes, or seek them out. In addition, markings may affect an animal's health or make it more conspicuous to predators.

The timing of such studies can affect the results. Few places yield abundant resources all year long, so many animal populations move between habitats as seasons change. In such cases, the capturing, marking, and then recapturing of individuals might be carried out more than once a year, for several years.

Figure 18.2 Animals marked for population studies. (**a**) Costa Rican owl butterfly (*Caligo*) and (**b**) Florida Key deer.

Each population has characteristic demographics: size, density, distribution pattern, and age structure.

Environmental conditions and species interactions shape these characteristics, which may change over time.

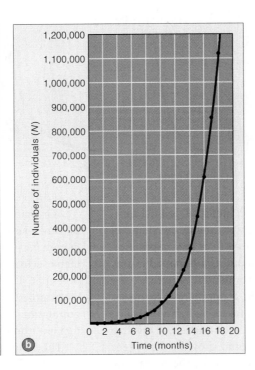

		Net Monthly Increase:		New Population Size:
$G = r \times$	3,920 =	1,568	=	5,488
$r \times$	5,488 =	2,195	=	7,683
$r \times$	7,683 =	3,073	=	10,756
$r \times$	10,756 =	4,302	=	15,058
$r \times$	15,058 =	6,023	=	21,081
$r \times$	21,081 =	8,432	=	29,513
$r \times$	29,513 =	11,805	=	41,318
$r \times$	41,318 =	16,527	=	57,845
$r \times$	57,845 =	23,138	=	80,983
$r \times$	80,983 =	32,393	=	113,376
$r \times$	113,376 =	45,350	=	158,726
$r \times$	158,726 =	63,490	=	222,216
$r \times$	222,216 =	88,887	=	311,103
$r \times$	311,103 =	124,441	=	435,544
$r \times$	435,544 =	174,218	=	609,762
$r \times$	609,762 =	243,905	=	853,667
$r \times$	853,677 =	341,467	=	1,195,134

(a)

(b)

Figure 18.3 *Animated!* (**a**) Data on net monthly increases in a population of field mice living in a cornfield. Start to finish, the numbers listed show a pattern typical of exponential growth. (**b**) Graph the data and you end up with a J-shaped growth curve.

18.2 Population Size and Exponential Growth

When resources are abundant, birth rates often exceed death rates and a population grows faster and faster. Each generation includes more breeding individuals than the prior one did, so each produces more offspring.

FROM ZERO TO EXPONENTIAL GROWTH

We measure changes in population size in terms of birth rates, death rates, and how many individuals are entering and leaving during a specified interval. The size increases by births and immigration. **Immigration** is a permanent addition of individuals that previously belonged to another population. Population size falls by deaths and emigration. **Emigration** is a permanent loss of individuals to another population. **Migration** is a round trip between regions. Many species migrate daily or seasonally, but because individuals return to the starting point, we need not consider this transient effect in our initial study of population size.

For our purposes, assume immigration is balancing emigration over time, so we may ignore the effects of both on population size. Doing so allows us to define **zero population growth** as an interval during which the number of births and the number of deaths are balanced. The population's size is stabilized during an interval, with no overall increase or decrease.

Births, deaths, and other variables that might affect population size can be measured as **per capita** rates, or rates per individual. *Capita* means heads, as in head counts. Imagine 2,000 mice in a cornfield. Twenty or so days after their eggs get fertilized, the female mice produce a litter, nurse offspring for a month or so, and get pregnant again. If they give birth to 1,000 mice every month, the birth rate will be 1,000/2,000 = 0.5 per mouse per month. If 200 of the 2,000 mice die in that interval, the death rate would be 200/2,000 = 0.1 per mouse per month.

If we assume the birth rate and death rate remain constant, we can combine both into one variable: the net reproduction per individual per unit time, or *r*. In this example, *r* is 0.5 − 0.1 = 0.4 per mouse per month. In equation form, $G = rN$, which simply means:

population growth per unit time	=	net population growth rate per individual per unit time	×	number of individuals

As the next month begins, 2,800 mice are scurrying about. With that net increase of 800 fertile mice, the reproductive base is larger. Assume *r* doesn't change. A net increase of 0.4 × 2,800 = 1,120 puts the population at 3,920 (Figure 18.3*a*). Over time, *r* stays constant, and a growth pattern emerges. In less than two years from the first count, the number of mice in the cornfield has increased from 2,000 to more than a million!

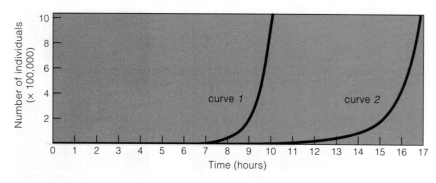

Figure 18.4 Effect of deaths on the rate of increase in two bacterial populations. Plot population growth for bacterial cells that reproduce every half hour and you get growth curve *1*. Plot the growth of a population of cells that divide every half hour, with 25 percent dying between divisions, and you get growth curve *2*. Deaths slow the rate of increase but can't directly prevent exponential growth.

The carrying capacity of an environment can vary over time. Drought may have accelerated the unraveling of the agriculturally based society on Easter Island, shown in the satellite image above. Long-term drought may have been a factor in the crop failures, deforestation, and increased social unrest. This volcanic island has few streams. Most of the rainfall seeps into underground basins. Even today, islanders face severe water shortages, often for extended times.

Plot the monthly population increases against time, and you will end up with a graph line in the shape of a J, as in Figure 18.3*b*. When growth of any population over increments of time plots out as a J-shaped curve, you know you are tracking exponential growth.

With **exponential growth**, a population increases by a *fixed* percentage of the total in each interval. The number of births and deaths are a function of the size of a population. With exponential growth, the bigger the reproductive base of a population, the more new individuals will be added during a specified interval. The rate of increase for the population gets higher and higher, even though the per capita rate of increase does not change.

Consider a lone bacterium in a culture flask with all the nutrients it requires for growth. Thirty minutes later, the cell divides in two. Thirty minutes pass, and each of these two cells divides, and so on every thirty minutes. If no cells die between divisions, population size will double in each interval—from 1 to 2, then 4, 8, 16, 32, and so on. The length of time it takes for a population to double in size is its **doubling time**.

The larger the population gets, the more cells there are to divide. After 9–1/2 hours (nineteen doublings), it has more than 500,000 bacterial cells. After 10 hours (twenty doublings), there are more than 1 million cells. Plot the size of this population against time, and you will get Figure 18.4, curve *1*.

To see the effect of deaths on the growth rates, start over. Assume 25 percent of the first cell's descendants die every half hour. It now takes about seventeen hours (not ten) for population size to reach one million. But the deaths did not stop the population from growing (Figure 28.4, curve *2*). As long as the birth rate exceeds the death rate, a population will grow exponentially.

WHAT IS THE BIOTIC POTENTIAL?

Every species has a **biotic potential**, a maximum rate of increase per individual under ideal conditions. This is the theoretical rate at which the population would increase if resources were unlimited and there were no predators, parasites, or pathogens around.

For many bacteria, the biotic potential can be 100 percent every half hour. For humans and other large mammals, it's 2 to 5 percent per year. The *actual* rate of increase in nature is nearly always lower than biotic potential. The degree of difference between a species' biotic potential and the observed rate of increase for a population in some environment shows the effect of that environment on that population.

The human population is not displaying its biotic potential. A woman is biologically able to bear twenty or more children over the course of her lifetime, but few do. Even so, our population size has been growing exponentially ever since the mid-eighteenth century, for reasons we will discuss later in this chapter.

> *During a specified interval, population size is generally an outcome of births, deaths, immigration, and emigration.*
>
> *With exponential growth, population size increases by a fixed percentage of the whole in each interval, so its reproductive base gets larger and larger over time. A plot of population size against time produces a J-shaped curve.*
>
> *As long as the per capita birth rate remains above the per capita death rate, a population will grow exponentially.*

18.3 Limits on the Growth of Populations

In nature, limits on resource availability usually prevent population size from expanding above a certain point. The quantity of resources varies among environments, and over time within an environment.

WHAT ARE THE LIMITING FACTORS?

Nearly all the time, environmental circumstances keep any population from fulfilling its biotic potential. That is why sea stars, the females of which could produce 2,500,000 eggs each year, do not fill up the oceans with

Figure 18.5 Amazing response to a scarcity of nesting sites, a limiting factor for weaver bird populations in dry habitats in Africa.

(**a**) African weavers construct densely woven, cup-shaped nests that are only wide and deep enough for a hen and her nestlings. Many nests hang from the same spindly limbs.

(**b**) This nest in Namibia is like an apartment house in a place where few trees are available. Between 100 and 300 pairs of sparrow weavers occupy their own flask-shaped nests, each with its own tubular entrance.

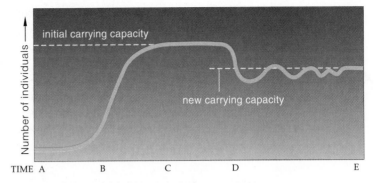

Figure 18.6 *Animated!* Idealized S-shaped curve characteristic of logistic growth. Growth slows after a phase of rapid increase (time B to C). The curve flattens out as the carrying capacity is reached (time C to D). S-shaped growth curves can show variations, as when changes in the environment lower the carrying capacity (time D to E).

sea stars. That also is why populations of humans will never fill up the planet.

To get a sense of what some of the constraints may be, start again with a bacterial cell in a culture flask, where you can control the variables. First you enrich the culture medium with glucose and other nutrients necessary for bacterial growth. Then you sit back and let bacterial cells reproduce for many generations.

At first the growth pattern seems to be exponential. Then growth slows, and population size is relatively stable. After the stable period, the size plummets until all the bacterial cells are dead. *What happened?* As the population grew ever larger, it used up more and more nutrients. Nutrient scarcity became an environmental signal for cells to stop dividing. When the supply was exhausted, the cells starved to death.

Any essential resource that is in short supply is a **limiting factor** on population growth. Food, mineral ions, refuge from predators, living space, and even a place to build a nest are examples (Figure 18.5). The number of limiting factors can be extensive, and their effects can vary. Even so, a scarcity of one factor alone often puts the brakes on population growth.

What if you had kept on freshening the supply of nutrients for the bacterial population? After growing exponentially, that population would have collapsed anyway. Like all other organisms, the bacteria produce metabolic wastes. Confined populations of bacteria drastically alter their own living conditions. By their own metabolic activities, they pollute experimentally

designed habitats and, in so doing, put a stop to any further exponential growth.

CARRYING CAPACITY AND LOGISTIC GROWTH

Now visualize a small population, with its individuals dispersed throughout the habitat. As its size increases, more and more individuals must share nutrients, living quarters, and other resources. As the share available to each diminishes, fewer individuals may be born, and more may die by starvation or nutrient deficiencies. The population's growth rate will slow until births are balanced (or outnumbered) by deaths. Ultimately, the *sustainable* supply of resources determines how large the population can become. **Carrying capacity** is the maximum number of individuals of a species that a given environment can sustain indefinitely.

The pattern of **logistic growth** shows how carrying capacity can affect population size. By this pattern, a small population starts growing slowly in size, then it grows rapidly, and finally its size levels off once the carrying capacity is reached. Figure 18.6 reveals how the pattern plots out as an S-shaped curve. Here is a way to represent this pattern as an equation:

population growth per unit time	=	maximum net population growth rate per individual per unit time	×	number of individuals	×	proportion of resources not yet used

LINK TO
SECTION
14.6

An S-shaped curve is only an approximation of what actually happens in nature. A population that is growing really fast may overshoot carrying capacity. If this happens, the death rate skyrockets, and the birth rate plummets. Population size may drop even below carrying capacity. The human population on Easter Island went through such a soar-and-crash experience. Figure 18.7 shows another example, the explosion and die-off of a reindeer population on an island.

DENSITY-DEPENDENT CONTROLS

The logistic growth equation just described deals with **density-dependent controls**. These controls are factors that come into play when population density increases. They adversely affect survival and reproduction.

When a small, rapidly growing population reaches its carrying capacity, dwindling resources work as a control to stabilize or decrease numbers. Besides the environmental factors, interactions among members of the population may keep the number of individuals below the maximum sustainable level.

Crowding often leads to interactions that increase death rates. Predators, parasites, and pathogens may exert greater effects when prey or host populations are at a high density. They usually reduce the numbers of prey or hosts. By thinning a population, they remove the condition that invited their controlling effect, and the population size may increase again.

Bubonic plague and *pneumonic plague*, two terrible diseases, are examples of density-dependent controls. Both are caused by a bacterium, *Yersinia pestis*. A large reservoir of *Y. pestis* persists inside rabbits, rats, and certain other small mammals. Fleas transmit the pest by biting new hosts. Bacterial cells reproduce rapidly in the flea gut. In time, their numbers and metabolic activities disrupt digestion and cause gnawing hunger sensations. They provoke the flea into feeding more often on more of its hosts, and so the disease spreads.

Bubonic plague devastated Europe in the fourteenth century. It swept through cities where many people were crowded together, sanitary conditions were poor, and rats were abundant. That one epidemic claimed 25 million lives. Plagues are still threats.

DENSITY-INDEPENDENT FACTORS

Sometimes events cause more deaths or fewer births regardless of a population's density. For instance, each year, millions of monarch butterflies fly down from Canada to Mexico's forested mountains, where they spend the winter (Figure 18.8). But logging cleared out many forest trees, which normally buffer temperature. In 2002 a sudden freeze, combined with deforestation, killed millions of butterflies. The events were **density-independent factors**, meaning that their effects came into play independently of population density.

Similarly, heavy applications of pesticides in your backyard may also kill most insects, mice, cats, birds, and other animals. They will do this regardless of how sparse or dense the populations are.

Resources in short supply put limits on population growth. Together, all of the limiting factors acting on a population dictate how many individuals can be sustained.

Carrying capacity is the maximum number of individuals of a population that can be sustained indefinitely by the resources in a given environment. The number may rise or fall with changes in resource availability.

The size of a low-density population may increase slowly, go through a rapid growth phase, then level off once the carrying capacity for the population is reached. This is a logistic growth pattern.

Density-dependent controls and density-independent factors can bring about decreases in population size.

Figure 18.7 Carrying capacity of a reindeer herd on an island. In 1944, twenty-nine reindeer were introduced to St. Matthew Island in the Bering Sea. With no predators, numbers grew into the thousands. Then, during the winter of 1963–64, a scarcity of food coupled with a heavy snowfall decimated the herd. In 1966, the island supported only forty-two reindeer. The curve reflects how the population overshot carrying capacity, and then crashed.

DESCRIBING POPULATIONS

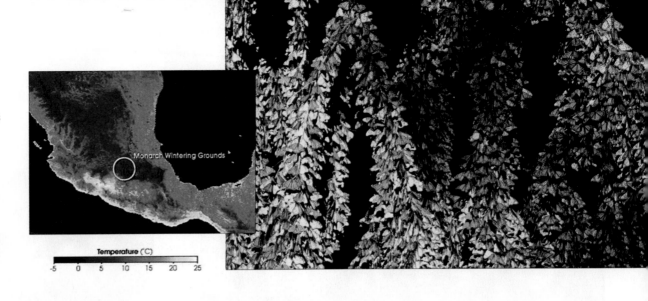

Figure 18.8 Monarch butterflies at their wintering ground in central Mexico. Each year these migratory insects travel hundreds of kilometers south to places that normally have been cool and humid in winter. If they were to stay in their northern breeding grounds, monarchs would risk being killed by more severe weather conditions.

Even so, deforestation and winter freezes in Mexico are now killing millions of them annually. The satellite image records the winter freeze in January 2002.

18.4 Life History Patterns

Researchers have identified age-specific adaptations that affect the survival, fertility, and reproduction of individuals of each species.

So far, we have looked at populations as if all of their members are identical during any given interval. For most species, however, the individuals of a population are at many different stages in a life cycle. Each stage may interact in a different way with other organisms and with the environment. Different stages may make use of different resources, as when larvae eat leaves and butterflies sip nectar. They may also be more or less vulnerable to predation or other threats.

In short, each species has a **life history pattern**, or a set of adaptations that influence longevity, fertility, and age at first reproduction. Each pattern reflects the individual's schedule of reproduction. In this section and the next, we look at a few of the environmental variables that underlie these age-specific patterns.

LIFE TABLES

Although each species has a characteristic life span, few of its individuals get to the maximum age possible. Death looms larger at some ages than others. Another variable: Individuals of a species tend to reproduce or emigrate during a characteristic age interval.

Age-specific patterns in populations are of interest to life insurance companies, as well as ecologists. Such investigators typically track a **cohort,** a collection of same-aged individuals, from the cohort's birth until all are dead. They also track the number of offspring born to all individuals during each interval. Life tables present this data as an age-specific death schedule.

Often the death data are converted into much cheerier "survivorship" schedules: the number of individuals reaching some specified age (*x*). Table 18.1 is a typical example. It lists data for the human population of the United States in 2001.

PATTERNS OF SURVIVAL AND REPRODUCTION

Evolutionarily speaking, we measure the reproductive success of individuals in terms of the number of their surviving offspring. But that number will vary among species, which differ in (1) how much energy and time are allocated to producing gametes, securing mates, and parenting, and in (2) the size of offspring. It seems

Table 18.1	Life Table for the Human Population of the United States in 2001			
Age Interval	Number at Start of Interval	Number Dying During Age Interval	Life Expectancy at Start of Interval	Reported Live Births
0–1	100,000	684	77.2	
1–5	99,316	132	76.7	
5–10	99,184	76	72.8	
10–15	99,108	96	67.9	7,315
15–20	99,012	330	62.9	525,493
20–25	98,682	468	58.1	1,022,106
25–30	98,214	471	53.4	1,060,391
30–35	97,743	554	48.6	951,219
35–40	97,189	801	43.9	453,927
40–45	96,388	1,154	39.2	95,788
45–50	95,234	1,682	34.7	5,244
50–55	93,552	2,373	30.3	263
55–60	91,179	3,474	26.0	
60–65	87,705	5,186	21.9	
65–70	82,519	7,397	18.1	
70–75	75,122	10,018	14.6	
75–80	65,104	13,284	11.5	
80–85	51,820	15,877	8.8	
85–90	35,943	16,147	6.5	
90–95	19,796	11,906	4.8	
95–100	7,890	5,845	3.6	
100+	2,045	2,045	2.7	

Figure 18.9 Three generalized survivorship curves. (**a**) Elephants and petunias are Type I populations. They both have high survivorship until some age (seventy years for an elephant, one for a petunia), then high mortality. (**b**) Snowy egrets are Type II populations. They have a fairly constant death rate. (**c**) Sea star larvae represent Type III populations, which show low survivorship early in life.

LINKS TO
SECTIONS
12.3, 12.5

that trade-offs have been made in response to selection pressures, such as prevailing conditions in the habitat and the type of species interactions.

A **survivorship curve** is the graphical equivalent of a life table. It shows the percentage of cohort members that remain alive at each age interval. Three general types of survivorship curves are possible. Each type of curve is associated with a particular way of allocating resources to reproduction.

Type I curves reflect high survivorship until fairly late in life, then a large increase in deaths. Such curves typify annual plants as well as large mammals that bear only one or a few large offspring at a time, then engage in extended parental care (Figure 18.9*a*). As an example, a female elephant gives birth to four or five calves and devotes several years to parenting each. Type I curves also are typical of human populations with widespread access to modern health care services.

Type II curves reflect a more or less constant death rate throughout the life span. These curves are typical of organisms that are just as likely to be killed or to die of disease at any age, such as many lizards, small mammals, and large birds (Figure 18.9*b*).

Type III curves signify a death rate that is highest early in life. We see this for species that produce many

small offspring and do little, if any, parenting. Most trees show this type of curve. So do nearly all marine invertebrates (Figure 18.9*c*). As one example, a sea star releases enormous numbers of eggs. The tiny larvae must feed, grow, and finish developing on their own without any support, protection, or guidance from parents. Most are devoured in the first days of life. The mortality rate drops as the sea stars increase in size and take on their adult spiny-skinned form.

At one time, it was thought that selection would tend to favor *either* early production of many small offspring *or* late production of a few large offspring. We now realize that the two patterns are extremes at opposite ends of a range of possible life histories. Both types of life history patterns, as well as intermediate ones, are sometimes evident in different populations of the same species, as the next example makes clear.

EVOLUTION OF LIFE HISTORY TRAITS

Any trait that has a genetic basis and varies among the members of a population has the potential to evolve in response to natural selection. Life history traits, such as age of first reproduction and number of offspring per reproductive event, are no exception.

LIFE HISTORY PATTERNS

In one long-term study, biologists David Reznick and John Endler documented the evolutionary effects of predation on the life history traits of one species of small fish, the guppy, *Poecilia reticulata*. Their study began with fieldwork in the mountains of Trinidad, an island in the southern Caribbean Sea. Here, guppies live in shallow freshwater streams (Figure 18.10).

Waterfalls in the streams act as barriers; they keep guppies from moving from one part of the stream to another, effectively isolating them genetically. Waterfalls also keep predators from moving from one part of a stream to another. As a result, different guppy populations are preyed upon by different predators.

The two kinds of guppy predators in this habitat, killifishes and pike-cichlids, differ in their body size. A killifish is a relatively small fish. It preys on the small, immature guppies but ignores big adults. Pike-cichlids are large fish. They tend to pursue big, mature guppies, and ignore small ones. Many of the streams have one of these predators, but not the other.

Reznick and Endler discovered that the guppies in streams with pike-cichlids grow faster and their body size is smaller at maturity, compared to the guppies in killifish-containing streams. Also, guppies in streams with pike-cichlids reproduce earlier, have more offspring at a time, and breed more often (Figure 18.11).

To determine whether these life history differences have a genetic basis, guppies from the two habitats were reared for several generations in aquariums. The results support the hypothesis that the differences are inherited. Descendants of guppies from streams where there were pike-cichlids showed different life history traits than descendants of guppies from streams with killifishes, even when both groups were raised under identical conditions.

Based on their field and laboratory observations, Reznick and Endler hypothesized that predation is a selective agent that shapes guppy life history patterns.

Figure 18.10 Evolutionary biologist David Reznick contemplating interactions among guppies (*Poecilia reticulata*) and their predators in a freshwater stream in Trinidad.

They set up a field experiment by introducing some guppies to a site upstream from one waterfall. Before the experiment, this waterfall had been an effective barrier to dispersal; it had prevented guppies and all predators except killifish from moving upstream. The guppies that were introduced to the experimental site were from a population that had evolved with pike-cichlids downstream from the waterfall. That part of the stream was designated the control site.

Eleven years later, researchers revisited the stream. They found that the guppy population at the upstream experimental site had evolved. Exposure to a novel predator had caused shifts in growth rate, age at first reproduction, and other life history traits. Guppies at the control site showed no such changes.

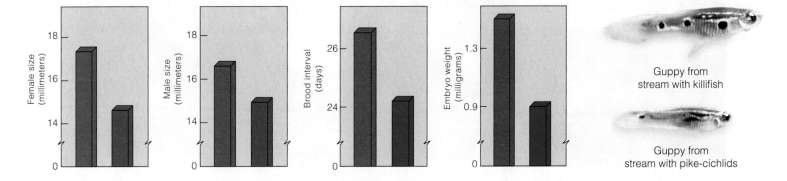

Figure 18.11 Variation in traits among guppies subject to different predation pressures. *Red* bars represent fish that evolved with killifish. *Green* bars are fish that evolved with pike-cichlids. Killifishes eat small guppies and pike-cichlids eat larger ones. The photos show males from two populations.

LINK TO
SECTION
16.13

The evolution of life history traits in response to predation pressure is not merely of theoretical interest. It can have enormous commercial importance. In the same way that predation caused evolution of guppies, fishing by humans may have caused rapid evolution in the North Atlantic codfish (*Gadus morhua*). From the mid-1980s to the early 1990s, as fishing pressure on cod increased, the fish matured at increasingly smaller ages and sizes.

In 1992, the Canadian government banned fishing for cod in some areas, but this and later closures came too late to prevent the population from plummeting. The population still has not recovered. Looking back, it seems that life history changes were an early sign that the cod population was in trouble. Had biologists recognized this, they might have been able to save the fishery and protect the livelihood of more than 35,000 fishers and associated workers. Ongoing monitoring of life history data for other economically important fishes may help prevent similar disasters in the future.

Tracking a group of individuals from birth until death provides data about age-specific survival and other life history characteristics. Like other traits, life history patterns can vary both between species and among populations of one species and can be altered by natural selection.

18.5 Human Population Growth

The world's human population has surpassed 6.4 billion. Now consider this: The annual additions to that base population will result in a larger absolute increase each year, well into the foreseeable future.

Figure 18.12 shows human population growth over time. Rapid population growth continues even though more than 1 billion people are already confronted by limits to growth. They are malnourished or starving. They are without clean water, adequate shelter, access to modern health care, or sewage treatment.

In 2004, the average rate of increase for the human population was 1.2 percent per year. As long as birth rate exceeds the death rate, these additions will result in a *larger* absolute increase each year.

WHY HAVE HUMAN POPULATIONS SOARED?

For most of its existence, the human population grew slowly. Then, during the past two centuries, growth rates skyrocketed. There are three main reasons. First, humans steadily developed their capacity to expand into many new habitats and into new climate zones.

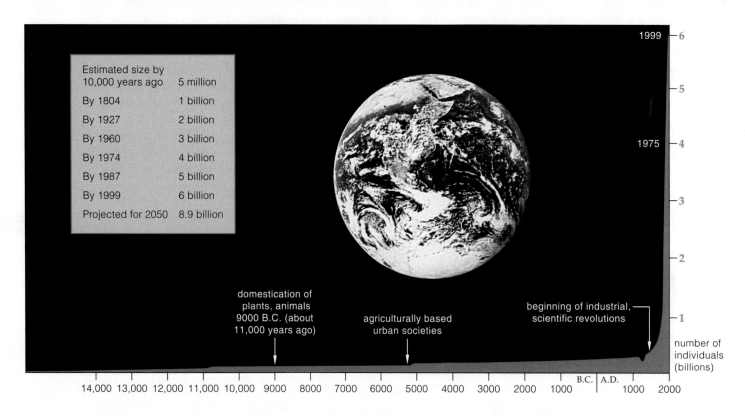

Estimated size by	
10,000 years ago	5 million
By 1804	1 billion
By 1927	2 billion
By 1960	3 billion
By 1974	4 billion
By 1987	5 billion
By 1999	6 billion
Projected for 2050	8.9 billion

domestication of plants, animals 9000 B.C. (about 11,000 years ago)

agriculturally based urban societies

beginning of industrial, scientific revolutions

number of individuals (billions)

14,000 13,000 12,000 11,000 10,000 9000 8000 7000 6000 5000 4000 3000 2000 1000 | B.C. | A.D. | 1000 2000

Figure 18.12 Growth curve (*red*) for the world human population. The *blue* box lists how long it took for the human population to increase from 5 million to 6 *billion*.

Second, humans developed new ways to increase the carrying capacity in existing habitats. Finally, human populations sidestepped limiting factors.

Early humans mostly gathered plant foods but also scavenged for meat. Small bands of hunter–gatherers started to disperse away from Africa about 2 million years ago. By 40,000 years ago, their descendants were established in much of the world (Section 16.12).

Most species can't expand into such a broad range of habitats. The early humans drew on a capacity for learning and memory to figure out how to make fires, shelters, and tools, and to plan out community hunts. Experience spread quickly from group to group by language. *The human population expanded into diverse environments very rapidly, as compared with the long-term geographic dispersal of other kinds of organisms.*

What about the second factor? Beginning 11,000 or so years ago, some human groups began to stay in one place and to domesticate various plants and animals. A pivotal factor was the domestication of wild grasses that were ancestral to modern wheat and rice.

Later in time, food supplies increased again by the use of fertilizers, herbicides, and pesticides. Methods of transportation improved; so did food distribution. *Thus, the management of food supplies through agriculture increased the carrying capacity for the human population.*

What about the third factor—the sidestepping of limiting factors? Think about what happened when medical practices and sanitary conditions improved. Until about 300 years ago, poor hygiene, malnutrition, and contagious diseases maintained high death rates. Diseases, a density-dependent control, swept through vermin-infested cities. Then came plumbing and water treatment. Vaccines and antibiotics were developed as weapons against pathogens. Death rates fell sharply, and rapid population growth was under way.

In the mid-eighteenth century, people discovered how to harness energy stored in fossil fuels, starting with coal. Industrialized societies began to form. *By controlling many disease agents and tapping into concentrated forms of energy (fossil fuels), humans have managed to sidestep major factors that previously limited their population growth.*

It took 2.5 million years for the human population size to reach 1 billion, then 123 years to reach 2 billion, then another 33 to reach 3 billion, then 14 more to get up to 4 billion, and just 13 more to get to 5 billion. It took only 12 more years to arrive at 6 billion!

We expect a decline in growth as birth rates drop or as death rates rise. Alternatively, our population growth might continue if technological breakthroughs are able to further increase carrying capacity. *However, human population growth cannot be sustained indefinitely.* Why? Density-dependent controls will kick in.

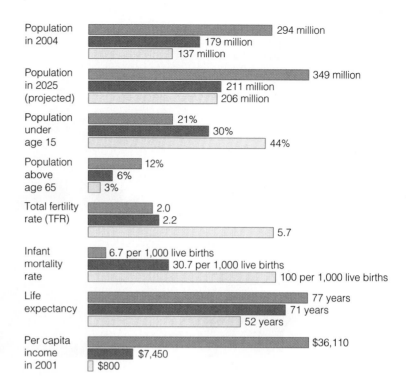

Figure 18.13 Key demographic indicators for three countries, mainly in 2003. The United States (*brown*) is highly developed, Brazil (*rust*) is moderately developed, and Nigeria (*gold*) is less developed.

FERTILITY RATES AND FAMILY PLANNING

Currently, most countries realize that their population growth, resource depletion, pollution, and quality of life are interconnected, as Figure 18.13 indicates. Many are attempting to lower long-term birth rates by providing information on methods of fertility control and access to these methods.

These attempts are having an impact; birth rates are declining worldwide. What about death rates? Overall, they are falling too, mainly because improved nutrition and health care are lowering infant mortality rates (the number of infants per 1,000 who die within the first year). However, the HIV/AIDS pandemic is causing death rates to soar in some African countries.

We still expect the world population to reach 8.9 billion by 2050. Think about the resources that will be required. We will have to boost food production and find more sources of energy and fresh water to meet even basic needs. These large-scale manipulations of resources will no doubt intensify pollution.

India, China, Pakistan, Nigeria, Bangladesh, and then Indonesia are expected to show the most growth, in that order. China (with 1.3 billion people) and India (with 1.1 billion) dwarf all other countries. Together, they make up 38 percent of the world population. The United States is next in line, with 295 million.

Figure 18.14 *Animated!* (**a**) General age structure diagrams for countries with rapid, slow, zero, and negative rates of population growth. The pre-reproductive years are *green* bars; reproductive years, *purple;* and the post-reproductive years, *light blue.* A vertical axis divides each graph into males (*left*) and females (*right*). Bar widths correspond to proportions of individuals in each age group. (**b**) Age structure diagrams for representative countries in 1997. Population sizes are measured in millions.

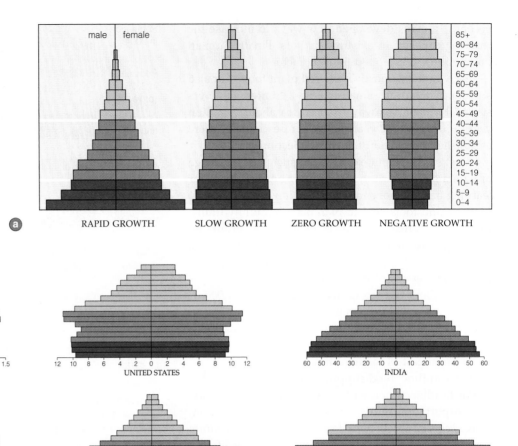

The **total fertility rate** (TFR) is the average number of children born to the women of a population during their reproductive years. TFR estimates are based on current age-specific rates. In 1950, the worldwide TFR averaged 6.5. By 2003, it had declined to 2.8 but was still above replacement levels, the number of children a couple must bear to replace themselves. At present, this replacement level is 2.1 for developed countries and as high as 2.5 in some developing countries. (It is higher in developing countries because more female children die before reaching reproductive age.)

The least developed countries continue to have the highest TFRs. In the African country of Niger, women average 7.5 births in their lifetime. About 12 percent of children born here die before their first birthday.

Comparing the age structure diagrams for different populations is revealing. In Figure 18.14, consider the size of the age groups that will be reproducing during the next fifteen years. The broader the base of an age structure diagram, the faster the population it depicts can be expected to increase in size. The United States

population has a relatively narrow base. It is growing more slowly than countries with younger populations.

Even if every couple decides to bear no more than two children, world population growth will not slow for sixty years, because 1.9 billion are about to enter the reproductive age bracket. *More than one-third of the world population is in the broad pre-reproductive base.*

ECONOMICS AND POPULATION GROWTH

We can correlate changes in population growth with changes that often unfold in four stages of economic development. These four stages are at the heart of the **demographic transition model**. By this model, living conditions are harshest during the *preindustrial* stage, before technological and medical advances become widespread. Birth and death rates are both high, so the growth rate is low. Next, in the *transitional* stage, industrialization begins. Food production and health care improve. Death rates drop. But birth rates stay high in agricultural societies. Why? Large families are

Stage 1 Preindustrial Stage 2 Transitional Stage 3 Industrial Stage 4 Postindustrial

Births and deaths (number per 1,000 per year)

relative population size

births

deaths

Change in population size

low   increasing   very high   decreasing   low   zero   negative

Growth rate over time ⟶

a supply of inexpensive labor able to help in the fields. The population continues to increase rapidly for some time. As living conditions improve, birth rates decline, and growth starts to level off (Figure 18.15).

During the *industrial* stage, when industrialization is in full swing, population growth slows dramatically. People move from the country to cities, and urban couples tend to want smaller families.

In the *postindustrial* stage, population growth rates become negative. The birth rate falls below the death rate, and population size slowly decreases.

The United States, Canada, Australia, and most of western Europe, Japan, and countries of the former Soviet Union are in the industrial stage. Their growth rate is slowly falling. Mexico and other less developed countries are in the transitional stage. If population growth continues to outpace economic growth, then death rates will likely rise. Thus many countries might get stuck in the transitional stage or go back to the harsh conditions of the preceding stage.

ENVIRONMENTAL IMPACTS OF POPULATIONS

The United States represents about 4.6 percent of the world population. Yet the people of the United States produce 21 percent of the world's goods and services. They also consume about 25 percent of all processed minerals and a big portion of nonrenewable sources of energy. They generate about 25 percent of the world's pollution and trash.

By contrast, India is home to a little over 15 percent of the world population. Yet it produces only about 1 percent of all goods and services. It uses 3 percent of minerals and nonrenewable energy resources, and it generates only about 3 percent of the total pollution and waste products.

Extrapolating from these numbers, G. Tyler Miller estimated that it would take 12.9 billion individuals living in India to have the same environmental impact as 284 million people in the United States. Think about this estimate when you reflect on possible solutions to the biodiversity crisis and environmental challenges sketched out in later chapters.

Through expansion into new habitats, cultural intervention, and technological innovation, the human population has temporarily skirted environmental resistance to growth. Its size has been skyrocketing ever since.

Birth rates fall as nations become more industrialized, but their per capita consumption of resources increases.

Figure 18.15 The demographic transition model for changes in population growth rates and in sizes, correlated with long-term changes in the economy.

Summary

Section 18.1 A population is a group of individuals of the same species occupying a given area. We typically measure its size, density, distribution, and age structure as well as characteristic ranges of heritable traits.

Biology ⊜ Now
Learn how to measure population size with the animation on BiologyNow.

Section 18.2 The growth rate for a population during a specified interval is determined by rates of birth, death, immigration, and emigration. In cases of exponential growth, the population increases by a fixed percentage of the whole with each successive interval. This trend plots out as a J-shaped curve. As long as a population's per capita birth rate is above its per capita death rate, it will show exponential growth.

Biology ⊜ Now
Observe the effects of exponential population growth with the animation on BiologyNow.

Figure 18.16 Saguaros growing very slowly in Arizona's Sonoran desert.

Section 18.3 Any essential resource that is in short supply is a limiting factor on population growth. The logistic growth equation describes how population growth is affected by density-dependent controls, such as disease or competition for resources. The population slowly increases in size, goes through a rapid growth phase, then levels off in size once carrying capacity is reached. Carrying capacity is the maximum number of individuals that can be sustained indefinitely by the resources available in their environment.

Biology ⑧ Now
Learn about logistic population growth with the animation on BiologyNow.

Section 18.4 Each species has a life history pattern. Life history traits, such as growth rate and age at first reproduction, are subject to selection pressures from predation and other environmental factors.

Section 18.5 The human population has surpassed 6.4 billion. Its rapid growth resulted from expansion into diverse habitats and agricultural and medical technologies that raised carrying capacity.

Worldwide, total fertility rate is declining. However, the broad pre-reproductive base of the human population means that population size will continue to increase for at least sixty years, even if all individuals have no more than a replacement number of children.

The demographic transition model is one way to correlate industrial and economic development with changes in population growth rates. Growth rates tend to decline as nations become more developed. But the per capita resource consumption of developed nations is many times higher.

Biology ⑧ Now
Compare the age structures of populations in different countries with the interaction on BiologyNow.

Self-Quiz

Answers in Appendix I

1. The rate at which a population grows or declines depends on the rate of _____ .
   a. births   c. immigration   e. a and b
   b. deaths   d. emigration   f. all of the above

2. Populations grow exponentially when _____ .
   a. population size expands by ever increasing increments through successive time intervals
   b. size of low-density population increases slowly, then quickly, then levels off once carrying capacity is reached
   c. a and b are characteristics of exponential growth

3. For a given species, the maximum rate of increase per individual under ideal conditions is the _____ .
   a. biotic potential   c. environmental resistance
   b. carrying capacity   d. density control

4. Resource competition, disease, and predation are _____ controls on population growth rates.
   a. density-independent   c. age-specific
   b. population-sustaining   d. density-dependent

5. An increase in the population of a prey species in some environment would most likely _____ the carrying capacity for that species' predators.
   a. increase   c. not affect
   b. decrease   d. stabilize

6. A life history pattern for a population is a set of adaptations that influence the individual's _____ .
   a. longevity   c. age when it starts reproducing
   b. fertility   d. all of the above

7. At present, the average natural rate of increase for the human population is about _____ percent per year.
   a. 0   c. 1.9   e. 3.8
   b. 1.2   d. 2.7   f. 4.0

8. Match each term with its most suitable description.
   ___ carrying capacity   a. maximum rate of increase per individual under ideal conditions
   ___ logistic growth   b. group of individuals born during the same period of time
   ___ exponential growth   c. population growth plots out as an S-shaped curve
   ___ biotic potential   d. largest number of individuals sustainable by the resources in a given environment
   ___ limiting factor   e. population growth plots out as a J-shaped curve
   ___ cohort   f. essential resource that restricts population growth when scarce

Additional questions are available on **Biology ⑧ Now™**

Critical Thinking

1. The Chittenango snail (*Novisuccinea chittenangoensis*) is New York State's most endangered animal. The dime-sized terrestrial snail lives only in the spray zone adjacent to a single waterfall. Water pollution and other disturbances threaten it. In 2002, a detailed census around the waterfall turned up 105 individuals. Each was marked and released. What type of information will marked-snail monitoring provide about this threatened species? How might this help save the snail from extinction?

2. Each summer, the giant saguaro cacti in deserts of the American Southwest produce tens of thousands of tiny black seeds apiece. Most die, but a few land in a sheltered spot and sprout the following spring. The saguaro is a CAM plant (Section 6.6) and growth is slow (Figure 18.16). After 15 years, the saguaro may be only knee high, and it will not flower until it is about age 30. The cactus may survive to age 200. Saguaros share their desert habitat with annuals, such as poppies, that sprout, produce seeds, and die in just a few weeks. How would you describe these two life history strategies? How could such different life histories both be adaptive in this environment?

3. A third of the world population is younger than fifteen. Describe the effect of this age distribution on the human population's future growth rate. If you think it will have a severe impact, what recommendations would you make to encourage these individuals to limit their family size? What are some of the factors that might prevent them from following the recommendations? How would you feel about following these recommendations yourself?

Fire Ants in the Pants

Two species of Argentine ants, *Solenopsis richteri* and *S. invicta*, entered the United States during the 1930s, most likely as cargo ship stowaways. They infiltrated communities throughout the Northern Hemisphere, starting with the southeastern states. *S. invicta*, the red imported fire ant shown at the right, has recently colonized Southern California and New Mexico.

The aptly named fire ants inflict hot, painful stings. Disturb one, and it will bite down with its mandibles and inject venom with the stinger on its abdomen. A pus-filled bump on the victim's skin follows the pain. At one time or another, about half the people who live in areas where fire ants are common have been stung. The venom has killed more than eighty of them.

Imported fire ants do more than bite humans. They attack anything that disturbs them, including livestock, pets, and wildlife. They outcompete native ant species and other animals that prey on insects. Their dispersal may be contributing to the decline of bobwhite quail, horned lizards, and other endemic species.

Invicta means invincible in Latin. So far, this insect pest is living up to its species name. Broad-spectrum

pesticides have done little to slow its march into new habitats. They actually may be helping the invader by killing off native competitors.

Ecologists are enlisting biological controls. Two fly species attack *S. invicta* in its native habitat. Both are parasitoids, a rather gruesome kind of parasite. A fly pierces an adult ant's protective cuticle and lays an egg in its soft tissues. The larva that hatches lives in and feeds on the ant's body. After it gets big enough, the fly larva moves into the ant's head and secretes an enzyme that makes the head fall off, as shown in the photo at the left. The larva metamorphoses into an adult inside the head.

In 1997, ecologists released one parasitoid fly species in Florida. In 2001, they released a second parasitoid species in the southern states. It is too soon to know if the flies work. As researchers wait for results, they are exploring other options. One is to use imported microbes that will infect *S. invicta* but not native ants. Another is to introduce another ant species that invades *S. invicta* colonies and kills their egg-laying queens.

As you will see in this chapter, every species is part of a complex web of relationships, some beneficial, some neutral, and some harmful. When species enter novel habitats far from home, they often shred this web, with unexpected consequences.

 How Would You Vote? *Currently, only a small fraction of the crates being imported to the United States are inspected for exotic species. Should the inspections increase? See BiologyNow for details, then vote online.*

 Key Concepts

SPECIES INTERACTIONS
A community includes all the populations of all species within some area. The structure of a community is determined by physical factors and by the interactions among species within it. Through their interactions, species exert selective pressure on one another.

STABILITY AND CHANGE
The kind and number of species in a community often change over time in response to species interactions, physical changes in the environment, and disturbances. Introduction of a new species into a community can have a dramatic effect on those already present.

ANALYZING BIODIVERSITY
Different communities include different numbers of species, and ecologists have identified some factors that shape global patterns of species diversity. Ongoing declines in species numbers in many communities are largely the result of human activities.

Links to Earlier Concepts

Section 1.1 introduced the concept of a biological community. Sections 13.8 and 13.9 described how species originate, change over time, and become extinct. Here you will learn how species interactions and human activities affect these processes, as well as their effect on population growth (18.3). You may wish to review some earlier examples of challenges faced by all species (17.2) and mutually beneficial interactions (3.6, 14.4, 15.6), as well as the evolution of infectious disease (14.6). Geologic forces also come into play (13.3), as does human population growth (18.5).

Figure 19.1 A sampling of the twelve or more fruit-eating pigeon species of New Guinea's tropical rain forests. Left to right, the turkey-sized Victoria crowned pigeon, the smaller superb crowned fruit pigeon, and the pied imperial pigeon.

Why are there so many? Wouldn't competition for fruit leave one the winner? Actually, the forest trees differ in the size of fruit and fruit-bearing branches. Big pigeons perch on fat branches and eat big fruit. Smaller pigeons, with smaller bills, cannot peck open big fruit. They eat small fruit hanging from branches too slender to hold big pigeons. They all coexist by *partitioning* the fruit supply.

Trees benefit, too. Tough-coated seeds in the fruit resist digestion when traveling through a pigeon gut. Pigeons fly about and dispense seed-containing droppings, often some distance from parent trees. Seeds dropped close to home cannot compete much with mature trees, which already have extensive roots and leafy crowns and have the edge in securing water, mineral ions, and sunlight.

19.1 Which Factors Shape Community Structure?

The type of place where each organism normally lives is its **habitat**. *Habitats have physical and chemical features, such as temperature, and an array of species. Directly or indirectly, populations of all species in each habitat associate with one another as a* **community**.

LINKS TO
SECTIONS
1.1, 3.6, 14.4, 15.6

Every community has a characteristic structure, which arises as an outcome of five factors. *First*, climate and topography interact to dictate temperatures, rainfall, type of soil, and other habitat conditions. *Second*, the kinds and quantities of food and other resources that become available through the year affect which species live there. *Third*, adaptive traits of each species help it survive and exploit specific resources in the habitat, as in Figure 19.1. *Fourth*, species interact by competition, predation, and mutually beneficial actions. *Fifth*, the overall pattern of population sizes affects community structure. This pattern reflects the history of changes in population size, the arrivals and disappearances of species, and physical disturbances to the habitat.

THE NICHE

All organisms of a community share the same habitat, the same "address." But each kind has a "profession" that makes it different from the rest. It is distinct in the sum of its activities and relationships as it goes about getting and using resources. The **niche** of a species is a description of the way that it utilizes its habitat.

The *fundamental* niche of any species describes all the ways that this species could utilize resources if it did not have to compete with any other species. In the real world, competition influences the dimensions of a more limited, *realized* niche. Unlike the fundamental niche, a realized niche can change as competition and resource availability vary.

CATEGORIES OF SPECIES INTERACTIONS

Dozens to hundreds of species interact even in simple communities. Some of the interactions are indirect. As one example, when Canadian lynx prey on snowshoe hares, they indirectly help the plants that these hares eat. The plants indirectly help the Canadian lynx by fattening hares and providing cover for ambushes.

We recognize five categories of direct interactions. **Commensalism** directly helps one species but does not affect the other much, if at all. For instance, some birds use tree branches for roosting sites. The trees get nothing but are not harmed. With **mutualism**, benefits flow both ways between the interacting species. Do not think of this as cozy cooperation. Benefits arise from two-way exploitation. With **interspecific competition**, disadvantages flow both ways between species. Finally, **predation** and **parasitism** are interactions that directly benefit one species (the predator or the parasite) and directly hurt the other (the prey or the host).

Commensalism, mutualism, and parasitism are all types of **symbiosis** ("living together"). Two species are symbiotic if they live in close proximity (often in or on one another) for an extended period or for a portion of their life cycle.

Close interactions between species, whether they are beneficial or detrimental, can lead to **coevolution**. Changes in one species exert selective pressure on the other species. Pollinator preferences select for changes in flower form, and these in turn select for discerning pollinators. Predation selects for swifter prey, then the swifter prey favor speedier predators, and so on.

Community structure arises from a habitat's physical and chemical features, resource availability over time, adaptive traits of its species, how its species interact, and the history of the habitat and its occupants.

A niche is the sum of all activities and relationships in which individuals of a species engage as they secure and use the resources necessary to survive and reproduce.

Direct and indirect interactions among species help shape community structure. Close interactions between species can lead to coevolution—evolution in close concert.

Mutually Beneficial Interactions

Chapter 15 described how flowering plants and pollinators take part in mutually beneficial interactions. In this and other mutualisms, two species form a close partnership that benefits both. In extreme cases, each partner requires the other to survive and reproduce.

Some mutualisms are *obligatory*; members of a species can complete their life cycle only with the assistance of a member of a different species. Yucca plants and yucca moths are one example (Figure 19.2).

A lichen is a case of obligatory mutualism between certain fungi and a photosynthetic alga or bacteria. So is a mycorrhiza, in which a fungus's hyphae penetrate root cells or form a dense, velvety mat around them. As earlier chapters explain, the fungus is efficient at absorbing mineral ions from soil, and the plant grows better when it gets extra ions from the fungus. The fungus in turn gets some of the sugars that the plant makes by way of photosynthesis. The fungus depends on the plant for reproductive success; it will not make spores if the plant stops photosynthesizing.

An anemone fish that needs a sea anemone to live and reproduce is another example (Figure 19.3). The anemone's nematocyst-laden tentacles protect the fish and its eggs. The anemone can survive and reproduce on its own, but having a fish partner lessens predation by other fishes on the anemone's tentacles.

Think back on how endosymbiosis gave rise to cell organelles (Section 3.8). Certain bacteria were able to

Figure 19.2 Mutualism on a rocky slope of Colorado's high desert. Only one yucca moth species pollinates each *Yucca* plant species. Each moth matures when yucca flowers blossom. The female moth collects yucca pollen and rolls it into a ball. She flies to another flower, pierces its ovary, and lays her eggs inside. As the moth crawls out, she pushes the ball of pollen onto the flower's pollen-receiving platform. Pollen tubes grow through the ovary tissues and deliver sperm to the plant's eggs. Seeds develop after fertilization. Meanwhile, the moth eggs develop into larvae that eat a few seeds, then gnaw their way out of the ovary. Seeds that larvae do not eat give rise to new yucca plants.

Figure 19.3 Mutualism near the Solomon Islands. The sea anemone (*Heteractis magnifica*) with a mutualistic pink anemone fish (*Amphiprion perideraion*). The anemone fish chases away butterfly fishes that try to eat bits of the sea anemone's tentacles. In return, the anemone fish gets protection and shelter for itself and for its eggs, which many reef dwellers consider tasty.

live inside other host cells. These symbionts got food from the host and reproduced on their own. In time, hosts came to rely on ATP produced by the symbionts, which were now called mitochondria and chloroplasts. Without these prokaryotic mutualists, no eukaryotic cells would be around today.

In cases of mutualism, each of the participating species reaps benefits from the interaction.

19.3 Competitive Interactions

Every habitat has only a limited supply of energy, nutrients, living space, and other essential resources. Often, many species must compete for access to them.

Intraspecific competition occurs between individuals of the same species. As Malthus and Darwin understood, these interactions can be fierce and they are important driving forces for natural selection (Section 12.1). In this chapter, our focus is on *interspecific* competition, the competition between members of different species. Usually this is not as intense. Why? *The requirements of two species may be similar, but they never can be as close as they are for individuals of the same species.*

With *interference* competition, the members of one species actively prevent members of another species from using a resource. For example, some chipmunks chase away other chipmunk species, thus preventing them from sharing seeds and shelter (Figure 19.4).

Plants can interfere with their competitors as well. Some plants, such as sagebrush, exude chemicals from their leaves that taint the surrounding soil and help to prevent potential competitors from taking root.

In *exploitative* competition, competing species have equal access to a resource and do not try to keep one another away. This is also called *scramble* competition. For example, blue jays, deer, and squirrels eat acorns in oak forests. The animals do not fight over acorns, but they do compete. By eating or storing acorns, each of these species reduces the number of acorns that is available for the others.

COMPETITIVE EXCLUSION

When species compete for some resource, each gets less of that resource than it would if it lived alone. In this way, competition has a negative impact on both competitors. The more alike two species are in their needs, the more intensely they will compete. If two species have identical needs, the superior competitor will drive the less competitive species to extinction in that environment. We call this **competitive exclusion**.

The concept of competitive exclusion grew out of experiments carried out by G. Gause in the 1930s. Gause grew two species of *Paramecium* separately and then in the same culture (Figure 19.5). Both of these protists required the same food—bacteria—and they competed intensely for it. In a few weeks, one species outgrew the other, driving it to extinction.

Gause also studied two other *Paramecium* species that did not overlap much in requirements. When he grew them together, one species tended to feed upon bacteria suspended in culture tube liquid. The other ate yeast cells at the tube's bottom. Population growth rates slowed for both of the species—but their overlap in resource use was not enough for one species to fully exclude the other. They continued to coexist.

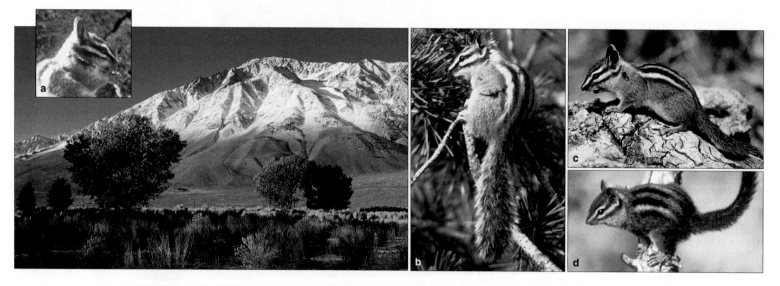

Figure 19.4 Example of interspecific competition in nature. On the slopes of the Sierra Nevada, competition helps keep nine species of chipmunks (*Tamias*) in different habitats. The alpine chipmunk (**a**) lives in the alpine zone, the highest elevation. Below it are the lodgepole pine, piñon pine, and then sagebrush habitat zones. Lodgepole pine chipmunks (**b**), least chipmunks (**c**), and other species live in the forest zones. Merriam's chipmunk (**d**) lives at the base of the mountains, in sagebrush. Its traits would allow it to move up into the pines, but the aggressive behavior of forest-dwelling species keep it from moving there. Food preferences keep the pine forest chipmunks out of the sagebrush habitat.

Paramecium caudatum

a Time (days)

b Time (days)

c Time (days)

Relative population density

P. aurelia

Figure 19.5 *Animated!* Results of competitive exclusion between two protist species that compete for the same food. (**a**) *Paramecium caudatum* and (**b**) *P. aurelia* were grown in separate culture tubes and established stable populations. The S-shaped graph curves indicate logistic growth and stability. (**c**) Then the two species were grown together. *P. aurelia* (*brown* curve) drove *P. caudatum* toward extinction (*green* curve in **c**). This experiment and others suggest that two species cannot coexist indefinitely in the same habitat *when they require identical resources.* If their requirements do not overlap much, one might influence the population growth rate of the other, but they may still coexist.

COEXISTING COMPETITORS

Think back on those twelve fruit-eating pigeon species in the same New Guinea forest. They all use the same resource: fruit. Yet they overlap only slightly in their use of it, because each prefers fruits of a certain size. They are a fine example of **resource partitioning**—the *subdividing* of some category of similar resources in a way that allows competing species to coexist.

A similar competitive situation arises among three species of annual plants in a plowed, abandoned field. Like other plants, all three require sunlight, water, and dissolved mineral ions. Yet each species is adapted to exploiting a different part of the habitat (Figure 19.6). The bristly foxtail grasses have a shallow, fibrous root system, and they are drought tolerant. Indian mallow plants have a taproot system and grow in deeper soil. Smartweed's taproot system branches in topsoil and soil below the roots of the other species.

Field studies allow ecologists to determine if two species are competing in nature. A typical experiment involves removing one possible competitor from some portion of the environment. For example, N. Hairston studied slimy salamanders (*Plethodon glutinosus*), and Jordan's salamander (*P. jordani*), which overlap in their ranges (Figure 19.7). He removed *P. glutinosus* from some plots and *P. jordani* from others. Control plots were unaltered. Five years later, salamander density in the control plots had not changed. But in the plots that lacked *P. jordani*, population density of *P. glutinosus* had soared. Similarly, *P. jordani* numbers had increased in the *P. glutinosus*-free plots. Hairston concluded that under natural circumstances, the species compete and that competition reduces the population size of both.

smartweed   Indian mallow   bristly foxtail

Soil depth (centimeters)

bristly foxtail roots

Indian mallow roots

smartweed roots

Figure 19.6 Resource partitioning among three annual plant species in an abandoned field. The plants differ in how they are adapted to secure soil water and mineral ions. The roots of each species tap into a different region of the soil.

Figure 19.7 Two competing species of salamander that coexist in the Great Smokey and Balsamic Mountains. (**a**) *Plethodon glutinosus* and (**b**) *P. jordani*.

Competition has negative effects on both participants. In some situations, one species actively prevents another from using a resource. In others, both species have equal access to the resource and each scrambles to get a bigger share.

Species that have dissimilar resource requirements are more likely to coexist than those having similar needs.

Two competing species also may coexist by partitioning a resource—accessing it at different times or locations.

LINK TO
SECTION
18.2

19.4 Predator–Prey Interactions

Directly or indirectly, interactions among coexisting populations organize the community to which they belong. Let's look at patterns of species abundances, species interactions, and local biodiversity, starting with the nature of interactions between predators and prey.

Predators are consumers that feed on living organisms —their **prey**—and typically cause their prey's death. Diverse carnivores and omnivores fall in this category. Herbivores are similar to predators, but they often do not cause death of a plant. Even so, many herbivores and the plants they eat coevolve in the same way that prey and predators do.

MODELS FOR PREDATOR–PREY INTERACTIONS

There are three general patterns of predator responses to increases or decreases in prey density (Figure 19.8). In the type I response, a predator removes a constant proportion of prey over time, regardless of levels of prey abundance. The number of prey killed depends only on prey density. This type of response is typical of passive predators, such as spiders. The more flies there are, the more get tangled in spider webs.

In a type II response, the ability of predators to eat and digest prey determines how many they capture. As prey density increases, the proportion caught rises steeply at first, then slows as predators are exposed to more prey than they can deal with at one time. Figure 19.8*b* gives an example. A wolf that has just killed a caribou will not hunt another until it has eaten and digested the first one.

In a type III response, predators capture only a small proportion of prey when the prey density is low. They capture the most at intermediate prey densities, and then the response levels off. This response occurs among predators that switch to other prey when low abundance makes a prey species difficult to locate.

Other factors besides individual predator response to prey density are at work. For example, predator and prey reproductive rates influence the interaction. So do hiding places for prey, the presence of other prey or predators, and carrying capacities (Section 18.3).

THE CANADIAN LYNX AND SNOWSHOE HARE

Often predator and prey populations will oscillate. For example, there is a ten-year oscillation in populations of one predator, the Canadian lynx, and the snowshoe hare that is its main prey (Figure 19.9). To identify the causes behind this cyclic pattern, the ecologist Charles Krebs and coworkers recorded hare densities for ten years in the Yukon River Valley in Alaska. They set up 1-square-kilometer control plots. They also set up a number of experimental plots. Some plots had electric fences that kept out predatory mammals. Other plots had extra food added or were fertilized to increase plant growth. The team captured and released more than a thousand snowshoe hares, lynx, squirrels, and coyotes after putting radio collars on each one.

In predator-free plots, hare density doubled. In plots with extra food, it tripled. In plots with extra food *and* fewer predators, hare density increased elevenfold.

Nevertheless, the experimental manipulations only delayed the cyclic declines in population density. They did not stop them. Why? The fence could not keep out owls and other raptors. Only 9 percent of the collared hares starved to death; predators devoured 83 percent

a Prey population density

b Caribou per square kilometer

Figure 19.8 *Animated!* (**a**) Three alternative models for responses of predators to prey density. Type I: Prey consumption rises linearly as prey density rises. Type II: Prey consumption is high at first, then levels off as predator bellies stay full. Type III: When prey density is low, it takes longer to hunt prey, so the predator response is low. (**b**) A type II response in nature. For one winter month in Alaska, B. W. Dale and coworkers observed four wolf packs (*Canis lupus*) feeding on caribou (*Rangifer taranuds*). This interaction fits the type II model for the response of predators to prey density.

SPECIES INTERACTIONS

Figure 19.9 (**a**) Correspondence between abundances of Canadian lynx (*dashed* line) and snowshoe hares (*solid* line), based on counts of pelts sold by trappers to Hudson's Bay Company during a ninety-year period. (**b**) This photograph reinforces Charles Krebs's observation that predation causes mass paranoia among snowshoe hares, which continually look over their shoulders during the declining phase of each cycle. (**c**) This photograph supports the Krebs hypothesis that there is a three-level interaction going on, one that involves plants.

The graph is a good test of whether you tend to readily accept someone else's conclusions without questioning their basis in science. Remember those sections in Chapter 1 that introduced the nature of scientific methods?

What other factors may have had impact on the cycle? Did the weather vary, with more severe winters imposing greater demand for hares (to keep lynx warmer) and higher death rates? Did the lynx compete with other predators, such as owls? Did the predators turn to alternative prey during low points of the hare cycle? When fur prices rose in Europe, did the trapping increase? When the pelt supply outstripped the demand, did trapping decline?

Graph labels: y-axis: Number of pelts taken (× 1,000); x-axis: Time (years); x-axis values: 1845, 1865, 1885, 1905, 1925; y-axis values: 0, 20, 40, 60, 80, 100, 120, 140, 160

of the rest. Krebs concluded that a simple predator–prey or plant–herbivore model could not explain the results of his experiments in the Yukon River Valley. Rather, in the Canadian lynx and snowshoe hare cycle, other variables are at work in a *three-level* interaction. Additional experiments may shed light on the causes of these cycles.

AN EVOLUTIONARY ARMS RACE

Predator and prey species often coevolve. Suppose a gene mutation that provides a novel means of defense arises in a prey species. Over generations, directional selection causes this gene to spread through the prey population. Suppose also that some members of the predator species have a gene that makes them better at thwarting the new defense. These predators, and their descendants, eat more often, hence are more likely to survive and reproduce. The predators exert selection pressure that favors better defenses by prey, which exert selection pressure on predators, and so it goes over generations. Let us now consider a few outcomes of such evolutionary arms races.

PREY DEFENSES You have already learned about some defensive adaptations. Many species have hard parts that enclose them or otherwise make them difficult to eat. Think back on the prickly elements in the bodies of sponges, the shells of mollusks, and the exoskeletons that protect arthropods and echinoderms.

Other prey species contain chemicals that taste bad or sicken predators. Some produce the toxins through their own metabolic processes. Others use chemical or physical weapons that they obtained from their prey, as when sea slugs store nematocysts from cnidarians (Figure 16.11*c*).

The leaves, stems, and seeds of many plants contain bitter, hard-to-digest, or poisonous substances that help to deter herbivores. Ricin-producing castor bean plants (Chapter 10) are a good example. Ricin sickens or kills animals that eat the seeds. Caffeine in coffee beans and nicotine in tobacco leaves are both defensive compounds for which humans have developed a taste.

Well-defended prey often show **warning coloration**, conspicuous patterns and colors that predators learn to recognize as *AVOID ME!* signals. An inexperienced bird that eats an orange-and-black patterned monarch

LINKS TO
SECTIONS
1.6, 17.2

a A dangerous model   **b** One of its edible mimics   **c** Another edible mimic   **d** And another edible mimic

Figure 19.10   An example of mimicry. Edible insect species often resemble toxic or unpalatable species that are not at all closely related. (**a**) A yellowjacket can deliver a painful sting. It may be the model for nonstinging wasps (**b**) beetles (**c**), and flies (**d**) that have a similar appearance.

Figure 19.11 Prey demonstrating the fine art of camouflage. (**a**) What bird??? When a predator approaches its nest, the least bittern stretches its neck (which is colored like the surrounding withered reeds), thrusts its beak upward, and sways gently like reeds in the wind. (**b**) Find the plants (*Lithops*) that are hiding in the open from herbivores with their stonelike form, pattern, and coloring.

butterfly learns quickly to associate this patterning with the vomiting of foul-tasting toxins.

Mimicry is an evolutionary convergence in body form, in which one species benefits by its resemblance to another. Section 1.6 described how the resemblance between two unrelated unpalatable butterflies protects both species against predation. In other cases, prey that are tasty and harmless evolve to resemble unpalatable or well-defended species (Figure 19.10). Predators may ignore the mimic after contacting the model's repelling taste, toxic secretion, or painful bite or sting.

Some prey engage in **camouflage**, or hiding in the open. Such species have a form, patterning, color—and often behavior—that allows them to blend into their surroundings and avoid detection. Figure 19.11 shows two examples. One, *Lithops*, is a desert plant that looks like a rock. It flowers only during a brief rainy season, when other plants grow profusely and there is plenty of drinking water, so herbivores are less interested in its juicy water-filled tissues.

When a prey animal is cornered or under attack, survival may turn on a last-chance trick. Some animals startle the predator, as by hissing, puffing up, showing teeth, or flashing big eye-shaped spots. The tail may detach from the body and wiggle a bit as a distraction. Opossums pretend to be dead. So do hognose snakes, which also secrete smelly fluid. Skunks squirt a foul-smelling, irritating repellent.

PREDATOR RESPONSES TO PREY   In a coevolutionary arms race, predators counter prey defenses with their own adaptations. Stealth, camouflage, and ingenious ways of avoiding the repellents are countermeasures. Consider edible beetles that direct sprays of noxious chemicals at assailants (Figure 19.12*a*). Grasshopper mice plunge the beetle's sprayer end into the ground, then munch on the unprotected head (Figure 19.12*b*).

Many prey can outrun predators when they get a head start. But consider the cheetah, the fastest land

Figure 19.12 Predator responses to prey defenses. (**a**) Some beetles spray noxious chemicals at attackers, which deters them some of the time. (**b**) At other times, grasshopper mice plunge the chemical-spraying tail end of their beetle prey into the ground and feast on the head end. (**c**) Find the scorpionfish, a venomous predator with camouflaging fleshy flaps, multiple colors, and profuse spines. (**d**) Where do the pink flowers end and the pink praying mantis begin?

animal. One cheetah was clocked at 114 kilometers per hour (70 mph). A cheetah's body is highly adapted for sprinting. Compared to the other large cats, it legs are longer relative to its body size. Nonretractable claws act like cleats that increase traction. Coevolution with fast prey contributed to the cheetah's need for speed. Its preferred prey, the Thomson's gazelle, is among the fastest land animals and can maintain a speed of 80 kilometers per hour.

Camouflage helps some predators catch their prey. Think how snow white polar bears stalk seals, striped tigers crouch in tall, golden grasses, and the mottled scorpionfish sits on the seafloor (Figure 19.12*c*). Many predatory insects are remarkably difficult to recognize (Figure 19.12*d*). Camouflaged predators may select for enhanced sensory systems in prey. The visual systems of many prey can detect even the slightest motion that could give away a predator.

> *Different kinds of predators show different behavioral responses to changes in prey density.*
>
> *Carrying capacities for predators and prey, the response of predators to changes in prey density, the availability of prey refuges, and the presence of alternative prey will all influence the stability of predator and prey populations.*
>
> *Predators and prey exert selective pressure on one another. The result is an evolutionary arms race.*

19.5 Parasites and Parasitoids

Parasites spend all or part of their life cycle in or on other living organisms, from which they draw nutrients. The host is harmed, but usually is not killed outright. Parasites help keep host populations in check, and we sometimes call them into service in our battle against pests.

Parasites have wide-reaching ecological impacts. By draining nutrients from hosts, they alter the amount of energy and nutrients that a host population demands from a habitat. Also, weakened hosts are usually more vulnerable to predation and less attractive to potential mates. Some parasite infections cause sterility. Others shift the ratio of host males to females. In such ways, parasitic infections lower birth rates, raise death rates, and affect intraspecific and interspecific competition.

Sometimes the gradual drain of nutrients during a parasitic infection indirectly leads to death. The host becomes so weak that it cannot fight off secondary infections. Nevertheless, in evolutionary terms, killing a host too quickly is bad for a parasite's reproductive success. A parasitic infection must last long enough to give the parasite time to produce some offspring. The longer it survives in the host, the more offspring. We therefore expect natural selection to favor individual parasites that have less-than-fatal effects on hosts.

Only two types of interactions typically kill a host. First, a parasite may attack a novel host, one with no coevolved defenses against it. Second, the host may be supporting too many individual parasites at the same time. The collective assault may overwhelm the body.

The previous chapters introduced diverse parasites. Some spend their entire life cycle in or on a single host. Others have different hosts during different stages of their life cycle. Insects and other arthropods are often **vectors** that convey a parasite from host to host.

LINK TO
SECTION
14.6

All viruses and some of the bacteria, protists, and fungi are parasites. Figure 19.13 describes a parasitic fungus that infects amphibians of tropical rain forests. Tapeworms and flukes are all parasites. So are some roundworms, insects and crustaceans, and all ticks.

Figure 19.13 Examples of parasites. (**a**) Harlequin frog (*Atelopus varius*) from Central America. (**b**) Section through skin of a harlequin frog affected by *chytridiomycosis*, an infectious disease caused by a fungus. *Arrows* point to flask-shaped cells of the fungal parasite. Each cell holds spores that will be released to the skin surface. They may be dispersed to other host frogs. (**c,d**) Dodder (*Cuscuta*) is also known as strangleweed or devil's hair. This parasitic flowering plant lacks chlorophylls. It winds around a host plant during growth. Modified roots penetrate the host's vascular tissues and draw water and nutrients from them.

Figure 19.14 A brown-headed cowbird nestling (*Molothrus ater*) with its foster parent. A female cowbird laid an egg in this smaller bird's nest. The host species does not recognize a cowbird nestling as an intruder and feeds it as if were its own offspring.

Figure 19.15 Biological control—a commercially raised parasitoid wasp about to deposit an egg in an aphid. This wasp cuts aphid numbers by crippling the egg laying capacity of its host even before the larva hatches inside and kills the aphid.

Even a few plants are parasitic. Nonphotosynthetic types, such as dodders, obtain sugars and nutrients from other plants (Figure 19.13c,d). The mistletoes and similar types carry out photosynthesis but tap into the nutrients and water in a host plant's tissues.

The larvae of **parasitoids** develop inside another species and devour it from the inside as they grow. The adults are free-living. Unlike parasites, parasitoids usually kill their hosts. Most parasitoids are insects. The ant-killing flies described earlier are one example.

Social parasites are animals that take advantage of the social behavior of a host species to carry out their life cycle. For example, cuckoos and cowbirds lay their eggs in the nests of other birds. In this way, they steal parental care from these hosts (Figure 19.14).

BIOLOGICAL CONTROL AGENTS

Parasites and parasitoids are commercially raised and released in target areas as *biological controls*. They are promoted as a workable alternative to pesticides. The chapter introduction and Figure 19.15 give examples.

Effective biological controls display five attributes. The agents are adapted to a specific host species and to the host's habitat; they are good at locating hosts; their population growth rate is high compared to the host's; their offspring are good at dispersing; and they make a type III functional response (Section 19.4), with little lag time after shifts occur in the host population size.

Use of biological control agents is not without risks. Releasing more than one species of biological control agent in an area may trigger competition among them and lessen their effectiveness. Introduced control agents sometimes parasitize nontargeted species in addition to, or instead of, species they were expected to attack.

For example, in Hawaii, introduction of parasitoids to control a non-native stink bug caused the extinction of the koa bug, the state's largest native bug. No koa bugs have been sighted since 1978. Apparently, the exotic parasitoids found the koa bugs a more tempting target than the pests they were imported to attack. An estimated 13 percent of the now established biological control agents imported into Hawaii harm native or beneficial species as well as their intended targets.

Parasites live on and draw nutrition from a living host. They usually weaken but do not kill the host.

Parasitoids are free-living as adults, but larvae live in and feed on another organism, which they usually kill.

Social parasites steal parental care from another species.

Parasites and parasitoids can be used to control pests.

SPECIES INTERACTIONS

19.6 Changes in Community Structure Over Time

You probably already realize that the array of species in a community changes over time. Where there are now meadow grasses, there may one day be trees. Or, after a fire, grasses may thrive where trees now stand.

ECOLOGICAL SUCCESSION

The transformation, over time, of a meadow to a forest is an example of **ecological succession**. By this process, one array of species is replaced by another, which is in turn replaced by another, and so on.

Primary succession occurs in habitats that lack soil and have few or no existing species. A newly formed volcanic island or a rocky area exposed by the retreat of a glacier will undergo primary succession (Figure 29.16). Pioneer species are the first to take hold. They include lichens, mosses, and some weedy annuals. The early arrivals are easily dispersed. They have a short life cycle and are able to tolerate intense sunlight, big swings in temperature, and nutrient-poor soil.

By their actions, pioneer species can help to create or improve the soil. In doing so, they may set the stage for their own replacement. Many are mutualists with nitrogen-fixing bacteria, so they can grow in nitrogen-poor habitats. Seeds of later species find shelter inside mats of the pioneers, which do not grow high enough to shade out the new seedlings.

Over time, organic wastes and remains accumulate. They add volume and nutrients to soil, thus creating conditions that make it easier for additional species to take hold. Later successional species often crowd out the early ones, whose spores and seeds then travel as fugitives on wind and water—destined, perhaps, for another new but temporary habitat.

If an existing community is altered by a natural or man-made disturbance, **secondary succession** occurs. We observe this kind of succession after a fire destroys a forest or a farm is abandoned.

INTERMEDIATE DISTURBANCE HYPOTHESIS

Early studies of primary succession led ecologists to the view that a community goes through predictable stages, determined by a habitat's typography, climate, and soil, and the interactions among species. By this view, succession ends with the formation of a stable *climax* community that does not change over time. In any particular habitat, succession always produces the same climax community, and if something destroys this community, the process of secondary succession will eventually restore it.

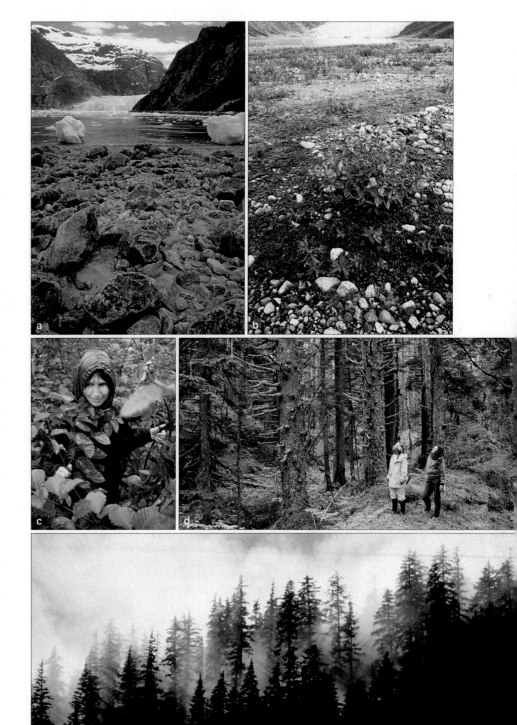

Figure 19.16 One of three pathways of primary succession observed in Alaska's Glacier Bay region. (**a**) A glacier is retreating from the sea. Meltwater leaches nitrogen and other minerals from the newly exposed soil. (**b**) The first invaders of nutrient-poor soil are lichens, horsetails, mosses, and fireweed and mountain avens. Some pioneer species are mutualists with nitrogen-fixing microbes. They grow and spread fast over the glacial till. (**c**) Alders, too, are symbionts with nitrogen-fixing microbes. Within fifty years, they have formed dense, mature thickets in which cottonwood, hemlock, and a few evergreen spruce grow quickly. (**d**) After eighty years, western hemlock and spruce trees crowd out the mature alders. (**e**) After a century, dense forests of spruce dominate. In the other successional pathways, different species of trees dominate at each stage.

moderate disturbances with an intermediate frequency. With this type of pattern, there are many opportunities for new species to enter the habitat after disturbances but not enough time between disturbances for many species to become competitively excluded from it.

SPECIES INTERACTIONS IN SUCCESSION

Facilitation, inhibition, and tolerance are interspecific interactions that influence succession. With *facilitation*, the presence of one species alters habitat conditions in a way that allows another to take root. For example, after Teresa Turner removed attached algae from plots in the intertidal zone in Oregon, surf grass could not move in. Algae has to be present to provide anchoring sites for the grasses.

With *inhibition*, a species alters the habitat in a way that discourages another species from growing there. As described earlier, sagebrush secretes chemicals into the soil that deter germination and growth of other plants. Wayne Sousa showed that the presence of one species of algae prevents another species of algae from growing on rocks along Southern California shores.

With *tolerance*, the presence of one species has no effect on the arrival of another species. For example, a fern that can tolerate a broad range of light levels will sprout regardless of whether or not trees are blocking its access to sunlight.

In 1980, an eruption of Mount Saint Helens leveled about 600 square kilometers of forest (Figure 19.17). This gave ecologists the chance to monitor succession firsthand. Researchers observed and recorded natural patterns of colonization. They also manipulated plots inside the blast zone. William Morris and David Wood showed that facilitation and inhibition affected the pattern of plant succession. By adding seeds of certain plant species to some of the plots and keeping other plots barren, they showed that some early colonizers helped different plant species become established and some kept different plant species out.

Figure 19.17 A natural laboratory for the study of succession. (**a**) Eruption of Mount Saint Helens in 1980. Nothing remained of the rich climax community at the base of this Cascade volcano. (**b**) In less than a decade, many pioneer species were established. (**c**) Twelve years after the eruption, seedlings of Douglas fir were taking hold in soils enriched with volcanic ash.

In recent times, ideas about community structure have changed. Experiments have shown that communities are not reconstituted after disturbances, and random factors—such the order in which species arrive—can determine community structure. Ecologists now regard communities as dynamic entities that are rarely, if ever, in a stable state. In this "non-equilibrium" view, fires, storms, freezes, droughts, and other disturbances play an important role in shaping community structure.

The **intermediate disturbance hypothesis** holds that the number of species in a community is influenced by how frequent and severe disturbances are. The highest species richness occurs in habitats that are subject to

By the process of ecological succession, one array of species replaces another in a sequential fashion. Primary succession occurs in habitats more or less devoid of life. Secondary succession occurs in areas where an existing community has been severely disturbed.

Traditionally, communities were expected to eventually reach a predictable stable state—the climax community. Ecologists now emphasize the role of random events and disturbances in determining the numbers and kinds of species in communities that are constantly changing.

The presence of one species in a habitat may aid, prevent, or have no effect on the success of later arriving species.

d Algal diversity in tidepools

e Algal diversity on rocks exposed at low tide

Figure 19.18 Effect of competition and predation on the community structure of an intertidal zone. (**a**) Grazing periwinkles (*Littorina littorea*) affect the number of algal species in different ways in different marine habitats. (**b**) *Chondrus* and (**c**) *Enteromorpha*, two kinds of algae in their natural habitat. (**d**) Periwinkles graze on *Enteromorpha*, the dominant alga in tidepools; less competitive algae that might otherwise be overwhelmed have an advantage. (**e**) Algal diversity is lower on exposed rock surfaces nearby. There, periwinkles do not graze on *Chondrus* and other kinds of red algae, the dominant species.

19.7 Forces Contributing to Community Instability

The importance of species interaction in community structure is illustrated by removing keystone species or by adding exotic invaders.

Community stability is an outcome of forces that have come into uneasy balance. Resources are sustained as long as populations do not reach carrying capacity. Predator and prey populations rise and fall together. Competitors continually engage in tugs-of-war over resources. In the short term, disturbances might slow the growth of some populations. Also, any long-term changes in climate typically have destabilizing effects.

If instability becomes great enough, the community might change in ways that will persist even when the disturbance ends or is reversed. If some of its species happen to be rare or do not compete as well as others, they might be driven to extinction.

THE ROLE OF KEYSTONE SPECIES

The uneasy balancing of forces in a community comes into focus when we observe the effects of a **keystone species**, a species that has a disproportionately large effect on a community relative to its actual abundance. Robert Paine was the first to describe the importance of a keystone species. He was studying interactions on the rocky shores of California's coast. Species in this rocky intertidal zone survive by clinging to rocks, and access to spaces to hold on to is a limiting factor. Paine set up control plots with the sea star *Pisaster ochraceus* and its prey—chitons, limpets, barnacles, and mussels. He removed all sea stars from his experimental plots.

With the sea stars gone, the mussels (*Mytilus*) took over Paine's experimental plots, and they crowded out seven other invertebrate species. The mussels are the preferred prey of the sea star, and the best competitors in its absence. Normally, predation by sea stars keeps the diversity of prey species high, because it restricts competitive exclusion by mussels. Remove all the sea stars, and the community shrinks from fifteen species down to eight.

The impact of a keystone species can vary between habitats that differ in their species arrays. Periwinkles (*Littorina littorea*) are alga-eating snails of intertidal zones. Jane Lubchenco showed that their removal can increase *or* decrease the diversity of algal species in different habitats (Figure 19.18).

In tidepools, the periwinkles prefer to eat the alga *Enteromorpha*, which can outgrow other algal species. By keeping *Enteromorpha* in check, periwinkles help less competitive algae survive. On rocks in the lower intertidal zone, *Chondrus* and other tough, red algae that periwinkles avoid are dominant. In this habitat, periwinkles prefer to graze on competitively weaker algal species. Thus, periwinkles *promote* algal diversity in tidepools but *reduce* it on exposed rock surfaces.

Figure 19.19 (**a**) Kudzu (*Pueraria lobata*) taking over part of Lyman, South Carolina. It has spread from East Texas to Florida and Pennsylvania. Ruth Duncan of Alabama, who makes 200 kudzu vine baskets a year, just can't keep up. (**b**) *Caulerpa taxifolia* suffocating yet another richly diverse marine ecosystem.

IMPACTS, ISSUES

The Texas horned lizard, or "horny toad," is a fierce-looking predator and the state's official reptile. The mainstay of its diet is harvester ants. However, these native ants cannot compete with imported fire ants, which are displacing them. That's bad news because the horned lizards do not eat the invaders. Imported fire ants also compete directly with the lizards for other insect prey.

HOW SPECIES INTRODUCTIONS TIP THE BALANCE

Sometimes, one member of an established community moves out from its home range and successfully takes up residence elsewhere. This directional movement, called **geographic dispersal**, happens in three ways.

First, over a number of generations, a population might expand its home range by slowly moving into outlying regions that prove hospitable. Second, some individuals might be rapidly transported across great distances. This *jump* dispersal often takes individuals across regions they could not cross on their own, as when tree snakes traveled from New Guinea to Guam inside a ship's cargo hold. Third, with imperceptible slowness, a population might be moved away from its home range over geologic time, as by continental drift.

Successful dispersal and colonization of a vacant adaptive zone are often rapid. Consider one of Amy Schoener's experiments in the Bahamas. She set out plastic sponges on the barren sand at the bottom of Bimini Lagoon. How fast did aquatic species take up residence in the artificial hotels? Schoener recorded occupancy by 220 species within thirty days.

When you hear someone bubbling enthusiastically about an exotic species, you can safely bet the speaker is not an ecologist. An **exotic species** is a resident of an established community that has dispersed from its home range and become established elsewhere. Unlike most imports, which never do take hold outside their home range, an exotic species permanently insinuates itself into a new community. Here, with no coevolved parasites, pathogens, or predators to keep the species in check, its numbers may soar.

More than 4,500 exotic species have established themselves in the United States after jump dispersal. These are just the ones we know about. We put some of the new arrivals, including soybeans, rice, wheat, corn, and potatoes, to use as food crops.

These and many other additions proved harmless or have had beneficial effects. However, other imports had adverse effects. Consider the following examples of how an imported plant, marine alga, and mammal wreaked havoc on community structure.

THE PLANTS THAT ATE GEORGIA   In 1876, a vine called kudzu (*Pueraria lobata*) from Japan was introduced to the United States. In temperate regions of Asia, it is a well-behaved legume with a well-developed root system. It *seemed* like a good idea to use it for forage and to control erosion. But kudzu grew faster in the Southeast, where herbivores, pathogens, and less competitive plants posed no serious threat to it.

With nothing to discourage it, kudzu shoots grew sixty meters a year. Its vines now blanket streambanks, trees, telephone poles, houses, and almost everything else in their path (Figure 19.19*a*). Attempts to dig up or burn it are futile. Grazing goats and herbicides help. But goats eat other plants too. Herbicides contaminate water supplies. Kudzu may spread as far north as the Great Lakes by the year 2040.

On the bright side, Asians use a starch extracted from kudzu in drinks, herbal medicines, and candy. A kudzu processing plant in Alabama may export this starch to Asia, where demand currently exceeds the supply. Also, kudzu may help save forests; it can be used as an alternative source for paper.

THE ALGA TRIUMPHANT They just looked so perfect in saltwater aquariums, those bright green, long, and feathery branches of the green alga *Caulerpa taxifolia*. The Stuttgart Aquarium in Germany even developed a hybrid strain that was sterile, and generously shared it with other marine institutions. Was it from Monaco's Oceanographic Museum that the strain escaped into the wild? Some say yes, Monaco says no.

The aquarium strain of *C. taxifolia* cannot reproduce sexually. It grows quickly, a few centimeters a day, and reproduces asexually by runners. Boat propellers and fishing nets help to disperse it. Since 1984, the alga has colonized more than 10,000 acres of coastal seafloor in the Mediterranean Sea (Figure 29.19b). It thrives along sandy or rocky shores and in mud, and survives ten days out of water. Unlike its tropical parents, this strain thrives in cool water, and tolerates pollutants. Its toxin repels herbivores. It displaces endemic algae, invertebrates, and fishes. It has been found in coastal waters of Southern Australia, where it has forced the closure of some waters to fishing and boating.

C. taxifolia has also invaded the coastal waters of the United States, but quick action may have beaten it back. In 2000, some scuba divers discovered the alga growing in areas off the coast of Southern California. Someone might have drained water from an aquarium into a storm drain or into the lagoon itself. In response to the discovery, a coalition of government and private groups sprang into action. They tarped over the area to shut out sunlight, pumped chlorine bleach into the mud to poison the alga, and used welders to boil it. The aggressive measures seem to have been effective. No sightings of the exotic invader have been reported in these waters since September of 2002.

Importing *C. taxifolia* or any closely related species of *Caulerpa* into the United States is now illegal. To protect native aquatic communities, aquarium water should never be dumped into storm drains or natural waterways. It should be discarded into a sink or toilet so that wastewater treatment can kill any algal spores.

THE RABBITS THAT ATE AUSTRALIA During the 1800s, British settlers in Australia just could not bond with koalas and kangaroos, and so they imported familiar animals from home. In 1859, in what would be the start of a major disaster, a northern Australian landowner imported and then released two dozen wild European rabbits (*Oryctolagus cuniculus*). Good food and sport hunting, that was the idea. An ideal rabbit habitat with no natural predators—that was the reality.

Six years later, the landowner had killed 20,000 rabbits and was besieged by 20,000 more. The rabbits displaced livestock and native wildlife. Now 200 to 300 million are hippity-hopping through the southern

half of the country. Their collective feeding is turning grasslands and shrublands into eroded deserts.

Rabbits have been shot and their warrens plowed under, fumigated, and dynamited. When a fence 2,000 miles long was built to keep rabbits out of western Australia, rabbits made it from one side to the other before workers could finish the job (Figure 19.20).

In 1951, the Australian government introduced a myxoma virus that infects South American rabbits and causes *myxomatosis*. Mosquitoes and fleas transmit the virus from host to host. Having no coevolved defenses against the import, European rabbits died in droves. But natural selection has since favored a rise in rabbit populations resistant to the imported virus.

In 1991, on an uninhabited island in Spencer Gulf, Australian researchers began an experimental study of a calicivirus. The virus kills rabbits by affecting their heart and lungs. In 1995, the virus made an unforeseen jump to the mainland, perhaps carried by insects. In 1996, it was deliberately released into other areas. By 2001, rabbit populations were 80 to 85 percent below their peak values. Grasses and shrubs have begun to rebound and native herbivores have begun increasing their numbers.

The jury is still out on the long-term impact of the viral releases. Some Australians fear it might infect animals other than rabbits—so far this seems unlikely.

The presence or absence of a single species can have a profound effect on the structure of a community. Competition from introduced species is driving many native species toward extinction.

Figure 19.20 Rabbit-proof fence? Hardly. This is part of a fence that was built to keep rabbits from spreading and destroying more of Australia's vegetation. The effort was a failure.

LINKS TO
SECTIONS
13.3, 18.1

19.8 Patterns of Species Diversity

You already learned that the frequency of disturbances affects the number of different species in a community. Ecologists have also documented other factors that cause species richness to vary from one region to the next.

In large part, the number of species in any region is an outcome of the evolutionary history of each member species and its resource requirements, physiology, and dispersal capacity. Rates of birth, death, immigration, and emigration also shape species numbers. In turn, those rates are shaped by conditions in the local and regional environments, and by species interactions.

The species richness of a region is also affected by geographic and climactic variables, such as sunlight intensity, rainfall, and temperature along gradients in latitude, elevation, and—for aquatic habitats—depth. Within broad regions, microenvironments give rise to local variations in the overall patterns.

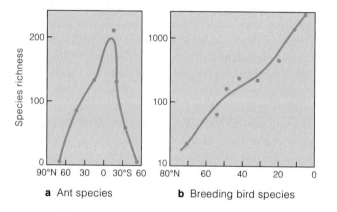

a Ant species   **b** Breeding bird species

Figure 19.21 Two patterns of species diversity corresponding to latitude, as revealed by the number of ant species (**a**) and birds (**b**) in Central and North America.

Figure 19.22 Surtsey, a volcanic island, at the time of its formation. The chart gives the number of colonizing species between 1965 and 1973. Newly formed, isolated islands are natural laboratories for ecologists.

MAINLAND AND MARINE PATTERNS

The most striking pattern of species richness relates to distance from the equator. For most groups of plants and animals, *the number of coexisting species on land and in the seas is greatest in the tropics, and it systematically declines from the equator to the poles.* Figure 19.21 shows examples of this pattern. Here we focus on the factors that help bring about the pattern and also maintain it.

First, for reasons explained in Chapter 21, tropical latitudes intercept more intense sunlight and receive more rainfall, and their growing season is longer. As one outcome, resource availability tends to be greater and more reliable in the tropics than elsewhere. This favors a degree of specialized interrelationships not possible where species are active for shorter periods.

Second, tropical communities have been evolving for a longer time than temperate ones, some of which did not begin forming until the end of the last ice age.

Third, species diversity might be self-reinforcing. Diversity of tree species in tropical forests is much greater than in comparable forests at higher latitudes. When more plant species compete and coexist, more kinds of herbivores also compete and coexist, partly because no single kind of herbivore can overcome all the chemical defenses of all the different plants. More predatory and parasitic species evolve in response to the larger array of prey and hosts. The same type of effect may promote diversity on tropical reefs.

ISLAND PATTERNS

Volcanic islands often serve as natural laboratories for biological studies. As an example, in 1965, a volcanic eruption produced a new island, called Surtsey, just southwest of Iceland. Within just six months, bacteria, fungi, seeds, flies, and some seabirds were established on it. One vascular plant appeared after two years, and mosses two years after that (Figure 19.22). The number of plant species rose after soils had become enriched. Seabirds began to nest on the island in 1970. Today, five species nest there and migratory birds pass through by the thousands. Some carry seeds in their feathers or bellies and so introduce new plant species.

As is the case for other islands, the number of new species on Surtsey did not increase indefinitely. Why not? Studies of community patterns on islands around the world provide us with two models.

First, any islands distant from a source of potential colonists receive few colonizing species, and the few that do arrive are adapted for long-distance dispersal (Figure 19.23). This is known as the **distance effect**. Second, larger islands tend to support more species than smaller islands that are the same distances from

ANALYZING BIODIVERSITY

a

Area (square kilometers)

Species richness (number of species)

Figure 19.23 *Animated!* (**a**) Two island patterns. Distance effect: Species richness on islands of a specified size declines with increasing distance from a source of colonizing species. *Green* circles signify islands less than 300 kilometers from the colonizing source. *Orange* triangles signify islands more than 300 kilometers from the source areas. Area effect: Among islands the same distance from a source of colonizing species, the larger ones support more species. (**b**) Wandering albatross, one travel agent for jump dispersals. Seabirds that island-hop long distances often have seeds stuck to their feathers. Seeds that successfully germinate in a new island community may give rise to a population of new immigrants.

a colonizing source. Figure 19.23*a* has examples of this **area effect**. Larger islands tend to offer more habitats with more variation. Most of them have more complex topography and higher elevations. These are the kinds of variations that favor species diversity. Also, being bigger targets, big islands intercept more colonists.

Most importantly, extinctions suppress biodiversity on small islands. There, populations are small and far more vulnerable to severe storms, volcanic eruptions, disease, and random shifts in birth and death rates.

Again, on any island, the number of species reflects a balance between immigration rates for new species and extinction rates for established ones. Small islands that are distant from a source of colonists have low immigration rates and high extinction rates. The result is that, over the long term, they support fewer species than do larger islands that lie closer to the mainland.

The number of species in a region—species richness—is affected by gradients in latitude, elevation, and (for the aquatic species) depth.

Species richness in a given area also is an outcome of the evolutionary history of each species, its requirements for resources, its physiology, its capacity for dispersal, and its rates of birth, death, immigration, and emigration.

Generally, communities in the tropics have the most species and those at the poles have the least. The species diversity on an island also depends on its size and distance from a source of potential colonists.

19.9 Conservation Biology

Based on many lines of evidence accumulated over the past few centuries, an estimated 99 percent of all species that have ever lived are extinct. Even so, the full range of species diversity is greater now than it has been at any time in the past.

Reflect on the story of life's evolution, as told in Unit Three. So immense were five of the mass extinctions that we use them as the major boundaries for geologic time. After each event, species diversity plummeted on land and in the seas. In the aftermath, other species embarked on adaptive radiations. Recovery to the same level of species richness was exceedingly slow, on the order of 20 million to 100 million years.

The five major mass extinctions and long recoveries overshadow other, less spectacular extinction events. For example, the K–T asteroid impact was the final blow for some lineages. But other factors had been at work. Many dinosaur lineages became extinct during the 10 million years that *preceded* the impact. Geologic changes may have been a factor. Pangea was breaking up even before the K–T impact. Eventually, the deep, coldwater currents from the south started moving into warm equatorial seas. The change in ocean circulation altered air circulation and climate. A wide variety of groups perished during that global reorganization.

ANALYZING BIODIVERSITY

LINKS TO
SECTIONS
13.8–13.10

ABOUT THE NEWLY ENDANGERED SPECIES

We recognize three levels of **biodiversity**—diversity encompassed in diverse genes, diverse species, and diverse habitats. All three components of biodiversity are now in decline. Human activities are causing this biodiversity crisis.

No giant asteroids have hit the Earth for 65 million years. Yet a sixth major extinction event is under way. Although mammals survived the asteroid impact that did in the dinosaurs, many have not been as lucky in their encounters with humans. Shortly after the first humans migrated into the Americas, three quarters of their large mammal species, known as the *megafauna*, disappeared. Climate changes may have played a role, but hunting was certainly a contributing factor.

The International Union for the Conservation of Nature (IUCN) defines an **endangered species** as one that has lost 80 percent of its population, has suffered a drastic reduction in its range, and is likely to become extinct in the next 20 to 50 years. By this definition, about 10 percent of mammals, 5 percent of birds, and 20 percent of amphibians are endangered. Nearly all are *endemic* species, that is, they evolved in one region and are found nowhere else.

THE MAJOR THREATS

Naturalist Edward O. Wilson defines **habitat loss** as physical reduction in suitable places to live, as well as the loss of suitable habitat through chemical pollution. Habitat loss is putting major pressure on more than 90 percent of the species that now face extinction.

Species-rich tropical forests are being decimated (Chapter 15). As other examples, 98 percent of tallgrass prairies, 50 percent of wetlands, and 85 to 95 percent of old-growth forests of the United States alone are gone. Even arid lands are not exempt from human intrusion. Figure 19.24 is a case in point.

Habitats often become fragmented, or chopped into patches, each with a separate periphery. Species at a patch's periphery are more vulnerable to temperature changes, winds, fires, predators, and pathogens. Also, fragmentation often creates patches that are not large enough to support the population sizes required for successful breeding. They may not have enough food or other resources to sustain a species.

Think back on the factors shaping species diversity on islands. Even though islands make up a fraction of the Earth's surface, about half the plants and animals driven to extinction since 1600 were island natives that just had nowhere else to go. R. MacArthur and E. O. Wilson used an island biogeography model to estimate the likelihood of extinctions in habitat islands—natural areas that are surrounded by a "sea" of habitat that has been degraded by human activities. The model is also used to estimate how large an area must be set aside as protected reserves to ensure survival of a species.

Some species help warn us of changes in habitats that could lead to local extinctions. Birds are one of the key **indicator species**. Different types live in all major land regions and climate zones, they react quickly to changes, and they are relatively easy to track.

Species introductions are another major threat to endemic species. Remember, each habitat has only so many resources, and its occupants must compete for a share. Introduced species often grow to large numbers and overrun endemic competitors. They have been a key factor in almost 70 percent of the cases where endemic species are being driven to extinction.

Also, in a sad commentary on human nature, the more rare a wild animal becomes, the more its value soars in the black market. Figure 19.25 gives you an idea of the money involved in illegal wildlife trading. As a final point, biodiversity also is threatened by overharvesting of species that have commercial value. Whales are a prime example. Humans kill whales for food, lubricating oil, fuel, cosmetics, fertilizer, even corset stays. Substitutes exist for all whale products. Yet fishermen of several nations ignore a moratorium on slaughtering large whales, and they routinely cross boundaries of sanctuaries set aside for recoveries.

Figure 19.24 Humans as competitor species—example of dominance in the competition for living space in a habitat.

ANALYZING BIODIVERSITY
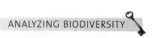

EMERGENCE OF CONSERVATION BIOLOGY

Awareness of the impending extinction crisis gave rise to **conservation biology**. We define this field of pure and applied research as (1) a systematic survey of the full range of biological diversity, (2) efforts to decipher biodiversity's evolutionary and ecological origins, and (3) attempts to identify methods that might maintain and use biodiversity to benefit the human population. Its goal is to conserve and use, in sustainable ways, as much biodiversity as possible.

THE ROLE OF SYSTEMATICS Reflect on Figure 19.26. Our current taxonomic knowledge is fairly complete for many groups, including flowering plants, conifers, and fishes, birds, mammals, and other vertebrates. We tend to know more about large organisms, especially the land dwellers, and the showy ones, such as birds and butterflies. Yet a staggering number of arthropods, fungi, protists, and prokaryotes are probably still out there, waiting to be identified.

For example, biologists recently surveyed a single hot spring in Yellowstone National Park. They turned up more archaean species than were already known for the rest of Earth—which gives you an idea of how much work is to be done.

When the inventorying of biodiversity is weighed against society's demands for, say, social services and protection from bioterrorism, guess which issue takes a back seat? Realistically, we just do not have enough time, money, or workers to complete a global survey of biodiversity. Possibly 99 percent of existing species have not even been described, and we do not know where most of them live. That is why the initial work has focused on identifying **hot spots**, or habitats of a large number of species that are found nowhere else and are in greatest danger of extinction.

At the first survey level, workers target a limited area, such as an isolated valley. A complete survey is out of the question, so they inventory flowering plants, birds, mammals, fishes, butterflies, and certain other indicator species within the habitat. At the next level, broader areas that are major or multiple hot spots are systematically explored. The widely separated forests of Mexico are an example of a key hot spot. Research stations have been set up at different latitudes and at different elevations across the region.

At the highest survey level, hot spot inventories are combined with existing knowledge of biodiversity in ecoregions. An **ecoregion** is a broad land or ocean region defined by climate, geography, and producer species. Figure 19.27 on the following page is a work in progress. It identifies 238 regions that are essential for global biodiversity: 142 land, 53 freshwater, and 43

Figure 19.25 Humans as predator species—confiscated tiger skins, leopard skins, rhinoceros horns, and elephant tusks. An illegal wildlife trade threatens the survival of more than 600 species, yet all but 10 percent of the trade escapes detection. A few of the black market prices: Live, mature saguaro cactus ($5,000–15,000), bighorn sheep, head only ($10,000–60,000), polar bear ($6,000), grizzly ($5,000), bald eagle ($2,500), peregrine falcon ($10,000), live chimpanzee ($50,000), live mountain gorilla ($150,000), Bengal tiger hide ($100,000), and rhinoceros horn ($28,600 per kilogram).

Figure 19.26 Current species diversity for a few major taxa. *Red* is the number of named species; *gold* is the estimated number of species not yet discovered.

marine. Initially, 25 are selected for rapid conservation efforts, which will be refined over the coming decades.

BIOECONOMIC ANALYSIS All nations have three forms of wealth—material, cultural, and biological. Until recently, many undervalued their biological wealth—*their biodiversity*, which is a source of food, medicine, and other goods. No one knows how much biological wealth they have already lost.

LINK TO
SECTION
18.5

For instance, one particularly observant Mexican college student located *Zea diploperennis*, a wild maize long believed to be extinct. It had disappeared from most of its range, but a relic population clung to life in a 900-acre region of mountainous terrain near Jalisco. Unlike domesticated corn, *Z. diploperennis* is perennial and resistant to most viruses. Gene transfers from the wild species into crop plants may boost production of corn for hungry people in Mexico and elsewhere.

In recognition of the potential value of this species, the Mexican government has set aside its mountainous habitat as a biological reserve, the first ever created to protect a wild relative of an important crop plant.

In the next fifty years, human population size may reach 8.9 billion, with most growth in the developing countries. Each person requires raw materials, energy, and living space. How many species will be crowded out? The countries with the greatest monetary wealth also have the least biological wealth and use most of the world's natural resources. Countries with the least monetary wealth and the most biological wealth have the fastest-growing populations.

People who are locked in poverty commonly must choose between saving themselves or an endangered species. Deprived of education and resources that the people in developed countries take for granted, they clear tropical forests. They hunt animals and dig up plants in "protected" parks and reserves. They try to raise crops and herds on marginal land.

Knowing that asking economically pressed people to stop is simplistic at best, conservation biologists are working to identify ways in which people might earn a living instead from their biological wealth—from the biodiversity in threatened habitats.

The systematic expansion of species inventories is a baseline for economic analysis, for assigning future value to ecoregions. Governments—and individuals—can protect a habitat and still withdraw resources in sustainable fashion. Threatened species can be studied for what they may offer. For example, one might be a source of new medicines and other chemical products. Most developing countries do not have laboratories for chemical analysis. However, large pharmaceutical companies do. The National Institute of Biodiversity of Costa Rica collects and identifies species that look promising and sends in chemical samples extracted from them to one company. If natural chemicals end up being marketed, Costa Rica will share the royalties, which are earmarked for conservation programs.

SUSTAINABLE DEVELOPMENT Identifying the possible uses of a country's biological wealth and putting them into practice is not easy. Yet there are successes, such as those in the next two examples.

Strip logging. Chapter 15 hinted at the destruction in once-great tropical rain forests, the richest homes of biodiversity on land. For now, simply think about one approach to minimizing the losses.

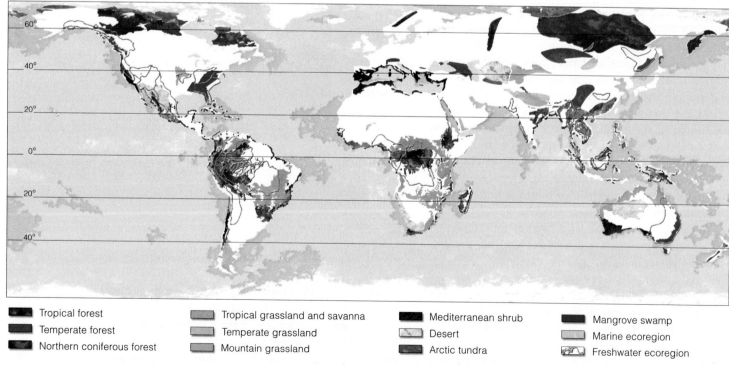

■ Tropical forest	▨ Tropical grassland and savanna
■ Temperate forest	▨ Temperate grassland
■ Northern coniferous forest	▨ Mountain grassland

■ Mediterranean shrub	■ Mangrove swamp
▨ Desert	▨ Marine ecoregion
■ Arctic tundra	▨ Freshwater ecoregion

Figure 19.27 Map of the most vulnerable regions of land and seas, compiled by the World Wildlife Fund.

Besides providing wood for local economies, trees of the tropical rain forests yield diverse, exotic woods prized by developed countries. What if they could be logged in a profitable, sustainable way that preserved biodiversity? Gary Hartshorn was the first to propose **strip logging** in portions of forests that are sloped and have many streams (Figure 19.28). The procedure is to clear a narrow corridor that parallels contours of the land, then cut a narrow road at the top to haul logs. In a few years, saplings grow in this cleared corridor. A second corridor is cleared above the road. Nutrients that leach from exposed soil here trickle into the first corridor and are taken up by the saplings and promote their growth. Later, a third corridor is cut above the second one—and so on in a profitable cycle of logging, which the habitat sustains over time.

Ranching in a riparian zone. Reflect on the map in Figure 19.27, and you see that the biodiversity crisis is not confined to developing countries. Now take a closer look at a riparian zone in the western United States. A **riparian zone** is a relatively narrow corridor of vegetation along a stream or a river, as in Figure 19.29. Its plants afford a major line of defense against flood damage by sponging up water during spring runoffs and summer storms. Shade cast by the canopy of taller shrubs and trees in the zone helps conserve precious water during droughts.

A riparian zone offers wildlife food, shelter, and shade, particularly in arid and semiarid regions. For instance, in the western United States, 67 to 75 percent of endemic species spend all or part of their life cycle in riparian zones. They include 136 songbird species, some of which nest only in riparian vegetation.

Many cattle are raised in the American West. In comparison with wild species, they have higher water needs and tend to congregate at rivers and streams. They trample and feed until grasses and herbaceous shrubs are gone. It takes only a few head of cattle to ruin a riparian zone. The damage can be reversed if cattle are excluded. Environmentally aware ranchers provide food and water away from streams and rotate grazing areas, so each area has time to recover.

> The current range of global biodiversity is an outcome of an overall pattern of extinctions and slow recoveries.
>
> For the past forty years, human activities have raised rates of extinction through habitat losses, species introductions, overharvesting, and illegal wildlife trading.
>
> Conservation biology entails a systematic survey of the full range of biodiversity, analysis of its evolutionary and ecological origins, and identification of methods to maintain and use biodiversity for the benefit of humans.

uncut forest

cut 1 year ago

dirt road

cut 3–5 years ago

cut 6–10 years ago

uncut forest

stream in watershed

Figure 19.28 Strip logging. This practice may sustain biodiversity even as it enhances the regeneration of economically valued forest trees.

Figure 19.29 Riparian zone restoration. This example comes from Arizona's San Pedro River, shown before and after restoration efforts.

ANALYZING BIODIVERSITY

Summary

Section 19.1 Each species occupies a certain habitat characterized by physical and chemical features and by the array of other species living in it. All populations of all species occupying a habitat are a community. Each species in the community has its own niche, or way of living.

Species in a community are linked by interactions of the types summarized in Table 19.1. Commensalism, mutualism, and parasitism are types of symbiosis: a relationship in which organisms live together for some part or all of the life cycle.

Section 19.2 Mutualism is a species interaction that benefits both participants. Some mutualists cannot complete their life cycle without their partner species.

Section 19.3 Two species that require the same limited resource tend to compete, as by exploiting it as fast or efficiently as possible or interfering with the other's use of it. Interspecific competition harms both participants. By the competitive exclusion concept, when two (or more) species require identical resources, they cannot coexist indefinitely. Species are more likely to coexist when they differ in their use of resources. Resource partitioning allows similar species to coexist by using different portions of shared resources.

Biology Now
Learn about competitive exclusion with the animation on BiologyNow.

Section 19.4 Predators are free-living and typically kill their prey. Predator and prey populations often show cyclic oscillations in their density. The carrying capacity for prey, likely refuges for prey, predator feeding efficiency, and often alternative prey sources influence the cycles.

Predators and prey exert selection pressure on each other. Evolved prey defenses include threat displays, chemical weapons, camouflage, and mimicry. Predators overcome defenses of prey by adaptive behavior, such as stealth and by camouflage.

Biology Now
Learn how predator and prey populations interact with the animation on BiologyNow.

Section 19.5 Parasites live in or on other living hosts and withdraw nutrients from host tissues for all or part of their life cycle. Hosts may or may not die as a result. Social parasites leave eggs and offspring in the care of a host species. Parasitoids are free-living as adults, but the larvae develop inside and feed on a host species, which they eventually kill. Parasites and parasitoids are used as biological control agents.

Section 19.6 Ecological succession is the sequential replacement of arrays of species within a community. Primary succession occurs in new habitats. Secondary succession occurs in disturbed ones. The first species in

Table 19.1 Types of Species Interactions

Type of Interaction	Direct Effect on Species 1	Direct Effect on Species 2
Commensalism	Benefits	None
Mutualism	Benefits	Benefits
Interspecific competition	Harmed	Harmed
Predation	Benefits	Harmed
Parasitism	Benefits	Harmed

a community are pioneer species. Their presence may help, hinder, or have no effect on other colonists.

The view that all communities eventually reach a stable climax state has been replaced by new models that emphasize continual change and acknowledge the importance of disturbances. Communities with an intermediate rate of disturbance tend to include more species than those disturbed more or less often.

Section 19.7 Community structure reflects an uneasy balance of forces that operate over time. Major forces are competition and predation, as by keystone species. Introduction of species that in one way or another expand their geographic range can change community structure in permanent ways.

Section 19.8 Global patterns of species diversity tend to follow environmental gradients. The species richness in a region is influenced by the evolutionary history of each species and its resource requirements, physiology, capacity for dispersal, and demographics. It also is influenced by species interactions.

Biology Now
Learn about the area effect and distance effect with the interaction on BiologyNow.

Section 19.9 The current range of biodiversity is an outcome of an overall pattern of mass extinctions and slow recoveries. Human activities are increasing rates of extinction. Conservation biologists seek ways to conserve and use biodiversity in a sustainable fashion.

Self-Quiz
Answers in Appendix I

1. A realized niche _____ .
   a. is smaller than a species' fundamental niche
   b. is affected by the presence of other species
   c. can shift over time
   d. all of the above

2. Which of the following interspecific interactions often lowers the population size of both participants?
   a. mutualism
   b. commensalism
   c. parasitism
   d. predation
   e. interspecific competition

Figure 19.30 Two phasmids: (**a**) A stick insect from South Africa. (**b**) A leaf insect from Java. (**c**) Many phasmid eggs resemble seeds.

Figure 19.31 One of the new realities: Casket in the ground or a concrete foundation for an artificial reef?

3. Which of the following is not a type of symbiosis?
   a. predation
   b. mutualism
   c. parasitism
   d. commensalism

4. With _____ both species benefit.
   a. predation
   b. interspecific competition
   c. mutualism
   d. commensalism

5. If one of two competing species is removed from a portion of a shared habitat, the population size of the species that remains will _____ over time.
   a. increase
   b. decrease
   c. remain the same

6. A predator population and prey population _____ .
   a. always coexist at relatively stable levels
   b. may undergo cyclic or irregular changes in density
   c. cannot coexist indefinitely in the same habitat
   d. both b and c

7. With _____ , one species modifies a habitat in a way that makes it easier for some other species to move in.
   a. tolerance
   b. facilitation
   c. inhibition
   d. a and c

8. The species diversity of an area is affected by _____ .
   a. climate and topography
   b. possibilities for dispersal
   c. evolutionary history
   d. both a and b
   e. a through c

9. Following mass extinctions, recovery to the same level of species diversity has taken many _____ of years.
   a. hundreds   b. millions   c. billions

10. Match the terms with the most suitable descriptions.
    ____ geographic dispersal
    ____ area effect
    ____ pioneer species
    ____ symbiotic species
    ____ keystone species
    ____ endemic species
    ____ riparian zone
    ____ niche

    a. early colonist of barren or disturbed places
    b. dominates community
    c. individuals leave home range, and get established elsewhere
    d. island size influences number of species it can support
    e. live together and interact for some or all of life cycle
    f. area near river or stream
    g. originated in one region and found nowhere else
    h. activities, interactions, and needs of one particular species

Additional questions are available on Biology ❸Now™

Critical Thinking

1. With antibiotic resistance rising, researchers are looking for ways to reduce use of antibiotics. Some cattle that once would have been fed antibiotic-laced food are now being given feed that contains cultures of bacteria. These *probiotic feeds* are designed to establish or bolster populations of helpful bacteria in the animal's gut. The idea is that if a large population of beneficial bacteria is in place, then the harmful bacteria cannot become established or thrive. Which ecological principle is guiding this research?

2. Insects called phasmids are extraordinary masters of disguise. Most resemble sticks or leaves. Even their eggs may be camouflaged (Figure 19.30). The phasmids are all herbivores. Most spend their days motionless, moving about and feeding only at night. If disturbed, a phasmid will fall to the ground and remain motionless.
   Speculate on the selective pressures that might have shaped phasmid morphology and behavior. Suggest an experiment that could help determine whether the cryptic appearance and behavior of one species have adaptive value.

3. Does focusing on species-rich hot spots cause us to undervalue regions where species diversity is lower? For example, broad bands of coniferous forests in the Northern Hemisphere have low species diversity, but photosynthesis here affects global carbon cycles. How can we ensure that regions that provide such biological services are protected?

4. Mentally transport yourself to a tropical rain forest of South America. Imagine you do not have a job. There are no jobs in sight, not even in the overcrowded cities some distance away. You have no money or contacts to get you anywhere else. And yet you are the sole supporter of a large family. Today a stranger approaches you. He tells you he will pay good money if you can capture alive a certain brilliantly feathered parrot in the forest. You know the parrot is rare; it is seldom seen. However, you have an idea of where it lives. What will you do?

5. Here's an alternative to formal burial on land. Ashes of cremated people are mixed with pH-neutral concrete and used to build *artificial reefs*. Each perforated concrete ball weighs 400 to 4,000 pounds and is designed to last 500 years. So far, 100,000 of these reefs are helping marine life in 1,500 locations (Figure 19.31). Proponents say that the burial costs less, does not waste land, and is ecologically friendly. Your thoughts?

Bye-Bye Bayou

Each Labor Day, the coastal Louisiana town of Morgan City celebrates the region's economic mainstays with the Louisiana Shrimp and Petroleum Festival. The state is the nation's third-largest petroleum producer and the leader in shrimp production. But the petroleum industry's success may be contributing to the decline of the state's fisheries.

The global air temperature is rising, and fossil fuel burning is a major culprit. Because warm air heats the surface waters of the ocean, and water expands when it is heated, the sea level is rising. Warmer air also is melting ancient glaciers and vast ice caps, and this meltwater further increases the ocean's volume.

If current trends persist, then low elevations along the United States coastline—including *14,720,000 acres* next to the Gulf of Mexico and Atlantic Ocean—will be one to three feet under water within fifty years!

With more than 40 percent of the country's coastal marshes, Louisiana has the most to lose. Its wetlands are already sinking, because dams and levees along its rivers interfere with the deposition of sediments that could replace those washed out to sea. The anticipated rise in sea level will obliterate about 70 percent of these wetlands.

Is an ecological and economic disaster unfolding? Consider: More than 3 billion dollars' worth of shellfish and fish are harvested from Louisiana's wetlands each year. About 40 percent of North America's migratory ducks—more than 5 million birds—overwinter here. Louisiana's wetlands also buffer low inland elevations from storm surges and hurricanes.

There's more. The warmer water will promote algal blooms that cause massive fish kills. It will favor the growth of bacterial pathogens, and more people will get sick after swimming in contaminated water or after eating contaminated shellfish.

Inland, heat waves and wildfires will become more intense, and deaths from heat stroke will climb. Warmer temperatures will allow disease-causing mosquitoes to extend their ranges farther inland.

Global warming is expected to increase evaporation and alter rainfall patterns, leading to droughts in some places and floods in others. Hurricanes are expected to increase in intensity. The damage from hurricanes in 2005 may be a preview of things to come.

This chapter will start you thinking about a major concept: The atmosphere, ocean, and land interact in ways that influence the living conditions of organisms throughout the biosphere. Driven by energy streaming in from the sun, these interactions give rise to globe-spanning cycles upon which life depends. The chapter also will start you thinking about a related concept of equal importance: We have become major players in these global interactions even though we still do not fully comprehend how they work.

 How Would You Vote? *Emissions from motor vehicles contain greenhouse gases. Should the government raise minimum fuel economy standards for cars and SUVs? See BiologyNow for details, then vote online.*

 ## Key Concepts

ECOSYSTEM STRUCTURE
Organisms and their physical environment make up an ecosystem. Energy flows in one way through organisms in an ecosystem. Producers capture energy from the sun and use inorganic compounds from the environment to build tissues. Consumers get energy and nutrients by eating other organisms. Nutrients are cycled among organisms.

GLOBAL CYCLES
Earth's waters are cycled on a vast scale through environmental reservoirs and the living components of ecosystems. So are nutrients, such as carbon, nitrogen, and phosphorus. Human activities are disrupting these global cycles.

Links to Earlier Concepts

Sections 1.1 and 1.2 introduced you to the idea of an ecosystem. Carbon fixation by plants (5.3) and protists (14.3), and the many ecological roles of prokaryotes (14.2) are relevant too. Laws governing energy (4.1) come into play as energy is transferred among organisms (19.4). We consider again the runaway growth of human populations (18.5).

20.1 The Nature of Ecosystems

In the preceding chapter, we focused on community structure, how species interact in a habitat. This chapter continues with those interactions, but at a higher level of organization. Each community, with its physical environment, works as an ecosystem.

OVERVIEW OF THE PARTICIPANTS

An **ecosystem** is an array of organisms together with their physical environment. They interact through a one-way flow of energy and the cycling of vital raw materials. It is an open system; it is unable to sustain itself without continual inputs of energy.

Diverse ecosystems abound on Earth's surface. In climate, landforms, soil, vegetation, animal life, and other features, deserts differ from hardwood forests, which differ from tundra and the prairies. Reefs differ from the open ocean, which differs from streams and lakes. Despite their many differences, all ecosystems are alike in certain aspects of structure and function.

Ecologists investigate these similarities and can use them to make predictions about how ecosystems will be affected by human-caused and natural changes.

With a few exceptions, ecosystems run on sunlight energy captured by photoautotrophs—mostly plants and phytoplankton. The photoautotrophs are **primary producers** (Figure 20.1). They convert sunlight energy to chemical bond energy, using it to synthesize organic compounds from inorganic materials.

All other organisms in the system are **consumers**. These heterotrophs feed on the tissues, products, and remains of other organisms. We describe consumers by their diets. *Herbivores* eat plants. *Carnivores* eat flesh. *Parasites* live in or on a host and feed on its tissues. **Detritivores**, such as crabs, eat detritus—particles of decomposing organic matter. Finally, the wastes and remains of all organisms are degraded to inorganic building blocks by the bacterial, protist, and fungal **decomposers**. Decomposers absorb most breakdown products, but some spill out into the environment.

Some consumers do not fall into simple categories. Many are *omnivores*, which dine on animals, plants, fungi, protists, and even bacteria. A red fox is like this (Figure 20.2). If it comes across a dead bird, the fox will scavenge it. A *scavenger* is an animal that ingests dead plants, animals, or both, all or some of the time. Vultures, termites, and hyenas are typical scavengers.

Recall from Section 4.1 that energy transfers are never 100 percent efficient. Over time, all the energy that was originally captured by producers will escape back to the environment, mostly as metabolic heat. Energy flow through an ecosystem is one-way because sunlight energy is transformed into heat energy, which cannot be captured by organisms and used to do more cellular work.

Nutrients are cycled in an ecosystem. The primary producers obtain hydrogen, oxygen, and carbon from water and carbon dioxide in the environment. They take up phosphorus, nitrogen, and other minerals as well. These are used as raw materials for biosynthesis. Later, decomposition returns the nutrients back to the environment, from which producers can take them up once again.

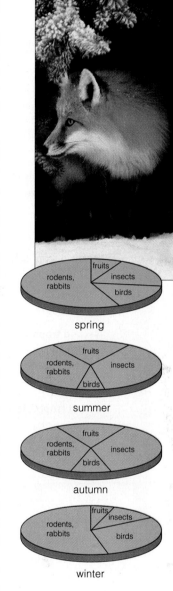

Figure 20.2 Seasonal shifts in the red fox diet.

Figure 20.1 Model for ecosystems. Energy flows in one direction: into an ecosystem, through living organisms, and out from it. Nutrients cycle among the producers and consumers. In nearly all ecosystems, energy flow starts with photoautotrophs that capture energy from the sun.

Energy input, from sun

Producers
Autotrophs (plants and other self-feeding organisms)

Nutrient Cycling

Consumers
Heterotrophs (animals, most fungi, many protists, many bacteria)

Energy output (mainly metabolic heat)

Organisms and their physical environment make up an ecosystem. A one-way energy flow and a cycling of materials interconnect them.

Primary producers capture sunlight energy and assemble organic compounds from inorganic raw materials. All other organisms are consumers; they get their carbon and energy secondhand by consuming other organisms, or their wastes or remains.

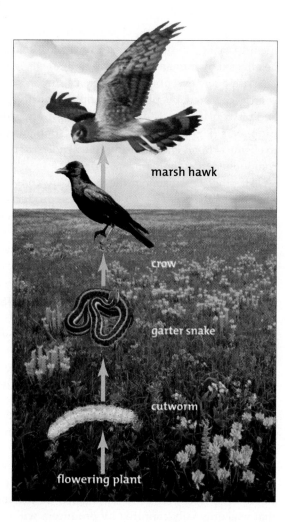

fifth trophic level

top carnivore
(fourth-level consumer)

marsh hawk

fourth trophic level

carnivore
(third-level consumer)

crow

third trophic level

carnivore
(second-level consumer)

garter snake

second trophic level

herbivore
(primary consumer)

cutworm

first trophic level

autotroph
(primary producer)

flowering plant

Figure 20.3 *Animated!* Example of a simple food chain and its corresponding trophic levels.

LINKS TO
SECTIONS
4.1, 19.4

TROPHIC INTERACTIONS

We classify all organisms of an ecosystem by exactly where they fit in a hierarchy of feeding relationships called **trophic levels** (*troph*, nourishment). "Who eats whom?" we ask. If organism **A** is eaten by organism **B**, energy is transferred from **A** to **B**. All organisms at a given trophic level are the same number of transfer steps away from the energy input into the ecosystem.

As one example, think about some organisms of a tallgrass prairie ecosystem. Plants that capture energy from the sun are at the first trophic level. These plants are eaten by herbivores, such as cutworms, which are at the next trophic level. The cutworms are among the primary consumers that are eaten by animals at the third trophic level, and so on up through successively higher tiers.

At each trophic level, organisms interact with the same sets of predators, prey, or both. Omnivores feed at several levels, so we would partition them among different levels or assign them to a level of their own.

A **food chain** describes a straight-line sequence of steps by which the energy and nutrients stored in the tissues of organisms at the lowest trophic level moves up to higher levels. In one tallgrass prairie food chain, energy from a plant flows to a cutworm that eats it, then to a garter snake that eats the cutworm, to a crow that eats the snake, and finally to a marsh hawk that eats the crow. Figure 20.3 shows this food chain.

Identifying a food chain is a simple way to begin thinking about how energy and nutrients are moved among the living components of an ecosystem. Keep in mind, however, that many different organisms are usually competing for food at each trophic level. Tallgrass prairie producers (mainly flowering plants) are eaten by various mammals, birds, and insects. Many species interact in nearly all ecosystems, especially at a lower trophic level. A number of food chains *cross-connect* with one another—as **food webs**.

FOOD WEBS

By untangling many food webs, ecologists discovered unifying patterns of organization. The patterns reflect both environmental constraints and the inefficiency of energy transfers from one trophic level to the next. Plants capture only a fraction of the energy from the sun. They store some of that energy in chemical bonds of organic compounds in new plant tissues. They lose the rest as metabolic heat. Consumers make use of the energy that became stored in plant tissues, remains, and wastes. Then they, too, lose metabolic heat. Taken together, all of these heat losses represent a one-way flow of energy out of the ecosystem.

Energy transfers can never be 100 percent efficient because a bit of energy is lost at each step. This puts a limit on the length of food chains. In most food chains, energy captured by the producers passes through no more than four or five trophic levels. Even ecosystems that include complex food webs, such as the one in Figure 20.4, do not have lengthy food chains.

Field studies and computer simulations of aquatic and land food webs suggest more patterns. Chains in food webs tend to be shortest where environmental conditions often vary widely over time. Food chains tend to be longer in habitats that are more stable, such as ocean depths. The most complex webs tend to have many herbivorous species, as happens in grasslands. By comparison, the food webs with fewer connections tend to have more carnivores. Also, the number and kind of species in a food web can change over time.

ECOSYSTEM STRUCTURE

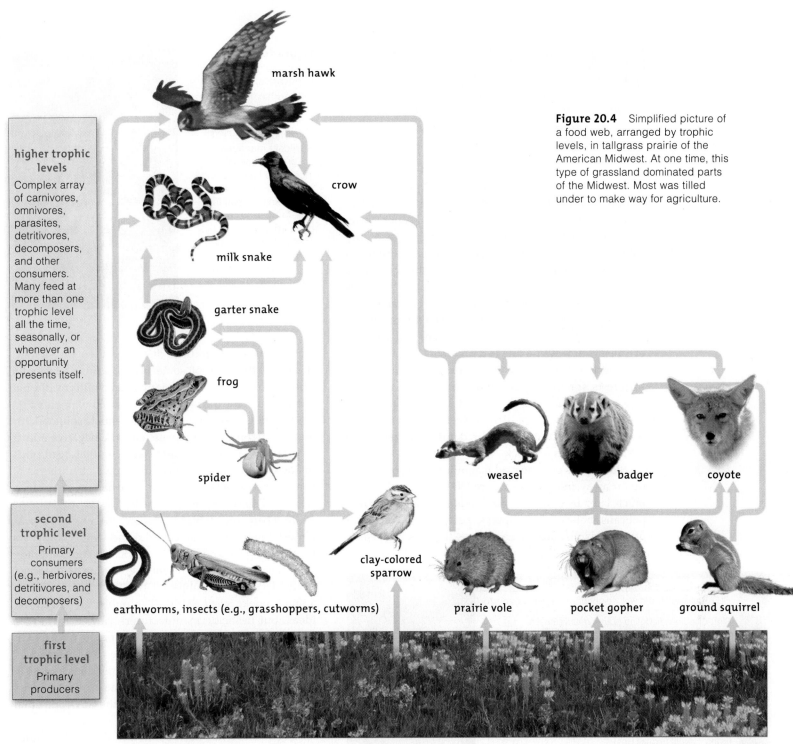

higher trophic levels
Complex array of carnivores, omnivores, parasites, detritivores, decomposers, and other consumers. Many feed at more than one trophic level all the time, seasonally, or whenever an opportunity presents itself.

second trophic level
Primary consumers (e.g., herbivores, detritivores, and decomposers)

first trophic level
Primary producers

marsh hawk

crow

milk snake

garter snake

frog

spider

weasel   badger   coyote

clay-colored sparrow

earthworms, insects (e.g., grasshoppers, cutworms)   prairie vole   pocket gopher   ground squirrel

grasses, composites

Figure 20.4 Simplified picture of a food web, arranged by trophic levels, in tallgrass prairie of the American Midwest. At one time, this type of grassland dominated parts of the Midwest. Most was tilled under to make way for agriculture.

Energy from producers flows in a single direction through two kinds of webs. In a **grazing food web**, energy stored by producers flows to herbivores, then to carnivores, and then to decomposers. In a **detrital food web**, energy from producers flows mainly into detritivores and decomposers.

In nearly all ecosystems, both kinds of webs cross-connect. For example, in a rocky intertidal ecosystem, energy captured by algae flows to snails, which are eaten by herring gulls as part of a grazing food web.

The same gulls also hunt crabs, which are among the primary consumers in the detrital food web.

An organism's trophic level describes how many steps it is from the ecosystem's energy source.

With each energy transfer, some of the energy is lost as heat. These losses limit the number of trophic levels in ecosystems to four or five.

20.2 Biological Magnification in Food Webs

We turn now to a premise that opened this chapter— that disturbances to one part of an ecosystem often have unexpected effects on other, seemingly unrelated parts.

DDT is a synthetic organic pesticide. It is nearly insoluble in water, so you may think it acts only where applied. But winds can carry DDT in vapor form; water can move fine particles of it. Being highly soluble in fats, DDT can collect in tissues, and so it can undergo **biological magnification**. By this process, a substance that degrades slowly or not at all becomes more and more concentrated in tissues of organisms at ever higher trophic levels of a food web (Figure 20.5).

Decades ago, DDT got into food webs and acted on organisms in ways no one had predicted. Where it was sprayed to control Dutch elm disease, songbirds died. In forests sprayed to kill budworm larvae, DDT got into small streams and fishes died. In fields sprayed to control one kind of pest, new pests moved in. DDT was also killing the natural predators that help keep pest populations in check.

Side effects of biological magnification also showed up far away from places where DDT was applied, and much later in time. Most vulnerable were the brown pelicans, bald eagles, ospreys, peregrine falcons, and other birds that are top carnivores (Figures 20.5, 20.6).

Figure 20.6
Peregrine falcon, a top carnivore of certain food webs. It almost became extinct as a result of biological magnification of DDT. A wildlife management program successfully increased its population sizes. Peregrine falcons were reintroduced into wild habitats. They also have adapted to cities. There, these raptors hunt pigeons, large populations of which are a messy nuisance.

DDT breakdown products disrupted their physiology. It caused the eggs of some birds to have unusually brittle shells. As a result, many chick embryos did not hatch, and some bird species even faced extinction.

Rachel Carson, a respected scientist, sounded the alarm. Carson began to investigate the negative effects of pesticides on wildlife after learning that dead birds inside a local wildlife sanctuary showed symptoms of some type of poisoning. The area had been sprayed with DDT to control mosquitoes.

Independent, unbiased research on pesticides was almost nonexistent. Carson, who had worked for years for the U.S. Fish and Wildlife Service, began compiling information about the harmful effects of widespread pesticide use. She published her findings in 1962 as a small book, *Silent Spring*. The public embraced her ideas. But pesticide manufacturers viewed them as a threat to sales and mounted a campaign to discredit her. At the time, Carson was in the final stages of a battle with terminal breast cancer. Yet she vigorously defended her position until her death in 1964. Later, her work became one impetus for the environmental movement in the United States.

Use of DDT has been banned in the United States since 1972. Its agricultural use is discouraged nearly everywhere, but some developing countries still use it to control disease-carrying mosquitoes. They may be trading one ill effect for another. Studies have shown that pregnant women exposed to DDT are at a higher risk for premature delivery and underweight infants.

DDT Residues (ppm wet weight of whole live organism)

Ring-billed gull fledgling (*Larus delawarensis*)	75.5
Herring gull (*Larus argentatus*)	18.5
Osprey (*Pandion haliaetus*)	13.8
Green heron (*Butorides virescens*)	3.57
Atlantic needlefish (*Strongylura marina*)	2.07
Summer flounder (*Paralichthys dentatus*)	1.28
Sheepshead minnow (*Cyprinodon variegatus*)	0.94
Hard clam (*Mercenaria mercenaria*)	0.42
Marsh grass shoots (*Spartina patens*)	0.33
Flying insects (mostly flies)	0.30
Mud snail (*Nassarius obsoletus*)	0.26
Shrimps (composite of several samples)	0.16
Green alga (*Cladophora gracilis*)	0.083
Plankton (mostly zooplankton)	0.040
Water	0.00005

Figure 20.5 Biological magnification in a Long Island, New York, estuary. In 1967, George Woodwell, Charles Wurster, and Peter Isaacson were aware that DDT exposure and wildlife deaths were connected. For instance, residues in birds that died from DDT poisoning were 41–295 parts per million (ppm) and 1–26 ppm in some fishes. In their study, some measured DDT levels were below lethal thresholds but still high enough to interfere with reproduction.

With biological magnification, a substance that degrades slowly or not at all gets ever more concentrated in tissues of organisms at ever higher trophic levels of a food web.

ECOSYSTEM STRUCTURE

20.3 Studying Energy Flow Through Ecosystems

Ecologists measure the amount of energy and nutrients that enter an ecosystem, the amounts captured, and the proportions stored in each trophic level.

ECOLOGICAL PYRAMIDS

Ecologists sometimes illustrate the trophic structure of an ecosystem in the form of an ecological pyramid. In such pyramids, the primary producers are shown as a base for successive tiers of consumers above them.

A **biomass pyramid** depicts the dry weight of all of an ecosystem's organisms at each tier (Figure 20.7). In most ecosystems, primary producers have the most biomass, and top carnivores the least. But in certain aquatic ecosystems the biomass pyramid is "upside-down." For example, in some marine ecosystems, the phytoplankton reproduce and are devoured so quickly that they never accumulate in large numbers.

An **energy pyramid** illustrates how the amount of usable energy diminishes as it is transferred through an ecosystem. Sunlight energy is captured at the base (first trophic level) and declines through successive levels to its tip (the top carnivores). Energy pyramids always have a large energy base, so they are always "right-side up."

PRIMARY PRODUCTIVITY

The rate at which producers capture and store energy in their tissues during a given interval is the **primary productivity** of an ecosystem. How much energy gets stored depends on (1) how many producers there are and (2) the balance between photosynthesis (energy trapped) and aerobic respiration (energy used). *Gross* primary production is all energy initially trapped by the producers. *Net* primary production is the fraction of trapped energy that producers funnel into growth and reproduction. **Net ecosystem production** is the gross primary production *minus* the energy used by producers and by detritivores and decomposers in soil. (In both cases, the subtracted amount of energy would not be available to organisms that eat plants.)

Many factors influence net production, its seasonal patterns, and its regional distribution (Figure 20.8). For example, energy acquisition and storage depend in part on the body size and body form of primary producers, the availability of minerals, the range of temperature, and the amount of sunlight and rain. The harsher environmental conditions are, the less new plant growth there is per season, and the lower the ecosystem's primary productivity.

Figure 20.7 Ecological pyramids for Silver Springs, Florida, a small aquatic ecosystem. *Above,* a biomass pyramid measured in grams per square meter. In this case, biomass decreases in successively higher tiers, but some biomass pyramids are widest on the top. *Below,* pyramid of energy flow measured in kilocalories per square meter per year. All energy pyramids are widest at the bottom because energy is lost with each transfer to a higher level.

Figure 20.8 Summary of satellite data on net primary productivity during 2002. Productivity is coded as *red* (highest) down through *orange, yellow, green, blue,* and *purple* (lowest). Although average productivity per unit of sea surface is lower than it is on land, total productivity on land and in seas is about equal, because most of Earth's surface is covered by water.

> *The trophic structure of an ecosystem can be represented by an ecological pyramid. Biomass pyramids may be top- or bottom-heavy depending on the ecosystem. In contrast, an energy pyramid always has the largest tier on the bottom.*
>
> *Gross primary productivity is an ecosystem's total rate of photosynthesis during a specified interval. The net amount is the rate at which primary producers store energy in tissues in excess of their rate of aerobic respiration. Heterotrophic consumption affects the rate of energy storage.*

20.4 Global Cycling of Water and Nutrients

Primary productivity depends on availability of fresh water and on nutrients dissolved in it. Global cycles distribute water and nutrients.

In a **biogeochemical cycle**, ions or molecules of some essential substance flow from the environment into organisms, then back to the environment. Some part or parts of the environment function as a reservoir. In most cases, the rate at which the substance moves into and out of environmental reservoirs is far slower than the rate at which it is transferred among organisms.

Nutrients move into and out of ecosystems as a result of geologic processes. Weathering of rocks will add nutrients to an ecosystem. Erosion and runoff can remove nutrients from the system. Usually, the quantity of a nutrient that cycles through an ecosystem per year is greater than the amount that enters or leaves.

By acting as decomposers, some bacteria help keep nutrients cycling within ecosystems. Other prokaryotic species play roles in transforming solids and ions into gaseous form and then back again. By their actions, they convert some essential nutrients to and from forms that can be taken up by the primary producers.

In the *hydrologic* cycle, oxygen and hydrogen move in the form of water molecules. We call this the global water cycle. In *atmospheric* cycles, some portion of the

nutrient occurs as an atmospheric gas. Nitrogen and carbon cycles are two examples. Phosphorus and other nutrients that do not commonly exist in gaseous forms move in *sedimentary cycles*. These nutrients move from land to the seafloor and eventually return to land by way of geological uplifting. Earth's crust serves as the main reservoir for nutrients with sedimentary cycles.

HYDROLOGIC CYCLE

Driven by ongoing inputs of solar energy, the Earth's waters move slowly, on a vast scale, from the ocean into the atmosphere, to land, and back to the ocean— the main reservoir. Figure 20.9 shows this **hydrologic cycle**. Water evaporating into the lower atmosphere stays aloft as vapor, clouds, and ice crystals, then falls mainly as rain and snow. Ocean circulation and wind patterns influence the cycle.

Water moves nutrients into and out of ecosystems. Long-term studies of watersheds demonstrated how. A **watershed** is any region where precipitation becomes funneled into a single stream or river. Watersheds may be an area drained by a small stream or as large as the Mississippi River watershed, which extends across one-third of the continental United States. Water entering a watershed mostly seeps into soil or joins surface runoff toward streams. Plants withdraw water and dissolved mineral ions from soil and lose water by transpiration.

You might think that surface runoff would quickly leach mineral ions. But research in New Hampshire's

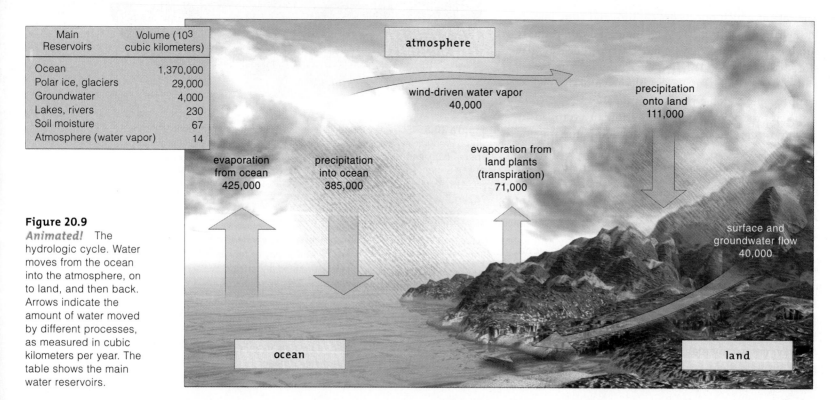

Main Reservoirs	Volume (10^3 cubic kilometers)
Ocean	1,370,000
Polar ice, glaciers	29,000
Groundwater	4,000
Lakes, rivers	230
Soil moisture	67
Atmosphere (water vapor)	14

atmosphere

wind-driven water vapor
40,000

precipitation onto land
111,000

evaporation from ocean
425,000

precipitation into ocean
385,000

evaporation from land plants (transpiration)
71,000

surface and groundwater flow
40,000

ocean

land

Figure 20.9
Animated! The hydrologic cycle. Water moves from the ocean into the atmosphere, on to land, and then back. Arrows indicate the amount of water moved by different processes, as measured in cubic kilometers per year. The table shows the main water reservoirs.

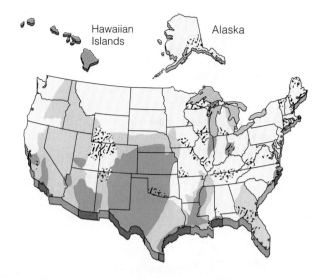

Figure 20.10 Experimental deforestation of New Hampshire's Hubbard Brook watershed. Researchers stripped vegetation from forest plots but did not disturb the soil. They applied herbicides for three years to prevent regrowth. They monitored surface runoff flowing over concrete catchments on its way to a stream below. Concentrations of calcium ions and other minerals were compared against concentrations in water passing over a control catchment, positioned in an undisturbed area.

Calcium losses were *six times* greater in deforested plots. Removing all vegetation from such forests clearly alters nutrient outputs in ways that can disrupt nutrient availability for an entire ecosystem.

Figure 20.11 Aquifer depletion, seawater intrusion, and groundwater contamination in the United States. *Green* signifies high overdrafts, *gold*, moderate overdrafts, and *pale yellow*, insignificant withdrawals. Shaded areas are sites of major groundwater pollution. *Blue* squares indicate saltwater intrusion from nearby seas.

Hubbard Brook watershed showed this is not the case. In undisturbed forests in this watershed, each hectare lost only about eight kilograms of calcium. The lost calcium was replaced by rainfall and by weathering of rocks. Also, tree roots mine the soil, so calcium was continually being stored in the biomass of tree tissues. The experimental deforestation of these forests caused a spike in nutrient losses (Figure 20.10).

THE WATER CRISIS

Our planet has a lot of water, but most of it is too salty to be of use. If all of Earth's waters fit into a bathtub, the freshwater portion would barely fill a teaspoon.

Agriculture accounts for two-thirds of our water usage, and irrigation alters the suitability of land for agriculture. The piped-in water commonly has high concentrations of mineral salts. Where the soil drains poorly, evaporation causes **salinization**, a buildup of salt in soil that stunts crop plants and decreases yields.

Soil and permeable rock layers called aquifers hold **groundwater**. About half the United States population taps this source for drinking water. Chemicals leached from landfills, hazardous waste dumps, and underground storage tanks contaminate it. Unlike flowing rivers and streams that recover quickly, groundwater pollution is difficult and expensive to clean up.

Groundwater overdrafts, or the amount that nature does not replenish, are high in many areas. Figure 20.11 shows some regions of aquifer depletion in the United States. Overdrafts have now depleted half of

the great Ogallala aquifer, which supplies irrigation water for 20 percent of the Midwest's croplands.

Contaminants, such as sewage, animal wastes, and agricultural chemicals, can make water unsuitable for drinking. These water pollutants also disrupt aquatic ecosystems, driving some species to local extinction.

If current rates of human population growth and water depletion continue, the amount of fresh water available for everyone will soon be 55 to 66 percent less than it was in 1976. In this past decade, thirty-three nations have already engaged in conflicts over reductions in water flow, pollution, and silt buildup in aquifers, rivers, and lakes.

Could **desalinization**—the removal of salt from seawater—meet our water needs? Salt can be removed by distillation or pushing water through membranes. The processes require fossil fuels, which makes them most feasible in Saudi Arabia and other countries with small populations and big fuel reserves. Most likely, desalinization will not be cost-effective for large-scale agriculture. It also produces mountains of waste salts.

> In the hydrologic cycle, water slowly moves on a global scale from the world ocean (the main reservoir), through the atmosphere, onto land, then back to the ocean.
>
> Plants absorb and store dissolved minerals. By doing so, they minimize the loss of soil nutrients in runoff.
>
> Aquifers that supply much of the world's drinking water are becoming polluted and depleted. Regional conflicts over access to clean, drinkable water are likely to increase.

Figure 20.12 *Animated!* Global carbon cycle.

Carbon's movement through typical marine ecosystems (**a**) and through ecosystems on land (**b**). *Gold* boxes show the main carbon reservoirs. The vast majority of carbon atoms are locked in sediments and rocks, followed by increasingly lesser amounts in ocean water, soil, and the atmosphere, and finally in biomass.

The sketch below shows the loop of ocean water that delivers carbon dioxide to carbon's deep ocean reservoir. It sinks in the cold, salty North Atlantic and rises in the warmer Pacific.

diffusion between atmosphere and ocean

combustion of fossil fuels

bicarbonate and carbonate dissolved in ocean water

photosynthesis   aerobic respiration

marine food webs
producers, consumers, decomposers, detritivores

incorporation into sediments

death, sedimentation

uplifting over geologic time

sedimentation

marine sediments, including formations with fossil fuels

a

20.5 Carbon Cycle

Most of the world's carbon is locked in sediments and rocks. Because most carbon moves into and out of ecosystems in gaseous form, its movement is said to be an atmospheric cycle.

Carbon moves through the lower atmosphere and all food webs on its way to and from the ocean's water, sediments, and rocks. Its global movement is called the **carbon cycle** (Figure 20.12). Carbon is released as cells engage in aerobic respiration, when fossil fuels burn, and when the molten rock in Earth's crust emits carbon during volcanic eruptions.

Carbon dioxide, or CO_2, is the most abundant form of carbon in the atmosphere. In the oceans, dissolved bicarbonate (HCO_3^-) and carbonate (CO_3^-) are the

main forms. All three carbon-oyxgen compounds have essential metabolic roles. For that reason, the cycle is sometimes called the *carbon–oxygen cycle*.

What keeps all the CO_2 dissolved in warm ocean surface waters from escaping into the air? Driven by winds and regional differences in water density, water makes a gigantic loop from the surface of the Pacific and Atlantic oceans down to the Atlantic and Antarctic seafloors. The CO_2 moves into deep storage reservoirs before the seawater loops back up (Figure 20.12).

Photosynthetic organisms lock billions of metric tons of carbon atoms into organic compounds each year (Chapter 5). The relative amount of time that a carbon atom spends in living and nonliving portions of an ecosystem varies. Organic wastes and remains decompose so quickly in tropical rain forests that not much carbon can accumulate at the soil surfaces. In swamps, bogs, marshes, and other anaerobic habitats, decomposers cannot degrade organic compounds into

atmosphere (mainly carbon dioxide)

volcanic action

combustion of fossil fuels

terrestrial rocks

photosynthesis

aerobic respiration

combustion of wood (for clearing land; or for fuel)

weathering

land food webs producers, consumers, decomposers, detritivores

deforestation

soil water (dissolved carbon)

death, burial, compaction over geologic time

peat, fossil fuels

leaching, runoff

b

inorganic subunits, so carbon gradually accumulates in peat and other forms of compressed organic matter.

Also, in ancient aquatic ecosystems, carbon became incorporated in staggering numbers of shells and other hard body parts of foraminiferans and other species. Those organisms died, sank through water, then were buried in sediments. Carbon stayed buried for many millions of years until geologic forces slowly uplifted part of the seafloor. Remember the carbon-rich white cliffs of Dover, shown in Figure 14.16?

As a final example, plants of Carboniferous swamp forests incorporated carbon into organic compounds. Those compounds were gradually converted into coal, petroleum, and gas reserves—which humans now tap as fossil fuels. Over the course of a few hundred years, we have been extracting and burning fossil fuels that took millions of years to form. In other words, these fuels are nonrenewable resources. Unfortunately, they are the energy source of choice for the majority of the

6.3 billion people now on Earth—mainly in developed countries that can afford them. At present, fossil fuel burning and some other human activities are releasing more carbon into the atmosphere than can be cycled naturally to the ocean's storage reservoirs.

Only about 2 percent of the excess carbon entering the atmosphere will become dissolved in ocean water. Most researchers now think the carbon buildup in the atmosphere is amplifying the greenhouse effect, which means it may be contributing to global warming. The next section will give you a closer look at this effect.

Most carbon is trapped in Earth's rocks and in its deep sea sediments. However, carbon moves between its main reservoirs and ecosystems mostly in the form of a gas— carbon dioxide. This is why the carbon cycle is considered to be one of the Earth's atmospheric cycles.

a Wavelengths in rays from the sun penetrate the lower atmosphere, and they warm the Earth's surface.

b The surface radiates heat (infrared wavelengths) to the atmosphere. Some heat escapes into space. But greenhouse gases and water vapor absorb some infrared energy and radiate a portion of it back toward Earth.

c Increased concentrations of greenhouse gases trap more heat near Earth's surface. Sea surface temperatures rise, so more water evaporates into atmosphere. The Earth's surface temperature rises.

Figure 20.13 *Animated!* The greenhouse effect.

LINK TO SECTION 18.5

20.6 Greenhouse Gases, Global Warming

The atmospheric concentrations of gaseous molecules help determine the average temperature near Earth's surface. That temperature has enormous impact on the global climate and ecosystems everywhere.

GREENHOUSE EFFECT

Countless molecules of carbon dioxide, water, ozone, methane, and other substances in the atmosphere affect global temperature. Collectively, these many molecules act somewhat like a pane of glass in a greenhouse—hence their name, "greenhouse gases." Incoming light energy streams past the molecules and reaches Earth's surface. This energy heats up the land, which radiates heat back into the atmosphere. Greenhouse gases slow the escape of this heat energy into space. They absorb heat and radiate it into the lower atmosphere, which warms up as a result (Figure 20.13). This warming of the lower atmosphere is called the **greenhouse effect**.

GLOBAL WARMING AND CLIMATE CHANGE

Without greenhouse gases, Earth's surface would be cold and lifeless. But there can be too much of a good thing. Primarily as the outcome of human activities, concentrations of greenhouse gases are now higher than they were in the past (Figure 20.14). Atmospheric CO_2 is at its highest level since 420,000 years ago. It may well be the highest for the past 20 million years.

In the 1950s, researchers in a laboratory on Hawaii's highest volcano set out to measure the atmospheric concentrations of greenhouse gases. That remote site is almost free of local airborne contamination. It is also representative of overall atmospheric conditions for the Northern Hemisphere. What did they find? Briefly, carbon dioxide concentrations follow annual cycles of primary production. The concentrations drop during the summer, when photosynthesis rates are high. They increase during the winter, when photosynthesis rates decline but aerobic respiration is still going on.

The alternating troughs and peaks along the graph line in Figure 20.14*a* reveal the annual lows and highs of atmospheric carbon dioxide concentrations. For the first time, scientists were able to study the integrated effects of carbon fluctuations for terrestrial and aquatic ecosystems of an entire hemisphere. Notice the overall trend. Notice especially that this line slants upward—*the concentration is steadily increasing.*

The enhanced greenhouse effect is linked to a long-term rise in the temperature near Earth's surface. We call this effect **global warming**. As an example, since direct atmospheric readings started in 1861, the lower atmosphere's temperature has risen by more than 1°F, with the biggest gains since 1946 (Figure 20.15). The trend is expected to accelerate, with temperature rising another 1°F by 2020. Polar ice is melting and glaciers are retreating. Sea level has risen an estimated twenty centimeters (eight inches) in the past century.

Greenhouse gases include carbon dioxide, methane chlorofluorocarbons (CFCs), and nitrous oxide (NO). They are among the many pollutants, substances that are new to ecosystems or are present in unnaturally

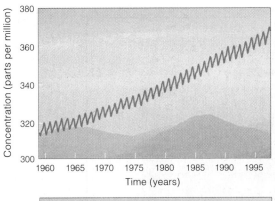

a Carbon dioxide (CO_2). Of all human activities, the burning of fossil fuels and deforestation contribute the most to increasing atmospheric levels.

c Methane (CH_4). Production and distribution of natural gas for use as fuel adds to the methane released by some bacteria that live in swamps, rice fields, landfills, and in the digestive tract of cattle and other ruminants.

b CFCs. Until restrictions were in place, CFCs were widely used in plastic foams, refrigerators, air conditioners, and industrial solvents.

d Nitrous oxide (N_2O). Denitrifying bacteria produce N_2O in metabolism. Also, fertilizers and animal wastes release enormous amounts; this is especially so for large-scale livestock feedlots.

Figure 20.14 Graphs of recent increases in four greenhouse gases.

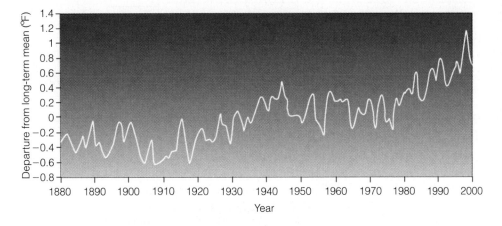

Figure 20.15 Recorded changes in global temperature between 1880 and 2000. At this writing, the hottest year on record was 2005.

Table 20.1 Per-Person CO_2 Emissions in 2002*	
Kuwait	5.97
United States	5.40
Saudi Arabia	4.77
Canada	3.87
Russian Federation	2.69
Germany	2.61
Japan	2.55
France	1.68
Mexico	1.19
China	0.60
India	0.29

* From burning fossil fuels; in metric tons.

large amounts. From a human perspective, pollutants are substances that become concentrated enough to cause adverse effects on our health or survival.

The burning of fossil fuels—including gasoline, diesel fuel, coal, and natural gas—is the main source of greenhouse gases, as well most other air pollutants. With a big population and per person CO_2 emissions exceeded only by Kuwait, the United States is the top CO_2 emitter (Table 20.1). China and India currently emit very little CO_2 on a per-person basis. Yet because of their huge populations, they are the second and fifth

greatest emitters of this gas. Both are industrializing rapidly and the resulting increase in their per person fuel use is expected to accelerate global warming.

Atmospheric concentrations of greenhouse gases trap heat and help keep Earth warm enough for life.

Fossil fuel burning and other human activities add to greenhouse gases and may lead to global warming.

Gaseous nitrogen in atmosphere

nitrogen fixation

food webs on land

fertilizers

uptake by autotrophs excretion, death, decomposition uptake by autotrophs loss by denitrification

ammonia, ammonium in soil nitrogen-rich wastes, remains in soil nitrate in soil

ammonification nitrification

loss by leaching nitrification nitrite in soil loss by leaching

Figure 20.16 *Animated!* The nitrogen cycle in an ecosystem on land. Activities of nitrogen-fixing bacteria make nitrogen available to plants. Other bacterial species cycle nitrogen to plants. They break down organic wastes to ammonium and nitrates.

LINK TO
SECTION
14.2

20.7 Nitrogen Cycle

Gaseous nitrogen makes up about 80 percent of the atmosphere. Successively smaller reservoirs are seafloor sediments, ocean water, soil, biomass on land, nitrous oxide in the atmosphere, and marine biomass.

Gaseous nitrogen (N_2) travels in an atmospheric cycle called the **nitrogen cycle**. Triple covalent bonds join two atoms ($N{\equiv}N$). Volcanic eruptions and lightning convert some gaseous nitrogen into forms that enter

food webs. Far more enters by **nitrogen fixation**. By this metabolic process, certain bacteria in the soil split all three bonds in gaseous nitrogen, and then use the nitrogen atoms to synthesize ammonia (NH_3). Later, the ammonia is converted to ammonium (NH_4^+) and nitrate (NO_3^-), two forms of nitrogen that most plants can easily take up (Figure 20.16).

Nitrogen-fixing bacteria are found in soil, in aquatic habitats, and as the photosynthetic partners of fungi in some lichens. Others are mutualists with plant roots, as when *Rhizobium* forms nodules on the roots of peas or other legumes. Collectively, nitrogen-fixing bacteria fix about 200 million metric tons of nitrogen each year.

The nitrogen incorporated into plant tissues moves through trophic levels of ecosystems and ends up in nitrogen-rich wastes and remains. By **ammonification**, bacteria and fungi break down nitrogenous materials, forming ammonium. They use some ammonium and the rest gets into the soil. Plants take it up, as do some nitrifying bacteria. In the first step of **nitrification**, the bacteria use oxygen to strip electron from ammonium and form nitrite (NO_2^-). Other nitrifying bacteria use nitrite and oxygen in reactions that form nitrate (NO_3^-).

Ecosystems lose nitrogen through **denitrification**. Denitrifying bacteria convert nitrate or nitrite to nitrous oxide (N_2O) or nitrogen gas, which can escape into the atmosphere. Most denitrifying bacteria are anaerobic; they live in waterlogged soils and aquatic sediments.

Ammonium, nitrite, and nitrate are also lost to land ecosystems as a result of leaching and runoff. Leaching, recall, is the loss of some nutrients to water draining through the soil. Leaching removes nitrogen from land ecosystems and adds it to aquatic ones.

Figure 20.17 Dead and dying trees in Great Smoky Mountains National Park. Forests are among the casualties of nitrogen oxides and other forms of air pollution.

HUMAN IMPACT ON THE NITROGEN CYCLE

Nitrogen losses through deforestation and grassland conversion for agriculture are huge. With each forest clearing and harvest, nitrogen stored in plant biomass is removed. Soil erosion and leaching take away more.

Many farmers counter nitrogen losses by rotating crops, as by alternating wheat with a crop that has nitrogen-fixing bacterial symbionts. These symbionts add some nitrogen to the soil. In developed countries, farmers also use nitrogen-rich fertilizers. They even select new strains of crop plants that have a greater metabolic capacity to take up fertilizers from soil. By such practices, crop yields per hectare have doubled and quadrupled over the past forty years.

However, heavy fertilizer applications alter the soil water in ways that can be harmful to plants. The surface charges of soil particles attract ions in soils. By a pH-dependent process called **ion exchange**, ions dissociate from soil particles, and others replace them. A change in the ion composition of soil water disrupts the number and the kinds of ions that are in an exchangeable form.

Calcium and magnesium are the most important exchangeable ions. When nitrogen-rich fertilizers are utilized, ammonium also assumes importance. Heavy applications of fertilizers make soil water more acidic. Hydrogen ions from the fertilizer fill up binding sites on soil particles, displacing magnesium and calcium ions, which leave the soil in runoff.

Humans introduce excess nitrogen into ecosystems in other ways. Nitrogen-rich sewage flows into rivers, lakes, and estuaries. Fossil fuel burned in power plants and vehicles releases nitrogen oxides. Winds carry the pollutants to forests at high elevations and northern latitudes, both of which have nitrogen-poor soils.

In forests that are adapted to nitrogen-poor soils, additional nitrogen makes trees grow faster and later into the fall. The late growth is vulnerable to damage from frost. Also, nitrogen-stimulated growth outstrips the availability of phosphorus and other nutrients, so trees suffer the effects of nutrient deficiencies. Leaves yellow and fall off, photosynthesis declines, and overall growth suffers. The weakened trees are more likely to die or be damaged by diseases, drought, insects, and other stresses (Figure 20.17).

In addition, the nitrogen in acid rain has the same effect on forest soils that fertilizers have on croplands. It increases the loss of potassium, calcium, and other ions that are essential plant nutrients. Thus, nitrogen pollution of forest soils can cause a growth spurt that is followed by a decline in growth as trees become starved for other nutrients.

The nitrogen cycle is an atmospheric cycle. Nitrogen fixation by bacteria converts nitrogen gas to forms plants can use. Other bacteria make the nitrogen in wastes and remains available to plants.

Ecosystems lose nitrogen naturally by the activities of denitrifying bacteria, which convert nitrite or nitrate to forms that escape into the atmosphere.

Ecosystems often lose or gain too much nitrogen through human activities, in ways that compromise plant growth.

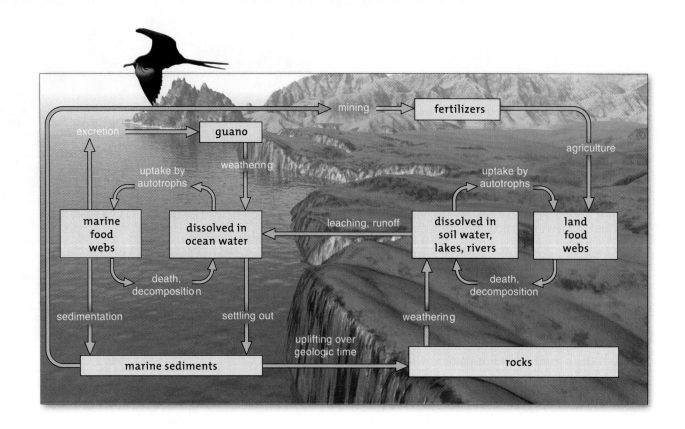

Figure 20.18
Animated!
Phosphorus cycle. In this sedimentary cycle, most of the phosphorus moves in the form of phosphate ions (PO_4^{3-}) and reaches the oceans. Phytoplankton of marine food webs take up some of it, then fishes eat the plankton. Seabirds eat the fishes, and their droppings (guano) accumulate on islands. Guano is collected and used as a phosphate-rich fertilizer.

LINK TO
SECTION
14.3

20.8 Phosphorus Cycle

Sedimentary cycles, again, move solid forms of nutrients into and out of all ecosystems. The phosphorus cycle is one of the most influential.

In the **phosphorus cycle**, phosphorus passes quickly through food webs as it moves from land to ocean sediments, then slowly back to land. Earth's crust is the largest phosphorus reservoir.

In rock formations, most phosphorus is in the form of phosphate ions (PO_4^{3-}). Weathering and erosion put phosphates into streams and rivers, then into ocean sediments (Figure 20.18). The ions combine with other minerals and accumulate on the submerged shelves of continents as insoluble deposits. Millions of years go by. Where movements of crustal plates uplift part of the seafloor, phosphates become exposed on drained land surfaces. In time, weathering and erosion release phosphates from exposed rocks once again.

All organisms use the ions to make phospholipids, NADPH, ATP, nucleic acids, and other compounds. Plants take up dissolved phosphates from soil water. Herbivores get needed phosphates by eating plants; carnivores get them by eating herbivores. All animals excrete phosphates as a waste in their urine and feces.

Bacterial and fungal decomposers release phosphates, then plants take them up again in this rapid cycle.

The hydrologic cycle helps move minerals through ecosystems. Water evaporates from the ocean and falls on land. On its way back to the sea, it transports silt and dissolved minerals, including phosphorus.

Of all mineral nutrients, phosphorus is most often the limiting factor in natural ecosystems. Only newly weathered, young soils are high in phosphorus. Older soils have lost much of this mineral to erosion and runoff. In aquatic habitats most phosphorus is tied up in the sediments.

Some human activities are lowering the levels of phosphorus in natural ecosystems. This is especially so for tropical and subtropical regions of developing countries, which often have weathered, phosphorus-poor soils. Phosphorus stored in biomass and slowly released from decomposing organic matter sustains undisturbed forests or grasslands. But people harvest trees and clear grassland for farms, thus depleting the soil phosphorus. Crop yields start out low and become nonexistent. Fields are soon abandoned, and regrowth of natural vegetation is sparse. About 1 to 2 billion hectares are already depleted of phosphorus, mostly where impoverished farmers are not engaging in soil management practices and cannot afford fertilizers.

In developed countries, fertilizer use has left many soils with phosphorus overloads. Phosphorus becomes

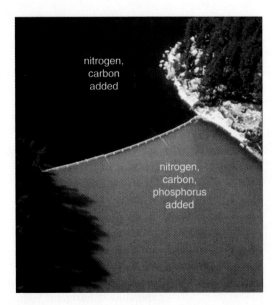

Figure 20.19 One of the eutrophication experiments. Researchers put a plastic curtain across a channel between two basins of a natural lake. They added nitrogen, carbon, and phosphorus on one side of the curtain (here, the *lower* part of the lake) and added nitrogen and carbon on the other side. Within months, the phosphorus-rich basin was eutrophic, with a dense algal bloom covering its surface.

concentrated in eroded sediments and runoff from agricultural fields. Phosphorus also is present in the outflows from sewage treatment plants and industries, in the runoff from cleared land, and even from lawns.

Dissolved phosphorus that enters streams, rivers, lakes, and estuaries can promote dense algal blooms. Like plants, photosynthetic algae cannot grow without phosphorus, nitrogen, and other mineral ions. In many freshwater ecosystems, nitrogen is not a limiting factor on algal growth because of the abundance of nitrogen-fixing bacteria. Phosphorus is. When decomposers act on the remains from algal blooms, they deplete water of oxygen, which kills fishes and other organisms.

Eutrophication refers to the nutrient enrichment of any ecosystem that is naturally low in nutrients. This is a natural process, but phosphorus inputs accelerate it, as field experiments demonstrate (Figure 20.19). The added phosphorus encourages algal growth.

Sedimentary cycles, in combination with the hydrologic cycle, move most mineral elements, such as phosphorus, through terrestrial and aquatic ecosystems.

Agriculture, deforestation, and other human activities upset the nutrient balance of an ecosystem because they accelerate the amount of minerals entering and leaving it.

Summary

Section 20.1 An ecosystem consists of an array of organisms and their environment. There is a one-way flow of energy through an ecosystem, and a cycling of materials among the organisms. All ecosystems have inputs and outputs of energy and, to some extent, nutrients.

Sunlight is the initial energy source for almost all ecosystems. Photoautotrophs (the primary producers) convert sunlight energy to ATP and assimilate nutrients that an ecosystem's consumers require.

All organisms in an ecosystem that are at the same trophic level are the same number of steps away from the energy source for that ecosystem. Energy transfers are inefficient and so most ecosystems support no more than four or five trophic levels.

Biology Now
Learn about trophic levels with the animation on BiologyNow.

Section 20.2 By biological magnification, a chemical substance is passed from organisms at one trophic level to those above, so that the substance becomes increasingly concentrated in body tissues.

Section 20.3 Primary productivity is the rate at which producers in an ecosystem capture and store energy. It varies with climate, season, and other factors. Energy pyramids show how usable energy lessens as it flows through an ecosystem. In contrast, a biomass pyramid is sometimes larger at the top than at the base.

Section 20.4 In a biogeochemical cycle, water or a nutrient moves through the environment, then through organisms, then back to an environmental reservoir. In the hydrologic cycle, waters move from the ocean into the atmosphere, to land, and back to the ocean— the main reservoir. Human actions are disrupting the cycle in ways that may lead to water shortages.

Biology Now
Learn about the hydrologic cycle with the interaction on BiologyNow.

Section 20.5 The global carbon cycle moves carbon from its main reservoirs in rocks and seawater, through its gaseous form (CO_2), and through living organisms. Deforestation and burning of fossil fuels are adding more carbon to the air than the oceans can absorb.

Biology Now
Learn about the carbon cycle with the animation and interaction on BiologyNow.

Section 20.6 The greenhouse gases trap heat in Earth's lower atmosphere and make life possible. But human activities are raising greenhouse gas levels and causing an increase in global temperatures.

Biology Now
Learn about the greenhouse effect and greenhouse gases with the interactions on BiologyNow.

Figure 20.20 Antarctica's Larson B ice shelf in (**a**) January and (**b**) March 2002. About 720 billion tons of ice broke from the shelf, forming thousands of icebergs. (**c**) These icebergs project about twenty-five meters above the sea surface.

Section 20.7 The atmosphere is the main reservoir for nitrogen, but plants cannot use gaseous forms. Soil bacteria and the fertilizer industry fix nitrogen (convert atmospheric nitrogen to forms plants take up). Human activities are disrupting the nitrogen cycle.

Biology ⊘ Now™
Learn about the nitrogen cycle with the animation on BiologyNow.

Section 20.8 Earth's crust is the largest reservoir of phosphorus. A sedimentary cycle moves the mineral from rocks and seawater into living organisms and back. Phosphorus is often a limiting factor for primary producers. Addition of phosphorus to aquatic systems contributes to eutrophication, the overloading of a system by excess nutrients. This can lead to harmful algal blooms.

Biology ⊘ Now™
Learn about the phosphorus cycle with the animation and interaction on BiologyNow.

Self-Quiz
Answers in Appendix I

1. Organisms at the lowest trophic level in a tallgrass prairie are all _____ .
   a. at the first step away from the original energy input
   b. autotrophs   d. both a and b
   c. heterotrophs   e. both a and c

2. Which of the following is a consumer?
   a. a carnivore   c. a detritivore
   b. an omnivore   d. all of the above

3. As a result of biological magnification, a substance will accumulate to its highest levels in the tissues of _____ .
   a. primary producers   c. decomposers
   b. herbivores   d. top carnivores

4. The _____ cycle is a sedimentary cycle.
   a. hydrologic   c. nitrogen
   b. carbon   d. phosphorus

5. Disruption of the _____ cycle is depleting aquifers.
   a. hydrologic   c. nitrogen
   b. carbon   d. phosphorus

6. Earth's largest reservoir for water is _____ .
   a. lakes and streams   c. atmospheric vapor
   b. oceans   d. polar ice and glaciers

7. Earth's largest carbon reservoir is _____ .
   a. the atmosphere   c. seawater
   b. sediments and rocks d. living organisms

8. What form(s) of nitrogen do plants easily take up?
   a. ammonium   d. nitrous oxide
   b. nitrate   e. both a and b
   c. nitrogen gas   f. both c and d

9. What process converts nitrogen gas to ammonia?
   a. ammonification   c. nitrification
   b. denitrification   d. nitrogen fixation

10. Match the terms with suitable descriptions.
   ____ producers   a. feed on plants
   ____ herbivores   b. feed on small bits of
   ____ decomposers   organic matter
   ____ detritivores   c. degrade organic wastes and
                          remains to inorganic forms
                       d. capture sunlight energy

Additional questions are available on **Biology ⊘ Now™**

Critical Thinking

1. Polar ice shelves are vast, thickened sheets of ice that float on seawater. In March 2002, 3,200 square kilometers (1,300 square miles) of Antarctica's largest ice shelf broke free from the continent and shattered into thousands of icebergs (Figure 20.20). Scientists knew the ice shelf was shrinking and breaking up, but this was the single largest loss ever observed at one time. Why should this concern those of us who live in more temperate climates?

2. Tissues of predatory marine fishes, such as swordfishes, tunas, marlins, and sharks, contain high levels of mercury. Many state governments have issued health advisories to pregnant women, suggesting they limit consumption of tainted species. Other fish from the same environments have lower mercury levels and are not on the warning list. Explain how species living in the same habitat can have very different mercury levels in their tissues.

3. Marguerite is growing vegetables in her garden in Maine. What are some variables that will affect the net primary production in her garden? Which of these can she alter and which are beyond her control?

4. List some of the manufactured goods and agricultural products you purchase that contribute to amplification of the greenhouse effect and to global warming. Which might you replace with products that are environmentally friendly? Which ones might you do without entirely?

5. Lawn fertilizers are rich in nitrogen and phosphorus. Why does adding these chemicals to the soil enhance plant growth? What effect will they have if they get into nearby streams and lakes?

6. Most of the body of any living organism is made up of water. Recall that carbon, nitrogen, and phosphorus make up large proportions of the remaining body weight. The importance of these elements in ecosystems stems from their roles in building the bodies of organisms. For each of these elements, list some biological molecules that include it.

Surfers, Seals, and the Sea

The stormy winter of 1997–1998 was a very good time for surfers on the lookout for the biggest waves. But it was a very bad time for seals and sea lions. While Ken Bradshaw was riding the monster wave shown below, sea lions on the Galápagos Islands were dying. Half of this population disappeared, including nine out of every ten pups. That same winter in California, the number of Northern fur seals plummeted and 75 percent of the newborns survived less than a few months. The animals were among the casualties of El Niño, an event that ushers in a spectacular seesaw in the world climate.

That winter, a massive volume of warm water from the southwestern Pacific moved east. It piled into coasts from California down through Peru, displacing currents that otherwise would have churned up nutrients from the deep. Without these nutrients, primary producers for marine food webs dwindled. Warm water and scarce producers drove out the migratory big fishes. Fishes and squids that could not migrate starved to death. So did many of the seals and sea lions—which stay alive by eating squids and fishes.

Marine mammals did not dominate the headlines, because the 1997–1998 El Niño gave humans plenty of other things to worry about. It battered Pacific coasts with fierce winds and torrential rains, massive flooding, and landslides. As another rippling effect, an ice storm in New York, New England, and central and eastern Canada crippled regional electrical grids. At the same time, the global seesaw caused drought-driven crop failures and raging wildfires in Australia and Indonesia.

The **biosphere**, again, is the sum of all places where we find life on Earth. With this chapter, we turn to the physical processes that help shape the distribution of organisms through the biosphere. With few exceptions, energy from the sun fuels the processes that underlie this distribution. As you will see, it drives the circulation of air and water in ways that start weather patterns. Some are highly predictable. Others, such as El Niño, are not yet fully understood.

 How Would You Vote? We cannot prevent El Niño events, but if we knew how to predict them, we might be able to minimize damage. Should the government fund more research that might help us predict El Niño? See BiologyNow for details, then vote online.

 ## Key Concepts

BIOGEOGRAPHIC PATTERNS
Sunlight energy warms Earth's atmosphere, land, and seas. This creates patterns of air and water circulation that shape climate and so determine the distribution of biodiversity.

MAJOR BIOMES
Climate varies with latitude and elevation, and different types of vegetation are adapted to different climates. As a result, we find similar types of ecosystems across broad bands of latitude and at similar elevations.

THE WATER PROVINCES
Gradients of brightness, nutrient availability, and temperature help shape freshwater and saltwater communities. Seasonal changes in these factors cause corresponding changes in primary productivity.

Links to Earlier Concepts

Section 2.4 introduced the concept of pH and 20.7 touched on acid rain. We revisit both those topics here, as well as the origin of life (14.1) and laws governing transfers of energy (4.1).

We also reconsider biodiversity hotspots (19.9), forests and deforestation (15.7), primary productivity (5.4, 20.3), the carbon cycle (20.5), eutrophication (20.8), and the ecology of protists (14.3) and corals (16.2).

21.1 Air Circulation and Climates

Within the biosphere are ecosystems ranging from continent-straddling forests to rainwater pools in cup-shaped clusters of leaves. Except for a few ecosystems at hydrothermal vents, climate influences all of them.

Light rays from the sun strike Earth's surface most directly at the equator and in the tropics.

To reach higher latitudes, light rays pass through more atmosphere, which scatters and reflects them. Rays that do reach the surface strike at an angle, so there is less incoming energy per unit area.

a

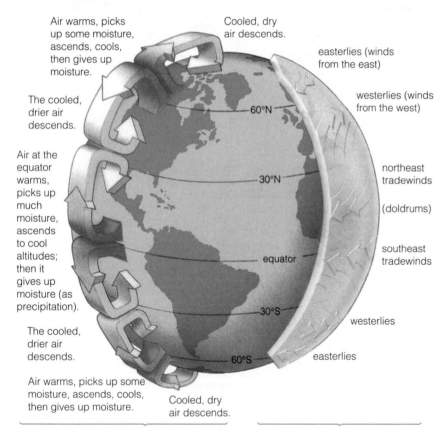

Air warms, picks up some moisture, ascends, cools, then gives up moisture.

Cooled, dry air descends.

The cooled, drier air descends.

Air at the equator warms, picks up much moisture, ascends to cool altitudes; then it gives up moisture (as precipitation).

The cooled, drier air descends.

Air warms, picks up some moisture, ascends, cools, then gives up moisture.

Cooled, dry air descends.

easterlies (winds from the east)

westerlies (winds from the west)

60°N

30°N

northeast tradewinds

(doldrums)

equator

southeast tradewinds

30°S

westerlies

60°S

easterlies

b Initial pattern of air circulation as air masses warm and rise, then cool and fall.

c Deflections in the initial pattern near Earth's surface.

Figure 21.1 *Animated!* Global air circulation patterns. Latitudinal differences in the amount of solar radiation reaching Earth (**a**) produce an initial north-south air circulation pattern (**b**). Earth's rotation and its curvature deflect the pattern in ways that produce prevailing east and west winds (**c**).

GLOBAL AIR CIRCULATION PATTERNS

Climate refers to average weather conditions, such as temperature, humidity, wind speed, cloud cover, and rainfall, over time. Many factors influence a region's climate. How much solar radiation a region gets varies with latitude. The global distribution of land masses and oceans affects climate. So does the topography of a region—whether it is a plain or mountainous. Complex interactions among these factors determine prevailing winds and the direction of ocean currents.

Warming effects of sunlight vary with latitude for two reasons. First, gases, bits of dust, and water vapor in the atmosphere absorb some wavelengths or reflect them back into space. Light rays that travel from the sun to the poles pass through more atmosphere than rays that travel to the equator, so fewer rays make it to Earth's polar surfaces. Second, light rays that do reach the surface at high latitudes are more spread out than rays that reach the equator (Figure 21.1a). As a result of these factors, Earth's surface is warmed the most at the equator and stays coolest at the poles.

Differential heating of Earth's surface sets up global air circulation patterns. Air at the equator heats up, expands, and rises. Convection causes this warm air to move north and south away from the equator, toward the cooler polar regions (Figure 21.1b).

Major winds do not blow directly north and south because Earth's rotation alters the initial movement. In the Northern Hemisphere, air masses become deflected rightward as Earth turns under them. In the Southern Hemisphere, this rotation deflects winds to the left of their initial course (Figure 21.1c).

Also, the distribution of land and water gives rise to pressure differences that produce regional winds. Land absorbs and gives up heat faster than water does, so air rises and falls more quickly over a land mass than over an ocean. Air pressure is lowest where air rises and is greatest where air sinks.

Air circulation affects rainfall patterns. Warm air can hold more moisture than cooler air. As air warms near the equator, it picks up moisture from the seas. It rises to cooler altitudes and gives up moisture as rain, which supports lush equatorial forests. Now drier, the air moves away from the equator. It descends at the latitudes of about 30°, where deserts often form. Still farther north and south, air again picks up moisture and ascends. This produces another rainy region at latitudes of about 60°. The air descends once again in polar regions. Here, frigid temperatures and minimal precipitation result in polar deserts.

Latitudinal variations in solar heating, combined with modified patterns of air circulation, help define the world's **temperature zones**, shown in Figure 21.2a.

The tilt of Earth's axis leads to seasonal variations in daylength and in the intensity of incoming sunlight, most notably at the poles and in temperate regions (Figure 21.2b). When it is summer in the Northern Hemisphere, this region is tilted *toward* the sun and sunlight strikes it most directly. In winter, this region is tilted *away* from the sun, so light rays come in at more of an angle. As described earlier, the greater the angle of incoming sunlight, the fewer light rays make it through the atmosphere to the surface, and the more widely those that do are spread out.

A FENCE OF WIND AND OZONE THINNING

Ozone (O_3) and oxygen (O_2) molecules in the upper atmosphere absorb most ultraviolet (UV) radiation in incoming sunlight. The concentration of ozone is the highest between 17 and 27 kilometers above sea level. This region is the "ozone layer" (Figure 21.3a).

In the mid-1970s, scientists began to notice that the ozone layer was getting thinner. The layer's thickness had always varied a bit with the season, but this was a steady decline. By the mid-1980s, springtime ozone thinning over Antarctica had become so pronounced that some scientists described it as an "ozone hole" over the South Pole (Figure 21.3b).

This quickly became a matter of great concern. With less ozone in the stratosphere, people would get more UV exposure. This could increase cases of skin cancers, cataracts, and weakened immunity. High UV levels also harm plants and wildlife, which do not have the option of staying inside or slathering on a bit more sunscreen. Higher UV levels might even affect the atmosphere's composition, by slowing photosynthetic oxygen production by plants and phytoplankton.

Chlorofluorocarbons, or CFCs, are the main ozone destroyers. These odorless gases were widely used as propellants in aerosol cans, as coolants in refrigerators and air conditioners, and in solvents and plastic foam. They slowly seep into the air, and resist breakdown. In the stratosphere, CFCs release free chlorine radicals that yank oxygen away from ozone. A single chlorine radical can break apart thousands of ozone molecules.

Ozone thins the most over the poles because during the dark polar winters, swirling winds fence in CFCs. Also within this dynamic enclosure are clouds of ice crystals, upon which CFCs are converted to different molecules. In the spring, as days get longer, UV light frees chlorine radicals from these molecules.

In response to the threat posed by ozone thinning, developed countries agreed in 1992 to phase out the production of CFCs and other compounds known to destroy ozone. But climate change, discussed later in the chapter, may also be contributing to ozone loss.

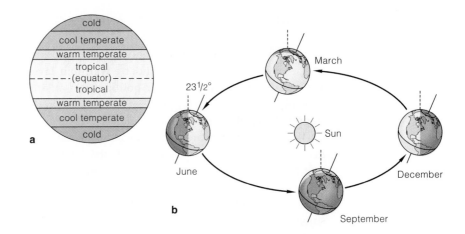

Figure 21.2 *Animated!* (**a**) World temperature zones. (**b**) Annual variation in incoming solar radiation. The northern end of Earth's fixed axis tilts toward the sun in June and away from it in December. This causes changes in the amount of light that reaches the surface and in daylength. Variations in sunlight intensity and daylength cause seasonal temperature variations.

Figure 21.3 (**a**) Earth's atmosphere. Most of the air circulation occurs in the lower atmosphere (troposphere), where temperature falls rapidly with altitude. The ozone layer in the stratosphere absorbs most of the ultraviolet wavelengths in the sun's rays. (**b**) Seasonal ozone thinning above Antarctica in 2001. *Darkest blue* indicates the area with the lowest ozone level, at that time the largest recorded.

> Heat energy from the sun warms the atmosphere and ultimately drives Earth's great weather systems.
>
> Ozone in the upper atmosphere is declining, exposing Earth's surface to higher levels of dangerous UV radiation.

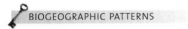 BIOGEOGRAPHIC PATTERNS

LINKS TO
SECTIONS
2.4, 4.1, 20.7

NO WIND, LOTS OF POLLUTANTS, AND SMOG

A **thermal inversion** occurs when weather conditions trap a layer of cool, dense air under a warm air layer. Thermal inversions can keep air pollutants in an area and contribute to smog (Figure 21.4). Pollutants in the trapped layer of air accumulate to dangerous levels.

Where the winters are cold and wet, *industrial* smog may develop as a gray haze over industrialized cities that burn coal and other fossil fuels. Burning releases dust, smoke, soot, ashes, asbestos, oil, bits of lead and other heavy metals, and sulfur oxides. Most industrial smog now forms in China, India, and eastern Europe.

In large cities in warm climates, *photochemical* smog forms as a brown haze, especially where land forms a natural basin. Los Angeles and Mexico City are classic examples. Nitric oxide, the main pollutant, escapes in vehicle exhaust. It reacts with atmospheric oxygen and other pollutants to yield photochemical oxidants, such as peroxyacyl nitrates (PANs) that irritate animal eyes and respiratory systems. PANs can also damage leaf surfaces and stunt plant growth.

WINDS AND ACID RAIN

Power plants, smelters, and factories that burn coal emit most sulfur dioxides. Vehicles, power plants that burn gas and oil, and nitrogen-rich fertilizers produce nitrogen oxides. In dry weather, these oxides can stay airborne for a bit, then fall as dry acid deposition. If air is moist, they form nitric acid vapor, sulfuric acid droplets, and sulfate and nitrate salts. Winds typically disperse them far from their source. They fall to Earth in rain and snow as wet acid deposition, or **acid rain**.

Figure 21.4 (**a**) Normal pattern of air circulation in smog-forming regions. (**b**) Air pollutants have become trapped under a thermal inversion layer.

The pH of normal rainwater is 5 or so. Acid rain can be 0 to 100 times more acidic, as potent as lemon juice! Acid rain corrodes metals, marble on buildings, rubber, plastics, nylon, and other manmade materials. It disrupts the physiology of organisms, the chemistry of ecosystems, and so endangers biodiversity.

Depending on soil type and vegetation cover, some regions are more sensitive to acid rain (Figure 21.5). Highly alkaline soil can neutralize acids before they enter the streams and lakes of watersheds. Water that is highly alkaline also can neutralize acids. But many watersheds of northern Europe, southeastern Canada, and regions throughout the United States have only a thin soil layer covering granite bedrock. These soils have little buffering capacity and are acidified quickly.

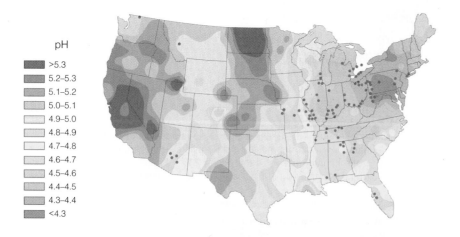

pH

■	>5.3
■	5.2–5.3
■	5.1–5.2
■	5.0–5.1
■	4.9–5.0
■	4.8–4.9
■	4.7–4.8
■	4.6–4.7
■	4.5–4.6
■	4.4–4.5
■	4.3–4.4
■	<4.3

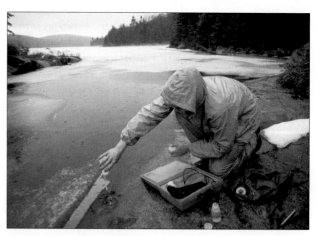

Figure 21.5 Average acidities of precipitation in the United States in 1998. *Red* dots pinpoint coal-burning power and industrial plants. *Right*, a biologist measuring the pH value of Woods Lake in New York's Adirondack Park. With a pH below 5, this lake is too acidic to support fish.

Figure 21.6 Harnessing solar energy. (**a**) A large array of electricity-producing photovoltaic cells in panels that collect sunlight energy. (**b**) A field of turbines harvesting wind energy.

Rain in much of eastern North America is thirty to forty times more acidic than it was several decades ago. Crop yields are declining. Fish populations have already disappeared from more than 200 lakes in New York's Adirondack Mountains. The air pollution from power plants and industry is increasing the acidity of rainfall. The change is contributing to the decline in forest trees and mycorrhizae that support new growth.

Coal-burning power plants are the single largest cause of acid rain. The United States produces about half of its electricity with such plants. Emissions from coal-burning plants in the midwest contribute to acid rain in Canada. Most pollutants in Scandinavia and the Netherlands, Austria, and Switzerland formed in the industrialized regions of western and eastern Europe. *Prevailing winds—and air pollutants they carry—do not stop at national boundaries.*

HARNESSING SOLAR AND WIND ENERGY

Paralleling the J-shaped curve of human population growth is a dramatic rise in total and per capita energy consumption. Some energy sources, most notably the fossil fuels, are not renewable. Others, such as direct solar energy and wind energy, are. The annual amount of incoming solar energy surpasses, by about 10 times, the energy in all the known fossil fuel reserves. That is 15,000 times more energy than human populations use on an annual basis. Maybe we should start to collect it in earnest, in something besides crops.

For example, when exposed to sunlight, electrodes in "photovoltaic cells" produce an electric current that splits water molecules into oxygen and hydrogen gas (H_2). This can be used directly as fuel to run cars, heat and cool buildings, and generate electricity. We have

the technology to obtain *solar–hydrogen energy* (Figure 21.6*a*). Such energy can be stored efficiently. It costs less to distribute H_2 than electricity, and water is the only by-product. Space satellites run on this energy.

Proponents of solar–hydrogen energy say that it could end smog, oil spills, and acid rain, without the risks associated with nuclear power. However, others see potential problems. As an example, hydrogen is a tiny molecule and highly corrosive; it is likely to leak from any pipeline or container. Leakage of hydrogen gas into the air might accelerate global warming.

We already are using winds to convert solar energy to mechanical energy. Where prevailing winds travel faster than 7.5 meters per second, we find *wind farms* (Figure 21.6*b*). Wind farms now generate 1 percent of California's electricity. The winds of North and South Dakota alone might meet all but 20 percent of the energy needs of the continental United States. Winds do not blow constantly, but when they do, energy can be stored in batteries or fed into utility grids to supply local and more distant power needs.

On the downside, the rotating blades of turbines at wind farms kill birds and bats. Also wind farms have been called a form of "visual pollution" that destroys scenic views and thus lowers property values.

Accumulation of CFCs in the upper atmosphere has led to thinning of the ozone layer. A ban on these chemicals is in place and is helping to reverse this effect.

Smog formation and acid deposition are two forms of air pollution in specific regions, but prevailing winds often distribute pollutants across regional and political boundaries.

Solar and wind energy are alternatives to fossil fuels. They could help meet rising energy needs.

The Ocean, Landforms, and Climates

The ocean, a continuous body of water, covers more than 71 percent of Earth. Driven by solar heat and wind friction, its upper 10 percent moves in currents. Those currents also influence regional climates.

OCEAN CURRENTS AND THEIR EFFECTS

Latitudinal and seasonal variations in sunlight warm and cool the ocean. When heated, that vast volume of water expands; where cooled, it shrinks. So sea level is about eight centimeters, or three inches, higher at the equator than it is at either pole. The volume of water in that "slope" is enough to get surface waters moving as a response to gravity, most often toward the poles. Along the way, water warms the portion of air just above it. At midlatitudes, it transfers *10 million billion calories* of heat energy per second to the air!

Rapid bulk flows of water, or *currents*, arise from the tug of trade winds and westerlies. Their direction and properties are influenced by Earth's rotation, the distribution of land masses, and the shapes of oceanic basins. The water circulates clockwise in the Northern Hemisphere, and it circulates counterclockwise in the Southern Hemisphere (Figure 21.7).

Swift, deep, and narrow currents of nutrient-poor waters move parallel with the east coast of continents. For instance, along North America's east coast, warm water moves north as the Gulf Stream. Slower, shallow, broad currents paralleling the west coast of continents move cold water toward the equator.

Why are Pacific Northwest coasts cool, foggy, and mild in summer? Winds nearing the coast give up heat to cold coastal water, which the California Current is moving down toward the equator. Why are Baltimore and Boston so muggy in summer? Air above the Gulf Stream gains heat and moisture, which southerly and easterly winds deliver to those cities. Why are winters milder in London and Edinburgh than in Ontario and central Canada, which are all at the same latitude? The North Atlantic Current picks up warm water from the Gulf Stream. As it flows past northwestern Europe, it gives up heat energy to the prevailing winds.

Figure 21.7 *Animated!* Major climate zones correlated with surface currents and drifts of the world ocean. Warm surface currents move from the equator toward the poles. Water temperatures differ with latitude and depth, which contributes to differences in air temperature and rainfall patterns. Which way a current flows depends on prevailing winds, Earth's rotation, the shapes of ocean basins, and the presence of land masses.

BIOGEOGRAPHIC PATTERNS

a Prevailing winds move moisture inland from the Pacific Ocean

b Clouds pile up and rain forms on side of mountain range facing prevailing winds

c Rain shadow on side facing away from the prevailing winds makes arid conditions

4,000/75
3,000/85
2,000/25
1,800/125
1,000/25
1,000/85
moist habitats
15/25

Figure 21.8 *Animated!* Rain shadow effect of the Sierra Nevada. On the side of mountains facing away from prevailing winds, rainfall is reduced. *Blue* numbers show the average yearly precipitation, in centimeters. The *white* numbers signify elevations, in meters.

RAIN SHADOWS AND MONSOONS

Mountains, valleys, and other topographical features affect regional climates. *Topography* refers to a region's variations in elevation. Consider an air mass that picks up moisture off California's coast. It moves inland to the Sierra Nevada, a mountain range paralleling the coast. The air cools as it rises to higher altitudes, then loses moisture as rain (Figure 21.8).

Belts of vegetation at different elevations reflect the temperature and moisture differences. Arid grasslands form at the western base of the range. Higher up, you see deciduous and evergreen species adapted to more moisture and cooler air. Higher still, a subalpine belt supports just a few evergreen species that withstand a rigorously cold habitat. Above the subalpine belt, only low, nonwoody plants and dwarfed shrubs can grow.

Air warms and holds more water after it flows over mountain tops. As the air descends, it retains moisture and pulls more water out of plants and soil. The result is a **rain shadow**, a semiarid or arid region of sparse rainfall on the side of mountains that faces away from the wind. The Himalayas, Andes, Rockies, and other great mountain ranges cause rain shadows. Tropical forests flourish on one side of Hawaii's high volcanic peaks, and arid conditions prevail on the other.

Monsoons are air circulation patterns with effects on continents north or south of warm oceans. The land heats intensely, so low-pressure air parcels form above it. The low pressure draws in moisture-laden air from the ocean, as from the Bay of Bengal into Bangladesh. Converging trade winds and the equatorial sun cause intense heating and heavy rain. Air moving north and south gives rise to alternating wet and dry seasons.

Recurring coastal breezes are like mini-monsoons. Water and coastal land have different heat capacities. In the morning, water does not warm as fast as the land. When warm air above land rises, cooler marine air moves in. After sunset, land loses heat faster than water. Then, land breezes flow in the reverse direction.

Surface ocean currents also influence regional climates and help distribute nutrients in marine ecosystems.

Air circulation patterns, ocean currents, and landforms interact in ways that influence regional temperatures and moisture levels. Thus they also influence the distribution and dominant features of ecosystems.

LINKS TO
SECTIONS
13.3, 19.9

21.3 Realms of Biodiversity

Differences in physical and chemical properties, and in evolutionary history, help explain why grasslands, forests, deserts, and tundra form where they do, and why some water provinces are richer in species than others.

Suppose you live in the coastal hills of California and decide to tour the Mediterranean coast, the southern horn of Africa, or central Chile. There you find many-branched, tough-leafed woody plants that resemble the shrubby chaparral plants you are familiar with at home. Great geographic and evolutionary distances separate the plants. Why are they alike?

You decide to compare their locations on a global map. You discover that American and African desert plants live about the same distance from the equator. Chaparral plants and their distant look-alikes all grow along the western and southern coasts of continents positioned between the latitudes of 30° and 40°. As Alfred Wallace, Charles Darwin, and other naturalists did so long ago, you have just stumbled onto one of many biogeographic patterns.

Those naturalists divided Earth's land masses into six **biogeographic realms**—vast expanses where they could expect to find communities of specific types of plants and animals, such as palm trees and camels in the Ethiopian realm (Figure 21.9).

Biomes are finer subdivisions of the great realms, but they are still identifiable on a global scale. Their distribution has been shaped partly by evolutionary processes and events, as when evolving species were slowly rafted about on different land masses after the splitting of Pangea (Section 13.3). They also owe their distribution to topography, climate, and interactions among species. Some biomes extend across more than one continent, and their communities, although similar in many ways, each include unique species. Notice, in

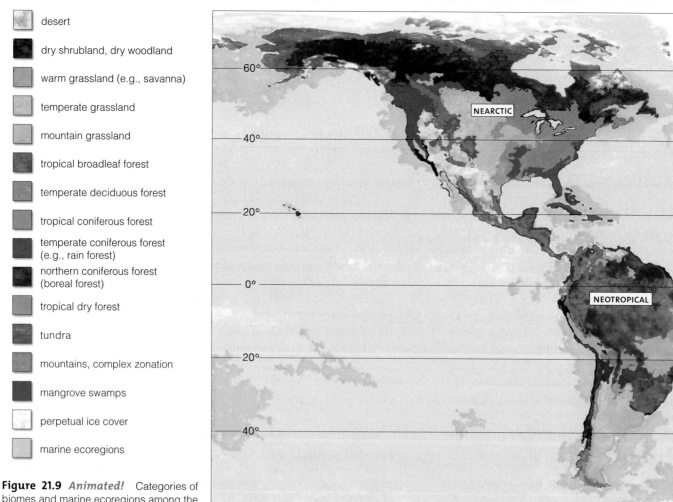

- desert
- dry shrubland, dry woodland
- warm grassland (e.g., savanna)
- temperate grassland
- mountain grassland
- tropical broadleaf forest
- temperate deciduous forest
- tropical coniferous forest
- temperate coniferous forest (e.g., rain forest)
- northern coniferous forest (boreal forest)
- tropical dry forest
- tundra
- mountains, complex zonation
- mangrove swamps
- perpetual ice cover
- marine ecoregions

Figure 21.9 *Animated!* Categories of biomes and marine ecoregions among the six biogeographic realms.

MAJOR BIOMES

Figure 21.9, the distribution of dry shrublands and woodlands (coded dark brown). Chaparral is part of that biome. Similarly, the temperate grassland biome occurs in North American prairie, South American pampa, southern African veld, and Eurasian steppe. In each case, the climax community is dominated by grasses that can tolerate strong winds and drought.

In general, communities that characterize a biome are arrays of species adapted to specific temperatures, patterns of rainfall, soil type, and to interacting with one another. Adaptations may include form, function, behavior, and life-history patterns.

Evolutionary events and environmental conditions also influence the distribution of marine ecosystems. Figure 21.9 shows the main marine ecoregions.

Conservationists are working to locate, inventory, and protect **hot spots**, those portions of biomes and ecoregions that are the richest in biodiversity and the most vulnerable to species losses. Figure 21.10 gives one example, Costa Rica's Monteverde cloud forest.

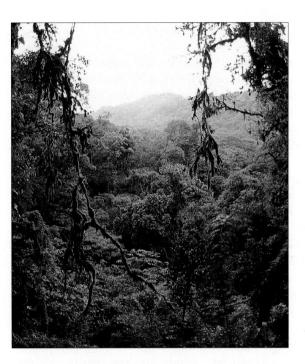

Figure 21.10 A Costa Rican hotspot. In the early 1970s, George Powell, a graduate student from the University of California at Davis, was studying this high elevation forest. When people began to move into and clear the forest, Powell got the idea to buy some land and set up a sanctuary. Powell's vision soon inspired individuals and conservation groups around the world to help protect this biological gem. Monteverde Cloud Forest Preserve now protects 5,000 hectares and is managed by a nonprofit organization. The reserve's species include 420 orchids, more than 100 species of mammals, 400 species of birds, and 120 species of amphibians and reptiles. It is one of the few habitats left for the jaguar, ocelot, and puma.

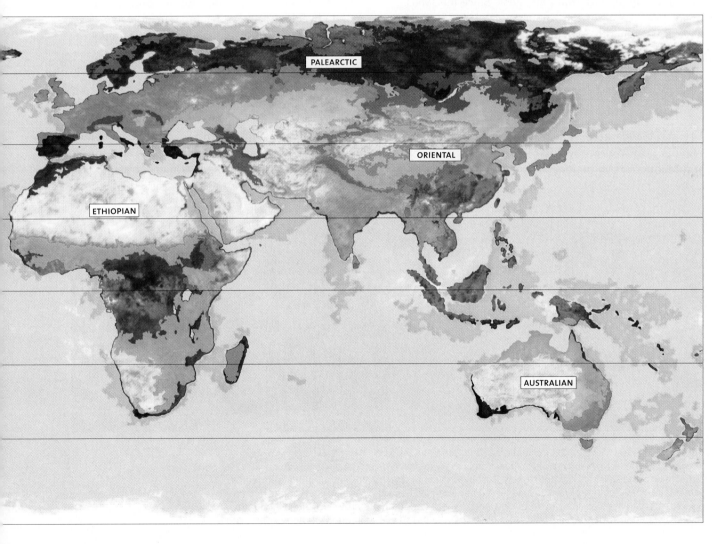

MAJOR BIOMES

Deserts form between latitudes of about 30° north and south. Here, annual rainfall is less than ten centimeters or so, and rates of evaporation are high. Heavy, brief, infrequent pulses of rainfall swiftly erode the topsoil. Humidity is so low that the sun's rays easily penetrate the air. The ground heats fast, then cools fast at night. These conditions do not favor large, leafy plants. The distribution of species within the biome is influenced by temperature and moisture gradients (Figure 21.11).

Long-term shifts in climate, and human activities, can convert grasslands to deserts. **Desertification** is the conversion of large regions of natural grasslands to desertlike conditions. Overgrazing by livestock and short-sighted agricultural practices contribute to the problem. Periodic droughts make matters worse.

Desertification has far-reaching ecological effects. For example, dust from northwestern Africa may be contributing to the decline of coral reefs far away, in the Caribbean Sea. Winds lift dust into the air and then transport it across the Atlantic. In that dust are fungal, bacterial, and viral pathogens. Since the 1980s, huge dust storms of the sort shown in Figure 21.12 have become more frequent in Africa—and so have disease outbreaks among the Caribbean corals.

DRY SHRUBLANDS, DRY WOODLANDS, AND GRASSLANDS

Dry shrublands get less than 25 to 60 centimeters of rain per year. We see them in California, South Africa, and Mediterranean regions, where they are known by

Figure 21.12 Satellite image of an immense dust storm that formed in the hot, dry Sahara Desert, then continued past the west coast of Africa, and is well on its way across the Atlantic.

Africa

such local names as fynbos and chaparral. California has 6 million acres of chaparral (Figure 21.13). In the summer, lightning-sparked fires sweep through these biomes. The shrubs have highly flammable leaves and quickly burn to the ground. But they grow back from root crowns and fire causes their seeds to germinate.

Dry woodlands dominate where annual rainfall is about 40 to 100 centimeters. The dominant trees can be tall, but they do not form a dense, continuous canopy. Eucalyptus woodlands of southwestern Australia and oak woodlands of California and Oregon are like this.

Grasslands form in the interior of continents in the zones between deserts and temperate forests. Warm temperatures prevail in the summer, and winters are extremely cold. Annual rainfall of 25 to 100 centimeters prevents deserts from forming, but it is not enough to support forests. Drought-tolerant primary producers handle strong winds, sparse and infrequent rainfall, and fast evaporation. Grazing and burrowing species are the dominant animals. Their activities, combined with periodic fires, help keep shrublands and forests from encroaching on the fringes of many grasslands.

Figure 21.11 Two views of the same biome—the Sonoran Desert of Arizona. Sunlight is just as intense in the desert lowlands (**a**) as in the uplands (**b**). Vegetation differences reflect water availability, temperature, and soils. The uplands rarely frost over. Also, they receive light but prolonged rain in winter, and heavy pulses of rain in summer. Winter and summer annuals as well as perennials are not as water-stressed as they are in the lowlands, where temperatures are exceedingly high and rainfall is sparse. Widely spaced creosote bush (*Larrea*) mixed with bursage (*Ambrosia*) dominates.

North America's main grasslands are *shortgrass* and *tallgrass* prairie. Most form on flat or rolling land, as in Figure 21.14. The roots of perennials extend profusely through topsoil. In the 1930s, the shortgrass prairie of the Great Plains was overgrazed and plowed under to grow wheat, which requires more water than the area sometimes receives. Strong winds, prolonged drought, and unsuitable farming practices turned much of the prairie into the Dust Bowl. John Steinbeck's historical novel, *The Grapes of Wrath,* eloquently describes the impact of this environmental disaster on people.

Tallgrass prairie, of the sort you read about in the preceding chapter, once extended west from deciduous forests in North America. Tall grasses, legumes, and composites, such as daisies, thrived in the continent's interior, which had richer topsoil and slightly more frequent rainfall. Farmers converted nearly all of the original tallgrass prairie to agriculture, but some areas are now being restored.

Between the tropical forests and the hot deserts of Africa, South America, and Australia are broad belts of grasslands with a smattering of shrubs and trees. These are savannas (Figure 21.14). Rainfall averages 90 to 150 centimeters a year, and extended, seasonal droughts occur. Where rainfall is sparse, fast-growing grasses dominate. Acacia and other shrubs grow in regions that get a bit more moisture. Where rainfall is higher, savannas grade into tropical woodlands with low trees and shrubs and tall, coarse grasses.

Monsoon grasslands form in southern Asia where heavy rains alternate with a dry season. Dense stands of tall, coarse grasses form, then die back and often burn during the dry season.

Figure 21.13 California chaparral. The dominant plants are multibranched, woody, and typically only a meter or so tall. The leaves of many produce oils that deter herbivores, but also make the plant highly flammable. The photo at right shows a firestorm racing through chaparral in a canyon above Malibu.

Figure 21.14 *Above*, bison in a shortgrass prairie to the east of the Rocky Mountains. At one time, bison were the dominant large herbivore of North America's grasslands. Abundant grasses of this biome supported 60 million bison before Europeans arrived.

Left, the African savanna, a warm grassland with scattered stands of shrubs and trees. More varieties and greater numbers of large ungulates live here than anywhere else. They include the migratory wildebeests shown here, and giraffes, Cape buffalos, zebras, and impalas.

MAJOR BIOMES

LINK TO
SECTION
15.7

BROADLEAF FORESTS

In forest biomes, tall trees grow closely together and form a fairly continuous canopy over a broad region. In general, there are three types of trees. Which type prevails depends partly on distance from the equator. *Evergreen broadleafs* dominate between the latitudes of 20° north and south. *Deciduous broadleafs* are typical in moist, temperate latitudes where the winters are mild. *Evergreen conifers*, the third category of forest tree, are dominant in high, cold latitudes and high mountains.

Evergreen broadleaf forests sweep across tropical zones of Africa, the East Indies, Malaysia, Southeast Asia, South America, and Central America. Annual rainfall can exceed 200 centimeters and never is less than 130 centimeters. One type, tropical rain forests, flourish with regular and heavy rainfall, an annual mean temperature of 25°C, and humidity of at least 80 percent (Figure 21.15). Evergreen trees of this highly productive forest grow new leaves and shed old ones all year long. Decomposition and mineral cycling are rapid, given the hot, humid climate, so litter does not accumulate. Soils are weathered, humus-deficient, and poor nutrient reservoirs.

Deciduous broadleaf forests start in regions of mild temperatures and regular seasonal rainfall. In tropical deciduous forests, the trees drop some or all of their leaves during a pronounced dry season, a moisture-preserving adaptation. Monsoon forests of India and Southeast Asia contain such trees. Farther north, in the temperate zone, rainfall is even lower. Winters are cold and dry, and temperate deciduous forests prevail (Figure 21.16). Decomposition is not as rapid as in the humid tropics, and many nutrients are conserved in accumulated litter on the forest floor.

CONIFEROUS FORESTS

Conifers (cone-bearing trees) are primary producers of coniferous forests. Most have thickly cuticled, needle-shaped leaves and recessed stomata, adaptations that help them conserve water during droughts and winter. They dominate the boreal forests, montane coniferous forests, temperate rain forests, and pine barrens.

Boreal forests stretch across northern Europe, Asia, and North America. Such forests also are called *taigas*, which means "swamp forests." Most form in glaciated

Figure 21.15 Tropical rain forests, biomes of great biodiversity. Among the millions of known species are jaguars and *Rafflesia*, a vine with a foul-smelling fly-pollinated flower that grows three meters across. No doubt many more astonishing rain forest species are still to be discovered. Who knows how many other biological marvels we are losing as these forests disappear?

Figure 21.16 Temperate deciduous forest in North America.

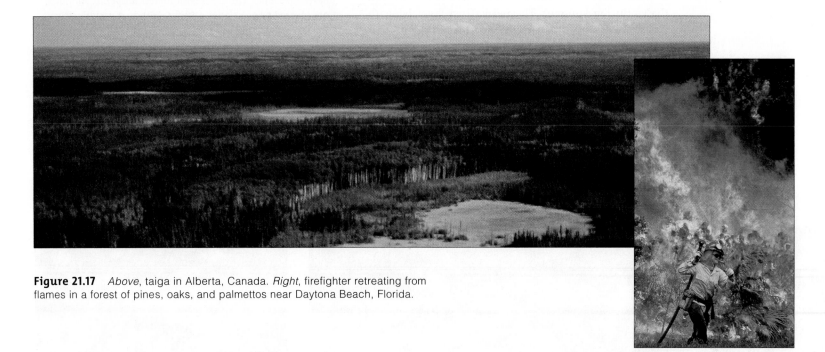

Figure 21.17 *Above*, taiga in Alberta, Canada. *Right*, firefighter retreating from flames in a forest of pines, oaks, and palmettos near Daytona Beach, Florida.

regions having cold lakes and streams (Figure 21.17). It rains mostly in summer, and evaporation is low in the cool summer air. The cold, dry winters are more severe in eastern parts of these biomes than in the west, where ocean winds moderate the climate. Spruce and balsam fir dominate the boreal forests of North America. Pine, birch, and aspen forests form in areas that were burned or logged. In poorly drained soil, acidic bogs dominated by peat mosses, shrubs, and stunted trees prevail. Boreal forests are less dense to the north, where they grade into arctic tundra.

In the Northern Hemisphere, montane coniferous forests extend southward through the great mountain ranges. Spruces and firs dominate in the north and at higher elevations. They give way to firs and pines in the south and at lower elevations. Coniferous forests thrive in certain temperate lowlands. One enormous temperate rain forest parallels the coast from Alaska into northern regions of California. It contains some of the world's tallest trees—Sitka spruce to the far north and redwoods to the south. Over the past century, extensive logging has destroyed much of this biome.

Southern pine forests dominate the coastal plains of the southern Atlantic and Gulf states. Pine species are adapted to dry, sandy, nutrient-poor soil and to frequent fires. Such fires do not get hot enough to kill big trees but they burn up shrubs and accumulated debris. Forests of pine, scrub oak, and wiregrass grow in New Jersey. Palmettos grow below loblolly and other pines in the Deep South. In one recent heat wave in Florida, fires destroyed more than 320,000 acres. Counties where buildings were near the forest had not allowed any controlled burns; the resulting buildup of dry material fed the fires (Figure 21.17).

LINKS TO
SECTIONS
20.3, 20.5

ARCTIC AND ALPINE TUNDRA

Tundra means "treeless plain" in Finnish and it is the coldest biome. *Arctic* tundra lies between the polar ice cap and broad belts of boreal forests in the Northern Hemisphere. The annual snowmelt and rain are less than 25 centimeters. Plants grow rapidly in the nearly continuous sunlight of a brief growing season (Figure 21.18*a,b*). Lichens and short, shallow-rooted plants are a base for food webs that include voles, arctic hares, caribou, arctic foxes, wolves, and polar bears. Many migratory birds nest here during summer, when the air becomes thick with mosquitoes and other insects.

Not much more than a meter or two of surface soil thaws during summer. The **permafrost** layer beneath this is always frozen solid. In some places, permafrost is more than 500 meters thick. It prevents drainage, so the surface soil that thaws is perpetually waterlogged. Cool, anaerobic conditions slow nutrient cycling.

The permafrost is a vast storehouse of carbon. With global warming, some permafrost has begun to melt, releasing stored carbon as methane. The release of this greenhouse gas may, in turn, increase global warming.

We find *alpine* tundra at high altitudes around the world (Figure 21.18*c*). At lofty elevations, where night temperatures usually fall below freezing, trees cannot grow. The soil is thin, nutrient-poor, and—unlike that of arctic tundra—fast draining, because there is no permafrost. The growing season is short and, even in summer, patches of snow persist. Strong winds whip the vegetation. Lichens, mosses, grasses, and small-leafed shrubs that form ground-hugging cushions and mats dominate the community.

Low rainfall produces deserts at latitudes between 30° north and south. Latitudes with somewhat higher rainfall are covered by grasslands and shrublands, especially in the interiors of continents.

At the equator, high rainfall supports forests dominated by evergreen broadleaf trees. In temperate zone forests, deciduous broadleaf trees dominate. High latitudes in the Northern Hemisphere are the realm of evergreen conifers.

The highest altitudes and latitudes where plants can survive support hardy tundra species.

Figure 21.18 Arctic tundra in summer (**a**) and winter (**b**). About 4 percent of the Earth's land mass is arctic tundra, blanketed with snow for as long as nine months of the year. Most arctic tundra is found in northern Russia and Canada, although some forms in Alaska and Scandinavia. For hundreds of thousands of years, small bands of humans have hunted, fished, and herded the caribou in these sparsely populated regions. These are very fragile biomes, where plants and animals must survive extreme cold in winter. They might be vulnerable to industrial pollution. Severe climactic changes that are expected to result from global warming are putting them at risk. (**c**) Compact, low-growing, hardy plants typical of the alpine tundra in Washington's Cascade Range.

21.4 The Water Provinces

Freshwater and saltwater provinces cover more of Earth's surface than all biomes combined. They include the world ocean, lakes, ponds, wetlands, and coral reefs. All aquatic ecosystems have gradients in light penetration, dissolved gases, and temperature.

LAKE ECOSYSTEMS

A lake is a body of standing fresh water, as in Figure 21.19. Over time, erosion and sedimentation alter its dimensions; most are eventually filled in.

A lake has littoral, limnetic, and profundal zones (Figure 21.20*a*). Its littoral zone extends all around the shore to the depth where rooted aquatic plants stop growing. This well-lit, shallow, warm region supports many species. Beyond the littoral zone is the limnetic zone, open sunlit waters that support photosynthesis. The limnetic waters teem with *phytoplankton*, such as cyanobacteria, green algae, and diatoms. They are food for *zooplankton* such as rotifers, copepods, freshwater shrimp, and other small invertebrates. Beneath this, the deeper waters of the profundal zone are dark and cold. Wastes and remains drift down through this zone to the lake bottom, where decomposers in the lake's sediments break them down and release nutrients.

SEASONAL CHANGES IN LAKES In temperate regions having warm summers and cold winters, lakes show seasonal changes in density and temperature from the surface to the bottom. A layer of ice often forms in midwinter. Water near the freezing point is the least dense and accumulates under the ice. Water at 4°C is the most dense. It collects in deeper layers that are a bit warmer than the surface layer in midwinter.

In spring, daylength increases and the air warms. Lake ice melts, the surface water warms to 4°C, and temperatures turn uniform throughout. Winds above the surface cause a **spring overturn**: Strong vertical movements deliver dissolved oxygen from the surface waters to the depths. They deliver nutrients released by decomposition from the sediments to the surface.

Figure 21.19 In the Canadian Rockies, the basin of Moraine Lake, formed by glacial action during the most recent ice age.

A *thermocline* has formed by midsummer. An abrupt drop in temperature at this midlayer blocks vertical mixing (Figure 21.20*b*). Being warmer and less dense, the surface water floats on the thermocline. In cooler water below, decomposers deplete dissolved oxygen. In autumn, the upper layer cools and becomes denser. It sinks, and the thermocline vanishes. During this **fall overturn**, water mixes vertically, so dissolved oxygen moves down and nutrients move up.

Primary productivity is seasonal. After the spring overturn, longer daylengths and the cycled nutrients favor higher rates of photosynthesis. Phytoplankton and rooted aquatic plants quickly take up phosphorus, nitrogen, and other nutrients. In the growing season, vertical thermocline mixing stops. Nutrients locked in the remains of organisms sink through deeper water, so photosynthesis slows. By late summer, shortages limit photosynthesis. A fall overturn cycles nutrients back to the surface, which supports a burst of primary productivity. But there are fewer daylight hours, and they cannot sustain a long-lasting burst. Not until the spring will primary productivity increase once more.

TROPHIC NATURE OF LAKES Each lake's topography, climate, and geologic history dictate the numbers and kinds of species, how they are dispersed in the lake,

Figure 21.20
(**a**) Lake zonation. A littoral zone extends all around the shore to a depth where rooted aquatic plants stop growing. The profundal is all water where light does not penetrate. Above the profundal are the open, sunlit waters of the limnetic zone. (**b**) Thermal layering that forms in many temperate zone lakes during the summer.

LINK TO
SECTION
20.8

and how they cycle nutrients. Soils of the surrounding region and of the lake basin contribute to the type and amount of nutrients available to support organisms.

Interplays among climate, soil, basin shape, and metabolic activities of the lake's residents contribute to conditions that range from oligotrophy to eutrophy. **Eutrophication**, recall, refers to any processes that enrich a body of water with nutrients (Section 20.8). *Oligotrophic* lakes typically are deep, clear, and nutrient poor. Primary productivity is not great. *Eutrophic* lakes often are shallower, nutrient rich, and high in primary productivity. These conditions develop as sediments accumulate in the lake basin. Over time, water slowly becomes cloudier and less deep. Phytoplankton come to dominate the aquatic communities. A completely filled-in basin is the final successional stage.

ESTUARIES AND MANGROVE WETLANDS

Rich ecosystems develop near the coasts of continents and of islands. Estuaries and mangrove wetlands are major types. Like freshwater ecosystems, they vary in their physical and chemical properties, including how deep light penetrates, and water temperature, depth, and salinity.

Estuaries are somewhat enclosed coastal regions where seawater slowly mixes with nutrient-rich fresh water from rivers, streams, and runoff (Figure 21.21a). The confined conditions, slow mixing of water, and tidal action combine to trap vital nutrients.

Primary producers are phytoplankton, plants that tolerate submergence at high tide, and algae. Detrital food webs abound. Because many invertebrates and fishes spend their early life stages here, estuaries are

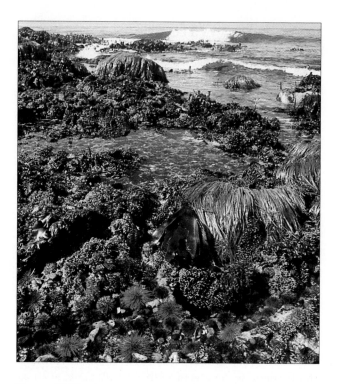

Figure 21.22 A rocky shore in the Pacific Northwest. Algae are the basis for a grazing food web.

called marine nurseries. Estuaries are like truck stops for migratory birds, places where they feed and rest.

Estuaries range from the broad shallow Chesapeake Bay, Mobile Bay, and San Francisco Bay to the narrow, deep fjords of Norway and equivalent bays in Alaska and British Columbia. All need a constant inflow of unpolluted, fresh water to remain healthy. Diversion of fresh water for human uses and contamination by pollutants threaten estuaries.

Figure 21.21 Wetlands. (**a**) In South Carolina, a salt marsh that interacts with the open sea. *Spartina*, a marsh grass, is the main producer. (**b**) From Florida's Everglades, a mangrove wetland with red mangroves (*Rhizophora*).

THE WATER PROVINCES

In tidal flats at tropical latitudes, we find nutrient-rich mangrove wetlands. "Mangrove" refers to forests of salt-tolerant plants in sheltered areas along tropical coasts. These plants have adventitious prop roots that extend out from the trunk (Figure 21.21b). Specialized cells at the surface of the roots allow oxygen in.

Mangroves are the base for a diverse aquatic food web, a nursery for aquatic organisms, and shelter for many birds. But the mangroves are under siege. Many tropical island nations where they grow are strapped for cash. Local people cut mangroves for firewood. Also, large stretches of coastline have been cleared to make way for shrimp farms that supply appetites in the United States, Europe, and Japan. At one time, about three quarters of the coastline of tropical seas was covered by mangrove forest. That amount has been halved and the losses are accelerating.

ROCKY AND SANDY COASTLINES

Some coastlines are nearly vertical, while others slope gently to the sea. Some are sandy and others are rocky. Still, organisms living along all shores share certain challenges. They must withstand the mechanical force of waves, as well as the physiological demands posed by alternating periods of submersion and drying out.

How high the water rises during a high tide varies over the course of a lunar cycle. An *upper* littoral zone is the part of the shore that is underwater only during the highest tide of a cycle. Biodiversity is higher in a *mid*littoral zone, which is regularly submerged by the high tide and exposed at low tide. Along rocky shores, red, brown, and green algae, snails, barnacles, crabs, nudibranchs, sea stars, and sea urchins typically occur here. Biodiversity is highest in the *lower* littoral zone. This region is underwater except during the lowest tide of the lunar cycle. In all three zones of a rocky shore, wave action prevents much detritus from piling up, so grazing food webs prevail (Figure 21.22).

Waves and currents continually rearrange stretches of loose sediments, the *sandy* shores. Few large plants or algae grow here so you will not find many grazing food webs. Instead, there are detrital food webs that start with organic debris from land or offshore.

CORAL REEFS

About 6,000 years ago, when the first dynasty in Egypt was gathering force, coral reefs began forming off the Florida coast. Today, countless numbers and kinds of marine invertebrates as well as fishes and algae live there (Figure 21.23a). Coral reefs form mainly in clear, warm waters in between the latitudes of 25° north and 25° south (Figure 21.23b). Calcium carbonate-rich coral

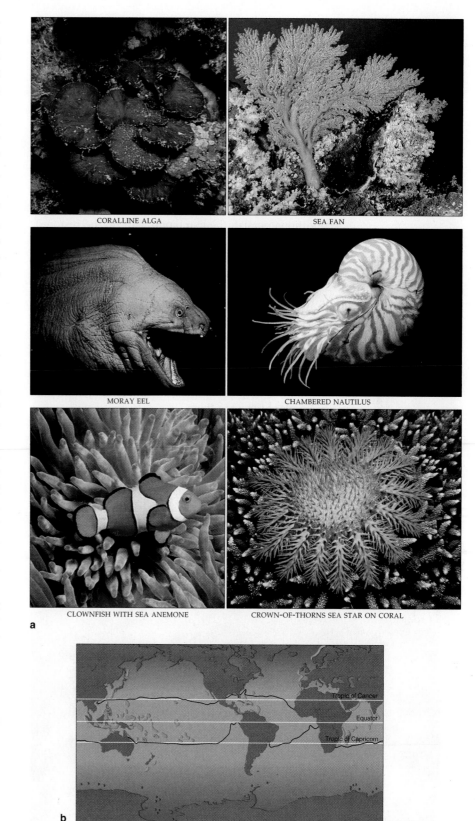

CORALLINE ALGA

SEA FAN

MORAY EEL

CHAMBERED NAUTILUS

CLOWNFISH WITH SEA ANEMONE

CROWN-OF-THORNS SEA STAR ON CORAL

a

b

Figure 21.23 (**a**) A sampling of biodiversity from coral reefs around the world. (**b**) Global distribution of reefs (*orange*), coral banks (*gold*), and solitary corals (*red*). Reef-building corals have photosynthetic symbionts in their tissues and live in shallow sunlit waters of warm seas (within dark lines). Coral banks are formed in deeper, colder water by corals without photosynthetic symbionts.

THE WATER PROVINCES

parts form the reef's spine. Some red algae contributed mineral-hardened cell walls to it. Secretions of many other organisms also help to cement a reef together.

Photosynthetic dinoflagellates live as mutualistic symbionts in tissues of most reef-building corals. They provide the polyps with oxygen and use their wastes. This helps reef-building corals survive in tropical seas, where oxygen and nutrient levels are typically low. If corals become stressed by environmental change, they expel their protistan symbionts. This response is called *coral bleaching*. If conditions improve fast, the symbiont population in the coral can recover. But when adverse conditions persist, the symbionts will not be restored, and the coral will die, leaving only bleached skeletons behind (Figure 21.24).

The frequency of coral bleaching and coral deaths is rising. A rise in water temperature can prompt corals to expel their symbionts. Water pollution kills corals directly and also can cause algal blooms that block out sunlight and release coral-killing toxins. For example, during 2002, a huge algal bloom ninety-six kilometers across drifted through the Florida Keys. The toxins that were released by this bloom killed about 70 percent of the corals exposed to them.

Florida's reefs are also increasingly suffering from disease. Remember the dust blowing across the ocean from eroded regions in Africa (Figure 21.12)? Spores of the fungus *Aspergillus sydowii* were blown along with this dust. The fungus kills sea fans (*Gorgonia*), a type of soft coral. Figure 21.23 shows one healthy specimen.

Figure 21.24 Coral bleaching. In 2002, water temperature around Australia's Great Barrier Reef was the highest recorded since 1870. Warmer water caused the most widespread bleaching ever observed on this barrier reef. Nearly 60 percent of the corals bleached out.

THE OPEN OCEAN

Beyond land's end are two vast ocean provinces. The ocean covers nearly three-fourths of Earth's surface. Its *pelagic* province is the full volume of ocean water (Figure 21.25a). Its neritic zone is all the water above continental shelves, and its oceanic zone is the water in the ocean basins. Sediments and rocks of the ocean floor constitute the *benthic* province. It extends from the lower intertidal zone to the greatest ocean depths.

Photosynthesis is seasonal in much of the ocean's upper surface (Section 5.4). Phytoplankton "pastures" are the first trophic level in upper ocean communities. Marine phytoplankton include photosynthetic bacteria, diatoms, dinoflagellates, coccolithophores, and algae.

In deep waters too dark to support photosynthesis, food webs start with **marine snow**. These tiny bits of organic matter drift down from communities above and are the food base for staggering biodiversity. The midoceanic water may be home to 10 million species!

Many zooplankton make a daily vertical migration, sometimes moving hundreds of meters. These species spend their days in deeper, darker, colder water where they are less visible to predators and their metabolic rate stays low. They move upward to feed as darkness falls, then migrate back down the next morning.

Carnivores at upper levels of marine food webs range from familiar types, such as sharks and squids, to the remarkable deep-sea fishes, such as angler fish (Figure 21.25b).

HYDROTHERMAL VENTS

Hydrothermal vents are like geysers on the ocean floor. They occur most commonly in areas where the sea floor is spreading. Water seeps into fissures and is heated by exposure to molten rock. Superheated water leaches mineral ions from these rocks. Iron, zinc, and copper sulfides, and magnesium and calcium sulfates dissolve in it. When mineral-rich hot water spews out through a vent and meets up with the cold water at the ocean floor, minerals come out of solution. Chemoautotrophic bacteria use these minerals as an energy source. These chemoautotrophs serve as primary producers for food webs that include giant tube worms (Figure 21.25c), shrimps, crabs, clams, and anemones. All spend their lives in perpetual darkness.

As Section 14.1 described, one hypothesis holds that life may have originated near such nutrient-rich places on the early seafloor. Conditions at the surface of early Earth were inhospitable. Radicals in the early atmosphere destroyed any organic molecules. There was no ozone layer and ultraviolet light would have broken bonds. Spaces in mineral-rich rocks near vents

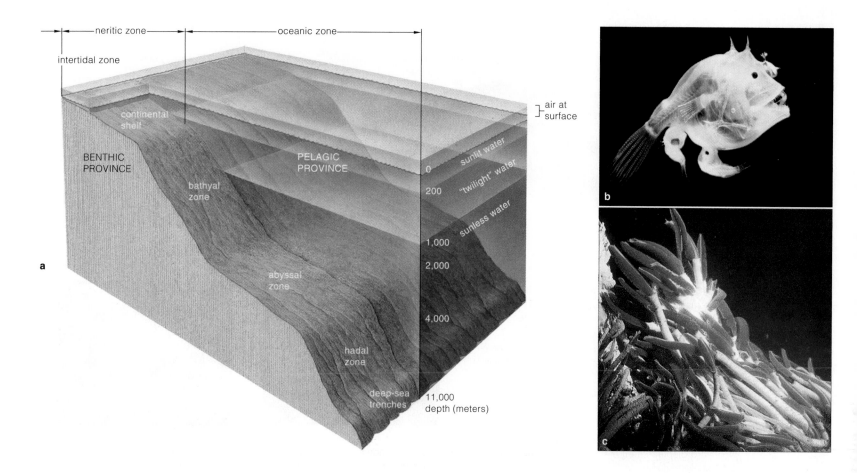

Figure 21.25 **Animated!** (**a**) Oceanic zones. The seafloor extends from the continental shelves to deep-sea trenches. Dimensions of each zone are not shown to scale. (**b**) A female deep-sea angler fish, with two smaller attached males on her ventral surface, is one of the many interesting inhabitants of the pelagic zone. (**c**) Tube worms that reach a meter in length cluster around a hydrothermal vent on the ocean floor.

may have shielded organic molecules, allowing the first cells to self-assemble.

UPWELLING ALONG COASTS

Prevailing winds that parallel the western coasts of continents tug on the ocean surface and get its upper waters moving. Earth's rotation deflects these waters away from the coasts. Cold, often nutrient-rich water moves in vertically in its place. We call this movement of water an **upwelling** (Figure 21.26).

Upwelling occurs in equatorial currents and along continents, and it cools air masses above it. When thick fogbanks form along California's north coast, an upwelling of cold water is interacting with warm air. Upwelling is disrupted by the weather pattern known as El Niño, introduced in the opening essay. We return to this subject again in the next section.

> *Freshwater and saltwater provinces are far more extensive than Earth's land biomes. All of the aquatic ecosystems in these water provinces display gradients in light intensity, temperature, and dissolved gases. Organisms have adapted to conditions along all of these gradients.*

a North wind starts coastal water moving

b Earth's rotational force deflects the moving water west

c Cold water moves up as replacement

Figure 21.26 **Animated!** Upwelling along the west coasts in the Northern Hemisphere. Winds that blow from the north start the ocean's surface water moving. Earth's rotational force deflects moving water westward. Cold, deeper, nutrient-rich water moves up and replaces it.

Figure 21.27 *Animated!* (**a**) Westward flow of cold, equatorial surface water in between ENSOs. (**b**) Eastward dislocation of warm ocean water during an ENSO.

Satellite data on primary productivity in the equatorial Pacific Ocean. The concentration of chlorophyll was used as the measure.

(**c**) During the 1997–1998 El Niño event, a massive amount of nutrient-poor water moved to the east, and photosynthetic activity was negligible.

(**d**) During the subsequent La Niña, upwelling and westward displacement of nutrient-rich water led to a huge algal bloom.

c Near-absence of phytoplankton in the equatorial Pacific during an El Niño.

d Immense algal bloom in the equatorial Pacific in the La Niña rebound event.

21.5 Applying Knowledge of the Biosphere

We now return to El Niño, a global event that affects primary productivity along coasts of the Northern Hemisphere and in the equatorial Pacific Ocean. That event has rippling effects on the weather and on land ecosystems around the world.

In the Southern Hemisphere, the health of commercial fisheries depends upon wind-induced upwelling along the coasts of Peru and Chile. Prevailing coastal winds blow in from the south and southeast, tugging surface water away from shore. Cold water delivered to the continental shelf by the Peru Current moves up near the surface. Huge quantities of nitrate and phosphate are pulled up and carried north by this current. These nutrients sustain phytoplankton that are the basis for one of the world's best fisheries.

Every three to seven years, warm surface waters of the western equatorial Pacific Ocean move eastward. This massive displacement of warm water affects the prevailing wind direction. The eastward flow speeds up so much that it influences the vertical movement of water along the coasts of Central and South America.

Water driven into a coastline is forced downward and flows away from it. Near Peru's coast, prolonged **downwelling** of nutrient-poor water displaces cooler waters of the Peru Current, which prevents upwelling. The warm current usually arrives around Christmas. Fishermen in Peru named it **El Niño** ("the little one," a reference to the baby Jesus).

El Niño episodes last about six to eighteen months. An El Niño Southern Oscillation, or **ENSO**, is defined by changes in the sea's surface temperature and in the air circulation patterns. *Between* ENSOs, warm waters and heavy rains move west (Figure 21.27*a*). *During* an ENSO, the prevailing surface winds in the western equatorial Pacific pick up speed and "drag" surface waters toward the east (Figure 21.27*b*). As a result, the westward transport of water slows, the temperature of the sea's surface water rises, evaporation increases, and air pressure falls. The changes have global effects, both in the seas and on land.

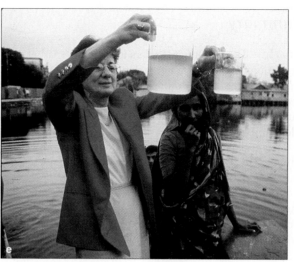

Figure 21.28 (a) *Vibrio cholerae*, the agent of cholera. (b) Copepods host a dormant stage of *V. cholerae* that waits out adverse environmental conditions that do not favor its growth and reproduction. (c) Typical Bangladesh waterway from which water samples were drawn for analysis. (d) Satellite data on rising sea surface temperatures in the Bay of Bengal correlated with cholera cases in the region's hospitals. *Red* signifies warmest summer temperatures. (e) Rita Colwell comparing unfiltered water with water that was filtered through folded sari fabric.

As you read in the chapter opening, 1997 ushered in an ENSO event. Average sea surface temperatures in the eastern Pacific rose 9°F (about 5°C). The unusually warm water extended 9,660 kilometers west from the coast of Peru.

The 1997–1998 El Niño event caused a tremendous change in the primary productivity of the equatorial Pacific. The massive eastward flow of nutrient-poor warm water caused phytoplankton populations to fall to almost undetectable levels (Figure 21.27c). During the following year, cooler, nutrient-rich water welled up and was displaced westward, along the equator in what is called a La Niña event. Satellite images taken during this period revealed a vast algal bloom that stretched across the equatorial Pacific (Figure 21.27d).

During the 1997–1998 El Niño event, the number of cholera cases soared. People already knew that water contaminated by *Vibrio cholerae* bacteria causes cholera epidemics (Figure 21.28a). The disease triggers severe diarrhea and the bacteria enter the water supply in feces. People who use the tainted water get infected.

What people did *not* know was where *V. cholerae* lurked between cholera outbreaks. It could not be found in humans or in water supplies. Then it would show up simultaneously in places that were often far apart—usually coastal cities, where the urban poor draw water from rivers flowing to the sea.

Marine biologist Rita Colwell had been thinking about cholera for some time. Colwell suspected that humans were not the host in between outbreaks. Was there an environmental reservoir for the pathogen? Maybe. But nobody had detected it in water samples that had been subjected to standard culturing.

Then Colwell had a flash of insight: Perhaps no one had found the pathogen because it sometimes enters a dormant stage that cannot be cultured.

Colwell discovered that she could use an antibody-based test to detect a protein on *V. cholerae*'s surface. Tests on samples from Bangladesh revealed a dormant form of the bacteria in fifty-one of fifty-two samples of water. Standard testing methods missed it in all but seven of these samples.

V. cholerae lives in rivers, estuaries, and the seas. As Colwell knew, plankton also thrive in these aquatic environments. Using her new method, she discovered *V. cholerae* on copepods, small marine crustaceans that graze on phytoplankton (Figure 21.28b).

Colwell knew that there was a seasonal pattern to cholera outbreaks in Bangladesh, and to temperatures in the Bay of Bengal (Figure 21.28c,d). She also knew that phytoplankton levels rise and fall seasonally. She hypothesized that the abundance of copepods—and of the *V. cholerae* living on them—rises and declines with phytoplankton abundance. Colwell decided to compare sea surface temperature data from 1990–1991 and 1997–1998 El Niño episodes with the incidence of cholera cases. As she expected, four to six weeks after the sea surface temperatures rose, so did the incidence of cholera cases.

Today, Colwell and Anwarul Huq, a Bangladeshi scientist, are looking into salinity and other factors that may also influence cholera outbreaks. Their goal is to design a more robust model for predicting where and when the next cholera outbreak will occur.

Colwell and Huq also devised a simple prevention method. They advised women in Bangladesh to use sari cloth as a filter to remove *V. cholerae* cells from the water (Figure 21.28e). Copepod hosts are too big to pass through the thin cloths, which can be rinsed in clean water, sun-dried, and used again and again. By using this simple and inexpensive method, seasonal cholera outbreaks have been reduced by 50 percent.

Summary

Section 21.1 "Climate" refers to average weather conditions over time. Climate depends upon differences in the amount of solar radiation reaching equatorial and polar regions, Earth's tilt and its annual path around the sun, the distribution of continents and seas, and the elevation of land masses.

The use of CFCs has depleted ozone in the upper atmosphere. Ozone thinning allows more damaging UV radiation to reach Earth's surface.

Smog, a form of air pollution, arises in areas where large amounts of fossil fuels are burned. Coal-burning power plants are also the main contributors to acid rain, which damages soils and kills plants and fish.

Solar energy and wind energy are renewable, clean sources of energy for the human population.

Biology ⊘ Now
Learn how sunlight energy drives global patterns of air circulation with the animation on BiologyNow.

Section 21.2 Latitudinal and seasonal variations in sunlight warm ocean waters and set currents in motion. The currents distribute heat energy and affect weather patterns. Ocean currents, air currents, and topography interact to shape regional climates.

Biology ⊘ Now
See the patterns of major ocean currents with the animation on BiologyNow.

Section 21.3 The world's land masses are realms of biodiversity. Each is more or less isolated from others and maintains a characteristic array of species.

Biomes are a category of major ecosystems on land. They form and are maintained by regional variations in climate, landforms, and soils.

Deserts form at latitudes about 30° north and south, where annual rainfall is sparse. Slightly wetter southern or western coastal regions support dry woodlands and dry shrublands. In midlatitude interiors of continents, vast grasslands form. Near the equator, high rainfall and mild temperatures support evergreen tropical forests. Deciduous broadleaf forests grow in regions with hot summers and cool winters. Where a cold, dry season alternates with a cool, rainy season, coniferous forests prevail. Tundra forms at high latitudes and high altitudes.

Biology ⊘ Now
Examine the distribution of Earth's biomes with the animation on BiologyNow.

Section 21.4 Water provinces cover more than 71 percent of the Earth's surface. They include standing fresh water, running fresh water, and the world oceans. All aquatic ecosystems show gradients in penetration of sunlight, water temperature, salinity, and dissolved gases. These factors vary daily and seasonally, and they influence primary productivity.

Wetlands, intertidal zones, rocky and sandy shores, tropical reefs, and regions of the open ocean are major marine ecosystems. Superheated, mineral-rich waters support communities at deep-sea hydrothermal vents.

Upwelling is an upward movement of deep, cool ocean water along coasts of continents that often carries nutrients to the surface.

Biology ⊘ Now
Learn about the oceanic zones with the animation on BiologyNow.

Section 21.5 An El Niño event disrupts upwelling and causes massive temporary changes in rainfall and other weather patterns with global effects.

Biology ⊘ Now
Observe how an El Niño event affects ocean currents and upwelling with the animation on BiologyNow.

Self-Quiz
Answers in Appendix I

1. Solar radiation drives the distribution of weather systems and so influences _____ .
   a. temperature zones   c. seasonal variations
   b. rainfall distribution   d. all of the above

2. _____ shields against the sun's UV radiation.
   a. A thermal inversion   c. The ozone layer
   b. Acid precipitation   d. The greenhouse effect

3. Regional variations in the global patterns of rainfall and temperature depend on _____ .
   a. global air circulation   c. topography
   b. ocean currents   d. all of the above

4. A rain shadow is a reduction in rainfall _____ .
   a. on the inland side of a coastal mountain range
   b. along eastern coasts in the Northern Hemisphere
   c. as a result of human activities
   d. near the poles

5. The Gulf Stream flows _____ along the east coast of the United States, and the California current flows _____ along its west coast.
   a. north, south   c. south, south
   b. north, north   d. south, north

6. Broadleaf evergreen trees dominate in _____ .
   a. chapparal   c. boreal forests
   b. tropical rain forests   d. deserts

7. The major cause of acid rain is operation of _____ .
   a. nuclear power plants   c. coal-burning plants
   b. automobiles   d. agricultural machinery

8. Biomes are _____ .
   a. water provinces   d. characterized by
   b. water and land zones   dominant plants
   c. broad land regions   e. both c and d

9. Biome distribution corresponds roughly with regional variations in _____ .
   a. climate   c. soils
   b. topography   d. all of the above

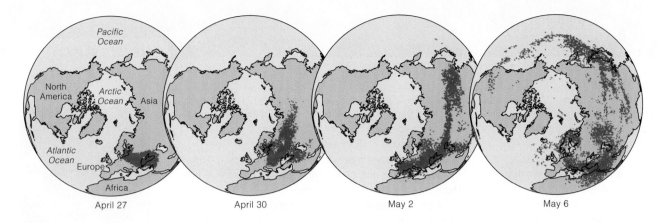

Figure 21.29 Global distribution of radioactive fallout after the 1986 meltdown of the Chernobyl nuclear power plant. The fallout put 300 million to 400 million people at risk for leukemia and other radiation-induced disorders. By 1998, the rate of thyroid abnormalities in children living immediately downwind from Chernobyl was nearly seven times as high as for those upwind; their thyroid gland concentrated the iodine radioisotopes.

10. Grasslands most often predominate _____ .
    a. near the equator   c. in interiors of continents
    b. at high altitudes   d. b and c

11. Permafrost underlies _____ .
    a. grasslands   c. temperate zone forests
    b. tundra   d. boreal forests

12. During _____ , deeper, often nutrient-rich water moves to the surface of a body of water.
    a. spring overturn   c. upwelling
    b. fall overturn   d. all of the above

13. Match the terms with the most suitable description.
    _____ tundra
    _____ chaparral
    _____ desert
    _____ savanna
    _____ estuary
    _____ boreal forest
    _____ tropical rain forest

    a. equatorial broadleaf forest
    b. area where freshwater and seawater mix
    c. type of grassland with trees
    d. low-growing plants at high latitudes or high elevations
    e. found at latitudes 30° north and 30° south
    f. conifers dominate
    g. dry shrubland

Additional questions are available on Biology ⒼNow™

Critical Thinking

1. As described in Section 14.3, the dry oak woodlands that dominate much of central California are threatened by *sudden oak death*, a fatal disease caused by a pathogenic oomycote. Trees already weakened by prolonged drought are especially vulnerable. If the oak woodlands disappear, what type of vegetation cover would you expect to replace them? Why?

2. Conifers have evergreen leaves and they maintain a higher rate of photosynthesis than broadleaf trees when temperatures are low. They are also less likely than broadleafs to form flow-interrupting bubbles in their vascular tissue as sap thaws. How do these traits help explain the distribution of boreal forests?

3. On April 26, 1986, in Ukraine, a meltdown occurred at the Chernobyl nuclear power plant. Nuclear fuel burned for nearly ten days and released up to 250 million curies of radioactive matter into the atmosphere. Inhaling as little as ten-millionths of a curie of plutonium can cause cancer. Thirty-one people died at once; others died from radiation poisoning in the following weeks. Winds carried radioactive fallout around the globe (Figure 21.29).

The Chernobyl accident stiffened opposition to nuclear power in the United States, but recent developments have some people reconsidering. Increasing the use of nuclear energy would diminish the country's dependence on oil from the politically unstable Middle East. Nuclear power does not contribute to global warming, acid rain, or smog. It does produce highly radioactive wastes. Investigate the pros and cons of nuclear power, and decide if you think the environmental benefits outweigh the risks. Would you feel differently if a nuclear power plant was going to be built ten kilometers upwind from your home?

4. Use of off-road recreational vehicles may double over the next twenty years. Many off-road enthusiasts would like increased access to government-owned desert areas. They argue that it is just the perfect place for off-roaders because "There's nothing there." How would you reply to this argument?

5. Write a short essay on the effect global warming might have on the spring overturn and thermocline formation in a Minnesota lake.

6. *Wetlands,* remember, are transitional zones between aquatic and terrestrial ecosystems in which plants are adapted to grow in periodically or permanently saturated soil. They include mangrove swamps and riparian zones. Wetlands everywhere are being converted for agriculture, home building, and other human activities. Does the great ecological value of wetlands for the nation as a whole outweigh the rights of the private citizens who own many of the remaining wetlands in the United States? Should private owners be forced to transfer title of their land to the government? If so, who should decide the value of a parcel of land *and* pay for it? Or would such seizure of private property violate laws of the United States?

My Pheromones Made Me Do It

A few years ago, as Toha Bererub (below) walked down a street near her Las Vegas home, she felt a sharp pain above her right eye. Then another. And another. A few seconds later, stinging bees encased her upper body. Firefighters wearing protective gear rescued her, but only after she had been stung more than 500 times.

Bererub's attackers were Africanized honeybees, a hybrid between the gentle European honeybee and an aggressive African strain. African bees were imported to Brazil in the 1950s for cross-breeding experiments. Breeders were after a gentle but zippier pollinator for commercial orchards. But some of the bees from Africa escaped and started mating with the European bees that had been imported earlier. Then, in a grand example of geographic dispersal, some descendants buzzed north all the way from Brazil, into Mexico, and then into the United States.

All honeybees live in big colonies and defend them by stinging. Each bee can sting only once. Compared with their milder-mannered relatives, the Africanized bees are more easily alarmed and more persistent in their defensive behavior. Some have chased a target for a quarter of a mile. Although dubbed "killer bees" by the media, attacks are painful but generally not fatal. A healthy adult usually survives even a huge number of stings. Bererub was seventy years old and recovered fully after spending a week in the hospital.

What makes Africanized bees more aggressive? Their reaction to isopentyl acetate, for one thing. This banana-scented chemical functions as a component of the honeybee alarm pheromone.

A **pheromone** is a chemical communication signal that conveys information among members of the same species. A honeybee releases alarm pheromone when she senses a threat or as she is stinging something. Other bees follow the gradient of pheromone molecules diffusing through the air to guide them to the target.

Researchers once tested colonies of Africanized and European honeybees to compare their responses to an alarm pheromone. They positioned a tiny target in front of each colony, then released an artificial pheromone. Africanized bees flew out and attacked the target faster than European bees did. They plunged six to eight times as many stingers into it.

These differences among honeybees lead us into the world of animal behavior—coordinated responses that animal species make to stimuli. We will start with the genetic basis of behavior, then look at examples of instinctive and learned mechanisms. Along the way, we also will consider the adaptive value of behavior.

 How Would You Vote? *Africanized honeybees are slowly expanding their range in North America. Should we fund more studies of their genes and behavior? See BiologyNow for details, then vote online.*

Key Concepts

THE BASIS OF BEHAVIOR
Behavior arises through interactions among gene products and the environment. Most behavior has innate components but can be modified by environmental factors.

NATURAL SELECTION AND BEHAVIOR
Like other traits, behavioral traits can evolve by natural selection. Behavior that enhances the reproductive success of an individual will be favored. For example, communication systems and mating systems have been shaped by natural selection.

GROUP LIVING
Life in social groups has reproductive benefits and costs. Self-sacrificing behavior has evolved among a few animals that live in large family groups.

Links to Earlier Concepts

Alternative gene splicing (Section10.6) can affect behavior.

Much behavior is shaped by directional selection (12.5) or sexual selection (12.6). We will revisit the concept of adaptation (12.2) and consider how predation (19.4) and parasitism (19.5) influence social behavior. You may wish to review information about the primate lineage (16.13).

22.1 So Where Does Behavior Start?

As you learned in earlier chapters, nervous and endocrine systems govern an animal's responses to stimuli. Because genes specify the substances required for constructing those systems, they provide the heritable foundation for animal behavior.

GENES AND BEHAVIOR

An animal's nervous system is set up in a way that allows it to detect, interpret, and issue commands for responses to stimuli—specific bits of information about the world that sensory receptors have received. One way or another, diverse gene products affect each step in the formation and operation of the animal's brain, nerve pathways, and sensory receptors. Do some gene products influence *behavioral* responses to stimuli?

Stevan Arnold showed that feeding preferences of different populations of California garter snakes have a genetic basis. The garter snakes that live in the damp forests near the coast preferentially feed on large land-dwelling slugs known as banana slugs (Figure 22.1*a,b*). Snakes that live inland prefer tadpoles and tiny fishes. Offer them a banana slug and they ignore it.

In one set of experiments, Arnold offered newborn captive garter snakes a bit of slug as their first meal. Coastal snake offspring usually ate it and flicked their tongue at cotton swabs that smelled like slug (Figure 22.1*c*). (Tongue-flicking helps snakes to sense smells.) Inland snake offspring ignored slug-soaked swabs and rarely ate slug meat. Since both groups of snakes had no prior experience with slugs, Arnold concluded that garter snakes are genetically programmed to accept or reject slugs. He hypothesized that coastal and inland populations might have allele differences that affect development of their odor-detecting mechanisms.

To test this hypothesis, Arnold crossbred coastal and inland snakes. He predicted that hybrid offspring of these snakes would make an *intermediate* response to slug chunks and odors. The test results matched his prediction. Compared with a typical newborn inland snake, most baby snakes of mixed parentage tongue-flicked more often at slug-scented cotton swabs—but less often than newborn coastal snakes did.

Researchers are sometimes even able to pinpoint exactly which gene affects a behavior. For example, in fruit flies (*Drosophila melanogaster*) the gene known as *fruitless* (*fru*) controls male courtship behavior. Both males and females have the gene and both express it, but differences in how transcripts are spliced (Section 10.6) make their final protein products differ.

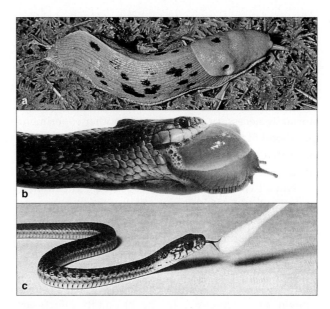

Figure 22.1 (**a**) Banana slug, food for (**b**) an adult garter snake of coastal California. (**c**) Newborn garter snake from a coastal population, tongue-flicking at a cotton swab drenched with tissue fluids from a banana slug.

As you will learn later in this chapter, specialized courtship displays precede mating in many animals, and fruit flies are no exception. A female fruit fly will mate only after her suitor performs a display in which he waves his wings and licks and taps her body. Male flies with a mutated version of the *fru* gene perform poorly at this task, so females remain unresponsive.

Perhaps even more interesting is what happened when researchers manipulated flies so that each ended up making the fruitless protein typical of the opposite sex. Males that expressed the female protein were less inclined to court females; but lavished attention on the normal males. Females with the male protein courted other females, displaying stereotypical male behaviors such as wing waving. The genetically altered females had little interest in ordinary males, but they did court males who had been made to smell like females.

The researchers concluded that the fruitless protein acts as a master switch. Which form of the protein is present affects the nervous system in ways that have far-reaching effects on complex behavior.

HORMONES AND BEHAVIOR

Gene products called hormones also guide behavior. For instance, all mammals make and secrete oxytocin, a hormone with roles in the birth process and lactation. In many species, oxytocin also affects social behavior, including pair bonding.

In the prairie vole (*Microtus ochrogaster*), oxytocin is the key that unlocks a female's heart. The female of these small, fluffy rodents bonds with a male after a night of repeated matings, and she mates for life. To test the effects of oxytocin, researchers placed a female vole with a male for a few hours but prevented mating. Then they injected oxytocin into the female. The result was love at first sight, a pair bond even without the normally necessary sex act. In yet another experiment, pair-bonded female prairie voles were injected with a drug that blocks oxytocin's effects. They immediately dumped their partners.

Not all vole species are monogamous. The number and distribution of oxytocin receptors may play a role. The brains of monogomous prairie voles are richer in oxytocin receptors than the brains of more promiscuous montane voles (*M. montanus*) (Figure 22.2).

The brains of monogamous vole species also have more receptors for antidiuretic hormone (ADH), also known as vasopressin. In the brain, its effects are similar to those of oxytocin; it facilitates bonding.

Researchers isolated the gene for an ADH receptor in monogamous prairie voles and transferred copies of it into forebrain cells of male meadow voles (*M. pennsylvanicus*). After the transfer, males of this usually promiscuous species showed an increased tendency to partner with just one female.

Male meadow voles used as a control group also got copies of the gene, but in a brain region not known to be involved in pair-bonding. Unlike the experimental group, these males retained their playboy ways.

Even among prairie voles, there is variation in how much time a male devotes to his mate and offspring. Breeding studies have shown that a male's tendency to remain around his family is inherited. Compared to

Figure 22.2 Hormones and prairie voles. Distribution of oxytocin receptors (*red*) in the brain of (**a**) a mate-for-life prairie vole and (**b**) a promiscuous montane vole.

less devoted males, a doting dad vole has more ADH receptors in his brain.

REGARDING INSTINCT AND LEARNING

Slug-munching snakes, courting fruit flies, and pair-bonding voles all demonstrate **instinctive behavior**. This term means the behavior is performed without having first been learned by actual experience in the environment. The nervous system of a newly born or hatched animal is prewired to recognize *sign stimuli* —one or two simple, well-defined environmental cues that trigger a suitable response. The snake's response is a stereotyped motor program. When a garter snake recognizes certain sign stimuli, a **fixed action pattern**

Figure 22.3 A few instinctive behaviors. (**a**) A social parasite, the European cuckoo, lays eggs in the nests of other birds. Even before this cuckoo hatchling opens its eyes, it responds instinctually to a round shape— a host's egg—and shoves it out of the nest. (**b**) The clueless foster parents show instinctual behavior as they respond to the usurper's gaping mouth, not to its size and other traits that characterize their own species. (**c**) A human baby instinctively imitates adult facial expressions.

follows. It is a program of coordinated muscle activity that will run to completion independently of feedback from the environment.

Cuckoos provide another example of instinctive behavior. Like the cowbirds described in Section 19.5, cuckoos are social parasites. The females deposit their eggs in the nests of other bird species. Cuckoos are blind when they first hatch, but contact with an egg or another round object stimulates a fixed action pattern. The cuckoo hatchling maneuvers the egg onto its back, then pushes the egg from the nest (Figure 22.3a). This ensures that the cuckoo receives undivided attention.

The cuckoo's "foster parents" are oblivious to the odd color and size of the usurper. They respond only to one sign stimulus—the gaping mouth of a chick—and continue to nurture the invader (Figure 22.3b).

Human infants display instinctive behavior. Three days after birth, a human infant already has a capacity to mimic the facial expressions of an adult who comes close (Figure 22.3c). This may seem easy, but the infant cannot watch its own face, nor can it feel which facial muscles the adult is using. Somehow it is able to open its mouth, protrude its tongue, or rotate its head the same way as the adult. Infants will also respond to a simplified stimulus—a flat, face-sized mask with two dark spots for eyes. One "eye" will not do the trick.

Animals also show **learned behavior**. They process information about experiences and use it to change or vary responses to stimuli. For instance, a young toad's nervous system commands it to flip its sticky tongue instinctively at any dark objects in its field of vision, which usually are edible insects. But what if the dark object is a wasp that can sting the toad's tongue? The toad will quickly learn to avoid all "black-and-yellow-banded objects that sting." For the warning coloration and mimicry described in Section 19.4 to evolve by natural selection, some predators must be able to learn to associate an unpleasant or painful experience with a particular prey coloration or pattern.

Imprinting is a classic example of learned behavior. This time-dependent form of learning is triggered by exposure to particular sign stimuli. Exposure usually takes place during a sensitive period when the animal is young. Imprinting of baby geese on their mother is a well-documented example (Figure 22.4).

Most behavior is shaped when genes interact with the "environment." Each animal's sensory experience, nutritional state, and many other variable factors that can directly or indirectly change gene activity are part of the environment.

For example, many male birds have an instinct to sing during the nesting season. Yet in different areas, songbirds may pick up variations, or dialects, of their species song. Peter Marler demonstrated that many male birds learn their full species song ten to fifty days after hatching by listening to other birds sing it. The male nervous system is prewired to recognize that song; his learning mechanism is primed to select and respond to acoustical input. What he hears during the sensitive period influences his rendition of the song.

In one study, Marler raised white-crown sparrow nestlings in soundproof chambers so they could not hear adult males. The captives sang when mature, but their songs were simpler than normal. Marler exposed some other isolated white-crown sparrow nestlings to recordings of adult males of their own species and

Figure 22.4 No one can tell these imprinted baby geese that Konrad Lorenz is not Mother Goose!

Imprinting is a time-dependent form of learning. It involves exposure to sign stimuli, most often early in development. In response to a moving object and probably to certain sounds, baby geese imprint on the mother and follow her during a short, sensitive period after hatching. They are neurally wired to learn crucial information —the identity of the individual that will protect them in the months ahead. Usually that individual is the mother or father. Imprinting occurs among many animals.

Konrad Lorenz, one of the early investigators of animal behavior, must have presented the baby geese in this figure with sign stimuli that made them form an attachment to him. When imprinted later in life, birds also direct sexual attention to members of the species they fixated upon when young.

Figure 22.5 Male marsh wren belting out a territorial song. Male birds start to imitate the species song when they are about fifteen days old, and they continue to learn songs throughout their life. The western marsh wren males learn as many as 200! Eastern marsh wrens have a more limited repertoire. The difference appears to have a genetic basis. Western birds raised in the east still mastered a far greater number of songs than native easterners.

another. At maturity, these captives sang the white-crown song only. Apparently, they had an instinctual ability to pay attention to this one song, and ignore others. These birds even picked up the dialect of their unseen same-species tutor. Marler carried out many such experiments; his results support the hypothesis that birdsong starts with *a genetically based capacity to learn* from specific acoustical cues. Many birds seem to have an inherited outline of their species song. They recognize this outline in an overheard song and then use that song as a guide to fill in their song's details.

This is not the whole story. In another experiment, Marler let young, hand-reared white-crowns interact with a "social tutor" of a different species, as opposed to listening to taped songs. The males tended to learn their tutor's song. In this case, birds were influenced by their *social experience* as well as acoustical cues.

Also, within a species there may be genetic variation in the capacity to learn different songs (Figure 22.5).

THE ADAPTIVE VALUE OF BEHAVIOR

Whenever forms of behavior vary and the variation has a genetic basis, they can be subject to evolution by natural selection. **Natural selection**, remember, is an outcome of differences in reproductive success among individuals of a species that show variation in their heritable traits. Over time, we would expect the alleles underlying the most adaptive behaviors to increase in frequency in the population, while alleles that cause less adaptive behaviors would become less frequent.

Natural selection theory can be used to develop and test possible explanations of why a behavior has endured. It helps explain how some actions bestow reproductive benefits that offset the reproductive costs (disadvantages) associated with them. If a behavior is adaptive, it must promote the *individual's* production of offspring. Here are five definitions to keep in mind:

1. *Reproductive success.* The individual produces some number of surviving offspring.

2. *Adaptive behavior.* Behavior that helps an individual promote propagation of its own genes; its frequency is maintained or increases in successive generations.

3. *Social behavior.* Interdependent interactions among individuals of the species.

4. *Selfish behavior.* Any form of behavior that improves an individual's chance to produce or protect its own offspring regardless of the impact on the population as a whole.

5. *Altruism.* Also called self-sacrificing behavior. Some individual behaves in a way that helps others in the population but lowers its chance to produce offspring.

When biologists speak of selfish or altruistic behavior, they do not mean an individual is conscious of what it is doing or of the behavior's reproductive goal. A lion does not need to know that capturing zebras will be good for its reproductive success. Its nervous system just shouts HUNTING BEHAVIOR! when the lion is hungry and sees a zebra. Hunting behavior persists in the lion population because the genes that make this behavior possible are passed along when lions reproduce.

To assess the adaptive value of behavior, biologists consider the ways that it might promote reproductive success. For example, many cavity-nesting birds, such as starlings (*Sturnus vulgaris*), decorate their nest with fresh sprigs of greenery. The birds particularly favor the leaves of strongly scented plants, such as the wild carrot (*Daucus carota*) (Figure 22.6).

Why do the birds do this? Larry Clark and Russell Mason suspected that nest-decorating behavior helps minimize the number of parasitic mites in their nests. Mites suck blood from nestlings and large numbers could affect growth and survival of the offspring.

The biologists decided to test their "nest protection hypothesis" by replacing the natural nests of starlings with experimental ones. Half of the replacement nests included wild carrot leaves and the other half had no aromatic plant matter.

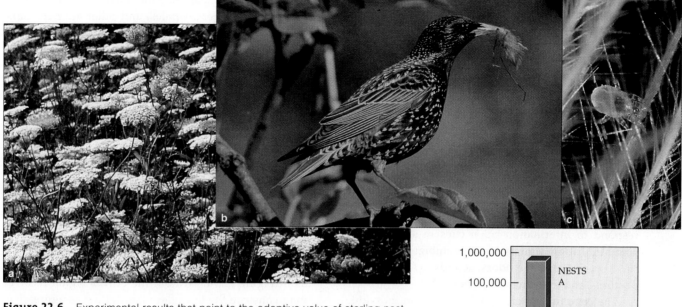

Figure 22.6 Experimental results that point to the adaptive value of starling nest-decorating behavior. (**a**) Wild carrot (*Daucus carota*) and (**b**) a nest-building starling (*Sturnus vulgaris*). Nests designated *A* were kept free of fresh sprigs of wild carrot and other plants that make aromatic compounds. The compounds prevent baby mites (*Ornithonyssus sylviarum*) from developing into (**c**) adult mites. Other nests, designated *B*, had fresh sprigs tucked in every seven days. (**d**) Counts of the number of mites infesting the nests were made at the time the starling chicks left the nest, twenty-one days after this experiment was under way.

After the starling chicks had left these nests, the researchers recorded the number of mites left behind (Figure 22.6*d*). The number of mites in sprig-free nests was far greater than the number in sprig-decorated ones. Why? Wild carrot shoots have highly aromatic compounds that repel herbivores and thus help these plants survive. By coincidence, one of the compounds also prevents mites from maturing sexually.

Mason and Clark concluded that decorating a nest with sprigs deters blood-sucking mites. They inferred that this nest-decorating behavior is adaptive because it promotes nestling survival, and thus increases the reproductive success of nest-decorating birds.

As is often the case in science, additional research suggests that the story may be a bit more complicated. Other researchers have not found the same deterrent effect of greenery on mite numbers. They suggest that nest decorating may be adaptive because the greenery deters other nest parasites, or because it enhances the immune function of the nestlings. Ongoing studies are testing these competing hypotheses.

Do not get the idea that all behavior is adaptive. A behavior may be adaptive under some circumstances but not others. Just as some morphological traits are a legacy of past selective pressures, certain behavioral characteristics persist even after the environment in which they evolved has been altered. Consider moths that exhaust themselves by fluttering against brightly lit window panes at night. Is this adaptive? No, but for millions of years, flying toward a bright, highly visible star allowed moths to navigate a straight line. Today, with the widespread use of electric lights at night, this behavior often turns out to be maladaptive. It draws the night-flying moths to windows, or even worse, to electric bug zappers.

Genes underlie animal behavior—coordinated responses to stimuli. Certain gene products are essential in constructing and operating the nervous system, which governs behavior. Genes that encode hormones or their receptors also affect many forms of behavior.

In instinctive behavior, animals make complex, stereotyped responses to specific, often simple environmental cues. In learned behavior, responses may vary or change as a result of individual experiences in the environment.

Whether instinctive or learned, behavior develops through interactions between genes and environmental inputs.

Like other genetically determined traits, behavioral traits are subject to natural selection. Adaptive behaviors increase an animal's reproductive success under current conditions.

22.2 Communication Signals

Competing for food, defending territory, alerting others to danger, advertising sexual readiness, forming bonds with a mate, caring for the offspring—such interactions among the members of a species require communication between two or more individuals.

THE NATURE OF COMMUNICATION SIGNALS

Intraspecific interactions involve mixes of instinctive and learned behaviors by which individuals send and respond to information-laden cues in stimuli. These **communication signals** have unambiguous meaning for a species. Specific odors, sounds, colors, patterns, postures, and movements are examples.

The communication signals sent by an individual, the signaler, are meant to change behaviors of others of the same species. A signal evolves or persists when it helps the reproductive success of both the sender *and* receiver. If a signal is usually disadvantageous to either party, then natural selection favors individuals that do not send the signal or do not respond to it.

Pheromones are chemicals released by one member of a species that influence another member of the same species. *Signaling* pheromones, such as honeybee alarm pheromone or a sex attractant, induce an immediate response. *Priming* pheromones act over a longer time frame, causing a change in the receipient's physiology. For instance, exposure to a pheromone in the urine of male mice triggers physiological changes that lead to estrus in female mice.

Acoustical signals abound in nature. Male birds, frogs, grasshoppers, whales, and other animals make sounds that attract females. Prairie dogs bark alarms. Wolves howl and kangaroo rats drum their feet on the ground when advertising possession of territory.

Some signals never vary. Zebra ears pressed flat to the head convey hostility; ears pointing up convey its absence. Other signals convey a message's intensity. A zebra that has ears laid back is not too riled up when its mouth is open a bit. When its mouth gapes, watch out. That combination is one type of *composite* signal, which has information encoded in two cues (or more).

Signals often take on different meaning in different contexts. A lion emits a spine-tingling roar to keep in touch with its pride *or* to threaten rivals. Also, a signal can convey information about signals to follow. Dogs and wolves solicit play behavior with their play bow, as in Figure 22.7*a*. Without the bow, a signal receiver may construe the behaviors that follow as aggressive, sexual, or even exploratory—but not playful.

EXAMPLES OF COMMUNICATION DISPLAYS

The play bow is a communication display, a pattern of behavior that is a social signal. The threat display is another common pattern. It unambiguously announces that a signaler is prepared to attack a signal receiver. When a rival for a receptive female confronts him, a dominant male baboon will roll his eyes upward and "yawn," which exposes his canines (Figure 22.7*b*). The signaler often benefits when the rival backs down, for he retains access to the female without having had to fight. The signal receiver benefits because he avoids a serious beating, infected wounds, and possibly death.

Figure 22.7 Communication displays. (**a**) A young male wolf using a play bow to solicit a romp. (**b**) Exposed canines are part of a male baboon's threat display. (**c**) Courtship display of king penguins.

NATURAL SELECTION AND BEHAVIOR

Communication displays can evolve when behavior that occurs in one context is performed in another and becomes *ritualized*. This means that over generations, aspects of the behavior may become exaggerated and stereotyped in ways that make it easier for a receiver to recognize and interpret. Feathers or other body parts may become colored or enlarged in ways that enhance the display's effectiveness. Ritualized behavior is very commonly a part of courtship displays (Figure 22.7c).

In tactile displays, a signaler touches the receiver in ritualized ways. After locating a source of pollen or nectar, a foraging honeybee returns to its colony, a hive, and performs a complex dance. It moves about in a circle, jostling in the dark with a crowd of workers. Other bees may follow and maintain physical contact with the dancer. Honeybees that do so get information about the general location, distance, and direction of pollen or nectar (Figure 22.8).

ILLEGITIMATE SIGNALERS AND RECEIVERS

Sometimes the wrong parties intercept communication signals. Termites will respond with defensive behavior if they catch a whiff of the scent from an invading ant. That scent is meant to identify the ant as a member of an ant colony, and it elicits cooperative behavior from other ants. The scent does have evolved functions, but announcing the ant's presence to a termite colony is not one of them. A termite might detect the scent and kill the ant. In that case, the termite is an *illegitimate receiver* of a signal that is meant for other ants.

Male tungara frogs make two kinds of calls; one simple, the other complex. The calls say "come hither" to female frogs, but they also mean "dinner here" to fringe-lipped bats. Complex songs are more attractive to females, but make it easier for the bats to locate the caller. When bats are in the area, the male frogs call less and are more likely to make the simpler call.

There are also *illegitimate signalers*. Certain assassin bugs hook dead termite prey on their back to acquire the victim's odor. By signaling that they "belong" to a termite colony, they can more easily hunt termite prey. As another example, when the female of a predatory firefly species sees the flash from a male of a different species, she flashes back. If she lures him into attack range, she eats him—an evolutionary cost of having an otherwise useful response to a come-hither signal.

A communication signal between individuals of the same species is an action having a net beneficial effect on both the signaler and receiver. Natural selection tends to favor communication signals that promote reproductive success.

When bee moves straight down comb, recruits fly to source directly away from the sun.

When bee moves to right of vertical, recruits fly at 90° angle to right of the sun.

When bee moves straight up comb, recruits fly straight toward the sun.

Figure 22.8 *Animated!* Dances of the honeybees, as examples of tactile displays. (**a**) Honeybees that visit food sources close to their hive perform a *round* dance on the honeycomb. Worker bees that maintain contact with the forager through the dance go out and search for food near the hive.

(**b**) Bees trained to visit feeding stations over 100 meters from the hive do a *waggle* dance. During the dance, a bee makes a straight run and waggles its abdomen. The waggle dancers also vary the dance speed to convey more information about the distance of a food source. For example, when a site is 150 meters away, the dance is executed much faster, with more waggles per straight run, compared with a dance concerning a food source that is 500 meters away.

(**c**) As Karl von Frisch discovered, a *straight* run's orientation varies, depending on the direction in which food is located. When he put a sugar solution on a direct line between a hive and the sun, foragers that located it returned to the hive and oriented their straight runs right up the honeycomb. When he put a sugar solution at right angles to a line between the hive and the sun, foragers made straight runs 90 degrees to vertical. Thus, a honeybee "recruited" into foraging orients its flight *with respect to the sun and the hive*. Using this code, bees waste less time and energy during their food-gathering expedition.

22.3 Mates, Offspring, and Reproductive Success

Sexual reproduction requires a partner. An individual must attract a member of the opposite sex with which to mate. In all animals except birds, any parental care is most often provided by the female alone.

SEXUAL SELECTION AND MATING BEHAVIOR

Competition among members of one sex for access to mates is common. So is choosiness in selecting a mate. Such activities, recall, are forms of sexual selection. This microevolutionary process favors traits that give the individual a competitive edge in attracting and holding on to mates (Section 12.6).

But *whose* reproductive success is it—the male's or female's? Males, recall, produce large numbers of tiny sperm, and females produce far larger but fewer eggs. For the male, success generally depends on how many eggs he fertilizes. For the female, it depends more on how many of her eggs survive and reach adulthood. In most cases, *the key factor influencing a female's sexual preference is the quality of the mate—not the quantity of partners*. Males often employ tactics that might help them fertilize as many eggs as possible.

Female hangingflies (*Harpobittacus apicalis*) choose males that offer superior food. A male kills a moth or some other insect, then releases a sex pheromone to attract females to his "nuptial gift" (Figure 22.9*a*). A female selects a male with a large calorie-rich offering. Only *after* she has been eating for five minutes or so does she start to accept sperm from her partner, which she stores in a special internal compartment. She lets the male continue inseminating her, but only as long as the gift holds out. Before twenty minutes are up, she can break off the mating at any point. When she does, she may well mate with another male and accept his sperm. Doing so dilutes the reproductive success of her first partner.

Some females shop around for good-looking males. For example, fiddler crabs abound along many muddy shores (Figure 22.9*b,c*). One of the male's two claws is so enlarged that it may make up more than 50 percent of his body weight. When spring tides are favorable, the males build elaborate mating burrows next to each other and stand beside them, waving their oversized claws. Females stroll by, checking out males and their burrows. If a female likes what she sees, she enters a male's burrow and engages in sex.

Many birds show a similar pattern of mating. Male sage grouse (*Centrocercus urophasianus*) congregate in a

Figure 22.9 (a) Male hangingfly dangling a moth as a nuptial gift for a future mate. Females of some hangingfly species choose sexual partners on the basis of the size of prey offered to them. (b) A male fiddler crab waves his oversized claw to attract the interest of a female (c). A male sage grouse (d) competes for female attention at a communal display ground.

NATURAL SELECTION AND BEHAVIOR

type of communal display ground called a lek. Each male stakes out a few square meters. With tail feathers erect and large neck pouches puffed out, the males emit booming calls and stamp about like wind-up toys (Figure 22.9d). Females tend to select and mate with one male only, then they go off to nest by themselves in the sagebrush. Many females often choose the same stand-out male, so most of the males never mate.

The females of some species cluster in defendable groups when they are sexually receptive. Where you come across such a group, you are likely to observe males competing for access to clusters. Competition for ready-made harems favors highly combative male lions, sheep, elk, elephant seals, and bison, to name a few examples (Figure 22.10).

PARENTAL CARE

If females are fighting over males, then males probably provide some service beyond sperm delivery, such as parental care. A midwife toad is a fine example. A male winds strings of fertilized eggs around his legs, where they stay until the eggs hatch (Figure 22.11a). With her eggs in the care of the male, a female is free to mate with other males—if she can find some unencumbered by eggs. As the season progresses, such males become rare, and the females fight for access to them. Females even attempt to pry mating pairs apart.

Parental behavior has reproductive costs. It drains time and energy that parents might otherwise spend on improving their own likelihood of living to reproduce another time. Yet, for many species, parental behavior improves the likelihood that offspring will live. The benefit of devoting time and energy to their offspring (immediate reproductive success) might outweigh the cost of reduced reproductive success at a later time.

Once young amphibians and reptiles have hatched, parental care is uncommon. Like birds, crocodilians are an exception. Parents build a nest where eggs are laid. The young start calling when they are ready to hatch. Parents dig them up and care for them for some time.

Many birds are monogamous and both parents often care for the young (Figure 22.11b). In mammals, males typically leave after mating. Females raise the young alone, and the males attempt to mate again or conserve energy for the next breeding season (Figure 22.11c). Mammalian species in which males do assist in care of the young tend to be monogamous.

Selection theory explains some aspects of mating behavior, for sexual selection favors behavioral traits that give the individual a competitive edge in reproductive success.

Figure 22.10 (**a**) Sexual competition between male elephant seals, which fight for access to a cluster of females. (**b**) Male bison locked in combat during the breeding season.

Figure 22.11 (**a**) Male midwife toad with developing eggs wrapped around his legs. (**b**) Male and female Caspian terns cooperate in the care of their chick. (**c**) A female grizzly will care for her cub for up to two years. The male takes no part in the cub's upbringing.

22.4 Costs and Benefits of Social Groups

Survey the animal kingdom and you'll find a range of social groups. Individuals of some species spend most of their lives alone or in small family groups. Others live in groups of thousands of related individuals. Still others live in social units with few close relatives, as humans do. Evolutionary biologists often study the costs and benefits of sociality in terms of individual reproductive success, measured by contributions to the next generation's gene pool.

Figure 22.12 Group defenses. Musk oxen adults (*Ovibos moshatus*) form a ring of horns, often around the young.

COOPERATIVE PREDATOR AVOIDANCE

By cooperatively responding to predators, some groups reduce the net risk to all. Vulnerable individuals, too, scan for predators, join in a counterattack, or engage in more effective defenses (Figures 22.12 and 22.13).

Vervet monkeys, meerkats, prairie dogs, and many other mammals give alarm calls, as in Figure 22.13*a*. A prairie dog barks different signals for eagle or coyote sightings. Others of its group then dive into burrows to escape an aerial attack or stand erect, the better to scan the horizon and pinpoint a coyote.

Ecologist Birgitta Sillén-Tullberg found evidence of group benefits for Australian sawfly caterpillars that live in clumps on branches. When something disturbs their clump, caterpillars collectively rear up from the branch and writhe about (Figure 22.13*b*). At the same time, they throw up partly digested eucalyptus leaves, which taste awful to most predators—including birds that prey on caterpillars.

As Sillén-Tullberg hypothesized, individual sawfly caterpillars benefit from their coordinated repulsion of predatory birds. She used her hypothesis to predict that birds are more likely to eat a solitary caterpillar.

To test her prediction, she gave young, hand-reared Great Tits (*Parus major*) a chance to eat the caterpillars. She offered caterpillars one at a time or in a group of twenty. When birds were offered caterpillars one at a time, they devoured an average of 5.6. Birds that were

offered caterpillars as a clump only ate an average of 4.1. As predicted, each individual caterpillar was a bit safer when in the midst of a fluid-spewing group.

THE SELFISH HERD

Simply by their physical position in the group, some individuals form a living shield against predation on others. They belong to a **selfish herd**, a simple society formed (not consciously) by reproductive self-interest. The selfish-herd hypothesis has been tested for male bluegill sunfishes, which build adjacent nests on lake bottoms. Females deposit eggs where males have used their fins to scoop out depressions in mud.

If a colony of bluegill males is a selfish herd, then we can predict competition for the "safe" sites—at the center of a colony. Compared to eggs at the periphery, eggs in nests at the center are less likely to be eaten by snails and largemouth bass. Competition does indeed occur. The largest, most powerful males tend to claim centermost locations. Other, smaller males assemble around them and bear the brunt of predatory attacks. Even so, they are better off in the group than on their own, fending off a bass single-handedly, so to speak.

COOPERATIVE HUNTING

Many predatory mammals, including wolves, lions, and wild dogs, live in social groups and cooperate in hunts (Figure 22.14). Are group hunts more successful than solitary hunts? Surprisingly, the answer is often no. Researchers observed a solitary lion that captured prey about 15 percent of the time. Two lions hunting together captured prey twice as often, but they had to share it, so the number of successful hunts per lion balanced out. When more lions joined the hunt, the success rate per lion actually fell. Studies of wolves revealed a similar pattern. Among many cooperative hunters, group living is probably best explained by

Figure 22.13 Sentries and slime. (**a**) Black-tailed prairie dogs bark an alarm call that warns others of predators. Does this put the caller at risk? Not much. Prairie dogs usually act as sentries only if they are done feeding and are standing beside their burrows. (**b**) Australian sawfly caterpillars form clumps and collectively regurgitate a fluid (yellow blobs) that is distasteful to most predators.

GROUP LIVING

Figure 22.14
Members of a wolf pack (*Canis lupus*) cooperate to hunt, care for young, and defend a territory. Benefits are not distributed equally. Only the highest-ranking individuals, the alpha male and alpha female, breed.

factors other than success in hunting. Individuals not only hunt together, they fend off scavengers, care for one another's young, and protect the group's territory.

DOMINANCE HIERARCHIES

In many social groups, resources are not distributed equally. Certain individuals are subordinate to others. Most wolf packs have one dominant male that breeds with a dominant female. The other pack members are nonbreeding brothers and sisters, aunts and uncles. They all hunt and bring food to individuals that stay in the den and guard the young.

Another example: Baboons live in large groups. A female spends her life within the group in which she was born, and she inherits her social standing from her mother. High-ranking females displace lower-ranked females from food, water, and grooming opportunities. As a result, offspring of higher-ranking mothers grow and mature faster than those of lower-ranking ones.

Why do subordinates bother staying in a group if they must give up resources and often do not breed? They might do so to avoid being hurt and to remain in the group. Maybe it is impossible for them to survive on their own. Maybe challenging a strong individual will invite injuries or death. Maybe a subordinate has a chance to reproduce if a predator or old age removes dominant peers. Subordinate wolves and baboons do move up the social ladder if the opportunity arises.

REGARDING THE COSTS

If social behavior can be advantageous, *then why are there relatively few social species?* In most environments, costs outweigh the benefits. For instance, when large numbers of individuals live together, they compete more. Cormorants and many other seabirds form huge breeding colonies (Figure 22.15). All must compete for a share of the same ecological pie.

Large groups are attractive to predators. Crowded social groups also invite parasites and diseases that are readily transmitted from host to host. Still another cost is the risk of being killed or exploited by others in the group. Given the opportunity, breeding pairs of herring gulls cannibalize a neighbor's eggs and any chicks that wander away from their nest.

Figure 22.15 The nearly uniform spacing among a densely packed cormorant colony. Imagine how easily pathogens and parasites can spread through such breeding colonies.

Living in a social group can provide benefits, as through cooperative defenses of shielding against predators.

Group living has costs, in terms of increased competition, increased vulnerability to infections, and exploitation by others of the group.

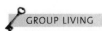

22.5 Why Sacrifice Yourself?

Extreme cases of sterility and self-sacrifice have evolved in only two insect groups and one group of mammals. How are genes of the nonreproducing individuals passed from one generation to the next?

SOCIAL INSECTS

Honeybees and fire ants are among the true social (eusocial) insects. Like termites, they stay together for generations in a group that has a reproductive division of labor. Permanently sterile individuals cooperatively care for the offspring of a few breeding individuals. Often these sterile workers are highly specialized in their body form and function (Figure 22.16).

Consider a honeybee colony, or hive. Its queen bee, the only fertile female, secretes a pheromone that other female bees distribute through the hive. This signaling molecule suppresses the development of ovaries in all other females, rendering them sterile. The queen bee is larger than the worker bees, partly because of her enlarged egg-producing ovaries (Figure 22.17a).

About 30,000 to 50,000 female workers feed larvae, clean and maintain the hive, and build honeycomb from waxy secretions. Adult workers live for about six weeks during the spring and summer. When foragers return to the hive after finding a rich source of nectar or pollen, they engage others in a dance. This tactile display recruits more foragers (Figure 22.8). Workers also cooperate by transferring food to one another. They guard the entrance to the hive and will sacrifice themselves to repel intruders.

Males, the stingless drones, develop only in spring and summer. They do not take part in the day-to-day work and subsist on the food gathered by their worker sisters. Drones live for sex. Each day, they fly out in search of a mate. If a drone is lucky, he will locate a virgin queen on her single flight away from her home colony. The sole function of her flight is to meet up with and mate with drones. A drone dies right after he inseminates a queen, which then starts a new colony. The queen stores and uses the drone's sperm for years, perpetuating his genes and those of the colony.

Like honeybees, termites live in huge family groups with a large queen specialized for producing eggs (Figure 22.17b). She mates repeatedly with a resident king. Their offspring are flightless sterile workers of both sexes and winged sexual forms that are produced and leave the colony seasonally.

SOCIAL MOLE-RATS

Vertebrates are not known for sterility and extreme self-sacrifice. The only eusocial mammals are African mole-rats. The best studied is *Heterocephalus glaber*, the naked mole-rat. Clans of this nearly hairless rodent occupy burrows in arid parts of East Africa.

One reproducing female dominates the clan and she mates with one to three males (Figure 22.17c). Other, nonbreeding members live to protect and care for the queen, her consorts, and their offspring. Sterile diggers excavate the underground tunnels and chambers that serve as living areas or waste-disposal sites. When a digger locates a tasty root or tuber, it hauls a bit of it back to the main chamber and emits a series of chirps. The chirps recruit others, which help carry the food back to the chamber. In this way, the queen, her retinue of males, and her offspring get fed.

Digger mole-rats also deliver food to other helpers that usually loaf about, shoulder to shoulder and belly to back, with the reproductive royals. These "loafers" actually spring to action when a snake or some other predator threatens the clan. Collectively, and at great risk, they chase away or attack and kill the predator.

  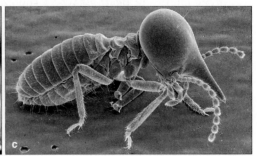

Figure 22.16 Serving and defending the colony with specializations in form and function. **(a)** An Australian honeypot ant worker. This sterile female is a living container for the colony's food reserves. **(b)** An army ant soldier (*Eciton burchelli*) displays its relatively enormous mandibles. **(c)** Eyeless soldier termites (*Nasutitermes*) bombard intruders with a stream of sticky goo from their nozzle-shaped head.

Figure 22.17 Three queens. (**a**) A queen honeybee with her court of sterile worker daughters. (**b**) A termite queen (*Macrotermes*) dwarfs her offspring and her mate. Her body pumps out thousands of eggs a day. (**c**) This queen naked mole-rat has twelve mammary glands, the better to feed her many offspring. In a laboratory colony at Cornell University, one female produced one litter of twenty-eight pups. She gave birth to more than 900 offspring during her lifetime.

Figure 22.18 Damaraland mole-rats in a burrow. Like their relatives, the naked mole-rats, they live in colonies having nonbreeding workers. But unlike the naked mole-rats, these fuzzy burrowers are not highly inbred.

INDIRECT SELECTION THEORY

None of the altruistic individuals of a honeybee hive, termite colony, or naked mole-rat clan directly passes genes to the next generation. So how are genes that underlie altruistic behavior perpetuated? According to William Hamilton's theory of **inclusive fitness**, genes associated with altruism can be favored by selection if they lead to behavior that will increase the number of offspring produced by an altruist's closest relatives.

A sexually reproducing, diploid parent caring for offspring is not helping exact genetic copies of itself. Each of its gametes, and each of its offspring, inherit one-half of its genes. Other individuals of the social group that have the same ancestors also share genes with their parents. Two siblings (brothers or sisters) are as genetically similar as a parent and its offspring. Nephews and nieces have inherited about one-fourth of their uncle's genes.

Sterile workers may be indirectly promoting genes for "self-sacrifice" through altruistic behavior that will benefit relatives. Members of a honeybee, termite, or ant colony are an extended family. The nonbreeding family members are working in support of siblings, a few of which are future kings and queens. Although a guard bee will die after driving her stinger into a bear, her death helps to ensure that relatives in the hive will live to pass along some of her genes.

Does close relatedness explain why the naked mole-rats are the only eusocial mammals? Through DNA fingerprinting studies, we know that all individuals in one naked mole-rat clan are *very* close relatives and differ genetically from individuals of other clans. Each clan is highly inbred through generations of brother–sister, mother–son, and father–daughter matings.

However, inbreeding may not be required for mole-rat eusociality. The organization of the Damaraland mole-rat (*Cryptomys damarensis*) is similar to that of *H. glaber* (Figure 22.18). Nonbreeding members of both sexes cooperatively assist one breeding pair. However, breeding pairs of wild Damaraland mole-rat colonies usually are unrelated.

Researchers are now searching for other factors that select for mole-rat eusocial behavior. According to one hypothesis, an arid habitat and patchy food supplies favor the mole-rat genes that underlie cooperation in digging burrows, searching for food, and fending off competitors and predators.

Altruistic behavior may persist when individuals pass on genes indirectly, by helping relatives survive and reproduce.

By the theory of inclusive fitness, genes associated with altruistic behavior that is directed toward relatives may spread through a population in certain situations.

LINK TO
SECTION
16.13

22.6 A Look at Primate Social Behavior

Primates, especially chimpanzees and bonobos, are our closest living relatives. All live in groups, and the social environment plays an important role in determining an individual's reproductive success.

In the 1960s, a 26-year-old primatologist named Jane Goodall began what would turn out to be a life-long study of chimpanzees in Tanzania. One of her earliest discoveries was the chimpanzee's ability to make and use simple tools by stripping leaves from branches to create "fishing sticks." These long flexible sticks are inserted into a large termite mound as shown in Figure 22.19*a*. After agitated termites climb aboard the stick, it is carefully withdrawn and the chimpanzee enjoys a high-protein snack. Chimpanzees use thicker sticks to poke holes in a mound so they can insert the fishing sticks. Different chimpanzee groups exhibit somewhat different tool-shaping and termite-fishing methods. In each group, youngsters learn by imitating the adults.

Male chimpanzees spend their lives in the group in which they are born and form strong social bonds. The females are often unrelated and interact little with one another. A female's status is dictated mostly by how she gets along with the males. Before the rainy season, mature females that are entering a fertile cycle undergo hormone-driven physiological and behavioral changes. External genitalia become swollen and vivid pink as the females become sexually active. The swellings are strong visual signals to males. They are flags of sexual jamborees—of great gatherings of highly stimulated chimpanzees in which any males present may have a turn at copulating with the same female.

Male chimpanzees cooperatively hunt for monkeys, small pigs, and antelopes. They may also cooperate in attacks on neighboring groups. Males sometimes even kill infants. By one hypothesis, infanticidal behavior by males may exert selection pressure for promiscuity by females. A female who mates with many males may protect her offspring by obscuring their exact paternity. A male would be expected to avoid killing any infant that may carry his genes.

Comparative studies of the closely related bonobo reveal contrasting sexual and social behavior (Figure 22.19*b*). As in chimpanzees, adult males in a group are related and females are not. Yet bonobo females form strong bonds. Unlike chimpanzees, female bonobos are receptive to sex at any time, not only when fertile. Male bonobos interact less than male chimpanzees do. They do not hunt together, and infanticidal behavior has not been observed among bonobo males.

What explains the differences? Does a higher level of interaction help female bonobos fend off potentially infanticidal males? Does unrestricted access to sexually receptive females interfere with male–male bonding or diffuse male aggression? We do not know. Hormones that affect pair-bonding may play a role. Like prairie voles and mountain voles, chimpanzees and bonobos differ in a regulatory region near a gene that encodes one ADH receptor. In voles, a longer sequence in this region correlates with more family-oriented behavior. Interestingly, the bonobo sequence for this region is about 360 bases longer than the chimpanzee sequence. What about humans? Our sequence in this region is nearly identical to that of the bonobos. And with this in mind we turn briefly to human behavior.

Chimpanzees and bonobos are our closest living relatives. Both live in social groups but the exact details of social organization, degree of female cooperation, and extent of male aggression vary between the two species.

Figure 22.19 (**a**) A group of chimpanzees (*Pan troglodytes*) using sticks as tools to extract protein-rich termites from a nest. (**b**) A bonobo (*Pan paniscus*) mother with her offspring. Like humans, bonobos are bipedal; they often walk about on two legs. Also like humans, and unlike chimpanzees, bonobo females have sexual organs that allow them to copulate face to face with their partners and are sexually receptive at any time of year. They use sex as a means of strengthening social bonds.

22.7 Human Social Behavior

Evolutionary forces have shaped the behavior of all animals, including humans. Only humans, however, make moral choices about their behavior.

HUMAN PHEROMONES

Do humans secrete and sense pheromones? Possibly. When women live together, as in a college dormitory, their menstrual cycles typically become synchronized. Studies by Martha McClintock and Kathleen Stern showed that the length of a woman's menstrual cycle could be lengthened or shortened by exposing her to sweat secreted by a woman who was in a different part of the cycle.

Men and women secrete different chemicals and also respond differently to them. One study using PET scans found that a chemical typically released in male sweat activated certain brain areas in women but not in most men. In contrast, a chemical released in female urine activated most male brains more than it did female ones. Intriguingly, male homosexuals showed the same response to male sweat that the women did.

In most mammals, pheromones bind to receptors in a vomeronasal organ, or VNO. Neurons connect it to parts of the brain that control behavior. In humans, the VNO is a tiny ductlike structure on the septum that divides the nose into two nostrils. Many scientists contend that the human VNO is a vestigial structure (no longer functional). Others suspect that the human VNO does still connect with the brain by way of some pathway that has not yet been discovered.

HORMONES AND BONDING BEHAVIOR

The hormones oxytocin and ADH, which affect pair-bonding in other mammals, may also influence human social attachment. Consider *autism*. A person who has this behavioral disorder does not form normal social relationships. Compared to unaffected individuals, an autistic child has significantly less oxytocin circulating in the blood. Also associated with autism is a shorter than normal regulatory sequence near the gene for one ADH receptor. This is the same sequence that is shorter in chimpanzees than in bonobos and humans.

Today, researchers are studying roles of hormones in mother–infant bonding and romantic attachments. Nursing stimulates the secretion of oxytocin. So does orgasm, or even a friendly massage. Whether or not these temporary elevations in oxytocin or any other hormonal responses contribute to what we perceive as love is an interesting and still open question.

EVOLUTIONARY QUESTIONS

If we are comfortable with analyzing the evolutionary basis of the behavior of termites, naked mole-rats, and other animals, would it not be rewarding to analyze such a basis of human behavior, too? Many resist the idea. Apparently, they believe that attempts to identify the adaptive value of a human trait are attempts to define its moral or social advantage. However, there is a clear difference between trying to explain something in terms of its evolutionary history and attempting to justify it. "Adaptive" does not mean "morally right." To biologists, adaptive only means valuable in terms of transmission of the individual's genes.

One example: Infanticide is morally repugnant. Is it unnatural? No. Many animals do it and it occurs in all human cultures. Male lions often kill offspring of other males when they take over a pride. Male chimpanzees are also known to kill infants. Doing so frees up the females to breed with them, increasing the infanticidal male's reproductive success.

A biologist may predict that unrelated human males are a threat, too. In fact, the presence of an unrelated male in the home increases the risk of injury to an American child under age two by seventy times.

What about parents who kill their own offspring? In her book on maternal behavior, primatologist Sarah Blaffer Hrdy cites a study of a village in New Guinea in which about 40 percent of newborns were killed by their parents. She argues that if resources are scarce or there is little support socially, a mother might increase her fitness by killing a newborn. The mother can then allocate child-rearing energy to her other offspring or save it for children she may have in the future.

Understanding infanticide's evolutionary roots may help us prevent it. Using our knowledge of conditions that make infanticide an adaptive choice, we can strive as a society to prevent those conditions from arising.

Summary

Section 22.1 Animal behavior starts with genes that specify products required for development of the nervous, endocrine, and skeletal-muscular systems. Hormones are among the products that affect behavior.

Instinctive behavior is performed without having been learned by experience in the environment. It is a prewired response to one or two simple, well-defined environmental cues.

An animal learns when it processes and integrates information from experiences, then uses it to vary or change how it responds to stimuli. Imprinting is a type of learning that can occur only during a sensitive period.

Behavior with a genetic basis is subject to evolution by natural selection. It evolved as a result of individual differences in reproductive success in past generations. A behavior is adaptive if its reproductive benefits are greater than the reproductive costs. A behavior that has evolved under one set of circumstances may become maladaptive if circumstances change.

Section 22.2 Signalers send communication signals meant to change the behavior of receivers of the same species. Pheromones are signaling molecules that have roles in social communication.

Visual signals are key components of courtship displays and threat displays. Acoustical signals are sounds that have precise, species-specific information. Tactile signals are specific forms of physical contact between a signaler and a receiver.

Predators and parasites sometimes intercept or mimic signals of their prey species.

Biology ⊜Now
Learn about the honeybee dance language with the animated interaction on BiologyNow.

Section 22.3 Sexual selection favors traits that give an individual a competitive edge in attracting and keeping mates. Female choice selects for males that have traits or engage in behaviors that attract females. When large numbers of females cluster in defensible areas, males may compete with one another to control these areas.

Parental care has reproductive costs in terms of future reproduction and survival. It is adaptive if the benefits to the current set of offspring offset these costs.

Section 22.4 Animals that live in social groups may benefit by cooperating in predator detection, defense, and rearing the young. The benefits of group living are often distributed unequally. Species that live in groups incur costs, including increased disease and parasitism, and increased competition for resources.

Section 22.5 Ants, termites, and some other insects as well as two species of mole-rats are eusocial. They live in colonies with overlapping generations and have a reproductive division of labor. Most colony members do not reproduce; they instead assist relatives.

According to the theory of indirect selection, such extreme altruism usually evolves because the altruistic individuals share some portion of their genes in common with their reproducing relatives. By aiding relatives, the altruists enhance the passage of shared genes—including the gene for altruism—on to the next generation.

Section 22.6 Our closest relatives, the chimpanzees and bonobos, both live in social groups, but they differ in their sexual behavior and social organization.

Section 22.7 Behavioral biologists ask questions about the mechanisms and the possible adaptive significance of human behavior. A behavior that is adaptive may still be judged by society to be morally wrong.

Self-Quiz

Answers in Appendix I

1. Genes affect the behavior of individuals by _____ .
   a. influencing the development of nervous systems
   b. affecting the kinds of hormones in individuals
   c. controlling what can be learned and when
   d. all of the above

2. A behavior is considered adaptive if it _____ .
   a. varies among individuals of a population
   b. has a genetic basis
   c. increases an individual's reproductive success
   d. is widespread across a species

3. The feeding preference of newly born coastal garter snakes for slugs is an example of _____ .
   a. indirect selection
   b. an instinctive response
   c. imprinting
   d. a composite display signal

4. Male sage grouse gather at a lek to _____ .
   a. display to and attract potential mates
   b. get protection from aerial predators
   c. share food resources with other males

5. In most mammals, the vomeronasal organ functions in _____ .
   a. hormone production   c. visual communication
   b. pheromone detection d. all of the above

6. Eusocial insects _____ .
   a. live in extended family groups
   b. are found among almost all insect orders
   c. show a reproductive division of labor
   d. a and c
   e. all of the above

7. In _____ , all sterile workers are females.
   a. honeybees   c. termites
   b. naked mole-rats   d. all of the above

8. Helping other individuals at a reproductive cost to oneself might be adaptive if those helped are _____ .
   a. members of another species
   b. competitors for mates
   c. close relatives
   d. illegitimate signalers

Figure 22.20 Behaviorally confused rooster.

Figure 22.21
A Nazca booby
tends its single
surviving chick.

9. Social behavior evolves because _____ .
   a. social animals are more advanced than solitary ones
   b. under some conditions, the costs of social life to individuals are offset by benefits to the species
   c. under some conditions, the benefits of social life to an individual offset its costs to the individual
   d. under some conditions, social life has no costs to an individual.

10. Match the terms with their most suitable description.

   _____ fixed action pattern
   _____ altruism
   _____ illegitimate receiver
   _____ imprinting
   _____ selfish herd

   a. time- and cue-dependent form of learning
   b. interceptor of signals
   c. stereotyped motor program that runs to completion independently of feedback from environment
   d. assisting another individual at one's own expense
   e. group in which individuals use others as a shield

Additional questions are available on Biology ᏜNow™

Critical Thinking

1. Sexual imprinting is common in birds. During a short sensitive period in early life, the bird learns the features it will later look for when it is ready to mate. The rooster in Figure 22.20 was observed wading into the water after ducks with amorous intent. Speculate on what might have caused this behavior.

2. Nazca boobies (*Sula granti*) lay two eggs, several days apart. No matter how much food is available, only one chick survives to adulthood (Figure 22.21). The first chick to hatch pushes its younger sibling from the nest, and that sibling dies of starvation and neglect. Formulate a hypothesis on how it might be adaptive for parents to lay two eggs if one sibling generally kills the other. Design an experiment to test your hypothesis.

3. In 2002, Svante Paabo suggested how one gene may have played a pivotal role in the evolution of *language*. Humans who have a single base-pair substitution in this gene, called *FOXP2*, are unable to speak in an intelligible way, to understand complex sentences, or even to make certain movements of the mouth and face.

All mammals have the *FOXP2* gene, which has a 715 base-pair sequence. The gene has changed very little over evolutionary time. The chimpanzee gene and the mouse gene differ by only one base pair. However, two additional base-pair changes appeared and became fixed in the much briefer interval since chimpanzees and humans diverged. The rapid spread of these two substitutions suggest that these mutations may have been highly adaptive and subject to directional selection. What effect do you think mutations that favored an ability to produce and comprehend more complex auditory signals had on social behavior in the lineage leading to humans?

4. A cheetah scent-marks plants in its territory with certain exocrine gland secretions. What evidence would you require to demonstrate that the cheetah's action is an evolved communication signal?

5. Among primates, differences in sexual behavior tend to be related to the size of a male's gonads. Gorillas have relatively tiny testicles; in a male weighing 450 pounds, they may weigh only about an ounce. Gorillas live in groups consisting of a male, a few females, and offspring. This is the most typical type of social group among the primates. When a female is ready to mate, there usually is only a single adult male around to inseminate her.

In contrast, a female chimpanzee advertises her fertile period and mates with many males (Section 22.6). A 100-pound chimpanzee male has testicles about four times as weighty as a gorilla's. By producing large amounts of sperm, this male increases the chance that his sperm, rather than a rival's, will fertilize a female's egg.

Although adult men are larger than chimpanzees, their testicles are only about half the weight. What might this suggest about female promiscuity and male competition to fertilize eggs in the lineage that led to humans?

6. In moths and many other insects, potential mates locate one another with the help of species-specific pheromones. These pheromones are usually mixes of chemicals derived from fatty acids. Sometimes a gene mutation causes a change in the mix of chemicals in a moth pheromone. How might such a mutation alter mating behavior and encourage speciation?

Epilogue

BIOLOGICAL PRINCIPLES AND THE HUMAN IMPERATIVE

Molecules, single cells, tissues, organs, organ systems, multicelled organisms, populations, communities, then ecosystems and the biosphere. These are architectural systems of life, assembled in increasingly complex ways over the past 3.8 billion years. We are latecomers to this immense biological building program. Yet within the relatively short span of 10,000 years, our activities have been changing the character of the land, ocean, and atmosphere, even the genetic character of species.

It would be presumptuous to think we alone have had profound effects on the world of life. As long ago as the Proterozoic, photosynthetic organisms were irrevocably changing the course of biological evolution by enriching the atmosphere with oxygen. During the past as well as the present, competitive adaptations led to the rise of some groups, whose dominance assured the decline of others. Change is nothing new. What *is* new is the capacity of one species to comprehend what might be going on.

We now have the population size, technology, and cultural inclination to use up energy and modify the environment at rapid rates. Where will this end? Will feedback controls operate as they do, for example, when population growth exceeds carrying capacity? In other words, will negative feedback controls come into play and keep things from getting too far out of hand?

Feedback control will not be enough, for it operates after deviation. Our patterns of resource consumption and population growth are founded on an illusion of unlimited resources and a forgiving environment. A prolonged, global shortage of food or the passing of a critical threshold for the global climate can come too fast to be corrected; in which case the impact of the deviation may be too great to be reversed.

What about feedforward mechanisms that might serve as early warning systems? For example, when sensory receptors near the surface of skin detect a drop in outside air temperature, each sends messages to the nervous system. That system responds by triggering mechanisms that raise the body's core temperature before the body itself becomes dangerously chilled.

Extrapolating from this, if we develop feedforward control mechanisms, would it not be possible to start corrective measures before we do too much harm?

Feedforward controls alone will not work, for they operate after change is under way. Think of the DEW line—the Distant Early Warning system. It is like a vast sensory receptor for detecting missiles launched against North America. By the time it does what it is supposed to, it may be too late to stop widespread destruction.

It would be naive to assume we can ever reverse who we are at this point in evolutionary time, to de-evolve ourselves culturally and biologically into becoming less complex in the hope of averting disaster. Yet there is reason to believe we can avert disaster by using a third kind of control mechanism—a capacity to anticipate events even before they happen. We are not locked into responding only after irreversible change has begun. We have the capacity to anticipate the future—it is the essence of our visions of utopia and hell. *We all have the capacity to adapt to a future that we can partly shape.*

For instance, we can stop trying to "beat nature" and learn to work with it. Individually and collectively, we can work to develop long-term policies that take into account biotic and abiotic limits on population growth. Far from being a surrender, this would be one of the most intelligent behaviors of which we are capable.

Having a capacity to adapt and using it are not the same thing. We have already put the world of life on dangerous ground because we have not yet mobilized ourselves as a species to work toward self-control.

Our survival depends on predicting possible futures. It depends on preserving, restoring, and constructing ecosystems that fit with our definition of basic human values and available biological models. Human values can change; our expectations can and must be adapted to biological reality. *For the principles of energy flow and resource utilization, which govern the survival of all systems of life, do not change.* It is our biological and cultural imperative that we come to terms with these principles, and ask ourselves what our long-term contribution will be to the world of life.

Appendix I Answers to Self-Quizzes *Italicized numbers refer to relevant section numbers*

Chapter 1

1. cell — *1.1*
2. d — *1.2*
3. metabolism — *1.2*
4. homeostasis — *1.2*
5. adaptive — *1.4*
6. mutations — *1.4*
7. domains — *1.3*
8. a — *1.2*
9. d — *1.2, 1.4*
10. c — *1.6*
11. c — *1.4*
    e — *1.4*
    d — *1.5*
    b — *1.5*
    a — *1.5*

Chapter 2

1. False — *2.1*
2. c — *2.2*
3. e — *2.3*
4. d — *2.4*
5. See Table 2.1
6. e — *2.7*
7. d — *2.5*
8. e — *2.6*
9. b — *2.7*
10. d — *2.8, 2.10*
11. d — *2.9*
12. c — *2.4*
    a — *2.8*
    g — *2.1*
    d — *2.3*
    f — *2.2*
    b — *2.1*
    e — *2.10*

Chapter 3

1. c — *3.3*
2. d — *3.5*
3. d — *3.8*
4. False; many have a cell wall outside the plasma membrane, and some prokaryotes have a capsule outside of that. — *3.4*
5. c — *3.1*
6. See Figures 3.12 and 3.13
7. e — *3.5*
   d — *3.5*
   a — *3.1*
   b — *3.5*
   c — *3.5*

Chapter 4

1. c — *4.1*
2. d — *4.1*
3. d — *4.2, 4.3*
4. d — *4.3*
5. a — *4.6*
6. d — *4.4*
7. b — *4.5*
8. d — *4.7*
9. d — *4.4*
10. c — *4.3*
    g — *4.5*
    a — *4.2*
    d — *4.2*
    e — *4.2*
    b — *4.3*
    f — *4.1*

Chapter 5

1. carbon dioxide; sunlight — *intro essay*
2. a — *5.1, 5.2*
3. b — *5.2*
4. c — *5.2*
5. c — *5.1, 5.3*
6. b — *5.3*
7. d — *5.2*
   b — *5.3*
   a — *5.3*
   c — *5.2*

Chapter 6

1. d — *6.2*
2. c — *6.2*
3. b — *6.3*
4. c — *6.3*
5. d — *6.4*
6. c — *6.4*
7. b — *6.4*
8. d — *6.5*
9. b — *6.2*
   c — *6.4*
   a — *6.3*
   d — *6.3*

Chapter 7

1. d — *7.1*
2. b — *7.1*
3. a — *7.3*
4. a — *7.3*
5. b — *7.2*
6. d, b, c, a — *7.3*
7. d — *7.5*
8. d — *7.5*
9. b — *7.5*
10. d — *7.5*
11. c — *7.5*
12. d — *7.5*
13. d — *7.3*
    a — *7.5*
    c — *7.3*
    b — *7.5*

Chapter 8

1. a — *8.1*
2. b — *8.1*
3. d — *8.1*
4. c — *8.5*
5. e — *8.8*
6. a — *8.5*
7. c — *8.4*
8. d — *8.9*
9. c — *8.9*
10. b — *8.1*
    d — *8.1*
    a — *8.1*
    c — *8.1*
11. c — *8.5*
    e — *8.8*
    d — *8.9*
    b — *8.8*
    a — *8.4*
    f — *8.5*

Chapter 9

1. c — *9.2*
2. d — *9.2*
3. c — *9.2*
4. a — *9.3*
5. d — *9.3*
6. c — *9.1*
   e — *9.4*
   a — *9.2*
   d — *9.3*
   b — *9.3*

Chapter 10

1. c — *10.2*
2. b — *10.2*
3. c — *10.1, 10.2*
4. c — *10.3*
5. a — *10.3*
6. d — *10.6*
7. d — *10.6*
8. a — *10.6*
9. b — *10.6*
10. b — *10.6*
11. c — *10.6*
12. e — *10.5*
    c — *10.4*
    a — *10.2*
    f — *10.3*
    d — *10.3*
    g — *10.2*
    b — *10.3*

Chapter 11

1. d — *11.5*
2. c — *11.1*
3. plasmid — *11.1*
4. b — *11.1*
5. b — *11.3*
6. d — *11.3*
7. b — *11.5, 11.7*
8. d — *11.4*
   c — *11.6*
   b — *11.2*
   e — *11.2*
   a — *11.7*

Chapter 12

1. populations — *12.3*
2. b — *intro essay*
3. a — *12.3*
4. b — *12.1*
5. b — *12.5*
6. c — *12.5*
7. b — *12.8*
8. c — *12.8*
   d — *12.1*
   a — *12.3*
   b — *12.7*

Chapter 13

1. a — *13.2*
2. Gondwana — *13.3*
3. a. — *13.4*
4. d — *13.7*
5. a — *13.8*
6. c — *13.10*
7. c — *13.10*
8. d — *13.10*
9. e — *13.10*
   f — *13.1*
   g — *13.1*
   c — *13.4*
   b — *13.10*
   d — *13.4*
   a — *13.9*

Chapter 14

1. a — *14.4*
2. d — *14.2*
3. b — *14.1, 14.2*
4. b — *14.3*
5. d — *14.4*
6. c — *14.3*
7. c — *14.4*
8. d — *14.5*
9. g — *14.3*
   d — *14.5*
   c — *14.2*
   e — *14.2*
   i — *14.4*
   f — *14.4*
   a — *14.3*
   j — *14.4*
   b — *14.3*
   h — *14.1*

Chapter 15

1. a — *15.1*
2. c — *15.4*
3. b — *15.2*
4. c — *15.3*
5. a — *15.3*
6. b — *15.5*
7. b — *15.4*
8. c — *15.6*
9. c — *15.5*
   d — *15.1*
   f — *15.3*
   e — *15.2*
   a — *15.1*
   b — *15.1*
   g — *15.6*

Chapter 16

1. c — *16.9*
2. d — *16.7*
3. f — *16.12*
4. f — *16.13*
5. a — *16.1*
6. c — *16.9*
7. a — *16.4*
8. d — *16.1*
9. b — *16.2*
   g — *16.2*
   l — *16.3*
   j — *16.6*
   e — *16.4*
   c — *16.7*
   d — *16.5*
   f — *16.8*
   k — *16.11*
   a — *16.10*
   h — *16.12*
   i — *16.12*

Chapter 17

1 a — *17.1*
2. c — *17.1*

3. d	17.2	f	18.3	3. d	20.2	13. d	21.3
4. e	17.2	b	18.4	4. d	20.8	g	21.3
5. c	17.3			5. a	20.4	e	21.3
6. d	17.5	**Chapter 19**		6. b	20.4	c	21.3
7. b	17.3	1. d	19.1	7. b	20.5	b	21.3
8. b	17.4	2. e	19.3	8. e	20.7	f	21.3
c	17.3	3. a	19.1	9. d	20.7	a	21.3
d	17.3	4. c	19.2	10. d	20.1		
a	17.5	5. a	19.3	a	20.1	**Chapter 22**	
e	17.3	6. b	19.4	c	20.1	1. d	22.1
g	17.3	7. b	19.6	b	20.1	2. c	22.1
f	17.3	8. e	19.8			3. b	22.1
		9. b	19.9	**Chapter 21**		4. a	22.3
Chapter 18		10. c	19.7	1. d	21.1	5. b	22.7
1. f	18.2	d	19.8	2. c	21.1	6. d	22.5
2. a	18.2	a	19.6	3. d	21.2	7. a	22.5
3. a	18.2	e	19.1	4. a	21.2	8. c	22.5
4. d	18.3	b	19.7	5. a	21.2	9. c	22.4
5. a	18.3	g	19.9	6. b	21.3	10. c	22.1
6. d	18.4	f	19.9	7. c	21.1	d	22.1, 22.5
7. b	18.5	h	19.1	8. e	21.3	b	22.2
8. d	18.3			9. d	21.3	a	22.1
c	18.3	**Chapter 20**		10. c	21.3	e	22.4
e	18.2	1. d	20.1	11. b	21.3		
a	18.2	2. d	20.1	12. d	21.4		

Appendix II   Answers to Genetics Problems

1. **a.** Both parents are heterozygotes (*Cc*). Their children may be albino (*cc*) or unaffected (*CC* or *Cc*).
   b. All are homozygous recessive (*cc*).
   c. Homozygous recessive (*cc*) father, and heterozygous (*Cc*) mother. The albino child is *cc*, the unaffected children *Cc*.

2. **a.** *AB*
   b. *AB, aB*
   c. *Ab, ab*
   d. *AB, Ab, aB, ab*

3. **a.** All offspring will be *AaBB*.
   b. 1/4 *AABB* (25% each genotype)
   1/4 *AABb*
   1/4 *AaBB*
   1/4 *AaBb*
   c. 1/4 *AaBb* (25% each genotype)
   1/4 *Aabb*
   1/4 *aaBb*
   1/4 *aabb*
   d. 1/16 *AABB* (6.25% of genotype)
   1/8 *AaBB* (12.5%)
   1/16 *aaBB* (6.25%)
   1/8 *AABb* (12.5%)
   1/4 *AaBb* (25%)
   1/8 *aaBb* (12.5%)
   1/16 *AAbb* (6.25%)
   1/8 *Aabb* (12.5%)
   1/16 *aabb* (6.25%)

4. A mating of two M^L cats yields 1/4 *MM*, 1/2 M^LM, and 1/4 M^LM^L. Because M^LM^L is lethal, the probability that any one kitten among the survivors will be heterozygous is 2/3.

5. **a.** Human males (XY) inherit their X chromosome from their mother.
   b. A male can produce two kinds of gametes. Half carry an X chromosome and half carry a Y chromosome. All the gametes that carry the X chromosome carry the same X-linked allele.
   c. A female homozygous for an X-linked allele produces only one kind of gamete.
   d. Fifty percent of the gametes of a female who is heterozygous for an X-linked allele carry one of the two alleles at that locus; the other fifty percent carry its partner allele for that locus.

6. Because Marfan syndrome is a case of autosomal dominant inheritance and because one parent bears the allele, the probability that any child of theirs will inherit the mutant allele is 50 percent.

7. Fred could use a testcross to find out if his pet's genotype is *WW* or *Ww*. He can let his black guinea pig mate with a white guinea pig having the genotype *ww*. If any F$_1$ offspring are white, then the genotype of his pet is *Ww*. If the two guinea pig parents are allowed to mate repeatedly and all the offspring of the matings are black, then there is a high probability that his pet guinea pig is *WW*. (For instance, if ten offspring are all black, then the probability that the male is *WW* is about 99.9 percent. The greater the number of offspring, the more confident Fred can be of his conclusion.)

8. **a.** Nondisjunction might occur during anaphase I or anaphase II of meiosis.
   b. One parent had a translocation in which part of chromosome 21 was exchanged with part of chromosome 14. A child receives the translocated chromosome 14 from that parent, in addition to a normal chromosome 21 from each parent. Although the somatic cells of the new individual have 46 chromosomes, the extra chromosome 21 material carried by the translocated chromosome 14 causes Down syndrome.

9. In the mother, a crossover at meiosis generates one X chromosome that carries neither mutant allele, and one that has both. A son that inherits the X chromosome with two normal alleles will be unaffected.

Appendix III Annotations to A Journal Article

This journal article reports on the movements of a female wolf during the summer of 2002 in northwestern Canada. It also reports on a scientific process of inquiry, observation and interpretation to learn where, how and why the wolf traveled as she did. In some ways, this article reflects the story of "how to do science" told in section 1.5 of this textbook. These notes are intended to help you read and understand how scientists work and how they report on their work.

(1) A R C T I C

(2) VOL. 57, NO. 2 (JUNE 2004) P. 196–203

(3) Long Foraging Movement of a Denning Tundra Wolf

(4) Paul F. Frame,[1,2] David S. Hik,[1] H. Dean Cluff,[3] and Paul C. Paquet[4]

(5) (Received 3 September 2003; accepted in revised form 16 January 2004)

(6) ABSTRACT Wolves (*Canis lupus*) on the Canadian barrens are intimately linked to migrating herds of barren-ground caribou (*Rangifer tarandus*). We deployed a Global Positioning System (GPS) radio collar on an adult female wolf to record her movements in response to changing caribou densities near her den during summer. This wolf and two other females were observed nursing a group of 11 pups. She traveled a minimum of 341 km during a 14-day excursion. The straight-line distance from the den to the farthest location was 103 km, and the overall minimum rate of travel was 3.1 km/h. The distance between the wolf and the radio-collared caribou decreased from 242 km one week before the excursion to 8 km four days into the excursion. We discuss several possible explanations for the long foraging bout.

(7) *Key words:* wolf, GPS tracking, movements, *Canis lupus*, foraging, caribou, Northwest Territories

(8) RÉSUMÉ Les loups (*Canis lupus*) dans la toundra canadienne sont étroitement liés aux hardes de caribous des toundras (*Rangifer tarandus*). On a équipé une louve adulte d'un collier émetteur muni d'un système de positionnement mondial (GPS) afin d'enregistrer ses déplacements en réponse au changement de densité du caribou près de sa tanière durant l'été. On a observé cette louve ainsi que deux autres en train d'allaiter un groupe de 11 louveteaux. Elle a parcouru un minimum de 341 km durant une sortie de 14 jours. La distance en ligne droite de la tanière à l'endroit le plus éloigné était de 103 km, et la vitesse minimum durant tout le voyage était de 3,1 km/h. La distance entre la louve et le caribou muni du collier émetteur a diminué de 242 km une semaine avant la sortie à 8 km quatre jours après la sortie. On commente diverses explications possibles pour ce long épisode de recherche de nourriture.

Mots clés: loup, repérage GPS, déplacements, *Canis lupus*, recherche de nourriture, caribou, Territoires du Nord-Ouest

Traduit pour la revue *Arctic* par Nésida Loyer.

(9) Introduction

Wolves (*Canis lupus*) that den on the central barrens of mainland Canada follow the seasonal movements of their main prey, migratory barren-ground caribou (*Rangifer tarandus*) (Kuyt, 1962; Kelsall, 1968; Walton et al., 2001). However, most wolves do not den near caribou calving grounds, but select sites farther south, closer to the tree line (Heard and Williams, 1992). Most caribou migrate beyond primary wolf denning areas by mid-June and do not return until mid-to-late July (Heard et al., 1996; Gunn et al., 2001). Conse-quently, caribou density near dens is low for part of the summer.

During this period of spatial separation from the main caribou herds, wolves must either search near the homesite for scarce caribou or alternative prey (or both), travel to where prey are abundant, or use a combination of these strategies.

Walton et al. (2001) postulated that the travel of tundra wolves outside their normal summer ranges is a response to low caribou availability rather than a pre-dispersal exploration like that observed in terri-torial wolves (Fritts and Mech, 1981; Messier, 1985). The authors postulated this because most such travel was directed toward caribou calving grounds. We report details of such a long-distance excursion by a breeding female tundra wolf wearing a GPS radio collar. We discuss the relationship of the excursion to movements of satellite-collared caribou (Gunn et al., 2001), supporting the hypothesis that tundra wolves make directional, rapid, long-distance movements in response to seasonal prey availability.

[1] Department of Biological Sciences, University of Alberta, Edmonton, Alberta T6G 2E9, Canada
[2] Corresponding author: pframe@ualberta.ca
[3] Department of Resources, Wildlife, and Economic Development, North Slave Region, Government of the Northwest Territories, P.O. Box 2668, 3803 Bretzlaff Dr., Yellowknife, Northwest Territories X1A 2P9, Canada; Dean_Cluff@gov.nt.ca
[4] Faculty of Environmental Design, University of Calgary, Calgary, Alberta T2N 1N4, Canada; current address: P.O. Box 150, Meacham, Saskatchewan S0K 2V0, Canada

196

Annotations (right margin):

1 Title of the journal, which reports on science taking place in Arctic regions.

2 Volume number, issue number and date of the journal, and page numbers of the article.

3 Title of the article: a concise but specific description of the subject of study—one episode of long-range travel by a wolf hunting for food on the Arctic tundra.

4 Authors of the article: scientists working at the institutions listed in the footnotes below. Note #2 indicates that P. F. Frame is the *corresponding author*—the person to contact with questions or comments. His email address is provided.

5 Date on which a draft of the article was received by the journal editor, followed by date one which a revised draft was accepted for publication. Between these dates, the article was reviewed and critiqued by other scientists, a process called peer review. The authors revised the article to make it clearer, according to those reviews.

6 ABSTRACT: A brief description of the study containing all basic elements of this report. First sentence summarizes the *background* material. Second sentence encapsulates the *methods* used. The rest of the paragraph sums up the *results*. Authors introduce the main *subject* of the study—a female wolf (#388) with pups in a den—and refer to later *discussion* of possible explanations for her behavior.

7 Key words are listed to help researchers using computer databases. Searching the databases using these key words will yield a list of studies related to this one.

8 RÉSUMÉ: The French translation of the abstract and key words. Many researchers in this field are French Canadian. Some journals provide such translations in French or in other languages.

9 INTRODUCTION: Gives the background for this wolf study. This paragraph tells of known or suspected wolf behavior that is important for this study. Note that (a) major species mentioned are always accompanied by scientific names, and (b) statements of fact or *postulations* (claims or assumptions about what is likely to be true) are followed by references to studies that established those facts or supported the postulations.

10 This paragraph focuses directly on the wolf behaviors that were studied here.

11 This paragraph starts with a statement of the *hypothesis* being tested, one that originated in other studies and is supported by this one. The hypothesis is restated more succinctly in the last sentence of this paragraph. This is the *inquiry* part of the scientific process—asking questions and suggesting possible answers.

Figure 1. Map showing the movements of satellite radio-collared caribou with respect to female wolf 388's summer range and long foraging movement, in summer 2002.

12 This map shows the study area and depicts wolf and caribou locations and movements during one summer. Some of this information is explained below.

13 STUDY AREA: This section sets the stage for the study, locating it precisely with latitude and longitude coordinates and describing the area (illustrated by the map in Figure 1).

14 Here begins the story of how prey (caribou) and predators (wolves) interact on the tundra. Authors describe movements of these nomadic animals throughout the year.

15 We focus on the denning season (summer) and learn how wolves locate their dens and travel according to the movements of caribou herds.

13 Study Area

Our study took place in the northern boreal forest–low Arctic tundra transition zone (63° 30′ N, 110° 00′ W; Figure 1; Timoney et al., 1992). Permafrost in the area changes from discontinuous to continuous (Harris, 1986). Patches of spruce (*Picea mariana, P. glauca*) occur in the southern portion and give way to open tundra to the northeast. Eskers, kames, and other glacial deposits are scattered throughout the study area. Standing water and exposed bedrock are characteristic of the area.

14 Details of the Caribou-Wolf System

The Bathurst caribou herd uses this study area. Most caribou cows have begun migrating by late April, reaching calving grounds by June (Gunn et al., 2001;

Figure 1). Calving peaks by 15 June (Gunn et al., 2001), and calves begin to travel with the herd by one week of age (Kelsall, 1968). The movement patterns of bulls are less known, but bulls frequent areas near calving grounds by mid-June (Heard et al., 1996; Gunn et al., 2001). In summer, Bathurst caribou cows generally travel south from their calving grounds and then, parallel to the tree line, to the northwest. The rut usually takes place at the tree line in October (Gunn et al., 2001). The winter range of the Bathurst herd varies among years, ranging through the taiga and along the tree line from south of Great Bear Lake to southeast of Great Slave Lake. Some caribou spend the winter on the tundra (Gunn et al., 2001; Thorpe et al., 2001).

15 In winter, wolves that prey on Bathurst caribou do not behave territorially. Instead, they follow the herd throughout its winter range (Walton et al., 2001; Musiani, 2003). However, during denning (May–

16 Other variables are considered—prey other than caribou and their relative abundance in 2002.

17 METHODS: There is no one scientific method. Procedures for each and every study must be explained carefully.

18 Authors explain when and how they tracked caribou and wolves, including tools used and the exact procedures followed.

19 This important subsection explains what data were calculated (average distance ...) and how, including the software used and where it came from. (The calculations are listed in Table 1.) Note that the behavior measured (traveling) is carefully defined.

20 RESULTS: The heart of the report and the *observation* part of the scientific process. This section is organized parallel to the Methods section.

21 This subsection is broken down by periods of observation. Pre-excursion period covers the time between 388's capture and the start of her long-distance travel. The investigators used visual observations as well as telemetry (measurements taken using the global positioning system (GPS)) to gather data. They looked at how 388 cared for her pups, interacted with other adults, and moved about the den area.

Table 1. Daily distances from wolf 388 and the den to the nearest radio-collared caribou during a long excursion in summer 2002.

Date (2002)	Mean distance from caribou to wolf (km)	Daily distance from closest caribou to den
12 July	242	241
13 July	210	209
14 July	200	199
15 July	186	180
16 July	163	162
17 July	151	148
18 July	144	137
19 July[1]	126	124
20 July	103	130
21 July	73	130
22 July	40	110
23 July[2]	9	104
29 July[3]	16	43
30 July	32	43
31 July	28	44
1 August	29	46
2 August[4]	54	52
3 August	53	53
4 August	74	74
5 August	75	75
6 August	74	75
7 August	72	75
8 August	76	75
9 August	79	79

[1] Excursion starts.
[2] Wolf closest to collared caribou.
[3] Previous five days' caribou locations not available.
[4] Excursion ends.

August, parturition late May to mid-June), wolf movements are limited by the need to return food to the den. To maximize access to migrating caribou, many wolves select den sites closer to the tree line than to caribou calving grounds (Heard and Williams, 1992). Because of caribou movement patterns, tundra denning wolves are separated from the main caribou herds by several hundred kilometers at some time during summer (Williams, 1990:19; Figure 1; Table 1).

16 Muskoxen do not occur in the study area (Fournier and Gunn, 1998), and there are few moose there (H.D. Cluff, pers. obs.). Therefore, alternative prey for wolves includes waterfowl, other ground-nesting birds, their eggs, rodents, and hares (Kuyt, 1972; Williams, 1990:16; H.D. Cluff and P.F. Frame, unpubl. data). During 56 hours of den observations, we saw no ground squirrels or hares, only birds. It appears that the abundance of alternative prey was relatively low in 2002.

17 Methods

Wolf Monitoring

18 We captured female wolf 388 near her den on 22 June 2002, using a helicopter net-gun (Walton et al., 2001). She was fitted with a releasable GPS radio collar (Merrill et al., 1998) programmed to acquire locations at 30-

minute intervals. The collar was electronically released (e.g., Mech and Gese, 1992) on 20 August 2002. From 27 June to 3 July 2002, we observed 388's den with a 78 mm spotting scope at a distance of 390 m.

Caribou Monitoring

In spring of 2002, ten female caribou were captured by helicopter net-gun and fitted with satellite radio collars, bringing the total number of collared Bathurst cows to 19. Eight of these spent the summer of 2002 south of Queen Maud Gulf, well east of normal Bathurst caribou range. Therefore, we used 11 caribou for this analysis. The collars provided one location per day during our study, except for five days from 24 to 28 July. Locations of satellite collars were obtained from Service Argos, Inc. (Landover, Maryland).

Data Analysis

19 Location data were analyzed by ArcView GIS software (Environmental Systems Research Institute Inc., Redlands, California). We calculated the average distance from the nearest collared caribou to the wolf and the den for each day of the study.

Wolf foraging bouts were calculated from the time 388 exited a buffer zone (500 m radius around the den) until she re-entered it. We considered her to be traveling when two consecutive locations were spatially separated by more than 100 m. Minimum distance traveled was the sum of distances between each location and the next during the excursion.

We compared pre- and post-excursion data using Analysis of Variance (ANOVA; Zar, 1999). We first tested for homogeneity of variances with Levene's test (Brown and Forsythe, 1974). No transformations of these data were required.

20 Results

Wolf Monitoring

21 **Pre-Excursion Period:** Wolf 388 was lactating when captured on 22 June. We observed her and two other females nursing a group of 11 pups between 27 June and 3 July. During our observations, the pack consisted of at least four adults (3 females and 1 male) and 11 pups. On 30 June, three pups were moved to a location 310 m from the other eight and cared for by an uncollared female. The male was not seen at the den after the evening of 30 June.

Before the excursion, telemetry indicated 18 foraging bouts. The mean distance traveled during these bouts was 25.29 km (± 4.5 SE, range 3.1–82.5 km). Mean greatest distance from the den on foraging

22

Figure 2. Details of a long foraging movement by female wolf 388 between 19 July and 2 August 2002. Also shown are locations and movements of three satellite radio-collared caribou from 23 July to 21 August 2002. On 23 July, the wolf was 8 km from a collared caribou. The farthest point from the den (103 km distant) was recorded on 27 July. Arrows indicate direction of travel.

bouts was 7.1 km (± 0.9 SE, range 1.7–17.0 km). The average duration of foraging bouts for the period was 20.9 h (± 4.5 SE, range 1–71 h).

The average daily distance between the wolf and the nearest collared caribou decreased from 242 km on 12 July, one week before the excursion period, to 126 km on 19 July, the day the excursion began (Table 1).

23 **Excursion Period:** On 19 July at 2203, after spending 14 h at the den, 388 began moving to the northeast and did not return for 336 h (14 d; Figure 2). Whether she traveled alone or with other wolves is unknown. During the excursion, 476 (71%) of 672 possible locations were recorded. The wolf crossed the southeast end of Lac Capot Blanc on a small land bridge, where she paused for 4.5 h after traveling for 19.5 h (37.5 km). Following this rest, she traveled for 9 h (26.3 km) onto a peninsula in Reid Lake, where she spent 2 h before backtracking and stopping for 8 h just off the peninsula. Her next period of travel lasted 16.5 h (32.7 km), terminating in a pause of 9.5 h just 3.8 km from a concentration of locations at the far end of her excursion, where we presume she encountered caribou. The mean duration of these three movement periods was 15.7 h (± 2.5 SE), and that of the pauses, 7.3 h (± 1.5). The wolf required 72.5 h (3.0 d) to travel a minimum of 95 km from her den to this area near caribou (Figure 2). She remained there (35.5 km2) for 151.5 h (6.3 d) and then moved south to Lake of the Enemy, where she stayed (31.9 km^2) for 74 h (3.1 d) before returning to her den. Her greatest distance from the den, 103 km, was recorded 174.5 h (7.3 d) after the excursion

22 The key in the lower right-hand corner of the map shows areas (shaded) within which the wolves and caribou moved, and the dotted trail of 388 during her excursion. From the results depicted on this map, the investigators tried to determine when and where 388 might have encountered caribou and how their locations affected her traveling behavior.

23 The wolf's excursion (her long trip away from the den area) is the focus of this study. These paragraphs present detailed measurements of daily movements during her two-week trip—how far she traveled, how far she was from collared caribou, her time spent traveling and resting, and her rate of speed. Authors use the phrase "minimum distance traveled" to acknowledge they couldn't track every step but were measuring samples of her movements. They knew that she went at least as far as they measured. This shows how scientists try to be exact when reporting results. Results of this study are depicted graphically in the map in Figure 2.

24 Post-excursion measurements of 388's movements were made to compare with those of the pre-excursion period. In order to compare, scientists often use *means*, or averages, of a series of measurements—mean distances, mean duration, etc.

25 In the comparison, authors used statistical calculations (F and df) to determine that the differences between pre- and post-excursion measurements were *statistically insignificant*, or close enough to be considered essentially the same or similar.

26 As with wolf 388, the investigators measured the movements of caribou during the study period. The areas within which the caribou moved are shown in Figure 2 by shaded polygons mentioned in the second paragraph of this subsection.

27 This subsection summarizes how distances separating predators and prey varied during the study period.

28 DISCUSSION: This section is the *interpretation* part of the scientific process.

29 This subsection reviews observations from other studies and suggests that this study fits with patterns of those observations.

30 Authors discuss a prevailing *theory* (CBFT) which might explain why a wolf would travel far to meet her own energy needs while taking food caught closer to the den back to her pups. The results of this study seem to fit that pattern.

---

began, at 0433 on 27 July. She was 8 km from a collared caribou on 23 July, four days after the excursion began (Table 1).

The return trip began at 0403 on 2 August, 318 h (13.2 d) after leaving the den. She followed a relatively direct path for 18 h back to the den, a distance of 75 km.

The minimum distance traveled during the excursion was 339 km. The estimated overall minimum travel rate was 3.1 km/h, 2.6 km/h away from the den and 4.2 km/h on the return trip.

24 **Post-Excursion Period:** We saw three pups when recovering the collar on 20 August, but others may have been hiding in vegetation.

Telemetry recorded 13 foraging bouts in the post-excursion period. The mean distance traveled during these bouts was 18.3 km (+ 2.7 SE, range 1.2–47.7 km), and mean greatest distance from the den was 7.1 km (+ 0.7 SE, range 1.1–11.0 km). The mean duration of these post-excursion foraging bouts was 10.9 h (+ 2.4 SE, range 1–33 h).

When 388 reached her den on 2 August, the distance to the nearest collared caribou was 54 km. On 9 August, one week after she returned, the distance was 79 km (Table 1).

Pre- and Post-Excursion Comparison

25 We found no differences in the mean distance of foraging bouts before and after the excursion period (F = 1.5, df = 1, 29, p = 0.24). Likewise, the mean greatest distance from the den was similar pre- and post-excursion (F = 0.004, df = 1, 29, p = 0.95). However, the mean duration of 388's foraging bouts decreased by 10.0 h after her long excursion (F = 3.1, df = 1, 29, p = 0.09).

26 *Caribou Monitoring*

Summer Movements: On 10 July, 5 of 11 collared caribou were dispersed over a distance of 10 km, 140 km south of their calving grounds (Figure 1). On the same day, three caribou were still on the calving grounds, two were between the calving grounds and the leaders, and one was missing. One week later (17 July), the leading radio-collared cows were 100 km farther south (Figure 1). Two were within 5 km of each other in front of the rest, who were more dispersed. All radio-collared cows had left the calving grounds by this time. On 23 July, the leading radio-collared caribou had moved 35 km farther south, and all of them were more widely dispersed. The two cows closest to the leader were 26 km and 33 km away, with 37 km between them. On the next location (29 July), the most southerly caribou were 60 km

farther south. All of the caribou were now in the areas where they remained for the duration of the study (Figure 2).

A Minimum Convex Polygon (Mohr and Stumpf, 1966) around all caribou locations acquired during the study encompassed 85 119 km^2.

27 **Relative to the Wolf Den:** The distance from the nearest collared caribou to the den decreased from 241 km one week before the excursion to 124 km the day it began. The nearest a collared caribou came to the den was 43 km away, on 29 and 30 July. During the study, four collared caribou were located within 100 km of the den. Each of these four was closest to the wolf on at least one day during the period reported.

28 ## Discussion

Prey Abundance

29 Caribou are the single most important prey of tundra wolves (Clark, 1971; Kuyt, 1972; Stephenson and James, 1982; Williams, 1990). Caribou range over vast areas, and for part of the summer, they are scarce or absent in wolf home ranges (Heard et al., 1996). Both the long distance between radio-collared caribou and the den the week before the excursion and the increased time spent foraging by wolf 388 indicate that caribou availability near the den was low. Observations of the pups' being left alone for up to 18 h, presumably while adults were searching for food, provide additional support for low caribou availability locally. Mean foraging bout duration decreased by 10.0 h after the excursion, when collared caribou were closer to the den, suggesting an increase in caribou availability nearby.

Foraging Excursion

30 One aspect of central place foraging theory (CPFT) deals with the optimality of returning different-sized food loads from varying distances to dependents at a central place (i.e., the den) (Orians and Pearson, 1979). Carlson (1985) tested CPFT and found that the predator usually consumed prey captured far from the central place, while feeding prey captured nearby to dependants. Wolf 388 spent 7.2 days in one area near caribou before moving to a location 23 km back towards the den, where she spent an additional 3.1 days, likely hunting caribou. She began her return trip from this closer location, traveling directly to the den. While away, she may have made one or more successful kills and spent time meeting her own energetic needs before returning to the den. Alternatively, it may have taken several attempts to make a kill, which she then fed on before beginning her return trip. We do not know if she returned food to the pups, but such behavior would be supported by CPFT.

31 Other workers have reported wolves' making long round trips and referred to them as "extraterritorial" or "pre-dispersal" forays (Fritts and Mech, 1981; Messier, 1985; Ballard et al., 1997; Merrill and Mech, 2000). These movements are most often made by young wolves (1–3 years old), in areas where annual territories are maintained and prey are relatively sedentary (Fritts and Mech, 1981; Messier, 1985). The long excursion of 388 differs in that tundra wolves do not maintain annual territories (Walton et al., 2001), and the main prey migrate over vast areas (Gunn et al., 2001).

Another difference between 388's excursion and those reported earlier is that she is a mature, breeding female. No study of territorial wolves has reported reproductive adults making extraterritorial movements in summer (Fritts and Mech, 1981; Messier, 1985; Ballard et al., 1997; Merrill and Mech, 2001). However, Walton et al. (2001) also report that breeding female tundra wolves made excursions.

Direction of Movement

32 Possible explanations for the relatively direct route 388 took to the caribou include landscape influence and experience. Considering the timing of 388's trip and the locations of caribou, had the wolf moved northwest, she might have missed the caribou entirely, or the encounter might have been delayed.

A reasonable possibility is that the land directed 388's route. The barrens are crisscrossed with trails worn into the tundra over centuries by hundreds of thousands of caribou and other animals (Kelsall, 1968; Thorpe et al., 2001). At river crossings, lakes, or narrow peninsulas, trails converge and funnel towards and away from caribou calving grounds and summer range. Wolves use trails for travel (Paquet et al., 1996; Mech and Boitani, 2003; P. Frame, pers. observation). Thus, the landscape may direct an animal's movements and lead it to where cues, such as the odor of caribou on the wind or scent marks of other wolves, may lead it to caribou.

33 Another possibility is that 388 knew where to find caribou in summer. Sexually immature tundra wolves sometimes follow caribou to calving grounds (D. Heard, unpubl. data). Possibly, 388 had made such journeys in previous years and killed caribou. If this were the case, then in times of local prey scarcity she might travel to areas where she had hunted successfully before. Continued monitoring of tundra wolves may answer questions about how their food needs are met in times of low caribou abundance near dens.

Caribou often form large groups while moving **34** south to the tree line (Kelsall, 1968). After a large aggregation of caribou moves through an area, its scent can linger for weeks (Thorpe et al., 2001:104). It is conceivable that 388 detected caribou scent on the wind, which was blowing from the northeast on 19–21 July (Environment Canada, 2003), at the same time her excursion began. Many factors, such as odor strength and wind direction and strength, make systematic study of scent detection in wolves difficult under field conditions (Harrington and Asa, 2003). However, humans are able to smell odors such as forest fires or oil refineries more than 100 km away. The olfactory capabilities of dogs, which are similar to wolves, are thought to be 100 to 1 million times that of humans (Harrington and Asa, 2003). Therefore, it is reasonable to think that under the right wind conditions, the scent of many caribou traveling together could be detected by wolves from great distances, thus triggering a long foraging bout.

Rate of Travel

Mech (1994) reported the rate of travel of Arctic **35** wolves on barren ground was 8.7 km/h during regular travel and 10.0 km/h when returning to the den, a difference of 1.3 km/h. These rates are based on direct observation and exclude periods when wolves moved slowly or not at all. Our calculated travel rates are assumed to include periods of slow movement or no movement. However, the pattern we report is similar to that reported by Mech (1994), in that homeward travel was faster than regular travel by 1.6 km/h. The faster rate on return may be explained by the need to return food to the den. Pup survival can increase with the number of adults in a pack available to deliver food to pups (Harrington et al., 1983). Therefore, an increased rate of travel on homeward trips could improve a wolf's reproductive fitness by getting food to pups more quickly.

Fate of 388's Pups

Wolf 388 was caring for pups during den observa- **36** tions. The pups were estimated to be six weeks old, and were seen ranging as far as 800 m from the den. They received some regurgitated food from two of the females, but were unattended for long periods. The excursion started 16 days after our observations, and it is improbable that the pups could have traveled the distance that 388 moved. If the pups died, this would have removed parental responsibility, allowing the long movement.

Our observations and the locations of radio-collared caribou indicate that prey became scarce in

31 Here our authors note other possible explanations for wolves' excursions presented by other investigators, but this study does not seem to support those ideas.

32 Authors discuss possible reasons for why 388 traveled directly to where caribou were located. They take what they learned from earlier studies and apply it to this case, suggesting that the lay of the land played a role. Note that their description paints a clear picture of the landscape.

33 Authors suggest that 388 may have learned in traveling during previous summers where the caribou were. The last two sentences suggest ideas for future studies.

34 Or maybe 388 followed the scent of the caribou. Authors acknowledge difficulties of proving this, but they suggest another area where future studies might be done.

35 Authors suggest that results of this study support previous studies about how fast wolves travel to and from the den. In the last sentence, they speculate on how these observed patterns would fit into the theory of evolution.

36 Authors also speculate on the fate of 388's pups while she was traveling. This leads to . . .

37 Discussion of cooperative rearing of pups and, in turn, to speculation on how this study and what is known about cooperative rearing might fit into the animal's strategies for survival of the species. Again, the authors approach the broader theory of evolution and how it might explain some of their results.

38 And again, they suggest that this study points to several areas where further study will shed some light.

39 In conclusion, the authors suggest that their study supports the hypothesis being tested here. And they touch on the implications of increased human activity on the tundra predicted by their results.

40 ACKNOWLEDGEMENTS: Authors note the support of institutions, companies and individuals. They thank their reviewers ad list permits under which their research was carried on.

41 REFERENCES: List of all studies cited in the report. This may seem tedious, but is a vitally important part of scientific reporting. It is a record of the sources of information on which this study is based. It provides readers with a wealth of resources for further reading on this topic. Much of it will form the foundation of future scientific studies like this one.

the area of the den as summer progressed. Wolf 388 may have abandoned her pups to seek food for herself. However, she returned to the den after the excursion, where she was seen near pups. In fact, she foraged in a similar pattern before and after the excursion, suggesting that she again was providing for pups after her return to the den.

37 A more likely possibility is that one or both of the other lactating females cared for the pups during 388's absence. The three females at this den were not seen with the pups at the same time. However, two weeks earlier, at a different den, we observed three females cooperatively caring for a group of six pups. At that den, the three lactating females were observed providing food for each other and trading places while nursing pups. Such a situation at the den of 388 could have created conditions that allowed one or more of the lactating females to range far from the den for a period, returning to her parental duties afterwards. However, the pups would have been weaned by eight weeks of age (Packard et al., 1992), so nonlactating adults could also have cared for them, as often happens in wolf packs (Packard et al., 1992; Mech et al., 1999).

Cooperative rearing of multiple litters by a pack could create opportunities for long-distance foraging movements by some reproductive wolves during summer periods of local food scarcity. We have recorded multiple lactating females at one or more tundra wolf dens per year since 1997. This reproductive strategy may be an adaptation to temporally and **38** spatially unpredictable food resources. All of these possibilities require further study, but emphasize both the adaptability of wolves living on the barrens and their dependence on caribou.

Long-range wolf movement in response to caribou **39** availability has been suggested by other researchers (Kuyt, 1972; Walton et al., 2001) and traditional ecological knowledge (Thorpe et al., 2001). Our report demonstrates the rapid and extreme response of wolves to caribou distribution and movements in summer. Increased human activity on the tundra (mining, road building, pipelines, ecotourism) may influence caribou movement patterns and change the interactions between wolves and caribou in the region. Continued monitoring of both species will help us to assess whether the association is being affected adversely by anthropogenic change.

40 Acknowledgements

This research was supported by the Department of Resources, Wildlife, and Economic Development, Government of the Northwest Territories; the Department of Biological Sciences at the University of Alberta; the Natural Sciences and Engineering Research Council of Canada; the Department of Indian and Northern Affairs Canada; the Canadian Circumpolar Institute; and DeBeers Canada, Ltd. Lorna Ruechel assisted with den observations. A. Gunn provided caribou location data. We thank Dave Mech for the use of GPS collars. M. Nelson, A. Gunn, and three anonymous reviewers made helpful comments on earlier drafts of the manuscript. This work was done under Wildlife Research Permit – WL002948 issued by the Government of the Northwest Territories, Department of Resources, Wildlife, and Economic Development.

41 References

BALLARD, W.B., AYRES, L.A., KRAUSMAN, P.R., REED, D.J., and FANCY, S.G. 1997. Ecology of wolves in relation to a migratory caribou herd in northwest Alaska. Wildlife Monographs 135. 47 p.

BROWN, M.B., and FORSYTHE, A.B. 1974. Robust tests for the equality of variances. Journal of the American Statistical Association 69:364–367.

CARLSON, A. 1985. Central place foraging in the red-backed shrike (*Lanius collurio* L.): Allocation of prey between forager and sedentary consumer. Animal Behaviour 33:664–666.

CLARK, K.R.F. 1971. Food habits and behavior of the tundra wolf on central Baffin Island. Ph.D. Thesis, University of Toronto, Ontario, Canada.

ENVIRONMENT CANADA. 2003. National climate data information archive. Available online: http://www.climate.weatheroffice.ec.gc.ca/Welcome_e.html

FOURNIER, B., and GUNN, A. 1998. Musk ox numbers and distribution in the NWT, 1997. File Report No. 121. Yellowknife: Department of Resources, Wildlife, and Economic Development, Government of the Northwest Territories. 55 p.

FRITTS, S.H., and MECH, L.D. 1981. Dynamics, movements, and feeding ecology of a newly protected wolf population in northwestern Minnesota. Wildlife Monographs 80. 79 p.

GUNN, A., DRAGON, J., and BOULANGER, J. 2001. Seasonal movements of satellite-collared caribou from the Bathurst herd. Final Report to the West Kitikmeot Slave Study Society, Yellowknife, NWT. 80 p. Available online: http://www.wkss.nt.ca/HTML/08_ProjectsReports/PDF/Seasonal MovementsFinal.pdf

HARRINGTON, F.H., and ASA, C.S. 2003. Wolf communication. In: Mech, L.D., and Boitani, L., eds. Wolves: Behavior, ecology, and conservation. Chicago: University of Chicago Press. 66–103.

HARRINGTON, F.H., MECH, L.D., and FRITTS, S.H. 1983. Pack size and wolf pup survival: Their relationship under varying ecological conditions. Behavioral Ecology and Sociobiology 13:19–26.

HARRIS, S.A. 1986. Permafrost distribution, zonation and stability along the eastern ranges of the cordillera of North America. Arctic 39(1):29–38.

HEARD, D.C., and WILLIAMS, T.M. 1992. Distribution of wolf dens on migratory caribou ranges in the Northwest

Territories, Canada. Canadian Journal of Zoology 70:1504–1510.

HEARD, D.C., WILLIAMS, T.M., and MELTON, D.A. 1996. The relationship between food intake and predation risk in migratory caribou and implication to caribou and wolf population dynamics. Rangifer Special Issue No. 2:37–44.

KELSALL, J.P. 1968. The migratory barren-ground caribou of Canada. Canadian Wildlife Service Monograph Series 3. Ottawa: Queen's Printer. 340 p.

KUYT, E. 1962. Movements of young wolves in the Northwest Territories of Canada. Journal of Mammalogy 43:270–271.

———. 1972. Food habits and ecology of wolves on barren-ground caribou range in the Northwest Territories. Canadian Wildlife Service Report Series 21. Ottawa: Information Canada. 36 p.

MECH, L.D. 1994. Regular and homeward travel speeds of Arctic wolves. Journal of Mammalogy 75:741–742.

MECH, L.D., and BOITANI, L. 2003. Wolf social ecology. In: Mech, L.D., and Boitani, L., eds. Wolves: Behavior, ecology, and conservation. Chicago: University of Chicago Press. 1–34.

MECH, L.D., and GESE, E.M. 1992. Field testing the Wildlink capture collar on wolves. Wildlife Society Bulletin 20:249–256.

MECH, L.D., WOLFE, P., and PACKARD, J.M. 1999. Regurgitative food transfer among wild wolves. Canadian Journal of Zoology 77:1192–1195.

MERRILL, S.B., and MECH, L.D. 2000. Details of extensive movements by Minnesota wolves (Canis lupus). American Midland Naturalist 144:428–433.

MERRILL, S.B., ADAMS, L.G., NELSON, M.E., and MECH, L.D. 1998. Testing releasable GPS radiocollars on wolves and white-tailed deer. Wildlife Society Bulletin 26:830–835.

MESSIER, F. 1985. Solitary living and extraterritorial movements of wolves in relation to social status and prey abundance. Canadian Journal of Zoology 63:239–245.

MOHR, C.O., and STUMPF, W.A. 1966. Comparison of methods for calculating areas of animal activity. Journal of Wildlife Management 30:293–304.

MUSIANI, M. 2003. Conservation biology and management of wolves and wolf-human conflicts in western North America. Ph.D. Thesis, University of Calgary, Calgary, Alberta, Canada.

ORIANS, G.H., and PEARSON, N.E. 1979. On the theory of central place foraging. In: Mitchell, R.D., and Stairs, G.F., eds. Analysis of ecological systems. Columbus: Ohio State University Press. 154–177.

PACKARD, J.M., MECH, L.D., and REAM, R.R. 1992. Weaning in an arctic wolf pack: Behavioral mechanisms. Canadian Journal of Zoology 70:1269–1275.

PAQUET, P.C., WIERZCHOWSKI, J., and CALLAGHAN, C. 1996. Summary report on the effects of human activity on gray wolves in the Bow River Valley, Banff National Park, Alberta. In: Green, J., Pacas, C., Bayley, S., and Cornwell, L., eds. A cumulative effects assessment and futures outlook for the Banff Bow Valley. Prepared for the Banff Bow Valley Study. Ottawa: Department of Canadian Heritage.

STEPHENSON, R.O., and JAMES, D. 1982. Wolf movements and food habits in northwest Alaska. In: Harrington, F.H., and Paquet, P.C., eds. Wolves of the world. New Jersey: Noyes Publications. 223–237.

THORPE, N., EYEGETOK, S., HAKONGAK, N., and QITIR-MIUT ELDERS. 2001. The Tuktu and Nogak Project: A caribou chronicle. Final Report to the West Kitikmeot/Slave Study Society, Ikaluktuuttiak, NWT. 160 p.

TIMONEY, K.P., LA ROI, G.H., ZOLTAI, S.C., and ROBINSON, A.L. 1992. The high subarctic forest-tundra of northwestern Canada: Position, width, and vegetation gradients in relation to climate. Arctic 45(1):1–9.

WALTON, L.R., CLUFF, H.D., PAQUET, P.C., and RAMSAY, M.A. 2001. Movement patterns of barren-ground wolves in the central Canadian Arctic. Journal of Mammalogy 82:867–876.

WILLIAMS, T.M. 1990. Summer diet and behavior of wolves denning on barren-ground caribou range in the Northwest Territories, Canada. M.Sc. Thesis, University of Alberta, Edmonton, Alberta, Canada.

ZAR, J.H. 1999. Biostatistical analysis. 4th ed. New Jersey: Prentice Hall. 663 p.

Appendix IV   A Plain English Map of the Human Chromosomes

Chromosome 1
- sweet taste receptors
- Rh blood type
- marijuana receptor
- (anorexia nervosa susceptibility)
- leptin receptor
- TSH β chain
- lamin A (progeria)
- Duffy blood group antigen

Chromosome 2
- LH/choriogonadotropin receptor (micropenis)
- CD8; cytotoxic T cell antigen
- antibody light chain
- lactase
- (cleft palate)
- glucagon

Chromosome 3
- oxytocin receptor
- HIV receptor
- rhodopsin
- (alkaptonuria)
- (sucrose intolerance)
- somatostatin

Chromosome 4
- (achondroplasia)
- (Huntington disease)
- (Ellis-van Creveld syndrome)
- alcohol dehydrogenase (susceptibility to alcoholism)
- red hair color

Chromosome 5
- Cri-du-chat syndrome
- bitter taste receptor
- growth hormone receptor (pituitary dwarfism)
- interleukin-4

Chromosome 6
- (gluten intolerance)
- HLA/MHC
- tumor necrosis factor
- α chains of HCG, FSH, LH, and TSH
- estrogen receptor

Chromosome 7
- cytochrome c
- elastin
- DLX 5/6 homeotic genes
- CFTR (cystic fibrosis)
- leptin (obesity)
- (blue-deficient colorblind)
- TCR β subunit

Chromosome 8
- gonadotropin releasing hormone
- helicase (Werner's syndrome)
- corticotropin releasing hormone

Chromosome 9
- (galactosemia)
- (cerebral palsy)
- (Friedreich ataxia)
- (fructose intolerance)
- ABO blood group

Chromosome 10
- vitamin B-12 receptor
- mannose binding protein
- perforin
- (gluten intolerance)

Chromosome 11
- hemoglobin β chain (sickle cell anemia)
- insulin
- parathyroid hormone
- catalase
- PAX6 (aniridia)
- FSH, β chain
- tyrosinase (albinism)

Chromosome 12
- CD4
- helper T cell antigen
- oncogene KRAS2 (lung cancer, bladder cancer, breast cancer)
- keratins
- lysozyme
- (phenylketonuria)
- aldehyde dehydrogenase (alcohol intolerance)

Chromosome 13
- ribosomal RNA
- BRCA 2 (breast cancer)
- (gastroesophageal reflux)

Chromosome 14
- ribosomal RNA
- presinilin (Alzheimer's)
- TSH receptor
- immunoglobulin heavy chains

Chromosome 15
- ribosomal RNA
- fibrillin 1 (Marfan syndrome)
- (Tay-Sachs disease)

Chromosome 16
- hemoglobin α chain
- DNAse I (lupus)

Chromosome 17
- (Canavan disease)
- p53 tumor antigen
- NF1 (neurofibromatosis)
- serotonin transporter
- BRCA 1 (breast, ovarian cancer)
- Growth hormone

Chromosome 18
- B cell apoptosis regulator (B cell lymphoma)
- myelin basic protein

Chromosome 19
- LDL receptor (coronary artery disease)
- insulin receptor
- brown hair color
- green/blue eye color
- (Warfarin resistance)
- HCG, β chain
- LH, β chain

Chromosome 20
- prion protein (Creutzfeld-Jacob disease)
- oxytocin
- GHRH (acromegaly)

Chromosome 21
- ribosomal RNA
- interferon receptors
- (bipolar disorder, early onset)

Chromosome 22
- ribosomal RNA
- immunoglobulin light chains
- myoglobin

Chromosome X
- dystrophin (muscular dystrophy)
- (anhidrotic ectodermal dysplasia)
- IL2RG (SCID-X1)
- XIST X chromosome inactivation control
- (hemophilia B)
- (hemophilia A)
- (red-deficient colorblind)
- (green-deficient colorblind)

Chromosome Y
- sex determining region Y (SRY)
- (no sperm)
- male stature

© 2002 Susan Offner/SK45176-02

Haploid set of human chromosomes. The banding patterns characteristic of each type of chromosome appear after staining with a reagent called Giemsa. The locations of some of the 20,065 known genes (as of November, 2005) are indicated. Also shown are locations that, when mutated, cause some of the genetic diseases discussed in the text.

For more information go to http://biology.brookscole.com/BTT2.

Glossary of Biological Terms

acid Any substance that releases hydrogen ions in water to lower the pH below 7.

acid rain Acidic precipitation; rain or snow with high levels of sulfur and nitrogen oxides.

active site Chemically stable crevice in an enzyme where substrates bind and a reaction can be catalyzed repeatedly.

active transport Pumping of a solute across a cell membrane against its concentration gradient, through the interior of a transport protein. Requires energy input, as from ATP.

adaptation Any heritable aspect of form, function, or behavior that helps an organism survive and reproduce.

adaptive zone A set of different niches that become filled by a group of related species.

aerobic respiration Metabolic pathway that uses oxygen to break down carbohydrates. Typical energy yield is 36 ATP per glucose.

age structure Of a population, the way that individuals are distributed among different age categories.

algal bloom Explosion in population size of phytoplankton after nutrient enrichment of an aquatic habitat; *e.g.*, a red tide.

alkylating agent Chemical that transfers small hydrocarbon groups to nucleotides in DNA. Causes mutation.

allele frequency Abundance of one allele relative to others in a population.

alleles Two or more molecular forms of a gene that encode slightly different versions of the same trait. Arise by mutation.

allopatric speciation Speciation model in which a physical barrier that arises separates two populations, so gene flow between them ends.

amino acid Monomer of polypeptide chains. Consists of a carboxyl group, an amino group, and a characteristic functional group (R).

ammonification Fungi and bacteria in soil degrade nitrogen-rich remains and wastes, then release ammonia and ammonium ions.

amniote Animal that produces amniote eggs; a reptile, bird, or mammal.

amphibian Ectothermic (cold-blooded) tetrapod vertebrate with a three-chambered heart and thin skin that functions in respiration.

anagenesis Pattern of speciation in which one species evolves into another over time without a branching event.

analogous structures Similar body parts in lineages that are not closely related.

anatomy The study of the body parts of an organism to determine their structure and positions relative to one another.

aneuploidy Chromosome abnormality in which body (somatic) cells have one extra or one less chromosome relative to the parental number.

animal Multicelled heterotroph that moves actively during part or all of the life cycle and passes through developmental stages.

annelid Soft-bodied, bilateral invertebrate with a coelom, organs, and many segments; a leech, polychaete, or oligochaete.

anthocyanin Type of accessory pigment that reflects red to blue light, depending on pH.

antibiotic A product of one microorganism that inhibits or kills another. Used as drugs.

anticodon Series of three nucleotide bases in a tRNA that can base-pair with mRNA codons.

antioxidant Enzyme or cofactor that prevents damage to biological molecules by neutralizing free radicals and other strong oxidizers.

apicomplexan A parasitic alveolate protist with a unique device at its anterior end that attaches to and penetrates a host cell.

apoptosis Programmed cell death.

Archaea Domain of prokaryotic species that share some traits with bacteria and other traits with eukaryotic species.

Archean Of the geologic time scale, the interval that extended from 3.8 billion years ago, to 2.5 billion years ago.

area effect Larger islands support more species than smaller ones the same distance from a source of colonizing species.

asexual reproduction Any reproductive mode by which offspring arise from one parent and inherit that parent's genes only.

atom Fundamental unit of matter; consists of protons, neutrons, and electrons.

ATP Adenosine triphosphate, a nucleotide that is the primary energy carrier between reaction sites in cells.

ATP synthase A type of membrane-bound active transport protein that catalyzes the formation of ATP.

ATP/ADP cycle How a cell regenerates ATP. ATP forms as ADP binds phosphate; ADP forms when ATP loses a phosphate.

australopith One of the early hominids; known only from fossils found in Africa.

autosome Of a sexually reproducing species, a chromosome that is the same in both males and females.

autotroph Organism that uses energy captured from sunlight or inorganic chemicals to make its own food from inorganic compounds.

Bacteria Domain of prokaryotic species; the most metabolically diverse organisms and the most common prokaryotes.

bacteriophage Virus that infects bacteria.

balanced polymorphism Two or more alleles for a trait persist in a population; the outcome of selection against homozygotes.

Barr body Inactivated X chromosome in the somatic cells of female mammals.

base Any substance that causes the pH of water to rise above 7.0. Also the nitrogen-containing component of a nucleotide.

base-pair substitution Mutation in which one nucleotide replaces another.

bell curve Idealized distribution of continuous variation of a trait in a population.

bilateral symmetry Body plan in which the main axis divides the body into two halves that are mirror images of one another.

biodiversity Biological diversity, measured by the number of genes, species, habitats.

biogeochemical cycle Slow movement of an element from environmental reservoirs, through food webs, then back.

biogeographic realm Vast expanse of land defined by a particular array of species.

biogeography Scientific study of patterns in the geographic distribution of species.

biological magnification Concentration of a contaminant in body tissues increases as it is passed along a food chain.

biological species concept Defines a sexually reproducing species as one or more populations that interbreed under natural conditions, produce fertile offspring, and are reproductively isolated from other such populations.

biomass pyramid Chart in which the size of successive tiers depicts the measured biomass (dry weight) of some ecosystem's producers, consumers, and decomposers.

biome Region that supports a characteristic type of vegetation; *e.g.*, desert, tundra.

biosphere All areas of Earth's land, water, and air that can sustain some form of life.

biotic potential Theoretical maximum rate of reproduction under ideal conditions.

bipedalism Habitually walking upright on two feet, as by ostriches and hominids.

bird Endothermic (warm-blooded), feathered amniote with a four-chambered heart; the dinosaurs and crocodilians are closest relatives.

bony fish Fish that has an endoskeleton composed mostly of bone tissue.

bottleneck Severe reduction in the size of a population, brought about by intense selection pressure or a natural calamity.

brain Of most nervous systems, the major integrating center that receives, processes, and stores sensory input, and issues coordinated commands for responses.

bryophyte Nonvascular land plant. The haploid stage dominates its life cycle; sperm must swim to eggs; a moss, hornwort, or liverwort.

buffer system Weak acid or base and its salt that together stabilize pH of a solution.

bulk flow Mass movement of molecules in one direction, usually in response to a pressure gradient.

C3 plant Type of plant in which three-carbon PGA is the first stable intermediate to form after carbon fixation.

C4 plant Type of plant in which four-carbon oxaloacetate is the first stable intermediate to form. Carbon is fixed twice, in two types of photosynthetic cells.

calcium pump Active transport protein; pumps calcium ions across a cell membrane against their concentration gradient.

Calvin–Benson cycle Cyclic pathway that forms sugars in the stroma of chloroplasts; also called light-independent reactions.

CAM plant Type of plant that conserves water by opening stomata only at night, when it fixes carbon by the C4 pathway.

camouflage Coloration, pattern, form, or behavior that helps an organism to blend with its surroundings.

cancer Mass of abnormally dividing cells that can invade and colonize other parts of the body. Also, malignant neoplasm.

capture–mark–recapture method Method of estimating population size of a mobile species.

carbohydrates Organic molecules of carbon, hydrogen, and oxygen typically in a 1:2:1 ratio. Structural materials or energy reservoirs.

carbon cycle Atmospheric cycle. Carbon moves from environmental reservoirs (sediments, the ocean, rocks), through the atmosphere (mostly as CO_2), food webs, and back to the reservoirs.

carbon fixation Process by which a cell incorporates carbon atoms from CO_2 into a stable organic compound.

carotenoid One of a class of accessory pigments in photosynthesis that reflect red, orange, or yellow light.

carrying capacity The maximum number of individuals of a species that some particular environment can sustain indefinitely.

cartilaginous fish Jawed fish having an endoskeleton of cartilage; *e.g.*, sharks.

catastrophism Idea that Earth's history was shaped by repeated violent, global catastrophes unlike any we see today.

cDNA DNA synthesized from mRNA.

cell Basic structural and functional unit of all organisms; smallest entity with the properties of life.

cell cortex Supportive mesh of crosslinked cytoskeletal elements just underneath the plasma membrane and attached to it.

cell cycle Of eukaryotic cells, a series of events from the time a cell forms until it reproduces.

cell differentiation In developing embryos, the process by which different cell lineages selectively express different fractions of their genome and so become specialized.

cell junction Of a tissue, a structure that connects adjoining cells, or a cell to matrix.

cell plate formation Cytoplasmic division mechanism of plant cells.

cell theory Theory that all organisms are made of one or more cells, cell is the smallest unit of organization that displays the properties of life, and all cells arise by division of another cell.

cell wall Of many cells (not animal cells), a semirigid but permeable structure that surrounds the plasma membrane.

central vacuole In many mature, living plant cells, an organelle that stores amino acids, sugars, and some wastes.

centromere Of eukaryotic chromosomes, constricted region where sister chromatids attach; binding site for spindles.

cephalization Of most bilateral animals, a concentration of sensory structures and nerve cells at the anterior end of the body.

chemical bond An attractive force that links two atoms.

chemical equilibrium Of reversible reactions, state in which the rate of the forward reaction equals that of the reverse.

chlorophyll *a* The primary photosynthetic pigment in plants and algae; appears green.

chlorophyll *b* Yellow–green photosynthetic accessory pigment.

chloroplast Organelle of photosynthesis in plants and algae. Two outer membranes enclose a semifluid interior, the stroma. A third, much-folded inner membrane (the thylakoid membrane) forms an inner compartment and functions in ATP and NADPH formation.

chordate Animal having a notochord, a dorsal hollow nerve cord, a pharynx with gill slits, and a tail extending past the anus in the embryo.

chromatin All of the DNA molecules and the associated proteins in a nucleus.

chromosome In eukaryotic cells, one of many linear DNA molecules with attached histones and other proteins. Prokaryotic cells have one circular chromosome.

chromosome number The sum of all of the chromosomes in cells of a given type.

ciliate A heterotrophic alveolate protist that has many cilia; *e.g., Paramecium*.

cilium, plural **cilia** A motile or sensory structure that projects from the plasma membrane of certain eukaryotic cells.

circadian rhythm Any biological activity repeated in cycles, each about twenty-four hours long, independently of any shifts in environmental conditions.

clade *See* monophyletic group.

cladogenesis Branching speciation pattern; genetic divergences occur after populations become reproductively isolated.

cladogram Evolutionary tree diagram that depicts relative relatedness among groups.

classification system A way to organize and retrieve information about species.

cleavage Process of animal cell division. Also, early stage of animal development in which mitosis divides a fertilized egg into smaller, nucleated cells (blastomeres).

climate Weather conditions that typically prevail in a region.

clone A genetically identical copy of DNA, a cell, or a multicelled organism.

cloning vector Any DNA molecule that can accept foreign DNA and be replicated inside a host cell.

club fungus Fungus that produces sexual spores in club-shaped cells.

cnidarian A radial invertebrate with epithelial tissues and a saclike gut; makes nematocysts.

codon Linear sequence of three nucleotides in mRNA; codes for an amino acid or a termination signal in protein synthesis.

coelom Of many invertebrates and all the vertebrates, a cavity lined with peritoneum that forms between the gut and body wall.

coenzyme An organic molecule that is a necessary participant in some enzymatic reactions; *e.g.*, a vitamin or NAD^+.

coevolution The joint evolution of two species interacting so closely that a change in one exerts selection pressure on the other.

cofactor A metal ion or a coenzyme that assists enzyme catalysis by alternately accepting and donating atoms, electrons, or functional groups.

cohesion The tendency of like molecules, particularly water molecules, to stick together.

cohort Group of individuals of the same age.

commensalism Interspecific interaction; one species benefits and the other is unaffected.

communication signal Body trait, chemical, or behavior that conveys information to other members of the same species.

community All populations of all species that occupy a given habitat.

comparative morphology Scientific study of comparable body parts in the embryonic and adult forms of major lineages.

compartmentalization Defensive response in some plants; an infected area is walled off by secretion of resins and toxins.

competitive exclusion A superior competitor drives a less competitive one with the same resource needs to local extinction.

compound Molecule consisting of two or more elements in unvarying proportions.

concentration Of a mixture, the number of molecules of a substance per unit volume.

condensation reaction A type of chemical reaction in which two molecules become covalently bonded into a larger molecule; water often forms as a by-product.

cone Reproductive structure of conifers; clusters of scales with exposed ovules.

conifer A cone-producing gymnosperm.

conjugation Among prokaryotes, transfer of a plasmid from one cell to another.

conservation biology The field of inquiry that surveys biodiversity, researches its evolutionary history, and determines how to maintain and use it to benefit humans.

consumer Organism that feeds on the tissues of other organisms; heterotroph.

continuous variation Of individuals of a population, a range of small differences in the phenotypic expression of a particular trait.

control group In experiments, a group used as a standard for comparison with one or more experimental groups.

covalent bond A type of chemical bond that forms when two atoms share a pair of electrons. Sharing is unequal in a *polar covalent bond*, and equal in a *nonpolar covalent bond*.

craniate A chordate with a cranium of bone or cartilage; a hagfish or vertebrate.

crossing over During prophase I of meiosis, exchange of segments between the nonsister chromatids of homologous chromosomes; puts novel combinations of alleles in gametes.

culture Sum of behavior patterns of a social group, passed between generations by learning and symbolic behavior.

cuticle Of plants, a protective surface layer secreted by epidermal cells. Of annelids, roundworms, and arthropods, a noncellular body covering. In the latter two groups it is periodically molted.

cycad A gymnosperm of subtropical or tropical habitats; has pollen-bearing and seed-bearing strobili on separate plants.

cyst Of many microbes, a resting stage with a secreted cover.

cytoplasm Semifluid matrix between the plasma membrane and nucleus or nucleoid.

cytoskeleton In a eukaryotic cell, the dynamic framework of protein filaments that support, organize, and move the cell and internal structures. Prokaryotic cells have a few similar protein filaments.

decomposer Prokaryotic or fungal heterotroph; obtains carbon and energy by breaking down organic wastes or remains.

deforestation Removal of all trees from a large tract of land.

deletion Loss of a chromosome segment; often leads to genetic disorders. Also, loss of one or more nucleotides from a gene.

demographic transition model Model that correlates changes in population growth with stages of economic development.

demographics Vital statistics of a population; *e.g.*, size, age structure.

denaturation Unravelling of a molecule's three-dimensional shape caused by high temperature, shifts in pH, or detergents.

denitrification Conversion of nitrate or nitrite to gaseous nitrogen by metabolic activity of certain bacteria in soil.

density-dependent control Any factor that comes into play in an overcrowded population; reduces birth rate or raises death and dispersal rates, as by intensified predation, parasitism, disease, competition.

density-independent factor A factor that causes fewer births or more deaths in a population regardless of its density; *e.g.*, a severe storm.

deoxyribonucleic acid *See* DNA.

derived trait A novel feature shared only by descendants of an ancestral species in which it originated.

desalinization Removal of salt from water.

desertification Conversion of grassland or cropland to desertlike conditions.

detrital food web Cross-connecting food chains in which energy flows primarily from plants through arrays of detritivores and decomposers (rather than herbivores).

detritivore Any animal that feeds on decomposing particles of organic matter.

deuterostome A member of the bilateral, coelomate, animal lineage in which the second indentation to form on the early embryo's surface becomes a mouth.

development Of complex, multicelled species, a series of stages by which an embryo becomes an adult organism.

diffusion Spontaneous dispersal of ions, molecules, or energy following a gradient.

dihybrid Individual that is heterozygous for two genes; *e.g.*, *AaBb*.

dihybrid experiment Cross between two identical dihybrids; *e.g.*, *AaBb* x *AaBb*.

dinoflagellate Flagellated, usually marine, protist; an alveolate with cellulose plates; photosynthetic and heterotrophic species.

dinosaur One of a group of reptiles that arose in the Triassic and became the dominant land vertebrates for 125 million years.

diploid Having two of each type of chromosome characteristic of the species.

directional selection Mechanism of natural selection by which forms at one extreme of a range of phenotypic variation are favored.

disease Illness caused by infection, diet, or environmental factors.

disruptive selection Mechanism of natural selection that favors different forms of a trait at both ends of a range of variation; intermediate forms are selected against.

distance effect Islands close to a source of colonists show greater species richness than islands farther from such a source.

DNA Deoxyribonucleic acid. Nucleic acid polymer of adenine, guanine, cytosine, and thymine bases, the sequence of which encodes primary hereditary information for all living organisms and many viruses.

DNA fingerprint The unique pattern of fragments of an individual's genomic DNA.

DNA polymerase Type of enzyme that catalyzes the addition of free nucleotides to new DNA strands during replication.

DNA replication Process by which a cell duplicates its DNA molecules before it divides into daughter cells.

DNA sequencing Method of determining the linear sequence of nucleotides in DNA.

dosage compensation Gene control mechanism that balances gene expression between the sexes; one X chromosome in female mammal somatic cells is inactivated.

doubling time Time it takes for a population to double in size.

downwelling Water is forced down and away from a coast after winds shift and make a surface current pile into the coast.

duplication Base sequence in DNA that has been repeated two or more times.

echinoderm A radial, predatory invertebrate with bilateral larvae and calcified spines or plates on the body wall; *e.g.*, sea stars.

ecological succession Replacement of one array of species in a community by another.

ecology Study of the ways in which organisms interact with one another and the environment.

ecoregion Broad region of land or aquatic habitat characterized by certain physical and biological factors.

ecosystem Community interacting with its physical environment.

ectoderm Outermost primary tissue layer of animal embryos; gives rise to the vertebrate nervous tissue and outer part of skin.

Ediacarans Collection of tiny precambrian species that may have been some of the earliest multicelled life forms.

effector Muscle (or gland); helps bring about movement (or chemical change) in response to neural or endocrine signals.

egg Mature female gamete, or ovum.

El Niño Periodic warming of waters in the western equatorial Pacific; disrupts ocean currents, upwelling, and global climate.

electron Negatively charged particle that occupies orbitals around an atom's nucleus.

electron transfer chain Array of enzymes and other molecules in a cell membrane that accept and give up electrons in sequence; operation of chain releases the energy of the electrons in small, usable increments.

electron transfer phosphorylation Final stage of aerobic respiration; electron flow through electron transfer chains embedded in the inner mitochondrial membrane sets up H+ gradients that drive ATP formation.

emigration Permanent move of one or more individuals out of a population.

endangered species A species endemic (native) to a habitat, found nowhere else, and highly vulnerable to extinction.

endocytosis The cellular uptake of external substances by formation of vesicles from patches of plasma membrane.

endoderm Inner primary tissue layer of animal embryos; source of the inner gut lining and organs derived from it.

endomembrane system Set of organelles that includes endoplasmic reticulum, Golgi bodies, and transport vesicles.

endoplasmic reticulum (ER) Organelle that extends from the nuclear envelope through the cytoplasm. New polypeptide chains become modified in the ribosome-studded rough ER. Membrane lipids are assembled, and fatty acids and toxins are broken down in the smooth ER.

endosymbiosis Intimate, permanent ecological interaction in which one species lives inside and reproduces in another's body.

energy The capacity to do work.

energy pyramid Diagram that depicts the energy stored in the tissues of organisms at each trophic level in an ecosystem. The lowest tier (producers) is always the largest.

ENSO El Niño Southern Oscillation. A warming and then cooling of waters in the western Pacific Ocean.

enzyme A type of protein or RNA that speeds up (catalyzes) a specific reaction, and is also unchanged by the reaction.

epithelial tissue *See* epithelium.

epithelium Animal tissue that covers external and internal body surfaces.

ER *See* endoplasmic reticulum.

eudicot Member of the most diverse group of flowering plants; has net-veined leaves, and an embryo with two cotyledons.

eukaryotic cell Type of cell that starts life with a nucleus and other organelles.

eutherian Placental mammal.

eutrophication Nutrient enrichment of water; promotes phytoplankton growth.

evaporation Process of conversion of a liquid to a gas; requires energy input.

evolution, biological Heritable change in a line of descent; occurs by microevolutionary events (mutation, natural selection, genetic drift, and gene flow).

evolutionary tree A treelike diagram in which each branch point represents a divergence from a shared ancestor.

exocytosis Fusion of a cytoplasmic vesicle with the plasma membrane and release of vesicle contents into extracellular fluid.

exon A base sequence in eukaryotic DNA that remains in mRNA as coding sequence.

exoskeleton An external skeleton.

exotic species Species established in a new community after dispersal from its home range.

experimental group In experiments, a group of objects or individuals that display or are exposed to a variable that is being investigated. Compared to a control group.

exponential growth An increase by the same percent of the whole in each interval.

extinction Irrevocable loss of a species.

extreme halophile Prokaryote adapted to life in an extremely salty habitat.

extreme thermophile Prokaryote adapted to life in an extremely hot aquatic habitat.

FAD Flavin adenine dinucleotide; type of nucleotide coenzyme.

fall overturn Vertical mixing of a body of water in fall as upper waters cool and sink.

fat Type of lipid with one, two, or three fatty acid tails attached to a glycerol.

fatty acid Organic compound that has a long carbon backbone with only single

bonds in *saturated* types, and one or more double bonds in *unsaturated* types.

feedback inhibition Mechanism by which a result of some cellular activity triggers a response that slows or stops the activity.

fermentation Type of metabolic pathway that breaks down carbohydrates without using oxygen. Net yield: 2 ATP per glucose. The end product of *alcoholic fermentation* is ethanol; that of *lactate fermentation* is lactate.

fern Seedless vascular plant that forms spores in sori on the underside of fronds.

fertilization Fusion of two gametes, the result being a single-celled zygote.

fin An appendage that helps stabilize, orient, and propel most fishes in water.

fitness Degree of adaptation to an environment, as measured by the relative genetic contribution to future generations.

fixation Of an allele, loss of all other alleles of a gene in a population.

fixed action pattern Instinctual program of coordinated, stereotyped movements, usually elicited by a simple stimulus.

flagellated protozoan A member of an ancient lineage of heterotrophic single-celled protists that have one or more flagella; *e.g.*, *Giardia*.

flagellum, plural **flagella** Long, whip-like motile structure that projects from a cell.

flatworm Simple bilateral animal with no coelom, saclike gut, three primary tissue layers.

flower Reproductive structure of plants; has nonfertile parts (sepals, petals), fertile parts (stamens, carpels), and a receptacle (modified base of floral shoot).

fluid mosaic model Model that describes the organization of membranes as a mixture of lipids and proteins, the interactions and motions of which impart fluidity to it.

food chain Linear sequence of steps by which energy stored in autotroph tissues enters higher trophic levels.

food web Cross-connecting food chains.

fossil Physical evidence of an organism that lived in the distant past.

fossilization How fossils form over time. An organism or evidence of it gets buried; and metal ions replace the minerals in bones and other hardened tissues.

founder effect A form of bottlenecking. By chance, a few individuals that establish a new population differ in allele frequencies relative to the original population.

free radical Uncharged molecule or atom with an unpaired electron; highly reactive.

functional group An atom or a group of atoms that imparts chemical properties to an organic compound.

fungus, plural **fungi** Eukaryotic consumer that obtains nutrients by extracellular digestion and absorption.

gametophyte Haploid, gamete-producing structure of plants and some algae.

gel electrophoresis Method of separating DNA fragments according to length. The fragments separate as they move through a gel matrix in response to an electric field.

gene Unit of heritable information; section of DNA that encodes one structural or functional component of an organism.

gene flow Microevolutionary process; alleles enter and leave a population by immigration and emigration. Counters mutation, natural selection, genetic drift.

gene pool All the alleles of all the genes in population; a pool of genetic resources.

genetic code The near-universal language of protein synthesis, the correspondence between nucleotide sequence and protein.

genetic disorder An inherited condition that causes mild to severe medical problems.

genetic divergence Accumulation of differences in the gene pools of two or more populations after gene flow stops and mutation, natural selection, and genetic drift occur in each.

genetic drift Change in allele frequencies in a population due to chance alone. Effects are most pronounced in small populations.

genetic equilibrium In theory, a state in which specific allele frequencies are not changing, so the population is not evolving with respect to that gene.

genome Complete complement of genetic information of an organism or species.

genomics The study of genes and gene function in humans and other organisms.

genus, plural **genera** Group of one or more related species that share a number of unique morphological traits.

geographic dispersal Individuals leave a home range and take root in a new community.

geologic time scale Time scale for Earth's history; major subdivisions correspond to mass extinctions.

gill A respiratory organ that has thin, moist, vascularized filaments for gas exchange.

global warming A long-term increase in the temperature of Earth's lower atmosphere.

glycolysis Set of metabolic reactions in which glucose is broken down to pyruvate; first stage of both aerobic respiration and fermentation. Yields 2 ATP per glucose.

gnetophyte A woody, vinelike or shrubby gymnosperm of the group most closely related to flowering plants.

Golgi body Organelle of the endomembrane system; enzymes in it modify new polypeptide chains and assemble lipids, then package both in vesicles.

Gondwana Paleozoic supercontinent that later became part of Pangea.

gradient Difference in a physical property, such as concentration, electric charge, or pressure, over distance. Drives diffusion.

gradual model, speciation Addresses the rate of speciation; morphological changes are said to accumulate over great time spans.

grazing food web Cross-connecting food chains in which energy flows from plants to an array of herbivores, then carnivores.

greenhouse effect Trapping of heat near Earth's surface by atmospheric gases. The gases absorb heat from Earth's sun-warmed surface, and then radiate some back to Earth.

groundwater Water in soil and porous rock layers called aquifers.

growth Of multicelled species, increases in the number, size, and volume of cells. Of single-celled prokaryotes, rise in the number of cells.

gut Sac or tube in which food is digested.

habitat Place where an organism or species normally lives; characterized by physical and chemical features and array of species.

habitat loss Reduction in suitable living space and closure of part of a habitat.

half-life Characteristic time it takes for half of a quantity of a radioisotope to decay.

haploid Having one of each type of chromosome characteristic of the species.

heterotroph Organism that obtains energy and carbon by feeding on other organisms, their wastes, or their remains.

heterozygous Condition of having nonidentical alleles at a given locus.

higher taxon, plural **taxa** One of ever more inclusive groupings of species; *e.g.*, family, order, class, phylum, kingdom.

homeostasis State of equilibrium in which an organism maintains its internal environment within a tolerable range.

homeotic gene Gene that, when expressed in a localized region of a developing embryo, helps determine the identity of a body part.

hominid All humanlike and human species; *e.g.*, australopiths, early *Homo*.

homologous chromosome One of a pair of chromosomes in diploid cells; except for nonidentical sex chromosomes, a pair has the same size, shape, and gene sequence.

homologous structures Similar body parts in related lineages; indicate shared ancestry.

homozygous dominant State of having a pair of dominant alleles; *e.g.*, AA.

homozygous recessive State of having a pair of recessive alleles; *e.g.*, Aa.

horsetail A seedless vascular plant with rhizomes, scale-like leaves, and hollow stems with silica-reinforced ribs.

hot spot A region where human activities are driving many species to extinction. Also, region where superplumes have ruptured Earth's crust.

human Primate of species *Homo sapiens*.

human gene therapy Transfer of normal or modified genes into a person, often with the intent to correct a genetic defect.

hybrid Heterozygous individual; also, the offspring of a cross between individuals of different species or different genera.

hydrogen bond Weak chemical bond that forms between a covalently bonded hydrogen atom and another atom taking part in a different polar covalent bond.

hydrologic cycle Driven by solar energy; water moves through atmosphere, on or through land, to the ocean and other bodies of water, and back to the atmosphere.

hydrolysis Cleavage reaction; an enzyme splits a molecule by adding —OH and —H from water to each of the two fragments.

hydrophilic Of a molecule, the property of dissolving easily in water. Also, polar.

hydrophobic Of a molecule, the property of not dissolving in water. Also, nonpolar.

hydrostatic pressure Pressure exerted by a volume of fluid against a cell wall or other enclosing structure. Also, turgor.

hypertonic Of two fluids, describes the one with the higher solute concentration.

hypha, plural **hyphae** Fungal filament with chitin-reinforced walls; part of a mycelium.

hypothesis Speculative explanation for, or generalization about, something observable.

immigration Migration of individuals into a population; adds to population size.

imprinting A form of learning triggered by exposure to a simple stimulus during a sensitive period, usually early in life.

in vitro fertilization Conception outside the body, "in glass" petri dishes or tubes.

inbreeding Mating among close relatives.

inclusive fitness theory Theory that genes for altruism can be favored by selection if they lead to behavior that increases the reproductive success of the altruist's close relatives.

independent assortment, theory of Theory that each homologous chromosome and its partner (and the genes they carry) are assorted into different gametes independently of the other pairs. Crossing over affects the outcome.

indicator species Any species which, by its abundance or scarcity, provides a measure of the health of its habitat.

infection Successful colonization of a host species by another organism.

insertion A mutation in which one or more extra bases enter a gene.

instinctive behavior A behavioral response that does not require learning.

integrator Internal control center that receives, processes, and stores sensory input, and then coordinates the responses; *e.g.*, a brain.

intermediate disturbance hypothesis Idea that species richness is greatest in habitats where disturbances are moderate.

intermediate filament Cytoskeletal element that mechanically strengthens some cells.

internal environment All body fluids *not* inside cells. Also, extracellular fluids.

interphase Of a eukaryotic cell cycle, interval between mitotic divisions.

interspecific competition Competition between members of different species.

interstitial fluid Fluid in spaces between the cells of all tissues except the blood (in which extracellular fluid is called plasma).

intron Noncoding DNA in eukaryotic chromosomes; excised from pre-mRNA.

inversion A chromosomal alteration; part of the DNA sequence gets oriented in the reverse direction, with no molecular loss.

invertebrate Animal without a backbone.

ion exchange pH-dependent process; ions dissociate from soil particles, then other ions dissolved in soil water replace them.

ionic bond Type of chemical bond that forms between ions of opposite charges.

isotonic Describes two fluids with the same solute concentrations.

isotope One of two or more forms of the same element (same number of protons) that differ in their number of neutrons.

jaw Paired, hinged cartilaginous or bony feeding structures of most chordates.

joint Area of contact between bones.

karyotype Preparation of an individual's diploid complement of chromosomes.

key innovation Modification in body structure or function that allows a species to exploit its environment more efficiently or in a novel way.

keystone species A species that influences community structure in disproportionally large ways relative to its abundance.

Krebs cycle The second stage of aerobic respiration in which many coenzymes and two ATP form as pyruvate from glycolysis is fully broken down to CO_2 and H_2O.

K–T asteroid impact theory Theory that a huge asteroid struck Earth 65 million years ago and caused a mass extinction.

lac operon Set of genes and gene controls for lactose metabolism in *E. coli* bacteria.

larva, plural **larvae** An immature stage between embryo and adult in the life cycle of animals that undergo metamorphosis.

lateral gene transfer Movement of genes from one cell to another cell, as by conjugation.

learned behavior Behavior that is modified as a result of experience.

library Collection of cells that have taken up different DNA fragments or assorted cDNAs from another species.

lichen Mutualism between a fungus and a cyanobacterium or alga.

life history pattern A description of when and how many offspring are produced in a lifetime.

light-dependent reactions First stage of photosynthesis in which photon energy is converted to the chemical energy of ATP.

light-independent reactions *See* Calvin–Benson cycle.

limiting factor Any essential resource that slows population growth when scarce.

lineage Line of descent.

lipid Nonpolar hydrocarbon; *e.g.*, a fat, oil, wax, sterol, or phospholipid. Cells use different kinds as storable forms of energy, structural materials, or signaling molecules.

lipid bilayer Major component of all cell membranes; mainly phospholipids arranged tail-to-tail in two layers.

logistic growth Of a population, a phase of nearly exponential growth; levels off as density-dependent controls slow growth.

lung Internal respiratory organ of all birds, reptiles, mammals, most amphians, some fish.

lycophyte A type of seedless vascular plant; living species are small with tiny leaves and branching stem and roots; *e.g.*, club moss.

lysis Gross damage to a cell wall or plasma membrane that causes cell death.

lysosome Vesicle filled with enzymes that functions in intracellular digestion.

macroevolution Large-scale patterns, rates of change, and trends among lineages.

magnoliid Least diverse of the three major flowering plant groups; *e.g.*, magnolia.

mammal Amniote with hair; feeds offspring with milk from the female's mammary glands.

marine snow Organic matter drifting down from ultraplankton to mid-oceanic water; supports food webs and marine biodiversity.

marsupial Pouched mammal, such as kangaroo.

mass extinction Catastrophic period in geologic time when major groups are lost.

megaspore Haploid spore that forms in female parts of seed-bearing plants; gives rise to the female gametophyte.

meiosis Nuclear division mechanism that halves the parental chromosome number, to a haploid (*n*) number. Precedes the formation of gametes and sexual spores.

mesoderm Of animals with a three-layered embryo, the middle primary tissue layer that gives rise to most internal organs and part of the skin.

messenger RNA mRNA. A single strand of ribonucleotides transcribed from DNA; the only type of RNA that carries protein-building information to ribosomes.

metabolic pathway A stepwise sequence of enzyme-mediated reactions.

metabolism Sum of all chemical processes by which cells acquire and use energy to grow, maintain themselves, and reproduce.

metamorphosis Major changes in body form of some animals. Hormone-controlled growth, tissue reorganization, and body part remodeling produces the adult form.

methanogen A prokaryote that makes methane gas as a by-product of anaerobic reactions.

microevolution Of a population, small-scale change in allele frequencies resulting from mutation, genetic drift, gene flow, natural selection, or a combination of them.

microfilament Cytoskeletal element that functions in cell contraction, movement, and structural support; forms cell cortex.

microspore A haploid spore that gives rise to pollen grains in seedbearing plants.

microtubule Cytoskeletal element; forms a dynamic framework that organizes and moves internal parts of the cell.

migration Of many animals, a recurring pattern of movement between two or more regions in response to seasonal change.

mimicry Resemblance between unrelated species; often a defense against predation.

mitochondrion Eukaryotic organelle of ATPformation; its inner membrane is the site of the second and third stages of aerobic respiration.

mitosis Nuclear division that maintains the parental chromosome number. The basis of growth in size, tissue repair, and often asexual reproduction of eukaryotes.

mitotic spindle Of eukaryotic cells, a dynamic array of microtubules that moves chromosomes during mitosis or meiosis.

mixture Two or more types of molecules intermingled in variable proportions.

model Theoretical explanation of a partially-understood process.

molecule Atoms joined by chemical bonds.

molecular clock Model used to estimate relative times of divergences by comparing the number of neutral mutations in the DNA of different species.

mollusk Only invertebrate with a mantle; most have an external or internal shell.

molting Periodic shedding and regrowth of cuticle, feathers, or other external parts.

monocot Flowering plant characterized by embryo sporophytes having one cotyledon; floral parts usually in threes (or multiples of three); and often parallel-veined leaves.

monohybrid Individual heterozygous for one gene; *e.g.*, *Aa*.

monohybrid experiment Cross between two identical monohybrids; *e.g.*, *Aa* × *Aa*.

monophyletic group Set of all species descended from a common ancestor. Also, clade.

monotreme Egg-laying mammal.

morphological convergence A pattern of macroevolution in which body parts of distant lineages evolve to become similar.

morphological divergence A pattern of macroevolution in which body parts of related lineages evolve to become different.

motor proteins Proteins that use ATP to move along a microfilament or microtubule; their concerted actions move a cell or cell structure.

multiple allele system Persistence of three or more alleles of a gene in a population.

mutation Heritable change in the sequence of DNA. Original source of new alleles and life's diversity, but most are harmful. *Lethal mutations* drastically affect phenotype, and thus usually cause death.

mutualism An interspecific interaction that benefits both species.

mycelium Of a multicelled fungus, a mesh of branching hyphal filaments that absorb food.

mycorrhiza "Fungus-root." A mutualism between a fungus and young plant roots.

NAD+ Nicotinamide adenine dinucleotide. Type of nucleotide coenzyme.

natural selection A microevolutionary process; differences in survival and reproduction among individuals of a population that differ in the details of their heritable traits.

negative control Gene control mechanism; a regulatory protein slows gene expression.

negative feedback mechanism A homeostatic mechanism by which an activity changes some condition, which in turn triggers a response that reverses the change.

neoplasm Abnormal mass of cells (tumor) that lost control over the cell cycle.

nerve cord A communication line that parallels the anterior–posterior axis of bilateral animals. In chordates, it develops as a hollow, neural tube and gives rise to the spinal cord and brain.

net ecosystem production All the energy the primary producers have accumulated during growth, reproduction in a specified interval (net primary production), *minus* energy that producers, and decomposers, have used.

niche Sum of activities and relationships in which individuals of a species engage as they secure and use resources.

nitrification Soil bacteria break down ammonia or ammonium to nitrite, then other bacteria break down nitrite to nitrate.

nitrogen cycle An atmospheric cycle in which nitrogen moves from its largest reservoir, the atmosphere, through the ocean waters and sediments, soils, and food webs, then back to the atmosphere.

nitrogen fixation Conversion of gaseous nitrogen to ammonia.

nondisjunction Failure of sister chromatids or homologous chromosomes to separate during meiosis or mitosis. Daughter cells receive too many or too few chromosomes.

nuclear envelope A double membrane that is the outer boundary of the nucleus.

nucleic acid Nucleotide polymer; used by all living cells and viruses to convey heritable information; *e.g.*, DNA, RNA.

nucleic acid hybridization Base-pairing between DNA or RNA from different sources.

nucleoid The portion of a prokaryotic cell where DNA is physically organized but not enclosed in a membrane.

nucleosome DNA wound twice around histone proteins; basic unit of structural organization in eukaryotic chromosomes.

nucleotide Small organic compound with a five-carbon sugar, a nitrogen-containing base, and phosphate. Monomers of nucleic acids are named after their component bases: adenine, guanine, cytosine, thymine, and uracil.

nucleus Defining organelle of eukaryotes; separates DNA physically and chemically from the cytoplasm.

oocyte Immature egg.

operator Part of an operon; DNA binding site for a regulatory protein.

organ Two or more tissues organized in a way that allows them to perform some task.

organ system A set of organs that interact chemically, physically, or both in a task.

organelle One of many membrane-bound compartments with special functions inside eukaryotic cells; *e.g.*, a nucleus.

organic compound A carbon-based molecule that also incorporates atoms of hydrogen and, often, oxygen, nitrogen, and other elements; *e.g.*, fats, proteins.

osmosis Diffusion of water across a selectively permeable membrane from a region where the water concentration is higher to a region where it is lower.

osmotic pressure Amount of pressure which, when applied to a hypertonic fluid, will stop osmosis into that fluid.

oxidation–reduction (redox) reaction Type of chemical reaction in which one molecule gives up electrons to another. Primary reaction in electron transfer chains.

Pangea Paleozoic supercontinent; the first land plants and animals evolved on it.

parapatric speciation A speciation model in which populations in contact along a common border evolve into new species.

parasite Organism that feeds on a living host, but usually does not kill it outright.

parasitism Symbiotic interaction in which one species draws nutrients from another, usually without killing it.

parasitoid A type of insect that, in a larval stage, grows inside a host (usually another insect), feeds on its soft tissues, and kills it.

passive transport Diffusion of a solute across a cell membrane, through the interior of a transport protein.

pathogen An organism that infects a host and disrupts its function, causing disease.

PCR Polymerase chain reaction. A method to rapidly copy specific DNA fragments.

pedigree Chart of genetic connections among individuals related by descent.

pellicle A thin, flexible, protein-rich body covering of some single-celled eukaryotes.

per capita Per individual.

permafrost An impermeable, perpetually frozen layer that underlies arctic tundra.

peroxisome Enzyme-filled vesicle that breaks down amino acids, fatty acids, and some toxic substances such as ethanol.

pH scale A scale of 0 to 14 that reflects the H^+ concentration of a solution. pH 7 is neutral.

pheromone A secreted signaling molecule that affects behavior and/or physiology of another individual of the same species.

phospholipid A lipid with a phosphate group in its hydrophilic head. The main constituent of cell membranes.

phosphorus cycle Sedimentary cycle that moves phosphates from land, through food webs, to ocean sediments, then back to land.

phosphorylation Enzyme-mediated transfer of a phosphate group to an organic compound.

photosynthesis The chemical process by which photoautotrophs capture and use light energy to make sugars from CO_2 and water.

photosystem In photosynthetic cells, a cluster of pigments and other molecules; converts light energy to chemical energy.

phycobilin One of a class of red or blue-green photosynthetic accessory pigments that occur in cyanobacteria and red algae.

physiology Study of how the multicelled body functions in its environment; more specifically, of mechanisms by which component parts grow, develop, and are maintained and reproduced.

pigment Molecule that absorbs light of certain energy; pigments are the bridge between light and photosynthesis.

plankton Aquatic community of mostly microscopic autotrophs and heterotrophs.

plant Multicelled photoautotroph, generally with roots, stems, and leaves. The primary producers on land.

plasma Liquid portion of blood; water with dissolved proteins and other solutes.

plasma membrane Outer cell membrane; the structural and functional boundary between cytoplasm and extracellular fluid.

plasmid Circular bacterial DNA molecule. Some are used as cloning vectors.

plate tectonics theory Theory that great slabs or plates of Earth's outer layer float on a hot, semi-molten mantle. The plates have rafted continents to new positions over time.

pleiotropy Multiple effects of one gene.

pollen grain Of seedbearing plants, a tiny structure formed from microspores; consists of a wall around a few cells that will develop into a mature, sperm-bearing, male gametophyte.

pollination The arrival of pollen on the female parts of a plant of the same species.

pollinator Any living agent that delivers pollen to egg-containing structures of the same species.

polypeptide chain Series of many amino acids linked by peptide bonds.

polyploidy A condition in which somatic cells have three or more of each type of chromosome.

population Group of individuals of one species living in a specified area.

population density Count of individuals of a population in a specified portion of a habitat.

population distribution The pattern in which the individuals of a population are dispersed through their habitat.

population size The total number of individuals that make up a population.

positive control Gene control mechanism by which a regulatory protein turns on or enhances gene expression.

positive feedback mechanism A homeostatic mechanism by which an activity changes some condition, which in turn triggers a response that intensifies the change.

predation Ecological interaction in which a free-living predator eats a prey organism.

predator A heterotroph that captures other living organisms (its prey), and feeds on them, usually killing them.

prediction A statement of what you would expect to observe if a hypothesis is valid.

pressure flow theory Theory that the flow of organic compounds through phloem is caused by pressure gradients between source and sink regions.

prey An organism that a predator eats.

primary producer An autotroph at the first trophic level of an ecosystem.

primary productivity The rate at which an ecosystem's primary producers secure and store energy in tissues in a given interval.

primary succession Establishment and replacement of one species array after another in an area where no community existed.

primary wall The first thin, pliable cell wall of young plant cells.

primate Placental mammal with nails rather than claws; adapted for grasping; a prosimian, tarsioid, or anthropoid.

primer Synthetic, single-stranded fragment of DNA used as an initiation site for DNA synthesis in PCR and DNA sequencing.

prion Protein particle that, in its infectious misfolded form, causes fatal brain diseases such as Creutzfeldt-Jakob disease.

probability The chance that one outcome of an event will occur; proportional to the number of possible outcomes.

probe Fragment of DNA complementary to a gene of interest, labeled with a tracer.

producer An organism that makes its own food from inorganic materials; autotroph.

prokaryotic cell A single-celled organism, often walled, without a nucleus or other organelles; a bacterium or archean.

promoter Short stretch of DNA to which RNA polymerase binds. Transcription then begins at the gene closest to the promoter.

protist One of the structurally simple single-celled or multicellular eukaryotes, which are now being classified into monophyletic groups.

proto-cell Membrane-bound sac enclosing metabolic machinery; presumed stage that preceded evolution of cells.

proton Positively charged particle in the nucleus of all atoms.

protostome A member of the bilateral, coelomate, animal lineage in which the first indentation to form on the early embryo's surface becomes a mouth; *e.g.*, mollusks, annelids, arthropods.

pseudopod A dynamic lobe of membrane-enclosed cytoplasm; functions in motility and phagocytosis by amoebas, amoeboid cells, and some white blood cells.

punctuation model, speciation Addresses the rate of speciation; morphological change is said to occur relatively quickly after two populations start to diverge.

Punnett-square method Method in which a matrix of genotypes is used to predict the probability of outcomes of a genetic cross.

pyruvate Three-carbon compound that forms as an end product of glycolysis.

quadrat One of a number of sampling areas of the same size and shape; data from these areas is used to estimate population density and other factors for a larger area.

radial symmetry Animal body plan with parts arrayed in a repeating pattern around a central axis, as in a sea anemone.

radiation Energy carried by a photon or subatomic particle. *Ionizing* radiation has enough energy to damage DNA and living tissues; *nonionizing* radiation does not.

radiometric dating Method of dating rocks and newer fossils in which the proportion of a radioisotope to its daughter element in a sample is used to calculate the sample's age.

rain shadow Reduction in rainfall on that side of a mountain range which faces away from prevailing winds; results in arid or semiarid conditions.

ray-finned fish A member of the most diverse group of bony fish; has fin supports derived from skin, a swim bladder, and thin flexible scales, *e.g.,* trout.

reaction Process of chemical change by which a cell builds, rearranges, or breaks down organic compounds.

receptor, molecular Protein that initiates a cellular response when it binds a hormone or other signalling molecule.

receptor, sensory Sensory cell or a specialized ending of one that detects a certain stimulus.

regulatory protein Protein that directly or indirectly adjusts gene expression.

reproductive base Of some population, the number of individuals already reproducing, plus those not yet reproductively mature.

reproductive isolating mechanism A heritable feature of body form, function, or behavior that prevents interbreeding between populations; sets the stage for genetic divergences.

reptile A vertebrate with lungs, waterproof skin and lungs; amniote eggs. Living forms are all ectothermic (cold-blooded). Not a monophyletic group; *e.g.,* dinosaur, turtle, crocodile, snake.

resource partitioning Use of the same resource in different places or at different times; permits species with similar resource needs to coexist.

restriction enzyme One of hundreds of proteins that recognize and cut specific base sequences in double-stranded DNA.

reverse transcriptase Viral replication enzyme that uses an RNA template to assemble a strand of DNA from nucleotides.

rhizoid A rootlike part of a bryophyte.

rhizome Horizontal-growing underground stem that gives rise to roots and shoots of many vascular plants.

ribosomal RNA rRNA. The RNA component of ribosomes.

ribosome In all cells, a structure that translates genetic information in mRNAs into proteins.

riparian zone The narrow strip of vegetation along a stream or river.

RNA Ribonucleic acid. Single-stranded nucleic acid polymer of adenine, guanine, cytosine, and uracil nucleotides. Central to the processes of transcription and translation.

RNA polymerase Enzyme that catalyzes transcription of DNA into RNA.

RNA world A hypothetical period in Earth history when RNA, not DNA, stored and transmitted genetic information.

roundworm Bilateral invertebrate with a false coelom, complete digestive system; cylindrical body is covered by a cuticle that is molted.

rubisco RuBP carboxylase. Carbon-fixing enzyme of the C3 photosynthesis pathway.

sac fungus Member of the largest fungal group; sexual spores form in sac-shaped cells.

salinization Salt buildup in soil.

salt Any compound that releases ions other than H^+ and OH^- in solution.

sampling error Difference between a statistical result derived from surveying a complete set of events or individuals, and a result derived from surveying a subset of it.

saprobe Heterotroph that extracts energy and carbon from organic remains and wastes.

scientific theory *See* Theory, scientific.

secondary succession Sequential changes in the array of species in a disturbed area.

secondary wall A rigid, permeable cell wall that forms inside the primary wall of many mature plant cells.

seed A mature ovule of a seed-bearing plant; contains a plant embryo.

segmentation Of animal body plans, existence of repeating units, some of which may be modified for special purposes.

segregation, theory of Mendel's theory that the two genes of a pair separate at meiosis to end up in different gametes.

selective permeability Capacity of a membrane to prevent or allow flow of specific substances across it.

selfish herd Social group held together by reproductive self-interest.

sensory receptor Cell or cell part that detects a specific form of energy.

sex chromosome One of the two chromosomes that determine gender.

sexual dimorphism A notable difference between the female and male phenotypes of a species; outcome of sexual selection.

sexual reproduction Reproductive process by which offspring arise through the combining of genetic material from two parents.

sexual selection Form of natural selection in which members of the opposite sex serve as the agents of selection.

shell model Model that depicts electron energy levels as a nested series of shells.

sister chromatids Two attached members of a duplicated eukaryotic chromosome.

six-kingdom classification system Grouping of all organisms into kingdoms Bacteria, Archaea, Protista, Fungi, Plantae, and Animalia.

sodium–potassium pump Protein that actively transports sodium out of a cell, and at the same time passively transports potassium into it.

solute Any dissolved substance.

speciation Macroevolutionary process by which a new species arises.

species One or more natural populations of individuals with at least one unique trait, derived from a common ancestor, that occurs in no other groups.

sperm Mature male gamete.

sponge Structurally simple, filter-feeding invertebrate with no body symmetry or tissues.

spore Structure of one or a few cells, formed sexually or asexually, often walled or coated, that protects and/or disperses a new generation. Some bacteria and protists, as well as fungi and plants, form spores.

sporophyte A diploid spore-producing body of a plant or multicelled alga; grows by mitotic cell divisions from a zygote.

spring overturn Of a large body of water, a downward movement of oxygenated surface water and an upward movement of nutrient-rich water from below during spring; increases primary productivity.

stabilizing selection Mode of natural selection; intermediate phenotypes are favored over extremes at both ends of the range of variation.

sterol Lipid with no fatty acids. Important component of eukaryotic cell membranes; some kinds are hormones.

stimulus A specific form of energy that activates a sensory receptor; *e.g.*, pressure.

stoma, plural **stomata** Tiny gap that forms across the epidermis of a leaf or primary stem when paired guard cells swell; allows water vapor and gases to cross the epidermis.

stratification Formation of sedimentary rock layers, or the layers themselves.

strip logging A way to minimize erosion from deforestation; a narrow corridor that parallels contours of sloped land is cleared; the upper part is used as a log-hauling road, then is reseeded from intact forest above it.

strobilus Of certain nonflowering plants, a cluster of spore-producing structures.

stroma Region between the thylakoid membrane and the outer membranes of a chloroplast; where the light-independent reactions of photosynthesis occur.

stromatolite Fossilized remains of bacterial mats that formed in shallow seas.

substrate A molecule that is specifically recognized and acted upon by an enzyme.

substrate-level phosphorylation Direct, enzyme-mediated transfer of a phosphate group from a substrate to another molecule.

surface-to-volume ratio Physical relationship that constrains the size of cells; the volume of a sphere increases with the cube of the diameter, but the surface area increases with the square.

survivorship curve Plot of age-specific survival of a cohort, from the time of birth until the last individual dies.

symbiosis An interspecific interaction in which members of different species interact closely for some or all of the life cycle; *e.g.*, mutualism, commensalism, parasitism.

sympatric speciation A speciation model. Occurs inside the home range of a species in the absence of a physical barrier.

syndrome Set of symptoms; characterizes a genetic abnormality or disorder.

systemic acquired resistance Of many plants, a mechanism in which infection of one region induces the synthesis and distribution of compounds that protect other regions from infection.

tandem repeats On a chromosome, many identical copies of a short base sequence, the number, position, and sequence of which are unique to an individual.

temperature Measure of molecular motion.

temperature zone Globe-spanning bands of temperature defined by latitude; *e.g.*, cool temperate, equatorial.

test Systematic determination of whether a prediction based on a hypothesis is valid; also, experiment.

tetrapod A vertebrate that is a four-legged walker or a descendant of one; an amphibian, reptile, bird, or mammal.

theory, scientific A rigorously tested, long-standing concept used to interpret a broad range of observations about some aspect of nature. Always open to revision.

thermal inversion A layer of dense, cool air trapped beneath a layer of warm air.

thermodynamics Science of the properties of energy. The *first law* states that energy cannot be either created or destroyed. The *second law*

states that energy disperses spontaneously; also, every time energy becomes converted from one form to another, some is lost.

three-domain system Classification system that groups all organisms into the domains Bacteria, Archaea, and Eukarya.

thylakoid Stack of interconnected discs formed from thylakoid membrane inside chloroplasts.

thylakoid membrane Much-folded inner membrane system of a chloroplast; forms a continuous compartment. In the first stage of photosynthesis, pigments and enzymes in the membrane form ATP and NADPH.

tissue Of multicelled organisms, organized group of cells and extracellular materials that interact in one or more tasks.

total fertility rate TFR. The average number of offspring born to a woman during her lifetime.

transcription Process by which an mRNA is assembled from nucleotides using a gene region in DNA as a template.

transfer RNA tRNA. One of a class of small RNA molecules that delivers amino acids to a ribosome. Its anticodon pairs with an mRNA codon during translation.

transgenic Describes an organism that has been genetically modified by experiment.

transition state The state in a chemical reaction at which the substrate bonds are at their breaking point.

translation Second stage of protein synthesis. At ribosomes, information encoded in an mRNA transcript guides the synthesis of a new polypeptide chain from amino acids.

translocation Chromosomal alteration in which a piece of chromosome becomes attached to a different chromosome. Also, process of fluid movement inside phloem.

transposon Transposable element. Small stretch of DNA that jumps spontaneously to another, random location in a genome.

triglyceride Lipid with three flexible fatty acid tails attached to a glycerol backbone. Most abundant lipid in the body.

trophic level All organisms that are the same number of transfer steps away from the energy input into an ecosystem.

uniformity Theory that Earth's surface has changed in uniformly repetitive ways by the typical forces of nature we see today.

upwelling Upward movement of deep, often nutrient-rich water near a coastline; replaces a mass of surface ocean water.

variable Of experimental tests, a specific aspect of an object or event that may differ over time or among individuals.

vector Organism that conveys a disease-causing pathogen from host to host.

vertebra, plural **vertebrae** One of a series of segments that make up a vertebrate backbone.

vertebrate A chordate that has a backbone.

viroid Infectious particle of short, tightly folded strands or circles of RNA.

virus A noncellular infectious agent that can be replicated only after its genetic material enters a host cell and subverts its metabolic machinery.

warning coloration Distinctive forms, colors, patterns, and other signals that predators learn to recognize and avoid.

watershed Region from which all water that falls as rain or snow drains into a particular stream or river.

wavelength Distance between the crests of two successive waves.

wax Type of lipid with long-chain fatty acid tails. Component of many waterproof, protective coatings on plants and animals.

X chromosome A type of sex chromosome that influences sex determination; *e.g.*, in mammals, an XX embryo becomes female; an XY embryo becomes male.

xanthophyll Photosynthetic accessory pigment; a type of carotenoid.

xenotransplantation Surgical transfer of an organ from one species to another.

x-ray diffraction image Film image of x-rays scattered by a crystalline sample.

zero population growth No net increase or decrease in population size during a interval.

zygomycetes Small fungal group in which diploid zygotes develop into zygospores, a type of thick-walled sexual spore in a thin, clear covering; e.g., black bread mold.

Art Credits and Acknowledgments

This page constitutes an extension of the book copyright page. We have made every effort to trace the ownership of all copyrighted material and secure permission from copyright holders. In the event of any question arising as to the use of any material, we will be pleased to make the necessary corrections in future printings. Thanks are due to the following authors, publishers, and agents for permission to use the material indicated.

Page iii Photographer Jay Barnes

TABLE OF CONTENTS: **Page vii** From top, © Peter Turnley/Corbis; Lisa Starr with PDB file from NYU Scientific Visualization Lab. **Page viii** From top, © Tony Brian, David Parker/SPL/Photo Researchers, Inc; © Paul Edmondson/Corbis; NASA. **Page ix** From top, © Professors P. Motta and T. Naguro/SPL/Photo Researchers, Inc; Micrograph, Dr. Pascal Madaule, France; © Science Photo Library/Photo Researchers, Inc. **Page x** From top, Photo courtesy of the College of Veterinary Medicine, Texas A&M University; PDB ID: 2AAI; Rutenber, E., Katzin, B.J., Ernst, S., Collins, E.J., Mlsna, D., Ready, M.P., Robertus, J.D.: Crystallographic refinement of ricin to 2.5 A. *Proteins* 10 pp. 240 (1991). UniProt copyright © 2002–2004 UniProt consortium; Dr. Jorge Mayer, Golden Rice Project. **Page xi** From top, © Bettmann/Corbis; NASA Galileo Imaging Team. **Page xii** From top, © Philippa Uwins/The University of Queensland; © Clinton Webb. **Page xiii** From top, P. Morris/Ardea London; Star Tribune/Minneapolis-St. Paul; © David Nunuk/Photo Researchers, Inc. **Page xiv** From top, John Kabashima; © Harold Naideau/Corbis. **Page xv** From top, © Hank Fotos Photography; © Scott Camazine; © Joseph Sohm, Visions of America/Corbis.

INTRODUCTION: NASA Space Flight Center

CHAPTER 1: **Page 1** © Peter Turnley/Corbis. **1.1** (b, above left) PDB file courtesy of Dr. Christina A. Bailey, Department of Chemistry & Biochemistry, California Polytechnic State University, San Luis Obispo, CA; (b, above center) PDB ID: 1BBB; Silva, M.M., Rogers, P.H., Arnone, A.; A third quarternary structure of human hemoglobin A at 1.7-A resolution; *J Biol Chem* 267 pp. 17248 (1992); (b above right) PDB file from Klotho Biochemical Compounds Declarative Database; (d) © Science Photo Library/Photo Researchers; (e) © Bill Varie/Corbis; (f–h) © Jeffrey L. Rotman/Corbis; (i) © Peter Scoones; (j–k) NASA. **Page 3** Bottom right, NASA. **1.3** Left, upper, PhotoDisc, Inc.; lower, Gary Head and Lisa Starr with photograph from David Neal Parks; right, © Y. Arthrus-Bertrand/Peter Arnold, Inc. **Page 5** Bottom right, NASA. **1.6** Page 6, Upper, left, © P. Hawtin, University of Southampton/SPL/Photo Researchers, Inc.; right, © CNRI/SPL/Photo Researchers, Inc.; lower, left, © R. Robinson/Visuals Unlimited, Inc.; right, © Dr. Harald Huber, Dr. Michael Hohn, Prof. Dr. K.O.Stetter, University of Regensburg, Germany. **1.6** Page 7, Upper left, clockwise from top, Lewis Trusty/Animals Animals; Emiliania Huxleyi photograph, Vita Pariente, scanning electron micrograph taken on a Jeol T330A instrument at Texas A&M University Electron Microscopy Center; Carolina Biological Supply Company; © Oliver Meckes/Photo Researchers, Inc.; Courtesy of James Evarts; upper right from left, © John Lotter Gurling/Tom Stack & Associates; Edward S. Ross; bottom from left, Robert C. Simpson/Nature Stock; Edward S. Ross; © Stephen Dalton/Photo Researchers, Inc. **1.7** Left, Photographs courtesy of Derrell Fowler, Tecumseh, Oklahoma; right, © Nick Brent. **Page 9** © Ed Kashi/Corbis. **Page 10** © SuperStock. **1.9** (a) © Chris D. Jiggins; (b) www.thais.it; (c) background, © Kevin Schafer/Peter Arnold, Inc.; (d) © Martin Reid. **Page 12** Gary Head. **Page 14** © Digital Vision/PictureQuest.

CHAPTER 2: **Page 15** Left, © John Collier. Great Britain, 1850–1934, "Priestess of Delphi," 1891, London, oil on canvas, 160.0 x 80.0cm. Gift of the Rt. Honourable, the Earl of Kintore, 1893; right, © Dr. John Brackenbury/Science Photo Library/Photo Researchers, Inc.. **2.1** © David Noble/FPG. **Page 17** © Michael S. Yamashita/Corbis. **2.3** (a) Above, Gary Head; below, Bruce Iverson. **Page 20** Left, © Dan Guravich/Corbis. **2.3** (c, above) PDB ID:1BNA; H.R. Drew, R.M. Wing, T. Takano, C. Broka, S. Tanaka, K. Itakura, R.E. Dickerson; Structure of a B-CAN Dodecamer, Conformation and Dynamics, PNAS; (c, center) PDB ID: ID:1BNA; H.R. Drew, R.M. Wing, T. Takano, C. Broka, S. Tanaka, K. Itakura, R.E. Dickerson; Structure of a B-CAN Dodecamer, Conformation and Dynamics, PNAS, V. 78 2179, 1981; (c, below) PDB ID:1BNA; H.R. Drew, R.M. Wing, T. Takano, C. Broka, S. Tanaka, K. Itakura, R.E. Dickerson; Structure of a B-CAN Dodecamer, Conformation and Dynamics, PNAS. **2.4** (a–c) PDB file from NYU Scientific Visualization Lab; (b) Right, © Steve Lissau/Rainbow. **2.6** (a) © Lester Lefkowitz/Corbis. **2.8** Michael Grecco/Picture Group. **2.9** PDB file from NYU Scientific Visualization Lab. **2.10** (a) PDB file from NYU Scientific Visualization Lab; (b) PDB file from Klotho Biochemical Compounds Declarative Database. **2.11** Above, left, © Eric and David Hosking/Corbis; right, © Wayne Bennett/Corbis. **2.16** © Steve Chenn/Corbis. © David Scharf/Peter Arnold, Inc. **2.18** PDB file courtesy of Dr. Christina A. Bailey, Department of Chemistry & Biochemistry, California Polytechnic State University, San Luis Obispo, CA. **Page 28** Top right, © Kevin Schafer/Corbis. **2.20** (a) PDB file courtesy of Dr. Christina A. Bailey. **2.21** (c) © Scott Camazine/Photo Researchers, Inc. **2.23** Below, PDB files from NYU Scientific Visualization Lab. **Page 32** Bottom right, © Corbis. **2.25** (a–b) PDB ID: 1BBB; Silva, M.M., Rogers, P.H., Arnone, A.; A third quarternary structure of human hemoglobin A at 1.7-A resolution; *J Biol Chem* 267 pp. 17248 (1992). **2.26** (a–b) PDB files from NYU Scientific Visualization Lab (c) Right, © Dr. Gopal Murti/SPL/Photo Researchers, Inc.; (d) Above left, Courtesy of Melba Moore. **2.29** PDB ID:1BNA; H.R. Drew, R.M. Wing, T. Takano, C. Broka, S. Tanaka, K. Itakura, R.E. Dickerson; Structure of a B-CAN Dodecamer, Conformation and Dynamics, PNAS. **2.30** © National Gallery Collection; By kind permission of the Trustees of the National Gallery, London/Corbis.

CHAPTER 3: **Page 38** © Dr. Tony Brian and David Parker/SPL/Photo Researchers, Inc. **3.2** Photographs, (a) Leica Microsystems, Inc. Deerfield, IL; (b) © George Musil/Visuals Unlimited. **3.3** (a–d) Jeremy Pickett-Heaps, School of Botany, University of Melbourne. **3.4** Art assembly, Gary Head; (hummingbird) Robert A. Tyrell; (humans) Pete Saloutos/Corbis; (redwood) Sally A. Morgan, Ecoscene/Corbis. (Left) With integrin: PDB ID:1JV2; Xiong, J.P., Stehle, T., Diefenbach, B., Zhang, R., Dunker, R., Scott, D.L., Joachimiak, A., Goodman, S.L., Arnaout, M.A.: Crystal Structure of the Extracellular Segment of Integrin; (right) Chris Keeney with Human growth hormone: PDB ID:1A22; Clackson, T., Ultsch, M. H., Wells, J. A., de Vos, A. M.: Structural and functional analysis of the 1:1 growth hormone: receptor complex reveals the molecular basis for receptor affinity. *J Mol Biol* 277 pp. 1111 (1998); HLA: PDB ID: 1AKJ; Gao, G.F., Tormo, J., Gerth, U.C., Wyer, J.R., McMichael, A.J., Stuart, D.I., Bell, J.I., Jones, E.Y., Jakobsen, B.K.; Crystal structure of the complex between human CD8alpha(alpha) and HLA-A2; *Nature* 387 pp. 630 (1997); glut1: PDB ID:1JA5; Zuniga, F.A., Shi, G., Haller, J.F., Rubashkin, A., Flynn, D.R., Iserovich, P., Fischbarg, J.: A Three-Dimensional Model of the Human Facilitative Glucose Transporter Glut1. *J.Biol.Chem.* 276 pp. 44970 (2001); calcium pump: PDB ID:1EUL; Toyoshima, C., Nakasako, M., Nomura, H., Ogawa, H.: Crystal Structure of the Calcium Pump of Sarcoplasmic Reticulum at 2.6 Angstrom Resolution. *Nature* 405 pp. 647 (2000); ATPase: PDB ID:18HE; Menz, R.I., Walker, J.E., Leslie, A.G. W.: Structure of Bovine Mitochondrial F1-ATPase with Nucleotide Bound to All Three Catalytic Sites: Implications for the Mechanism of Rotary Catalysis. *Cell* (Cambridge, Mass.) 106 pp. 331 (2001). **3.8** (a) Lisa Starr; (b) © K. G. Murti/Visuals Unlimited; (c) © R. Calentine/Visuals Unlimited; (d) Gary Gaard and Arthur Kelman. **3.9** Photographs, (a) Stephen L. Wolfe; (c) Don W. Fawcett/Visuals Unlimited, computer enhanced by Lisa Starr; page 47 (d) Don W. Fawcett/Visuals Unlimited, computer enhanced by Lisa Starr; (e) Gary Grimes, computer enhanced by Lisa Starr. **3.10** Keith R. Porter. **3.11** © Dr. Jeremy Burgess/SPL/Photo Researchers, Inc. **3.14** (a) © Stanley M. Awramik; (b–c) Andrew H. Knoll, Harvard University; (d) P.L. Walne and J.H. Arnott, *Planta*, 77:325–354, 1967. **Page 51** © CNRI/Photo Researchers, Inc. **3.16** (a) © J.W. Shuler/Photo Researchers, Inc.; (b) Courtesy Dr. Vincenzo Cirulli, Laboratory of Developmental Biology, The Whittier Institute for Diabetes University of California, San Diego, La Jolla, California; (c) Courtesy Mary Osborn, Max Planck Institute for Biophysical Chemistry, Goettingen, FRG. **3.17** (a) John Lonsdale, www.johnlonsdale.net; (b) David C. Martin, Ph.D. **3.19** (a) © CNRI/SPL/Photo Researchers, Inc.; (b) © Don W. Fawcett/Photo Researchers, Inc.; (c) © Mike Abbey/Visuals Unlimited. **3.21** (a) George S. Ellmore; (b, left) Science Photo Library/Photo Researchers, Inc.; (b, right) Bone Clones, www .boneclones.com. **3.23** (a–b) From *Tissue and Cell*, Vol. 27, pp. 421–427, Courtesy of Bjorn Afzelius, Stockholm University.

CHAPTER 4: **Page 58** © Paul Edmondson/Corbis. **4.1** Above, © Craig Aurness/Corbis; below, © William Dow/Corbis. **4.4** (a–b) From B. Alberts et al. *Molecular Biology of the Cell*, 1983, Garland Publishing. **4.5, 4.6** Art, Lisa Starr; photograph, © Randy Ury/Corbis. **4.7** PDB ID: 1DBF; Putnam, C.D., Arvai, A.S., Bourne, Y., Tainer, J.A.: Active and Inhibited Human Catalase Structures: Lingand Nadph Binding and Catalytic Mechanism, *J Mol Biol*. 296, p. 295 (2000). **4.10** Photograph, Scott McKiernan/ZUMA Press. **4.14** After David H. MacLennan, William J. Rice and N. Michael Green, "The Mechanism of Ca2+ Transport by Sarco (Endo)plasmic Reticulum Ca2+-ATPases." *JBC* Vol. 272, No. 46, November 14, 1997, pp. 28815–28818. **Page 68** © Hubert Stadler/Corbis. **4.16** Photographs, M. Sheetz, R. Painter, and S. Singer, *Journal of Cell Biology*, 70:193 (1976) by permission, The Rockefeller University Press. **4.19** (b) © Juergen Berger/Max Planck Institute/SPL/Photo Researchers, Inc. **4.21** Frieder Sauer/Bruce Coleman.

CHAPTER 5: **Page 73** NASA. **5.2** Larry West/FPG. PDB files from NYU Scientific Visualization Lab. Harindar Keer, Thorsten Ritz Laboratory at UC Irvine, Dept. of Physics and Astronomy, using VMD proprietary software. **Page 75** Bottom right, E.R. Degginger. **Page 76** Top left, © Richard Uhlhorn Photography. **5.7** (a–d) Light Harvesting Complex PDB ID: 1RWT; Liu, Z., Yan, H., Wang, K., Kuang, T., Zhang, J., Gui, L., An, X., Chang, W.: Crystal Structure of Spinach Major Light-Harvesting Complex at 2.72 A Resolution. *Nature* 428 pp. 287 (2004). **5.8** Above left, © Bill Beatty/Visuals Unlimited; above right, Micrograph, Bruce Iverson, computer-enhanced by Lisa Starr; Art, Gary Head. **Page 78** Bottom left, NASA. **5.9** (a, above) 2001 PhotoDisc; (a, below) Micrograph, © Ken Wagner/Visuals Unlimited, computer-enhanced by Lisa Starr. **5.10** © Chris Hellier/Corbis. **5.12** (a) Herve Chaumeton/Agence Nature; (b) Douglas Faulkner/Sally Faulkner Collection.

CHAPTER 6: **Page 82** Left, © Louise Chalcraft-Frank and FARA; right, © Professors P. Motta and T. Naguro/SPL/Photo Researchers, Inc. **6.4** (a) © Professors P. Motta and T. Naguro/SPL/Photo Researchers, Inc. **6.6** (a) Adrian Warren/Ardea, London; (b–c) © Foodpix/Ben Fink. **6.8** Photograph, Gary Head. **Page 91** © Lois Ellen Frank/Corbis. **6.9** Lisa Starr. **6.10** Thomas D. Mangelsen/Images of Nature.

CHAPTER 7: Page 94 Left, Courtesy of the family of Henrietta Lacks; right, Micrograph, Dr. Pascal Madaule, France. **7.1** Micrograph, C.J. Harrison et al., *Cytogenetics and Cell Genetics*, 35: 21–27 1983 S. Karger, A.G. Basel. **7.2** Above, © Tim Pannell/Corbis. **7.3** © L. Willatt, East Anglian Regional Genetics Service/SPL/Photo Researchers, Inc. **7.5** Pages 98–99, Photographs, © Jennifer W. Shuler/Science Source/Photo Researchers, Inc.; Art, Raychel Ciemma. **7.6** Micrograph, (a) © D.M. Phillips/Visuals Unlimited; (b) © R. Calentine/Visuals Unlimited. **Page 100** Bottom left, © Jennifer W. Shuler/Science Source/Photo Researchers, Inc. **7.8** Image courtesy of Carl Zeiss MicroImaging, Thornwood, NY. **7.14** © Francis Leroy, Biocosmos/Science Photo Library/Photo Researchers, Inc. **7.15** Lennart Nilsson from *A Child Is Born*, 1966, 1977 Dell Publishing Company, Inc. **7.16** © Phillip B. Carpenter, Department of Biochemistry and Molecular Biology. University of Texas-Houston Medical School. **7.17** © Science Photo Library/Photo Researchers, Inc. **7.19** © Ron Neumeyer, www.microimaging.ca.

CHAPTER 8: Page 112 Left, © Simon Fraser/RVI, New Castle-Upon-Tyne/SPL/Photo Researchers, Inc.; right, © Science Photo Library/Photo Researchers, Inc. **8.2** The Moravian Museum, Brno. **8.3** (a) Jean M. Labat/Ardea London. **8.9** David Scharf/Peter Arnold, Inc. **8.10** Upper left and right, William F. Ferguson; below, © Francesc Muntada/Corbis. **Page 118** Left, © Science Photo Library/Photo Researchers, Inc. **8.11** From left, Michael Stuckey/Comstock, Inc.; Michael Stuckey/Comstock, Inc.; Bosco Broyer, photograph by Gary Head. **8.12** © Bettman/Corbis. **8.13** From top, Frank Cezus/FPG; Frank Cezus/FPG; 2001 PhotoDisc, Inc.; Ted Beaudin/FPG; Stan Sholik/FPG. **8.15** Courtesy of Ray Carson, University of Florida News and Public Affairs. **8.17** Left, Pamela Harper/Harper Horticultural Slide Library; (a–c) From Prof. Otto Wilhelm Thomé, Flora von Deutschland Österreich und der Schweiz. 1885, Gera, Germany. **Page 122** Right, © Gideon Mendel/Corbis. **8.18** © Omikron/Photo Researchers, Inc. **8.19** Left, upper, 2001 PhotoDisc, Inc./Getty Images; lower, 2001 EyeWire, Inc. **8.21** © Stapleton Collection/Corbis. **8.22** Dr. Victor A. McKusick. **8.23** Steve Uzzell. **8.24** © Frank Trapper/Corbis Sygma. **8.25** Left, © Lois Ellen Frank/Corbis. **8.26** Below, Photo by Gary L. Friedman, www.FriedmanArchives.com. **8.28** © Bettman/Corbis. **8.29** Courtesy G.H. Valentine. **8.31** © CNRI/Photo Researchers, Inc. **8.32** (b) © Lauren Shear/Photo Researchers, Inc. **8.33** UNC Medical Illustration and Photography. **8.34** © Saturn Stills/SPL/Photo Researchers, Inc. **8.35** From Lennart Nilsson, *A Child is Born*, © 1966, 1977 Dell Publishing Company, Inc. **8.36** Left, Fran Heyl Associates © Jacques Cohen, computer-enhanced by © Pix Elation; right, © Matthew Alan/Corbis. **8.39** Upper, © Tom and Pat Leeson/Photo Researchers, Inc.; lower, Rick Guidotti, Positive Exposure. **8.40** Leslie Faltheisek. Clacritter Manx.

CHAPTER 9: Page 137 Photo courtesy of the College of Veterinary Medicine, Texas A&M University. **9.2** Bottom, © Eye of Science/Photo Researchers, Inc. **9.3** A.C. Barrington Brown, 1968 J.D. Watson. **9.5** Right, © SPL/Photo Researchers, Inc. **9.6** PDB ID: 1BBB; Silva, M.M., Rogers, P.H., Arnone, A.; A third quarternary structure of human hemoglobin A at 1.7-A resolution; *J Biol Chem* 267 pp. 17248 (1992). **Page 143** Fran Heyl Associates © Jacques Cohen, computer-enhanced by © Pix Elation. **9.8** © James King-Holmes/SPL/Photo Researchers, Inc. **9.9** (a–b) Photos courtesy of the College of Veterinary Medicine, Texas A&M University; (c–d) Photos by Victor Fisher, courtesy Genetic Savings & Clone.
CHAPTER 10: Page 146 Left, © Vaughan Fleming/SPL/Photo Researchers, Inc.; (right) PDB ID:2AAI; Rutenber, E., Katzin, B.M., Ernst, S., Collins, E.J., Mlsna, D., Ready, M.P., Robertus, J.D.: Crystallographic refinement of ricin to 2.5A. *Proteins* 10 pp. 240 (1991). **10.6** Above, Model by Dr. David B. Goodin, The Scripps Research Institute. **10.7** (a) Model, courtesy of Thomas A. Steitz from *Science*. **10.10** Left, Nik Kleinberg; right, P. J. Maughan. **10.11** © John W.

Gofman and Arthur R. Tamplin. From *Poisoned Power: The Case Against Nuclear Power Plants Before and After Three Mile Island*, Rodale Press, PA, 1979. **10.13** PDB ID: 1CJG; Spronk, C.A.E.M., Bonvin, A.M.J.J., Radha, P.K., Melacini, G., Boelens, R., Kaptein, R.: The Solution Structure of Lac Repressor Headpiece 62 Complexed to a Symmetrical Lac Operator. *Structure* (London) 7 pp. 1483 (1999). Also PDB ID: 1LBI; Lewis, M., Change, G., Horton, N.C., Kercher, M.A., Pace, H.C., Schumacher, M.A., Brennan, R.G., Lu, P.: Crystal structure of the lactose operon repressor and its complexes with DNA and inducer. *Science* 271 pp. 1247 (1996); lactose PDB files from the Hetero-Compound Information Centre-Uppsala (HIC-Up). **10.15** (a) © Lisa Starr; (b) © Visuals Unlimited; (c) © Gehring, P. Callaets and G. Halder; courtesy Walter J. Ghering/University of Basel, Switzerland; (d) Courtesy of Edward B. Lewis, California Institute of Technology. **10.16** (a) Dr. Karen Dyer Montgomery. **10.17** Jack Carey. **10.19** © Dr. M.A. Ansary/Science Photo Library/Photo Researchers, Inc. **Page 161** Lower right (both), Photo courtesy of the College of Veterinary Medicine, Texas A&M University.

CHAPTER 11: Page 162 Dr. Jorge Mayer, Golden Rice Project. **11.2** (a) © Professor Stanley Cohen/SPL/Photo Researchers, Inc.; (b) With permission of © QIAGEN, Inc. **11.7** Above, © TEK IMAGE/Photo Researchers, Inc. **11.8** Left, © David Parker/SPL/Photo Researchers, Inc.; right, Cellmark Diagnostics, Abingdon, UK. **11.10** © Volker Steger/SPL/Photo Researchers, Inc. **11.11** (a) Courtesy Calgene LLC; (b) Dr. Vincent Chaing, School of Forestry and Wood Projects, Michigan Technology University. **11.12** Photo left, © Lowell Georgis/Corbis; right, Keith V. Wood. **11.13** (a) Transgenic goat produced using nuclear transfer at GTC Biotherapeutics. Photo used with permission; (b) © Matt Gentry/Roanoke Times. **11.14** © Adi Nes, Dvir Gallery Ltd. **11.15** © Jeans for Gene Appeal. **Page 174** © Work of Atsushi Miyawaki, Qing Xiong, Varda Lev-Ram, Paul Steinbach, and Roger Y. Tsien at the University of California, San Diego.

CHAPTER 12: Page 176 © Bettmann/Corbis. **12.1** (a, c) © Wolfgang Kaehler/Corbis; (b) © Earl & Nazima Kowall/Corbis; (d–e) Edward S. Ross. **12.2** Left, Gary Head; right, above, © Bruce J. Mohn; below inset, Phillip Gingerich, Director, University of Michigan. Museum of Paleontology. **12.3** (a) Courtesy George P. Darwin, Darwin Museum, Down House; (b) Christopher Ralling; (c) Dieter & Mary Plage/Survival Anglia. **12.4** (a) © Karen Carr Studio/www.karencarr.com; (b) © Joe McDonald/Corbis. **12.5** (a) Heather Angel; (b) David Cavagnaro; (c) Dr. P. Evans/Bruce Coleman, Ltd.; (d) Alan Root/Bruce Coleman Ltd. **12.6** (a) Gary Head; (b) John W. Merck Jr., University of Maryland. **12.7** (a) Giorgio Gualco/Bruce Coleman, Ltd.; (b) © David Parker/SPL/Photo Researchers, Inc. **Page 182** Bottom, from left, © Peter Bowater/Photo Researchers, Inc.; © Sam Kleinman/Corbis; © Owen Franken/Corbis; © Christopher Briscoe/Photo Researchers, Inc. **12.8** Alan Solem. **12.11** Courtesy of © Hopi Hoekstra, University of California, San Diego. **12.12** © Reuters NewMedia, Inc./Corbis. **Page 182** Bottom left (Barn owl), © Lanz Von Horsten; Gallo Images/Corbis. **12.15** Thomas Bates Smith. **12.16** Bruce Beehler. **12.17** Left, © Michael Freeman/Corbis; (a–b) After Ayala and others. **12.18** Adapted from S.S. Rich, A.E. Bell, and S.P. Wilson, "Genetic drift in small populations of Tribolium," *Evolution* 33:579–584, Fig. 1, p. 580, 1979. Used by permission of the publisher. **Page 190** Bottom left, © Steve Bronstein/The Image Bank/Getty Images. **12.19** Frans Lanting/Minden Pictures. **12.20** Left, David Neal Parks; right, W. Carter Johnson. **12.21** © Elliot Erwitt/Magnum Photos, Inc.

CHAPTER 13: Page 194 NASA Galileo Imaging Team. **13.1** Left, H. P. Banks; right, Jonathan Blair. **13.2** © Jonathan Blair/Corbis. **Page 196** Bottom left, © David A. Kring, NASA/Univ. Arizona Space Imagery Center. **13.3** (b) © PhotoDisc/Getty Images. **13.5** © Corbis. **13.6** (a) NASA/GSFC. **13.7** Art, (a–d) After A.M.

Ziegler, C.R. Scotese, and S.F. Barrett, "Mesozoic and Cenozoic Paleogeographic Maps," and J. Krohn and J. Sundermann (Eds.), *Tidal Frictions and the Earth's Rotation II*, Springer-Verlag, 1983; (e) © Martin Land/Photo Researchers, Inc.; (f) © John Sibbick. **13.9** (a) J. Scott Altenbach, University of New Mexico, computer enhanced by Lisa Starr; (b) Frans Lanting/Minden Pictures, computer enhanced by Lisa Starr; (c) © Stephen Dalton/Photo Researchers, Inc. **13.10** (a) Above, Tait/Sunnucks Peripatus Research; below, © Jennifer Grenier, Grace Boekhoff-Falk and Sean Carroll, HMI, University of Wisconsin-Madison; (b) Above, Herve Chaumeton/Agence Nature; below, © Jennifer Grenier, Grace Boekhoff-Falk and Sean Carroll, HMI, University of Wisconsin-Madison; (c) Above, © Peter Skinner/Photo Researchers, Inc.; below, Courtesy of Dr. Giovanni Levi; (d) Dr. Chip Clark. **Page 204** © TEK IMAGE/Photo Researchers, Inc. **13.14** After F. Ayala and J. Valentine, *Evolving*, Benjamin-Cummings, 1979. **13.15** (a) John Alcock, Arizona State University; (b) © Alvin E. Staffan/Photo Researchers, Inc.; (c) G. Ziesler/ZEFA. **13.16** Left, © Digital Vision/PictureQuest; (a) © Joe McDonald/Corbis. **13.17** (a) © Graham Neden/Corbis; (b) © Kevin Schafer/Corbis; center, Ron Blakey, Northern Arizona University; (c) © Rick Rosen/Corbis SABA. **13.18** Po'ouli, Bill Sparklin/Ashley Dayer; All other photographs, © Jack Jeffrey Photography. **13.19** Left, above and below, © Below Water Photography/www.belowwater.com; right, Steve Gartlan. **13.20** After W. Jensen and F.B. Salisbury, *Botany: An Ecological Approach*, Wadsworth, 1972. **13.21** Courtesy of Dr. Robert Mesibov. **13.22** Courtesy of Daniel C. Kelley, Anthony J. Arnold, and William C. Parker, Florida State University Department of Geological Science. **13.23** Carnegie Museum of Natural History. **13.24** From left, © Science Photo Library/Photo Researchers, Inc.; © Dr. Dennis Kunkel/Visuals Unlimited; © Galen Rowell/Corbis; © Kevin Schafer/Corbis; Courtesy of Department of Entomology, University of Nebraska-Lincoln; Bruce Coleman, Ltd. **13.26** From left, © Hans Reinhard/Bruce Coleman, Inc.; © Phillip Colla Photography; © Randy Wells/Corbis; © Cousteau Society/The Image Bank/Getty Images; © Robert Dowling/Corbis. **13.28** © Gulf News, Dubai, UAE.

CHAPTER 14: Page 218 © Philippa Uwins/The University of Queensland. **14.1** Jeff Hester and Paul Scowen, Arizona State University, and NASA. **14.2** (a) © Eiichi Kurasawa/Photo Researchers, Inc.; (b) © Dr. Ken MacDonald/SPL/Photo Researchers, Inc.; (c) © Michael J. Russell, Scottish Universities Environmental Research Centre. **14.3** (a) Sidney W. Fox; (b) From Hanczyc, Fujikawa, and Szostak, Experimental Models of Primitive Cellular Compartments: Encapsulation, Growth, and Division; www.sciencemag.org. Science, 24 October 2003; 302;529, Figure 2, page 619. Reprinted with permission of the authors and AAAS. **14.4** Above, © Chase Studios/Photo Researchers, Inc.; below left, © John Reader/SPL/Photo Researchers, Inc.; below right, © Sinclair Stammers/SPL/Photo Researchers, Inc. **14.6** (a) © P. Hawtin, University of Southampton/SPL/Photo Researchers, Inc.; (b) © Dr. Dennis Kunkel/Visuals Unlimited. **14.8** (a) © Martin Miller/Visuals Unlimited; (b) © Alan L. Detrick, Science Source/Photo Researchers, Inc.; (c) © Dr. Harald Huber, Dr. Michael Hohn, Prof. Dr. K.O. Stetter, University of Regensburg, Germany. **Page 226** Bottom left, Courtesy of Agriculture Canada. **14.9** (a) © Dr. Jeremy Burgess/SPL/Photo Researchers, Inc.; (b) P.W. Johnson and J. MeN. Sieburth, Univ. Rhode Island/BPS. **14.10** Dr. Terry J. Beveridge, Department of Microbiology, University of Guelph, Ontario, Canada. **14.11** © Dr. Manfred Schloesser, Max Planck Institute for Marine Microbiology. **14.12** © Stem Jems/Photo Researchers, Inc.; inset, California Department of Health Services. **14.14** © Dr. Dennis Kunkel/Visuals Unlimited. **14.15** (a) © Oliver Meckes/Photo Researchers, Inc.; (b) Photo by Stephen Durr. **14.16** (a) Neal Ericsson; (b) Courtesy of Allen W.H. Bé and David A. Caron; (c, left) © Wim van Egmond; (c, right) © G. Shih and R. Kessel/Visuals Unlimited. **14.17** (a) © Andrew Syred/SPL/Photo Researchers,

Inc.; (b) Courtesy © James Evarts. **14.18** (a) © Dr. David Phillips/Visuals Unlimited; (b) C.C. Lockwood; (c) © Wim van Egmond/Micropolitan Museum; (d) © Frank Borges Llosa/www.frankley.com. **14.19** From left, © Sinclair Stammers/Photo Researchers, Inc.; micrograph center, Steven L'Hernault; © London School of Hygiene & Tropical Medicine/Photo Researchers, Inc.; micrograph right, © Moredum Animal Health, Ltd./Photo Researchers, Inc. **14.20** Science Museum of Minnesota; below, Emiliania Huxleyi. Photograph by Vita Pariente. Scanning electron micrograph taken on a Jeol T330A instrument at the Texas A&M University Electron Microscopy Center. **14.21** (a) © Lewis Trusty/Animals Animals; (b) © Wim van Egmond; art (left) T. Garrison, *Oceanography: An Invitation to Marine Science*, Brooks/Cole, 1993. **14.23** (a) © Andrew J. Martinez/Photo Researchers, Inc.; (b) Courtesy of Professor Astrid Saugestad; (c) Courtesy Microbial Culture Collection, National Institute for Environmental Studies, Japan. **14.24** (a) © Astrid Hanns-Frieder Michler/SPL/Photo Researchers, Inc.; (b) M. Claviez, G. Gerish, and R. Guggenheim; (c–d) Carolina Biological Supply Company; (e) Courtesy Robert R. Kay from R.R. Kay, et al., Development, 1989 Supplement, pp. 81–90. **14.26** Left, Micrograph Garry T. Cole, University of Texas, Austin/BPS; art, After T. Rost, et al., *Botany*, Wiley, 1979. **14.27** (a–b) Robert C. Simpson/Nature Stock. **14.28** (a) Jane Burton/Bruce Coleman, Ltd.; (b) © Chris Worden. **14.29** (a) © Andrew Syred/SPL/Photo Researchers, Inc.; (a, above) Micrograph Ed Reschke; (b) John Hodgin; (c) © North Carolina State University, Department of Plant Pathology; (d) © Michael Wood/mykob.com; (e) Garry T. Cole, University of Texas, Austin/BPS. **14.30** (a) © Mark E. Gibson/Visuals Unlimited; (b) Edward S. Ross; (c) © 1977 Sherry K. Pittam; (d) Mark Mattock/Planet Earth Pictures; (e) After Raven, Evert, and Eichhorn, *Biology of Plants*, fourth edition, Worth Publishers, New York, 1986. **14.31** (a) © 1990 Gary Braasch; (b) F.B. Reeves. **14.32** (a) © Dr. P. Marazzi/SPL/Photo Researchers, Inc.; (b) © Harry Regin. **14.33** (e) © CAMR/A.B. Dowsett/SPL/Photo Researchers, Inc.; (f) © Dr. Linda Stannard, UCT/SPL/Photo Researchers, Inc. **14.34** Above left, © Science Photo Library/Photo Researchers, Inc. **14.35** Above, © NIBSC/SPL/Photo Researchers, Inc.; below, © Sercomi/Photo Researchers, Inc. **14.36** With permission of © FSIS/USDA/FDA.

CHAPTER 15: **Page 244** Left, © T. Kerasote/Photo Researchers, Inc.; right, © Clinton Webb. **15.1** (a) © Reprinted with permission from Elsevier; (b) Patricia G. Gensel; (c) Illustration by Zdenek Burian, © Jeri Hochman and Martin Hochman. **15.3** After E.O. Dodson and P. Dodson, *Evolution: Process and Product*, third edition, p. 401, PWS. **15.5** (a) © Craig Wood/Visuals Unlimited; (b) © John D. Cunningham/Visuals Unlimited; (c) National Park Services, Paul Stehr-Green; (d) National Park Services, Martin Hutten. **15.6** Jane Burton/Bruce Coleman Ltd. **Page 248** Lower left, above, © Jean Miele/Corbis; below, © Thomas D. Mangelsen/Peter Arnold, Inc. **15.7** Photo, Field Museum of Natural History, Chicago (Neg. #7500C). **15.8** (a) © Ed Reschke/Peter Arnold, Inc.; (b) © Colin Bates; (c) W.H. Hodge. **15.9** A. & E. Bomford/Ardea, London. **15.10** © Klein Hubert/Peter Arnold, Inc. **15.11** (a) © Ralph Pleasant/FPG/Getty Images; (b) © Earl Roberge/Photo Researchers, Inc.; (c) © George Loun/Visuals Unlimited; (d) Courtesy of Water Research Commission, South Africa. **15.12** (a) © Dave Cavagnaro/Peter Arnold, Inc.; (b) Robert & Linda Mitchell Photography; (c) © E. Webber/Visuals Unlimited; (d) © Michael P. Gadomski/Photo Researchers, Inc.; (e) © Sinclair Stammers/Photo Researchers, Inc.; (f) Courtesy of Wayside Gardens/www.waysidegardens.com; (g) © Gerald & Buff Corsi/Visuals Unlimited; (h) © Fletcher and Baylis/Photo Researchers, Inc. **15.13** Left, Robert Potts, California Academy of Sciences; (a) Robert & Linda Mitchell Photography; (b) © R.J. Erwin/Photo Researchers, Inc. **15.15** (a) © Michelle Garrett/Corbis; (b) © Gregory G. Dimijian/Photo Researchers, Inc.; (c)

© Darrell Gulin/Corbis; (d) © Darrell Gulin/Corbis. **15.17** Gerry Ellis/The Wildlife Collection. **15.18** © William Campbell/TimePix/Getty Images.

CHAPTER 16: **Page 259** P. Morris/Ardea London. **16.5** After Eugene Kozloff. **16.6** (c) © Wim van Egmond/Micropolitan Museum; (d) © Brandon D. Cole/Corbis; (e) © Jeffrey L. Rotman/Corbis. **16.8** Right, © Andrew Syred/SPL/Photo Researchers, Inc. **16.9** (a) © Cabisco/Visuals Unlimited; (b) © Science Photo Library/Photo Researchers, Inc. **16.10** (a) Jon Kenfield/Bruce Coleman Ltd.; (b) Adapted from Rasmussen, *Ophelia*, Vol. 11, in Eugene Kozloff, Invertebrates, 1990; (c–d) J.A.L. Cooke/Oxford Scientific Films. **16.11** (b) Frank Park/ANT Photo Library; (c) Alex Kirstitch; (d) J. Grossauer/ZEFA. **16.12** (a) © Sinclair Stammers/SPL/Photo Researchers, Inc.; (b) © L. Jensen/Visuals Unlimited; (c) Dianora Niccolini. **16.13** Jane Burton/Bruce Coleman, Ltd. **16.14** (c) Dr. Chip Clark; (d) © Angelo Giampiccolo/FPG/Getty Images. **16.15** (a) © Corbis; (b) John H. Gerard; (c) © D. Suzio/Photo Researchers, Inc.; (d) © Andrew Syred/Photo Researchers, Inc. **16.17** (a) © Jeff Hunter/The Image Bank/Getty Images; (b) © Peter Parks/Imagequestmarine.com; (c) © Science Photo Library/Photo Researchers, Inc. **16.18** (a–c, i–j) Edward S. Ross; (d) © Mark Moffett/Minden Pictures; (e) Courtesy of Karen Swain, North Carolina Museum of Natural Sciences; (f) Chris Anderson/Darklight Imagery; (g) Joseph L. Spencer; (h) David Maitland/Seaphot Limited. **16.19** Left, Herve Chaumeton/Agence Nature; (a) Herve Chaumeton/Agence Nature; (b) © Fred Bavendam/Minden Pictures; (c) © George Perina, www.seapix.com; (d) Jan Haaga, Kodiak Lab, AFSC/NMFS. **16.20** Right, Runk & Schoenberger/Grant Heilman, Inc. **16.21** Above, Peter Parks/Oxford Scientific Films/Animals Animals; below, © California Academy of Sciences. **16.22** (a) © Ron & Valerie Taylor/Bruce Coleman, Inc.; (b–c) Adapted from A.S. Romer and T.S. Parson, *The Vertebrate Body*, sixth edition, Saunders, 1986; (d) Left, Photo by Lisa Starr; right, Courtesy of John McNamara, www.paleodirect.com. **16.23** (a) © Jonathan Bird/Oceanic Research Group, Inc.; (b) © Gido Braase/Deep Blue Productions. **16.24** (b) © Ivor Fulcher/Corbis; (c) © Roger Archibald. **16.25** (a) © Norbert Wu/Peter Arnold, Inc.; (b) Wernher Krutein/photovault.com; (c) © Alfred Kamajian. **16.26** (a) Bill M. Campbell, MD; (b) © Stephen Dalton/Photo Researchers, Inc.; (c) © John Serraro/Visuals Unlimited. **16.27** Pieter Johnson. **16.28** (a) © Illustration by Karen Carr; (b) © Kevin Schafer/Corbis; (c) Kevin Schafer/Tom Stack & Associates; (e) © Z. Leszczynski/Animals Animals; (e) © S. Blair Hedges, Pennsylvania State University. **16.29** (b) © Kevin Schafer/Corbis. **Page 279** Bottom right, above, P. Morris/Ardea London; below, James Reece, Nature Focus, Australian Museum. **16.30** (a) Sandy Roessler/FPG/Getty Images; (b) © Jean Phillipe Varin/Jacana/Photo Researchers, Inc.; (c) Corbis Images/PictureQuest; (d) © Merlin D. Tuttle/Bat Conservation International; (e) © Mike Johnson. All rights reserved, www.earthwindow.com. **16.31** (a) Larry Burrows/Aspect Photolibrary; (b) © Art Wolfe/Photo Researchers, Inc.; (c) © Dallas Zoo, Robert Cabello; (d) Allen Gathman, Biology Department, Southeast Missouri State University; right, monkey skeletons, Bone Clones®, www.boneclones.com; human, Gary Head. **16.32** Gerry Ellis/The Wildlife Collection; inset, Utah's Hogle Zoo. **16.33** (a) From left, © MPFT/Corbis Sygma; National Museum of Ethiopia, Addis Ababa. © 1985 David L. Brill; Original housed in National Museum of Ethiopia, Addis Ababa. © 1999 David L. Brill; Original housed in National Museum of Ethiopia, Addis Ababa. © 1999 David L. Brill; National Museum of Tanzania, Dar es Salaam. © 1985 David L. Brill; Transvaal Museum, Pretoria. © 1985 David L. Brill; (b) Dr. Donald Johanson, Institute of Human Origins; (c) Louise M. Robbins; (d) Kenneth Garrett/National Geographic Image Collection. **16.34** (a) National Museum of Kenya, Nairobi. © 1985 David L. Brill; (b) Jean Paul Tibbles. **16.35** Left, © Elizabeth Delaney/Visuals Unlimited; all others, John Reader © 1981. **16.37** Left, American Museum of Natural

History. © 1996 David L. Brill; right, Musée de l'homme, Paris. © 1985 David L. Brill. **16.38** Left, Housed in National Museum of Ethiopia, Addis Ababa. © 2001 David L. Brill; right, NASA. **16.39** © Pete & Judy Morrin/Ardea London.

CHAPTER 17: **Page 289** Star Tribune/Minneapolis-St. Paul. **17.1** Top, Courtesy of Charles Lewallen; others, © Bruce Iverson. **17.2** Left, © 2000 PhotoDisc, Inc, with art by Lisa Starr; upper right, © CNRI/SPL/Photo Researchers, Inc.; lower right, Dr. Robert Wagner/University of Delaware, www.udel.edu/Biology/Wags. **17.3** Upper, © Pat Johnson Studios Photography; lower, © Darrell Gulin/The Image Bank/Getty Images. **17.4** Left, © Cory Gray; right, © PhotoDisc/Getty Images. **17.5** (a) Heather Angel; (b) © Biophoto Associates/Photo Researchers, Inc. **17.7** © VVG/Science Photo Library/Photo Researchers, Inc. **17.8** Top, © Niall Benvie/Corbis. **17.9** (a) © Kennan Ward/Corbis; (b) G. J. McKenzie (MGS); (c) Frank B. Salisbury. **17.10** Slim Films. **17.11** (a–b) Courtesy of Dr. Kathleen K. Sulik, Bowles Center for Alcohol Studies, the University of North Carolina at Chapel Hill; (c) © John DaSiai, MD/Custom Medical Stock Photo. **17.12** (a) Courtesy of Dr. Consuelo M. De Moraes; (b–d) Andrei Sourakov and Consuelo M. De Moraes.

CHAPTER 18: **Page 300** © David Nunuk/Photo Researchers, Inc. **18.1** Left to right, © Amos Nachoum/Corbis; A.E. Zuckerman/Tom Stack & Associates; © Corbis. **18.2** (a) © Tom Davis; (b) © Cynthia Bateman, Bateman Photography. **18.3** © Jeff Lepore/Photo Researchers, Inc. **Page 304** Left, upper, © David Nunuk/Photo Researchers, Inc.; lower, NASA. **18.5** (a) © Lilliam Lampas/Omni-Photo; (b) © Christine Pemberton/Omni-Photo. **18.7** © Jacques Langevin/Corbis Sygma. **18.8** Left, NASA; right, © Danny Lehman/Corbis. **18.9** (a) © Joe McDonald/Corbis; (a, inset) © Whispers Appleblossom, courtesy of Goldsmith Seeds, Inc.; (b) © Wayne Bennett/Corbis; (c) © Douglas P. Wilson/Corbis. **18.10** Helen Rodd. **18.11** Right, David Reznick/University of California-Riverside; computer enhanced by Lisa Starr. **18.12** Photo, NASA. **18.15** Left, © Adrian Arbib/Corbis; right, © Don Mason/Corbis. **18.16** © John Alcock/Arizona State University.

CHAPTER 19: **Page 315** Left, USDA; right, above, Photo by Daniel Wojak/USDA; right, below, John Kabashima. **19.1** From left, Donna Hutchins; © Martin Harvey, Gallo Images/Corbis; © Len Robinson, Frank Lane Picture Agency/Corbis; © B.G. Thomson/Photo Researchers, Inc. **19.2** Left, Harlo H. Hadow; right, Bob and Miriam Francis/Tom Stack & Associates. **19.3** Thomas W. Doeppner. **19.4** (mountain peak) © Richard Cummins/Corbis; (a, d) Don Roberson; (b) © Kennan Ward/Corbis; (c) © D. Robert Franz/Corbis. **19.5** Left, above, © Michael Abbey/Photo Researchers, Inc.; below, © Eric V. Grave/Photo Researchers, Inc. **19.6** Art, After N. Weldan and F. Bazazz, *Ecology*, 56:681–688, © 1975 Ecological Society of America; Photographs from left, © Tony Wharton, Frank Lane Picture Agency/Corbis; © Hal Horwitz/Corbis; © Joe McDonald/Corbis. **19.7** Stephen G. Tilley. **19.8** (a) After Rickleffs & Miller, *Ecology*, fourth edition, p. 459 (Fig. 23.12a) and p. 461 (Fig. 23-14); photograph, © W. Perry Conway/Corbis. **19.9** (b) © Ed Cesar/Photo Researchers, Inc.; (c) © Kennan Ward. **19.10** (a–c) Edward S. Ross; (d) © Nigel Jones. **19.11** (a) JH Pete Carmichael; (b) W. M. Laetsch. **19.12** (a–b) Thomas Eisner, Cornell University; (c) Douglas Faulkner/Sally Faulkner Collection; (d) © Bob Jensen Photography. **19.13** (a) Courtesy of Ken Nemuras; (b) CDC; (c) © The Samuel Roberts Noble Foundation, Inc.; (d) Courtesy of Colin Purrington, Swarthmore College. **19.14** © E.R. Degginger/Photo Researchers, Inc. **19.15** © Peter J. Bryant/Biological Photo Service. **19.16** (a) © Doug Peebles/Corbis; (b) © Pat O'Hara/Corbis; (c–d) © Tom Bean/Corbis; (e) © Duncan Murrell/Taxi/Getty Images. **19.17** (a) R. Barrick/USGS; (b) © 1980 Gary Braasch; (c) Don Johnson/Photo Nats, Inc. **19.18** (a, c) Jane Burton/Bruce Coleman, Ltd.; (b) Heather Angel;

(d–e) Based on Jane Lubchenco, *American Naturalist*, 112:23–29, © 1978 by The University of Chicago Press. **19.19** Left, The University of Alabama Center for Public TV; (a) © Angelina Lax/Photo Researchers, Inc.; (b) © Pr. Alexande Meinesz, University of Nice-Sophia Antipolis. **Page 328** Left, upper, John Kabashima; lower, © Joe McDonald/Corbis. **19.20** John Carnemolla/Australian Picture Library/Westlight. **19.21** After W. Dansgaard, et al., *Nature*, 364:218–220, 15 July 1993; D. Raymond, et al., *Science*, 259:926–033, February 1993; W. Post, *American Scientist*, 78:310–326, July–August 1990. **19.22** Left, After S. Fridriksson, *Evolution of Life on a Volcanic Island*, Butterworth, London, 1975; photograph right, © Pierre Vauthey/Corbis Sygma. **19.23** (a) © Susan G. Drinker/Corbis; (b) Frans Lanting/Minden Pictures (computer modified by Lisa Starr). **19.24** © James Marshall/Corbis. **19.25** Left, © Bagla Pallava/Corbis Sygma; right, © A. Bannister/Photo Researchers, Inc. **19.28** Art, Lisa Starr with photographs © 2000 PhotoDisc, Inc. **19.29** Bureau of Land Management. **19.30** (a) © Anthony Bannister, Gallo Images/Corbis; (b) © Bob Jensen Photography; (c) Cedric Vaucher. **19.31** Courtesy of Eternal Reefs, Inc., www.eternalreefs.com.

CHAPTER 20: Page 338 Right, above, © Harold Naideau/Corbis; below, © Chase Swift/Corbis. **20.1** Art, Gary Head and Lisa Starr; photographs, above, © PhotoDisc, Inc.; below, David Neal Parks. **20.2** Above, photograph, Alan and Sandy Carey; art, After R.L. Smith, *Ecology and Field Biology*, fifth edition. **20.3, 20.4** Lisa Starr with photographs, flowering plants, © Frank Oberle/Stone/Getty Images; hawk, © J. Lichter/Photo Researchers, Inc; crow, © Ed Reschke; milk snake, Mike Pingleton/www.pingleton.com; garter snake, Michael Jeffords; frog, John H. Gerard; yellow spider, Michael Jeffords; weasel, courtesy Biology Department, Loyola Marymount University; sparrow, © Rod Planck/Photo Researchers, Inc; cutworm, © Nigel Cattlin/Holt Studios International/Photo Researchers, Inc.; gopher and prairie vole, © Tom McHugh/Photo Researchers, Inc.; all others © 2000 PhotoDisc, Inc. **20.5** Gary Head. **20.6** © David T. Grewcock/Corbis. **20.8** NASA/GSFC. **20.10** Left, After G.E. Likens and F.H. Bormann, "An Experimental Approach to New England Landscapes," in A.D. Hasler (ed.), *Coupling of Land and Water Systems*, Chapman & Hall, 1975; right, © Dr. Elizabeth Hane. **20.11** Water Resources Council. **Page 345** Right, upper, © Harold Naideau/Corbis; lower, © Jose Luis Pelaez, Inc./Corbis. **20.12** Left, Based on NASA photograph from JSC Digital Image Collection; (a–b) After Paul Hertz, photograph, © 2000 PhotoDisc, Inc. **20.13** Based on NASA photograph from JSC Digital Image Collection. **20.14** (a) Compilation of data from

Mauna Loa Observatory, Keeling and Whorf, Scripps Institute of Oceanography; (b) Compilation of data from Prinn, et al., CDIAC, Oak Ridge National Laboratory; World Resources Institute; Khalil and Rasmussen, Oregon Graduate Institute of Science and Technology, CDIAC DB-1010; Leifer and Chan, CDIAC DB-1019; (c–d) Compilation of data from World Resources Institute; Law Dome ice core samples, Etheridge, Pearman, and Fraser, Commonwealth Scientific and Industrial Research Organisation; Prinn, et al., CDIAC, Oak Ridge National Laboratory; Leifer and Chan, CDIAC DB-1019. **20.16** Photograph, © Jeff Vanuga/Corbis. **20.17** © Frederica Georgia/Photo Researchers, Inc. **Page 351** Right, upper, © Harold Naideau/Corbis; lower, © Ian Harwood; Ecoscene/Corbis. **20.18** Photograph, © 2000 PhotoDisc, Inc. **20.19** Fisheries & Oceans Canada, Experimental Lakes Area. **20.20** (a–b) Courtesy of NASA's Terra satellite, supplied by Ted Scambos, National Snow and Ice Data Center, University of Colorado, Boulder; (c) Courtesy of Keith Nicholls, British Antarctic Survey.

CHAPTER 21: Page 355 Left, © Hank Fotos Photography; right, © Wolfgang Kaehler/Corbis. **21.3** Photographs, NASA. **21.5** Left, Adapted from *Living in the Environment* by G. Tyler Miller, Jr., p. 428. © 2002 by BrooksCole; right, © Ted Spiegel/Corbis. **21.6** Alex MacLean/Landslides. **21.7** Art based on NASA photograph. **21.8** Below left, © Sally A. Morgan, Ecoscene/Corbis; below right, © Bob Rowan, Progressive Image/Corbis. **21.9** Lisa Starr, after The World Wildlife Fund and other sources, using NASA photograph. **21.10** Hans Renner. **21.11** (a–b) Courtesy of Jim Deacon, University of Edinburgh. **21.12** © Orbimage Imagery. **21.13** Left, Jack Wilburn/Animals Animals; right, AP/Wide World Photos. **21.14** Above, © Tom Bean Photography; below, Jonathan Scott/Planet Earth Pictures. **21.15** From left, © 1991 Gary Braasch; © Adolf Schmidecker/FPG/Getty Images; Edward S. Ross. **21.16** © James Randklev/Corbis. **21.17** Left, © Raymond Gehman/Corbis; right, © Nigel Cook/Dayton Beach News Journal/Corbis Sygma. **21.18** (a) © Darrell Gulin/Corbis; (b) © Paul A. Souders/Corbis; (c) © Pat O'Hara/Corbis. **21.19** D.W. MacManiman. **21.21** (a) © Annie Griffiths Belt/Corbis; (b) © Douglas Peebles/Corbis. **21.22** © Nancy Sefton. **21.23** (a) Clockwise from top left, Douglas Faulkner/Sally Faulkner Collection; © Corbis; Alex Kirstitch; Douglas Faulkner/Sally Faulkner Collection; © Amos Nachoum/Corbis; Jeff Rotman; (b) T. Garrison, *Oceanography: An Invitation to Marine Science*, third edition, Brooks/Cole, 2000. All rights reserved. **21.24** Greenpeace/Grace. **21.25** (b) © Peter David/FPG/Getty Images; (c) Robert

Vrijenhoek, MBARI. **21.27** (c–d) NASA–Goddard Space Flight Center Scientific Visualization Studio. **21.28** (a) © Eye of Science/Photo Researchers, Inc.; (b) © Douglas P. Wilson; Frank Lane Picture Agency/Corbis; (c) Courtesy of Dr. Anwar Huq and Dr. Rita Colwell, University of Maryland; (d) CHAART, at NASA Ames Research Center; (e) © Raghu Rai/Magnum Photos. **21.29** After M.H. Dickerson, "ARAC: Modeling and Ill Wind," in *Energy and Technology Review*, August 1987. Used by permission of University of California, Lawrence Livermore National Laboratory and U.S. Dept. of Energy.

CHAPTER 22: Page 378 Left, © Ralph Fountain; right, © Scott Camazine. **22.1** (a) Eugene Kozloff; (b–c) Stevan Arnold. **22.2** Left, © Robert M. Timm & Barbara L. Clauson, University of Kansas; (a–b) Reprinted from *Trends in Neuroscience*, Vol. 21, Issue 2, 1998, L.J. Young, W. Zuoxin, T.R. Insel, "Neuroendocrine bases of monogamy," pp. 71–75, ©1998, with permission from Elsevier Science. **22.3** (a) Eric Hosking; (b) © Stephen Dalton/Photo Researchers, Inc.; (c) © Jennie Woodcock, Reflections Photolibrary/Corbis. **22.4** Nina Leen/TimePix/Getty Images; inset, © Robert Semeniuk/Corbis. **22.5** © James Zipp/Photo Researchers, Inc. **22.6** (a) John Bova/Photo Researchers, Inc.; (b) Robert Maier/Animals Animals; (c) Jack Clark/Comstock, Inc.; (d) From L. Clark, *Parasitology Today*, 6(11), Elsevier Trends Journals, 1990, Cambridge, UK. **22.7** (a) © Monty Sloan, www.wolfphotography.com; (b) © Tom and Pat Leeson, leesonphoto.com; (c) © Kevin Schafer/Corbis. **22.8** Above, Stephen Dalton/Photo Researchers, Inc. **22.9** (a) John Alcock, Arizona State University 588; (b–c) © Pam Gardner, Frank Lane Picture Agency/Corbis; (d) © D. Robert Franz/Corbis. **22.10** (a) © Ingo Arndt/Nature Picture Library; (b) Michael Francis/The Wildlife Collection. **32.11** (a) © B. Borrell Casals; Frank Lane Picture Agency/Corbis; (b) © Frank Lane Agency/Bruce Coleman, Inc.; (c) © John Conrad/Corbis. **22.12** © Paul Nicklen/National Geographic/Getty Images. **22.13** (a) © Tom and Pat Leeson, leesonphoto.com; (b) John Alcock, Arizona State University. **22.14** © Jeff Vanuga/Corbis. **22.15** © Eric and David Hosking/Corbis. **22.16** (a) © Australian Picture Library/Corbis; (b) © Alexander Wild; (c) © Professor Louis De Vos. **22.17** (a) © Kenneth Lorenzen; (b) © Peter Johnson/Corbis; (c) Nicola Kountoupes/Cornell University. **22.18** © Dr. Tim Jackson, University of Pretoria. **22.19** (a) © Steve Bloom Images/www.stevebloom.com; (b) © Gallo Images/Corbis. **Page 393** © Matthew Alan/Corbis. **22.20** F. Schutz. **22.21** © Brad Bergstrom. **Page 395** © Joseph Sohm, Visions of America/Corbis.

Index

The letter i designates illustration; t designates table; **bold** designates defined term; ■ highlights the location of applications contained in text.